GALAXY INTERACTIONS AT LOW AND HIGH REDSHIFT

INTERNATIONAL ASTRONOMICAL UNION

UNION ASTRONOMIQUE INTERNATIONALE

GALAXY INTERACTIONS AT LOW AND HIGH REDSHIFT

PROCEEDINGS OF THE 186TH SYMPOSIUM OF THE
INTERNATIONAL ASTRONOMICAL UNION,
HELD AT KYOTO, JAPAN, 26–30 AUGUST 1997

EDITED BY

J. E. BARNES

and

D. B. SANDERS

Institute for Astronomy,
University of Hawaii,
Honolulu, HI, U.S.A.

SPRINGER SCIENCE+BUSINESS MEDIA, B.V.

A C.I.P. Catalogue record for this book is available from the Library of Congress.

ISBN 978-0-7923-5833-6 ISBN 978-94-011-4665-4 (eBook)
DOI 10.1007/978-94-011-4665-4

Printed on acid-free paper

CONTENTS

6. NUCLEAR ACTIVITY

7. GROUPS & CLUSTERS

PREFACE

Galaxy interactions, once an obscure backwater of extragalactic astronomy, have moved front and center after several decades of observational and theoretical research. Interactions are now seen to play a key role in the formation of galaxies, triggering of starbursts and active galactic nuclei, formation of star clusters, formation of dwarf galaxies, metal enrichment of the intergalactic medium, and galaxy evolution in clusters and the field. The goal of this Symposium was to further the development of *a new cosmological perspective tracing the effects of interactions from the current epoch back toward high redshifts*.

The ongoing torrent of data from ground-based telescopes (e.g. Keck, VLA, IRAM, NRO, & OVRO) and space-borne observatories (e.g. *HST*, *ROSAT*, *ASCA*, *IRAS*, & *ISO*) is now revealing interacting galaxies at nearly every wavelength from radio to X-rays. In the past few years, *HST* has imaged nearby merging systems in unprecedented detail, clarified the nature of blue cluster galaxies at intermediate redshifts, provided tantalizing hints of interactions in some high-redshift QSOs, and produced the clearest look yet at the formation and evolution of galaxies at $z \sim 3$ and beyond in the Hubble Deep Field (HDF). On the ground, radio and millimeterwave arrays are probing the spatial and kinematic distribution of the gas and dust in violently interacting galaxies and starburst systems. Theorists have responded by developing sophisticated models treating the dynamics of both stars and gas, pushing beyond pairs and small groups to study interactions in cluster environments, and striving to understand how interactions, mergers, and galactic activity at high redshift are connected to the properties of galaxies at $z \sim 0$.

IAU Symposium 186 on *Galaxy Interactions at Low and High Redshifts* pooled the expertise of observers, numerical simulators, and theoreticians to review the new observations and relate them to theoretical models. The Symposium took place 26–30 August, 1997, as part of the XXIII General Assembly in Kyoto, Japan. Over 300 astronomers attended the meeting, and nearly 200 of them presented results either as oral papers or posters.

Nearly all of these contributions are collected in these Proceedings. Within each of the nine topical sections, reviews and contributed papers are presented in the order given at the Symposium, and related poster papers follow in alphabetical order.

The selection of topics and talks for the Symposium was largely driven by recent advances in observation and modeling. In outline, the general order of presentation followed the title of the Symposium in moving from redshifts $z \sim 0$ out to the high redshift Universe. After a series of introductory talks reviewing galactic interactions (here collected in § 1), the first few sessions were largely devoted to evidence of interactions in the Local Group (§ 2), ongoing tidal interactions in well-studied galaxies (§ 3), and mergers both ancient and recent involving nearby galaxies (§ 4). Subsequent sessions began looking toward higher redshifts, examining how interactions and mergers trigger starbursts (§ 5) and active galactic nuclei (§ 6). Large-scale structure and environmental effects in groups and clusters were incorporated next (§ 7). Finally, studies of galaxy evolution and deep-field surveys (§ 8) bridged the entire redshift range available to observations.

IAU Symposium 186 was sponsored by Commission 28, with the support of Commissions 33, 34, 37, and 44. We thank the Local Organizing Committee in Kyoto for their gracious hospitality and for ensuring that the meeting was both a technical and scientific success.

J. E. Barnes *D. B. Sanders*

Scientific Organizing Committee

J.E. Barnes
F. Combes
G. Efstathiou
K.C. Freeman
I. Karachentsev
I.F. Mirabel
M. Noguchi
D.B. Sanders (Chair)
F. Schweizer
J. vanGorkom
V. Trimble
S.D.M. White
Y. Taniguchi (LOC representative)

PARTICIPANTS

Susanne Aalto, Onsala Observatory, Chalmers Univ. of Technology, Sweden
Roberto Abraham, Royal Greenwich Observatory, Cambridge, U.K.
Luis Aguilar, Institute de Astronomía, UNAM, Ensenada, Mexico
Joanna Anosova, Dept. of Astronomy, University of Texas, Austin TX, U.S.A.
Phil Appleton, Dept. of Astronomy, Iowa State University, Ames, IA, U.S.A.
A. Aragon-Salamanca, Institute of Astronomy, Cambridge, U.K.
Eliani Ardi, Dept. of Astronomy, Kyoto University, Japan
Taft Armandroff, National Optical Astronomy Observatory, Tucson, AZ, U.S.A.
Lia Athanassoula, Observatoire de Marseille, Marseille, France
Amanda Baker, Service d'Astrophysique, CEA Saclay, Gif-sur-Yvette, France
Josh Barnes, Institute for Astronomy, Univ. of Hawaii, Honolulu, U.S.A.
Peter Barthel, Kapteyn Institute, Groningen, The Netherlands
Kenji Bekki, Astronomical Institute, Tohoku University, Japan
Juan C.V. Beltran, Padova Astronomical Observatory, Padova, Italy
Francisco Bertola, Dept. di Astronomia, Università di Padova, Italy
Roger Blandford, California Institute of Technology, Pasadena, CA, U.S.A.
Elias Brinks, Universidad de Guanajuato, Guanajuato, Mexico
Jean Brodie, Lick Observatory, U.C. Santa Cruz, CA, U.S.A.
Gustavo Bruzual, Centro de Investigaciones de Astronomia, Mèrida, Venezuela
Martin Bureau, Mount Stromlo and Siding Spring Observatories, Australia
Natalija Bystrova, Astronomical Institute, St. Petersburg University, Russia
Dave Carter, Liverpool John Moores Univ., Astrophysical Research Inst., U.K.
Tapann Chatterjee, Facultad de Ciencias, Universidad A. Puebla, Mexico
Dave Clements, Institute d'Astrophysique Spatial, Orsay, France
Luis Colina, Space Telescope Science Institute, Baltimore, MD, U.S.A.
Francoise Combes, Observatoire de Paris, DEMIRM, France
Enrico Corsini, Dept. di Astronomia, Università di Padova, Italy
Lawrence Cram, Dept. of Astronomy, University of Sydney, Australia
Zugan Deng, Dept. of Physics, USTC, China
Pierre-Alain Duc, European Southern Observatory, Garching, Germany
Deborah Dultzin-Hacyan, Instituto de Astronomìa UNAM, Mexico
Richard Ellis, Institute of Astronomy, Cambridge, U.K.
Eiichi Egami, Max-Planck-Inst. fur Extraterrestrische Physik, Garching, Germany
Aaron Evans, California Institute of Technolugy, Pasadena, CA, U.S.A.
Duncan Forbes, School of Physics and Astronomy, University of Birmingham, U.K.
Marian Franx, University of Groningen, The Netherlands
Ken Freeman, Mount Stromlo and Siding Spring Observatories, Australia
Carlos Frenk, University of Durham, Durham, U.K.
U Fritze-v.Alvensleben, Universitats-Sternwarte Gottingen, Gottingen, Germany
Mitsuaki Fujimoto, Uedayama 4-901, Tempaku, Nagoya 468, Japan

Jun Fukue, Astronomical Institute, Osaka Kyoiku University, Japan
Carme Gallart, The Observatories-CIW, Pasadena, CA, U.S.A.
Yu Gao, Dept. of Astronomy, University of Illinois, Urbana, IL, U.S.A.
Doug Geisler, National Optical Astronomy Observatories, Tucson, AZ, U.S.A.
Ortwin Gerhard, Astronomisches Institut, Basel, Sweden
Jeroen Gerritsen, Kapteyn Institute, Groningen, The Netherlands
Karl Glazebrook, Anglo-Australian Observatory, Epping, Australia
Wlodzimierz Godlowski, Obser. Astron., Univ. Jagiellonski, Krakow, Poland
Paul Goudfrooij, Space Telescope Science Institute, Baltimore, MD, U.S.A.
Eva Grebel, Astronomical Institute, University of Wurzburg, Germany
Yasuhiro Hashimoto, Dept. of Astronomy, Yale University, New Haven, CT, U.S.A.
George Hau, Institute of Astronomy, Cambridge, U.K.
Paul Hickson, Dept. of Physics and Astronomy, UBC, Vancouver, Canada
John Hutchings, Dominion Astrophysical Observatory, Victoria, Canada
C.-Y. Hwang, Inst. of Astronomy and Astrophysics, Academia Sinica, Taipei, R.O.C.
Rodrigo Ibata, European Southern Observatory, Garching, Germany
Angela Iovino, Oss. Astron. di Brera, Milano, Italy
Cathrine Ishida, Institute for Astronomy, Univ. of Hawaii, Honolulu, U.S.A.
Masanori Iye, National Astronomical Observatory, Tokyo, Japan
Chanda Jog, Dept. of Physics, Indian Institute of Science, Bangalore, India
Shardha Jogee, Dept. of Astronomy, Yale University, New Haven, CT, U.S.A.
Bob Joseph, Institute for Astronomy, Univ. of Hawaii, Honolulu, U.S.A.
Rafik Kandalian, Byurakan Astrophysical Observatory, Byurakan, Armenia
Igor Karachentsev, Special Astrophysical Observatory, R.A.S., Russia
Guinevere Kauffmann, Max-Planck-Inst. fur Astrophysik, Garching, Germany
Jeff Kenney, Dept. of Astronomy, Yale University, New Haven, CT, U.S.A.
Wolfram Kollatschny, Universitats-Sternwarte Gottingen, Gottingen, Germany
David Koo, Lick Observatory, U.C. Santa Cruz, CA, U.S.A.
V. Korchagin, Institute of Physics, Rostov-on-Don, Russia
Oliver Kurth, Universitats-Sternwarte Gottingen, Gottingen, Germany
George Lake, Dept. of Astronomy, Univ. of Washington, Seattle, WA, U.S.A.
Michael Ledlow, Inst. for Astroph., Univ. New Mexico, Albuquerque, NM U.S.A.
Hee-won Lee, Dept. of Astronomy, Seoul National Univ., Seoul, Korea
Myung Gyoon Lee, Dept. of Astronomy, Seoul National Univ., Seoul, Korea
Claus Leitherer, Space Telescope Science Institute, Baltimore, MD, U.S.A.
Stephane Leon, Observatoire de Paris, DEMIRM, Paris, France
Ute Lisenfeld, Universidad de Granada, Granada, Spain
Fred Lo, Inst. of Astronomy and Astrophysics, Academia Sinica, Taipei, R.O.C.
Marcio Maia, Observatorio Nacional, Rio de Janeiro, Brazil
Paola Mazzei, Dept. di Astronomia, Università di Padova, Padova, Italy
Pat McCarthy, The Observatories-CIW, Pasadena, CA 91101, U.S.A.
Duilia de Mello, Space Telescope Science Institute, Baltimore, MD, U.S.A.
T.K. Menon, Dept. of Physics and Astronomy, UBC, Vancouver, Canada
Chris Mihos, Dept. of Astr., Case Western Reserve Univ., Cleveland, OH, U.S.A.
Felix Mirabel, Service d'Astrophysique, CEA Saclay, Gif-sur-Yvette, France
Kazutami Misaki, Dept. of Physics, Nagoya University, Nagoya, Japan
Vasilis Missoulis, Astronomical Institute, National Obs. of Athens, Athens, Greece
Keiko Miyahata, Dept. of Earth and Space Science, Osaka University, Osaka, Japan
Tadayuki Murai, Dept. of Physics, Nagoya University, Nagoya, Japan
Izumi Murakami, National Institute for Fusion Science, Toki, Japan
Masahiro Nagashima, Dept. of Earth and Space Science, Osaka University, Japan
Takao Nakagawa, Institute of Space and Astronautical Science, Tokyo, Japan
Kouichiro Nakanishi, Dept. of Astronomy, Kyoto University, Kyoto, Japan
Alistair Nelson, Dept. of Physics and Astronomy, University of Wales, Cardiff, U.K.
Yuen Ng, Padova Astronomical Observatory, Padova, Italy

Shingo Nishiura, Astronomical Institute, Tohoku University, Sendai, Japan
Masafumi Noguchi, Astronomical Institute, Tohoku University, Sendai, Japan
Kouji Ohta, Dept. of Astronomy, Kyoto Univ., Kyoto, Japan
Youichi Ohyama, Astronomical Institute, Tohoku Univ., Sendai, Japan
G.G.C. Palumbo, Dept. di Astronomia, Univ. di Bologna, Bologna, Italy
Daniel Pfenniger, Geneva Observatory, Univ. of Geneva, Sauverny, Switzerland
Sterl Phinney, California Institute of Technology, Pasadena, CA, U.S.A.
Francisco Prada, Instituto de Astronomía, UNAM, Ensenada, Mexico
Vladimir Reshetnikov, Astronomical Institute, St. Petersburg University, Russia
Huub Rottgering, Sterrewacht Leiden, Leiden, The Netherlands
Boudewijn Roukema, National Astronomical Observatory, Tokyo, Japan
Renzo Sancisi, Kapteyn Institute, Groningen, The Netherlands
Dave Sanders, Institute for Astronomy, Univ. of Hawaii, Honolulu, U.S.A.
Daniel Schaerer, Space Telescope Science Institute, Baltimore, MD, U.S.A.
David Schiminovich, Astronomy Dept., Columbia University, New York, NY, U.S.A.
Francois Schweizer, CIW-DTM, Washington DC,, U.S.A.
Nick Scoville, California Institute of Technology, Pasadena, CA, U.S.A.
Zhengyi Shao, Shanghai Astronomical Observatory, Shanghai, China
M. Shimada, Astronomical Institute, Tohoku University, Sendai, Japan
Masaaki Shimada, Dept. of Astronomy, Kyoto University, Kyoto, Japan
Yasuhiro Shioya, Astronomical Institute, Tohoku University, Sendai, Japan
Ignas Snellen, Sterrewacht Leiden, Leiden, The Netherlands
Phil Solomon, Dept. of Physics and Astronomy, SUNY Stony Brook, NY, U.S.A.
Peter Stockman, Space Telescope Science Institute, Baltimore, MD, U.S.A.
Alan Stockton, Institute for Astronomy, Univ. of Hawaii, Honolulu, U.S.A.
Curtis Struck, Dept. of Physics and Astronomy, Iowa State Univ., Ames, IA, U.S.A.
Hajime Sugai, Dept. of Astronomy, Kyoto University, Kyoto, Japan
Jason Surace, Institute for Astronomy, Univ. of Hawaii, Honolulu, U.S.A.
Marian Suran, Astronomical Inst. of the Romanian Academy, Bucharest, Romania
Tsutomu Takeuchi, Dept. of Astronomy, Kyoto University, Kyoto, Japan
Ichi Tanaka, Astronomical Institute, Tohoku University, Sendai, Japan
Yoshi Taniguchi, Astronomical Institute, Tohoku University, Sendai, Japan
Y. Tawara, Dept. of Astrophysics, Nagoya University, Nagoya, Japan
Bob Thompson, Dept. of Physical Sciences, University of Hertfordshire, U.K.
Dario Trevese, Instituto Astronomico, University of Rome, Italy
Virginia Trimble, Astronomy Dept., Univ. of Maryland, College Park, MD, U.S.A.
Toshio Tsuchiya, Dept. of Astronomy, Kyoto University, Kyoto, Japan
Alexander Turnbull, Dept. of Physical Sciences, University of Hertfordshire, U.K.
Roeland VanDerMarel, Space Telescope Science Institute, Baltimore, MD, U.S.A.
Jacqueline VanGorkom, Astronomy Dept., Columbia Univ., New York, NY, U.S.A.
Paul van der Werf, Sterrewacht Leiden, Leiden, The Netherlands
Wim vandriel, Unitè Scientifique Nancay, Observatoire de Paris, France
Hugo van Woerden, Kapteyn Institute, Groningen, The Netherlands
Sylvain Veilleux, Astronomy Dept., Univ. of Maryland, College Park, MD, U.S.A.
Laurent Vigroux, Service d'Astrophysique, CEA Saclay, Gif-sur-Yvette, France
Steve Vine, Institute of Astronomy, Cambridge, U.K.
Takehiko Wada, Dept. of Earth Science and Astronomy, University of Tokyo, Japan
Joachim Wambsganss, Astrophysikalisches Institut Potsdam, Potsdam, Germany
Hidenori Watarai, Dept. of Physics, Nagoya University, Nagoya, Japan
Nick White, Dept. of Physics and Astronomy, University of Wales, Cardiff, U.K.
Simon White, Max-Planck-Inst. fur Astrophysik, Garching, Germany
Brad Whitmore, Space Telescope Science Institute, Baltimore, MD, U.S.A.
Barbara Williams, Physics and Astro. Dept., Univ. of Delaware, Newark, DE, U.S.A.
Peter Williams, Dept. of Physics and Astronomy, University of Wales, Cardiff, U.K.

Hong Wu, Beijing Astronomical Observatory, Beijing, China
Xiaoyang Xia, Dept. of Physics, Tianjin Normal University, China
Toru Yamada, Astronomical Institute, Tohoku University, Sendai, Japan
Akira Yoshizawa, Astronomical Institute, Tohoku University, Sendai, Japan
Judy Young, Dept. of Physics and Astro., Univ. of Mass., Amherst, MA, U.S.A.
Min Yun, National Radio Astronomy Observatory, Socorro, NM, U.S.A.
Ann Zabludoff, Lick Observatory, U.C. Santa Cruz, CA, U.S.A
Dennis Zaritsky, Lick Observatory, U.C. Santa Cruz, CA, U.S.A.
Steve Zepf, Dept. of Astronomy, Yale University, New Haven, CT, U.S.A.
Xiaolei Zhang, Harvard-Smithsonian Center for Astrophysics, Cambridge, MA, U.S.A.

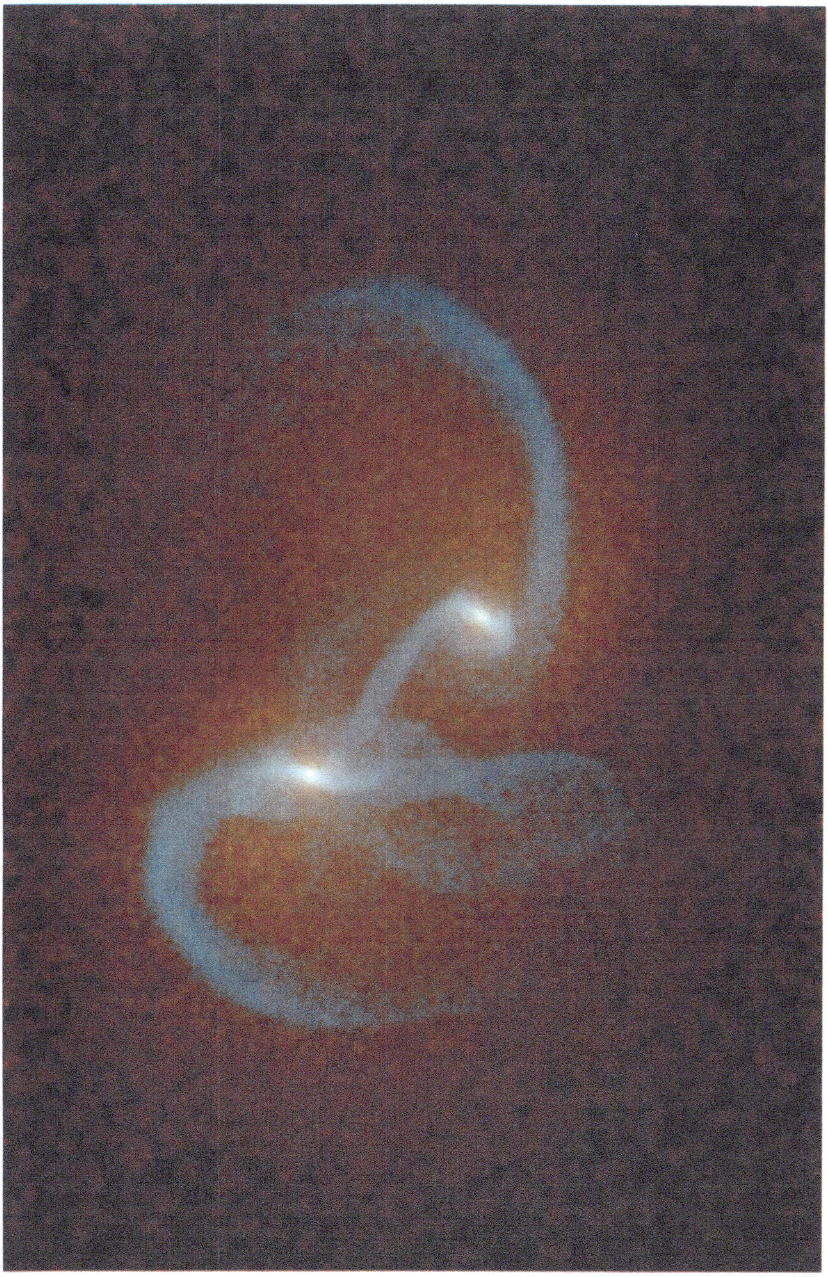

Plate 1. Poster illustration: a parabolic encounter of two disk galaxies, simulated self-consistently with $N = 262144$ particles. In this image, the central bulges are shown in yellow, the disks in blue, and the dark halo material in red.

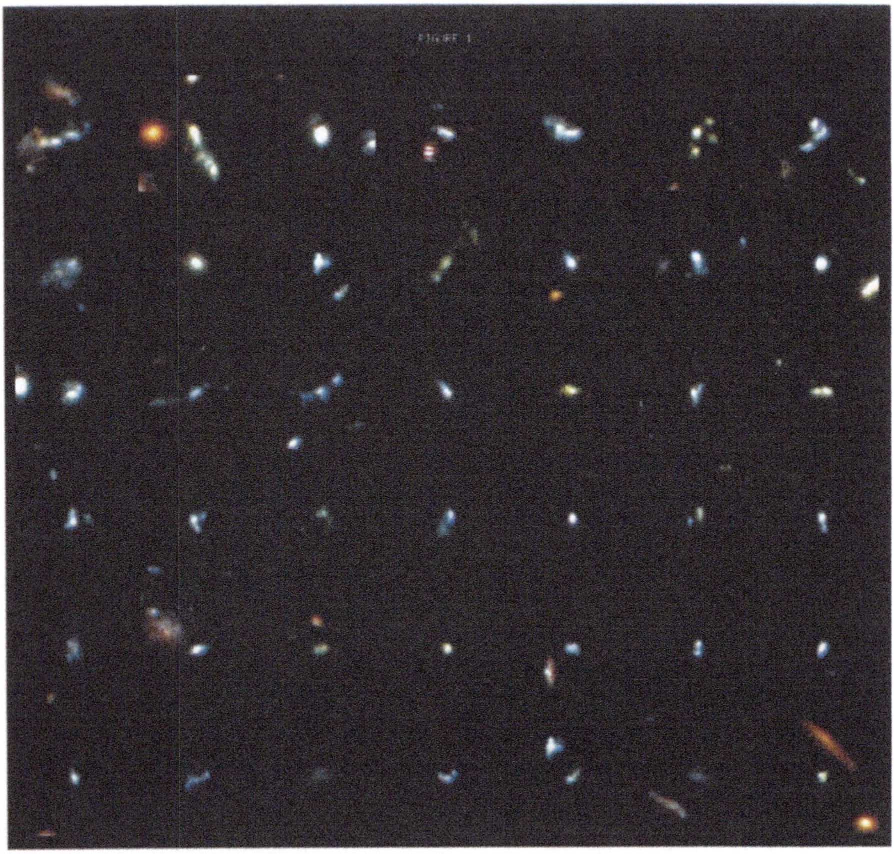

Plate 2. Candidate U_{300}-band dropout Lyman-limit systems in the Hubble Deep Field, with $I < 25$ mag, taken from van den Bergh et al. (1996). "True-color" images were constructed by combining the U_{300}, V_{606}, and I_{814}-band observations. [see Abraham, p. 16]

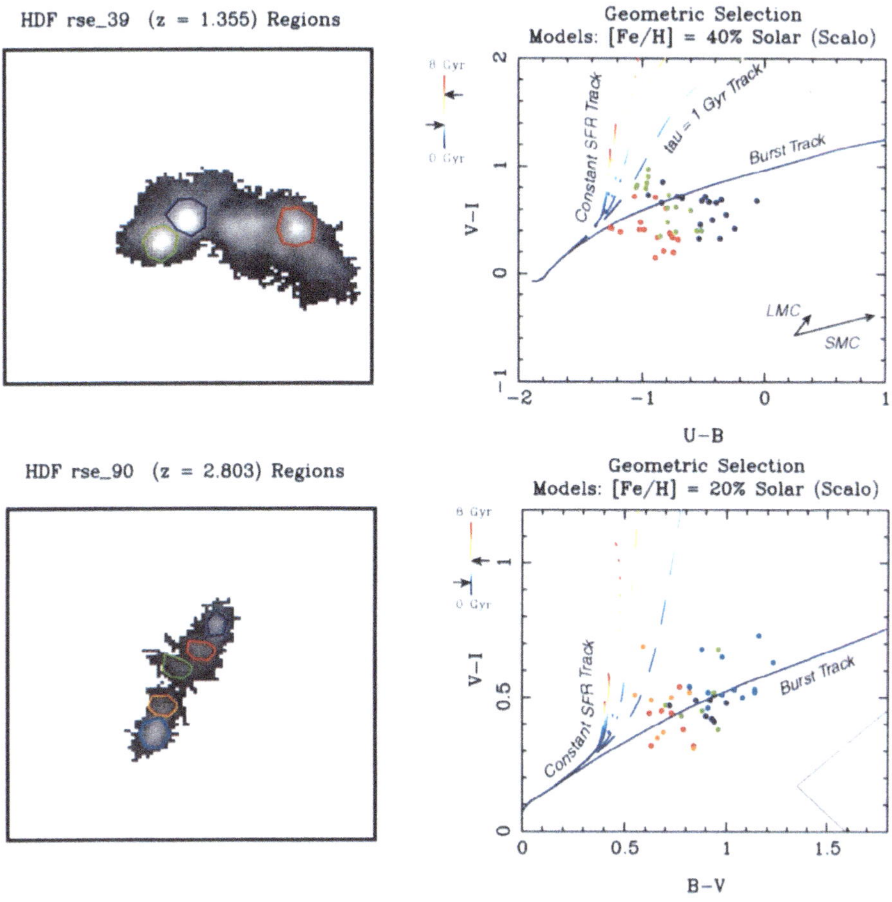

Plate 3. High redshift "chain galaxies" from the HDF (left), and corresponding pixel-by-pixel resolved color-color diagrams (right). Apertures on the images at left enclose subsets of pixels at right. Star-formation history tracks are shown as solid lines, keyed to the colored age bar. Note that both objects are consistent with pure protogalactic starburst tracks with ages < 0.2 Gyr. Synchronization in star-formation activity in the system at $z = 1.355$ (with the ages of bursts changing monotonically with age from a seed knot of star-formation) is seen in many chainlike galaxies. Arrows on the age bar correspond to the age of the universe in an $\Omega = 1$ and a low-Ω Universe. Arrows at the bottom-right corner of the right-hand panels are dust vectors (from Abraham et al., in preparation). [see Abraham, p. 19]

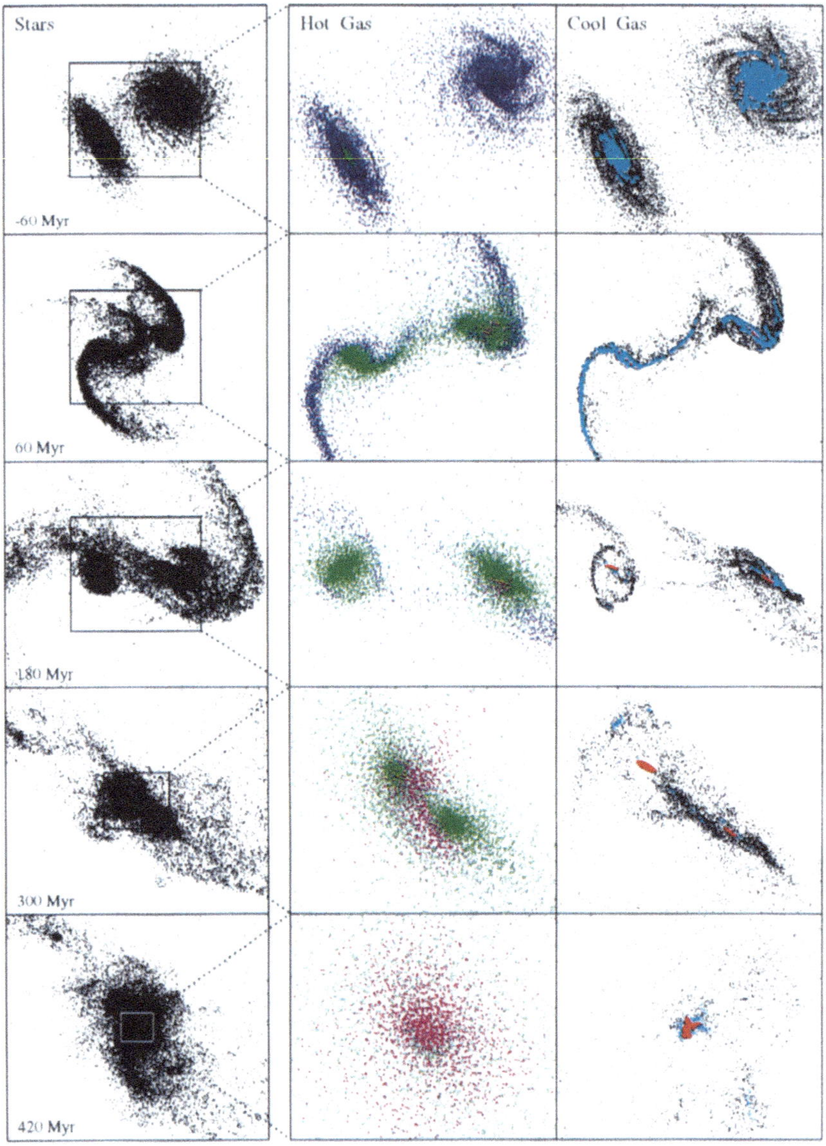

Plate 4. A parabolic encounter of two gas-rich disk galaxies. The stellar distribution is shown on the left; each frame is about 80×96 kpc. Times are given with respect to pericenter at $t = 0$. Hot gas is shown in the middle, enlarged with respect to the stellar frame as indicated; color codes temperature, with dark blue, green, and purple indicating factor-of-ten increases up to 2×10^6 K. Cool gas is shown on the right, on the same scale as the hot gas; here color codes local smoothed density, with black, light blue, and red indicating successive factor-of-hundred increases up to 10^2 cm^{-3}. [see Barnes, p. 142]

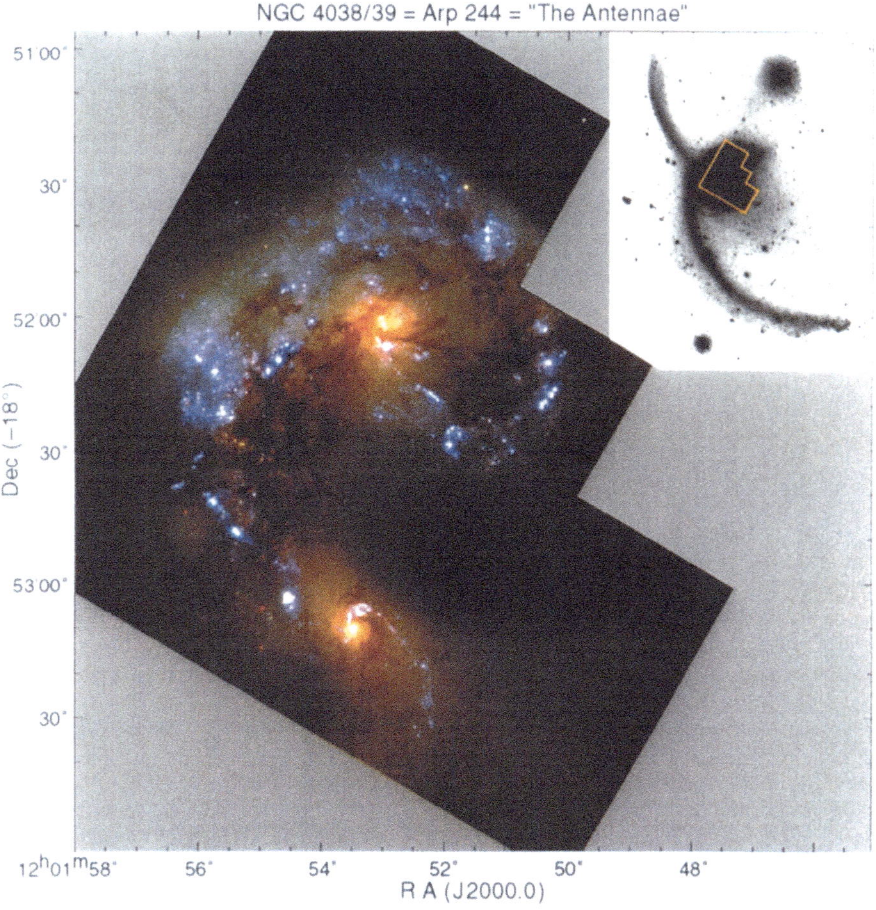

Plate 5. *HST*/WFPC2 image of NGC 4038/39 ("The Antennae") from Whitmore et al. (1998). The insert shows a deep exposure ground-based optical image (courtesy of F. Schweizer) that illustrates the extent of the tidal tails. [see Whitmore, p. 254]

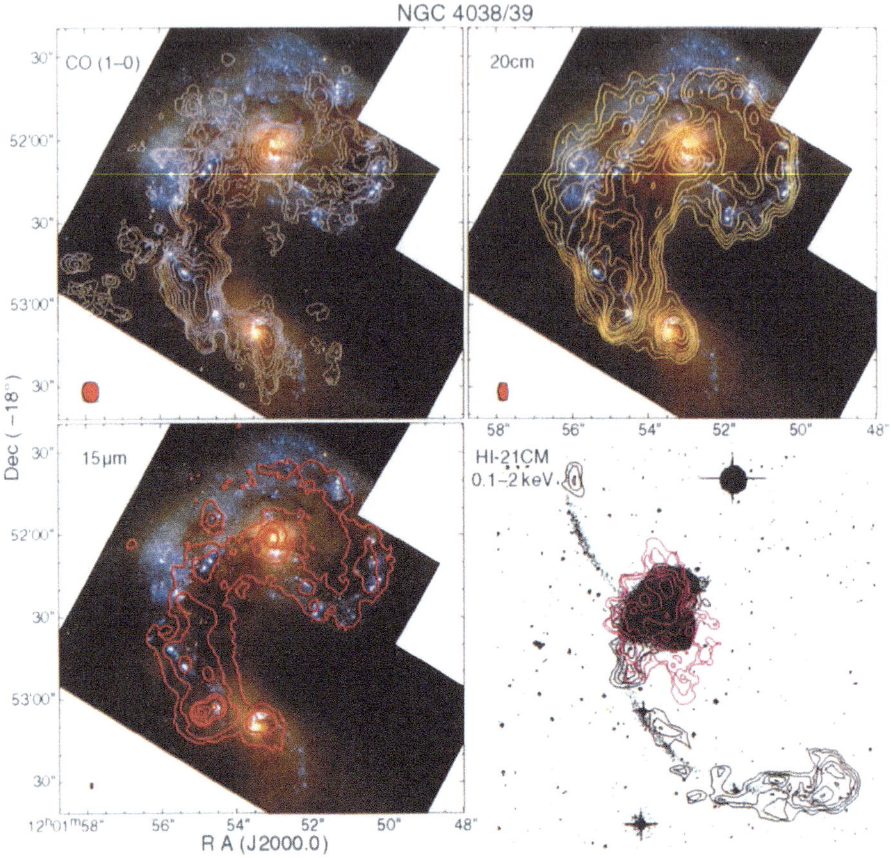

Plate 6. Multiwavelength maps of NGC 4038/39. (upper right) VLA 20 cm radio continuum map. The contour levels are 12,18,24,30,40,60,80,100,150,200 sigma where $1\,\sigma$ is 0.04 mJy/beam. (upper left) Berkeley-Illinois-Maryland-Array (BIMA) CO(1→0) map. The radio continuum and CO data were kindly provided by Fred Lo, Yu Gao, Siow-Wang Lee, and Robert Gruendl in advance of publication. (lower left) *ISO* 15μm continuum map. The ISOCAM contours are 0.4, 1, 3, 5, 10, and 15 mJy (Mirabel et al. 1998, *A&A*,**333**,L5). (lower right) VLA 21 cm-line data (Hibbard & van Gorkom 1996,*A.J.*,**111**,655), and contours of soft X-Ray emission (0.1–2 KeV) emission detected with *ROSAT* (Read et al. 1995, *MNRAS*,**277**,397). The molecular gas originally in the pre-merger disks of NGC 4038 and NGC 4039 has responded to the merger encounter by producing a large North-South concentration to the East of the two galaxy nuclei, in the "overlap" region from which extend the two long tidal tails. Additional concentrations of molecular gas are found around each nucleus and as part of the "ring" of star formation in the western part of the NGC 4038 disk. The most intense mid-infrared peak produces ~15% of the total 12.5–18 μm emission, and corresponds to an optically obscured knot of ~50 pc radius located in the southern portion of the major concentration of CO emission in the overlap region. This multiwavelength view of NGC 4038/39 illustrates the fact that caution must be applied when trying to interpret galaxy morphology or star-formation rates based on UV-optical data alone. [see Sanders et al. p. 291]

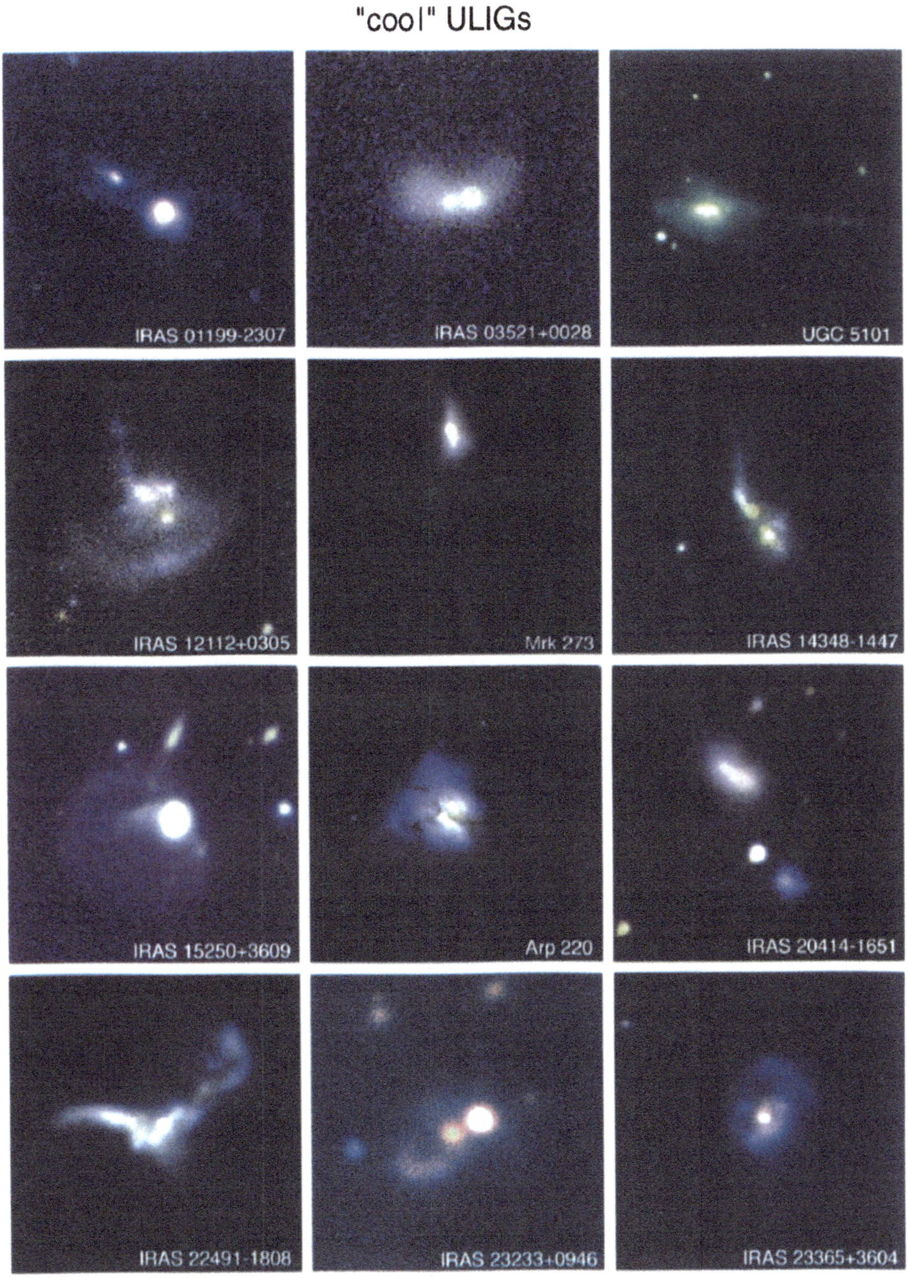

Plate 7. Near truecolor images of a complete sample of "cool" ($f_{25}/f_{60} < 0.2$) ULIGs constructed from B& I-band ground-based data (Surace 1998). [see Sanders et al., p. 290]

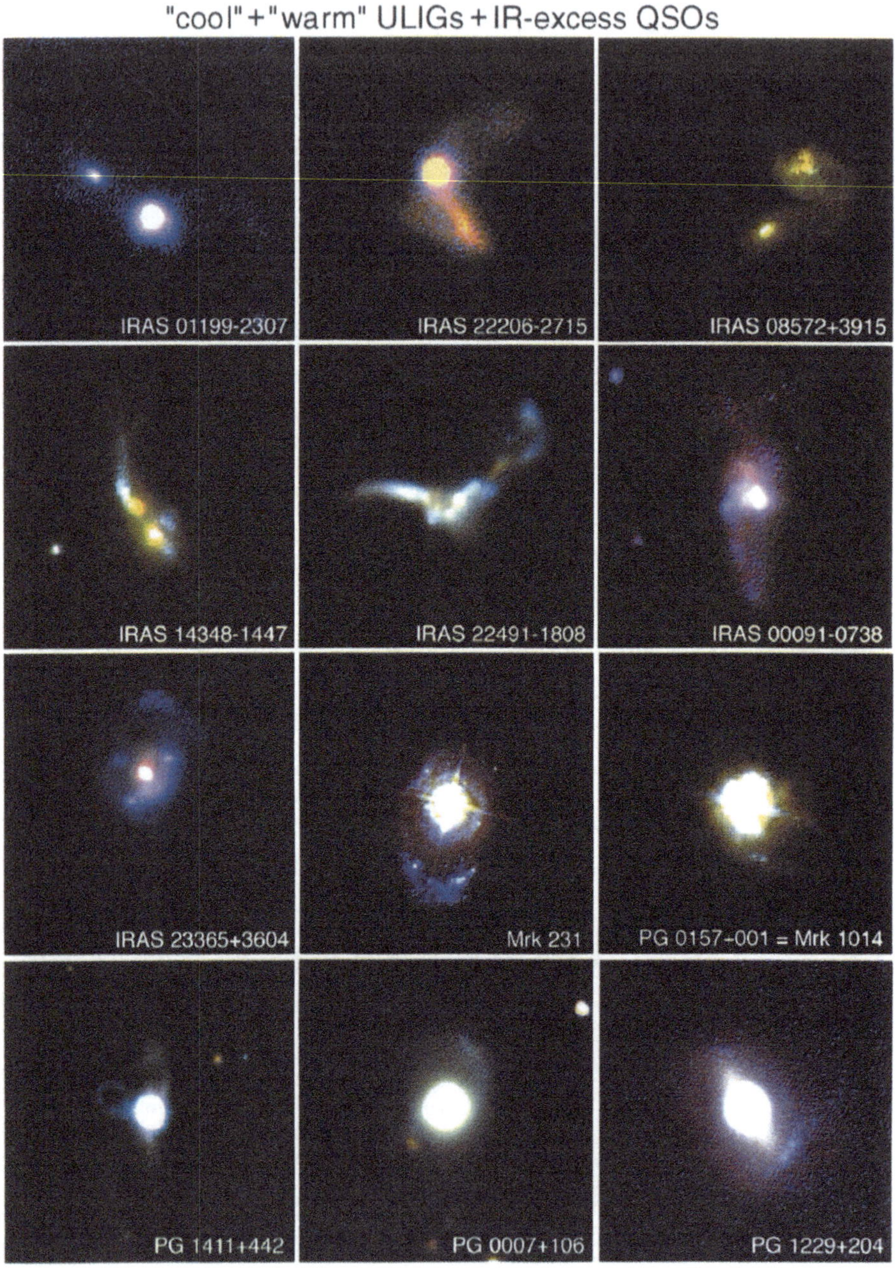

Plate 8. Near truecolor images of ULIGs arranged according to FIR color temperature, with (from upper left) "cool" ($f_{25}/f_{60} < 0.2$) ULIGs (panels 1-5), "warm" ($f_{25}/f_{60} > 0.2$) ULIGs (panels 6-8), and IR-excess ($L_{\rm ir}/L_{\rm B} > 1$) PGQSOs (panels 9-12). [see Sanders et al., p. 290, and Surace & Sanders, p. 363]

OVERVIEW: LOW-z OBSERVATIONS

Of Interacting and Merging Galaxies

FRANCOIS SCHWEIZER
Carnegie Institution of Washington
Department of Terrestrial Magnetism
Washington, DC 20015, USA

Abstract. Gravitational interactions and mergers affect the morphologies and dynamics of galaxies from our Local Group to the limits of the observable universe. Observations of interacting galaxies at low redshifts ($z \lesssim 0.2$) yield detailed information about many of the processes at work. I briefly review these processes and the growing evidence that mergers play a major role in the delayed formation of elliptical and early-type disk galaxies both in the field and in clusters. Low-z observations clearly contradict the notion of a single epoch of E formation at $z \gtrsim 2$; instead, E and S0 galaxies continue forming to the present. The different rates of E and S0 formation inferred from observations of distant and nearby clusters may partially reflect the dependence of dynamical friction on mass: Major, E-forming mergers may tend to occur earlier than minor, S0-forming mergers because the dynamical friction is strongest for equal-mass galaxies.

1. Introduction

Three published figures illustrate the dramatic progress in our understanding of galaxy structure and evolution made during the past six decades. The first of these is Hubble's (1936) empirical "Sequence of Nebular Types" (tuning-fork diagram), which begged the question: What determines the position of a galaxy along this sequence? And, more specifically, why are galaxies at one end of the sequence disk-shaped and at the other end ellipsoidal? The second figure is Toomre's (1977) sketch of "Eleven NGC Prospects for Ongoing Mergers," which illustrated how this puzzling shape dichotomy might result from disk galaxies merging to produce ellipticals (Toomre & Toomre 1972, hereafter TT for short). The third figure is Barnes's (1992, Fig. 9) display of the simulated final passages and merger of two disk galaxies, each represented with its own live halo, disk, and bulge. This figure not only validates TT's earlier hypothesis, but also shows that ellipticals formed via disk mergers bear signatures of the former disks in their fine structure.

The early work by TT and others on gravitational interactions led to many of the broader questions that preoccupy us at present: What dynamical processes drive galaxy evolution? What is the relative importance of tidal interactions, minor accretions, and major mergers in shaping the various types of galaxies? What fractions of stars formed quiescently versus in violent episodes (Schechter & Dressler 1987)? How fast do galaxies assemble? And how has the mix of galaxy types evolved over time?

Although we should not expect this Symposium to yield definitive answers to all these questions, the subject of galaxy interactions is progressing rapidly at present, and the moment seems opportune to review recent advances and chart new courses for addressing these questions.

2. Tidal Interactions

In essence, the tidal nature of galactic "bridges" and "tails" was deciphered during the early 1970s (e.g., TT; for a review, see Barnes & Hernquist 1992). Because of our human preference for the spectacular, there has been a strong observational bias toward studies of near-equal-mass collisions and mergers. Therefore, our knowledge of unequal-mass collisions (say, $m/M \lesssim 0.3$) and of the cumulative effects of weak, relatively distant interactions remains fragmentary. For example, of many galaxy deformations observed in the Local Group, only perhaps the Magellanic Stream (e.g., Gardiner & Noguchi 1996; Lin et al. 1995) and the elongated dwarf galaxy Sgr I (see below) are reasonably well understood. We still do not know the exact causes of the Milky Way's and M33's warps, M31's misaligned bulge, or NGC 205's tidal deformation.

Yet, the eight years since the Heidelberg meeting (Wielen 1990) have brought considerable progress in our general understanding of tidal interactions. Foremost perhaps is a steadily growing appreciation of the many phenomena associated with gas transport and induced star formation, as discussed briefly in §5 below and at length in the reviews by Kennicutt, Schweizer, & Barnes (1998). And certainly a high point has been the discovery of the Sagittarius dwarf, hitherto hidden behind the Milky Way's bulge and the first clear case of accretion observed in our own galaxy (Ibata et al. 1995). This apparently disintegrating companion strongly supports the hypothesis that our halo formed gradually from accreting fragments (Searle & Zinn 1978).

The ability of the *Very Large Array* to now routinely map the H I kinematics of interacting galaxies is rejuvenating the study of these systems.

Because tidal bridges and tails form from the gas-rich outskirts of disk galaxies, they contain H I along their full optically visible extent and often significantly beyond (Hibbard et al. 1994; Hibbard & van Gorkom 1996). The resulting velocity maps contain a wealth of optically inaccessible information and yield kinematic constraints that are invaluable for modeling these systems through N-body simulations (e.g., Hibbard & Mihos 1995). Even the tail lengths alone constrain the ratio of dark to luminous matter in disk galaxies to $M_{\rm d}/M_{\rm l} \lesssim 10$ (Dubinski et al. 1996; Mihos et al. 1998). In the long run, detailed modeling of interacting systems with mapped H I kinematics should yield not only this ratio for individual galaxies, but also the radial variation of it.

Evidence continues to grow that some dwarf galaxies form in tidal tails, as originally proposed by Zwicky (1956). Major clumps of stars and gas have been found in many tails by now (Mirabel et al. 1991; Duc & Mirabel 1994; Hunsberger et al. 1996), and two of these clumps have been shown to probably be self-gravitating entities from their measured H I velocity dispersions (Hibbard et al. 1994).

Many issues remain to be addressed. For example, despite assiduous work by observers and numerical simulators alike, we still do not have any fully successful models for the tidally generated spiral structures of M51 and M81. A new puzzle are the observed displacements between the stars and HI in some tidal tails, occasionally exceeding 2 kpc and perhaps reflecting non-gravitational forces acting upon the gas (Hibbard & van Gorkom 1996; Schiminovich et al. 1995). Finally, although there are many new observations of collisional ring galaxies (for a review, see Appleton & Struck-Marcell 1996), none have addressed yet the interesting issue of the nature of ring-galaxy remnants: Into what kind of galaxies do they evolve?

3. Mergers and the Formation of Ellipticals

The essence of this subject can be encapsulated in the following three questions: (1) What fraction of elliptical galaxies formed via major disk–disk (DD) mergers? (2) Did cluster ellipticals form via such DD mergers, via multiple minor mergers, or in a single collapse? And (3), what is the age distribution of elliptical galaxies?

There is now strong evidence that at least some DD-merger remnants are present-day protoellipticals, as envisaged by TT. My own two favorites are NGC 3921 and NGC 7252. Both feature double tidal tails, but single main bodies of $M_V \approx -23$ with $r^{1/4}$-type light distributions indicative of violent relaxation (Schweizer 1996, 1982). Their power-law cores, central luminosity densities, and *UBVI* color gradients are typical of ellipticals, and their central velocity dispersions fit the Faber–Jackson relation for

E's well (Lake & Dressler 1986). Their "E + A" spectra indicate recent ($\lesssim$1 Gyr ago) starbursts of strength $b \approx$ 10%–30%. During these major starbursts, the globular-cluster populations appear to have increased by $\gtrsim$40% in NGC 3921 (Schweizer et al. 1996) and ~80% in NGC 7252 (Miller et al. 1997). Within 5–7 Gyr, both remnants will have specific globular-cluster frequencies typical of field ellipticals. Therefore, in all their observed properties these two remnants appear to be 0.5–1 Gyr old protoellipticals.

A notable success has been the observational confirmation of the theoretical prediction by Barnes (1988) that most of the tidally ejected material must fall back. This phenomenon creates a strong connection between DD mergers and field ellipticals. In NGC 7252, and likely also in NGC 3921, the H I in the lower parts of the tails is observed returning toward the central remnant (Hibbard et al. 1994; Hibbard & Mihos 1995; Hibbard & van Gorkom 1996). It seems hardly coincidental, then, that many E and S0 galaxies feature inclined gas disks (van Gorkom & Schiminovich 1997), H I absorption in radio ellipticals always indicates infall (van Gorkom et al. 1989), and some well-known E and S0 galaxies like NGC 5128, NGC 1052, and NGC 5266 possess *two*, often nearly orthogonally rotating, H I disks (Schiminovich et al. 1994; Plana & Boulesteix 1996; Morganti et al. 1997). There can be little doubt that at least these kinds of field galaxies formed through major DD mergers.

Observations of fine structure in field E and S0 galaxies suggest that not just a few, but *most* of these galaxies formed through that same mechanism (Schweizer & Seitzer 1992). Roughly 70% of the E's and over 50% of the S0's show fine structure (mainly ripples and plumes) indicative of past disk mergers, and recent N-body simulations demonstrate that this fine structure is a natural byproduct of tidal material falling back after a major merger (Hernquist & Spergel 1992; Hibbard & Mihos 1995). Photometry of these structures suggests that there is more luminous matter in them than can be accounted for by the proverbial "gas-rich dwarf" that supposedly fell in (Prieur 1990). Therefore, given the high detection rates of fine structure and the limited duration of a significant flux of returning material ($\lesssim$5 Gyr), it seems now likely that the vast majority, and perhaps all, field E and S0 galaxies formed through DD mergers.

For cluster ellipticals the situation is less clear because they are poorer in H I and fine structure. Yet, there are at least three arguments to support the view that these ellipticals formed through DD mergers as well. First, cluster ellipticals are structurally indistinguishable from field ellipticals, for which the evidence for past DD mergers is strong. Second, many cluster ellipticals possess oddly rotating stellar cores, which seem to form naturally in DD mergers (Hernquist & Barnes 1991; Bender 1996) and point toward two, rather than multiple, merged components. And third, the bimodal

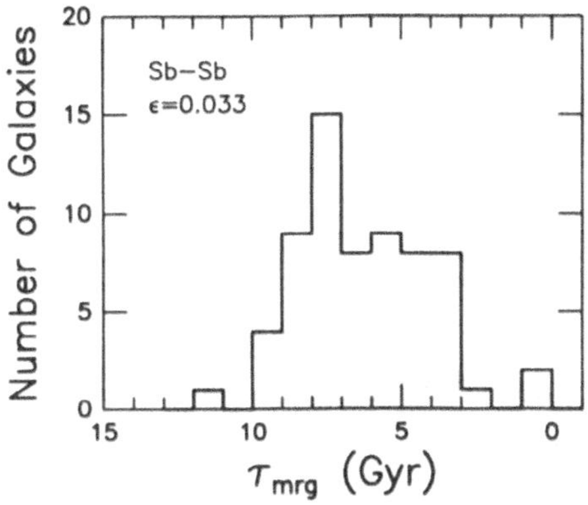

Figure 1. Distribution of merger ages for 65 field E + S0 galaxies. Ellipticals are shaded, age "0" marks present time. Ages are computed for presumed Sb–Sb mergers with star-formation efficiency $\epsilon = 0.033$. Note prolonged period of E + S0 formation suggested by these ages. After Schweizer & Seitzer (1992).

color distributions of globular clusters in cluster ellipticals like M87 and M49 (Whitmore et al. 1995; Geisler et al. 1996) imply a second major cluster-forming event, probably a merger (Ashman & Zepf 1992). Although the remarkably small scatter in the color–luminosity relations of cluster E and S0 galaxies is often taken as evidence against major DD mergers, it may simply be a natural consequence of such mergers having taken place in clusters relatively early (see §6).

If ellipticals did form through delayed DD mergers (Toomre 1977), their formation ages should show a wide spread. The evidence for a significant age spread is increasing, despite the fact that ages of individual E's remain uncertain and often controversial (O'Connell 1980, 1994; Schweizer & Seitzer 1992; González 1993; Faber et al. 1995; Davies 1996). One problem lies in the different possible definitions of "age." In an elliptical galaxy, one must distinguish at least three ages: (1) the true mean age $\langle\tau_*\rangle$ of the stars, (2) the luminosity-weighted mean age $\langle\tau_*\rangle_{\rm lum}$ of stars, which is what observers using *single*-burst population models for interpretation typically measure, and (3) the merger age $\tau_{\rm mrg}$, reckoned since the merger began or the starburst peaked. In general, $\langle\tau_*\rangle > \langle\tau_*\rangle_{\rm lum} > \tau_{\rm mrg}$. Figure 1 shows merger ages for 65 E and S0 galaxies based on *UBV* colors and a simple *two*-burst model of star formation. These formation ages spread over most of the age of the universe. The above model and Kauffmann's (1996) similar calculations also clarify why age-dating ellipticals from spectra is difficult: Because the bulk of stars formed before the final mergers, ellipticals *look* more uniform and old than they really are.

In summary, the combined evidence from low-z observations strongly indicates that *there was no single "epoch of E formation" at* $z \gtrsim 2$, as some have inferred from high-z observations. Instead, this evidence favors the view that E and S0 galaxies continue forming to the present and will do so into the future.

4. Mergers in Disk Galaxies

Although equal-mass mergers are the most spectacular, there must also be unequal-mass ("minor") mergers that affect disk galaxies without completely destroying their disks. Three central questions are: (1) Can bulges form through minor mergers? (2) What fraction of bulges formed in this manner? And (3) how fragile are galaxy disks?

Theoretical work on the fragility of disks is undergoing rapid revisions. As recently as 1992, Tóth & Ostriker argued that disks are very fragile and would be disrupted by any infalling companion more than a few percent the mass of the main disk. Yet, N-body simulations suggest that disk galaxies can survive minor mergers of up to $m/M \approx 0.3$, albeit with increases in bulge mass and a thickened disk (Walker et al. 1996).

Recent observations suggest that the effects of minor mergers on disk galaxies can be surprisingly complex. For example, S0 galaxies with polar rings were thought to have accreted their ring gas during a merger. Yet, new observations show that the H I content of polar rings is often large and typical of late-type disk galaxies, and that many of the S0's feature poststarburst spectra (e.g., Richter et al. 1994; Reshetnikov & Combes 1994). Thus it looks as if it is the central S0 galaxies (rather than the polar rings) that may have formed from a gas-rich companion falling into a spiral galaxy nearly over its poles. If so, these central S0 bodies represent *failed bulges* (Schweizer 1995; Arnaboldi et al. 1997).

Various observations suggest that quite in general early-type disk galaxies may be the remnants of minor mergers. From the statistics of counterrotating, skewedly rotating, and corotating ionized gas disks, one can conclude that 40%–70% of all S0 galaxies experienced minor mergers (Bertola et al. 1992). The phenomenon of counterrotating gas disks is observed—with decreasing frequency—into Hubble types S0/a, Sa, and Sab.

Another powerful kinematic signature of past mergers are subpopulations of stars counterrotating in disk galaxies of types S0 – Sb. The best known example of this phenomenon occurs in NGC 4550, an E/S0 galaxy with half of its disk stars rotating one way and the other half the other way (Rubin et al. 1992). In the Sa galaxy NGC 4138, the split between normal- and counterrotating disk stars is 75/25% (Jore et al. 1996), while in the Sb galaxy NGC 7217 it is 70/30% (Kuijken 1993). Finally, the whole bulge seems to counterrotate to the disk in the well-known Sb galaxy NGC 7331 (Prada et al. 1996). Some first N-body simulations suggest that minor and not-so-minor dD mergers can indeed produce the counterrotations observed in these systems (Thakar et al. 1997; Pfenniger 1998).

A connection between mergers in disk galaxies and bulge formation is also suggested by fine structure indicative of past mergers observed in many S0 and Sa galaxies (Schweizer & Seitzer 1988). Even NGC 4594, the "Sombrero," sports a faint fan of luminous material and an opposite tail signaling a not too ancient merger (Malin & Hadley 1997). Our present understanding of the effects of mergers on disk galaxies is then as follows.

Minor mergers do occur in disk galaxies and seem to move them toward earlier Hubble types. It appears that disks, especially those with a significant fraction of gas, are not nearly as fragile as thought just a few years ago. However, we must remember that bulgeless (e.g., M33) and lopsided disk galaxies do constrain the rate of minor mergers. For mergers of $m/M \approx 0.1$, this rate is estimated to be $\lesssim 0.07 - 0.25$ events/Gyr (Zaritsky & Rix 1997). At present, we can merely state that the fraction of bulges built through mergers is clearly >0, but its value remains unknown. A challenge for the future is to determine how unique or varied the possible paths to, say, a present-day Sb galaxy are. Does the disk form first or the bulge, and can each grow episodically and perhaps even by turns?

5. Interaction-Induced Processes

The *IRAS* sky survey opened our eyes to the fact that gas plays a disproportionately large role in interacting and merging galaxies. Due to lack of space, I can here only briefly sketch recent progress in our understanding of the four major interaction-induced processes.

Interaction-induced *starbursts* are fierce episodes of "galaxy building," during which 5%–20% of the luminous matter gets converted from gas into stars over periods of $\sim 10^8$ years. Fueled by molecular gas in quantities of up to $\sim 2 \times 10^{10}\, M_\odot$, these starbursts appear to be self-limiting: the star-formation rate does not exceed $\sim 0.7\, M_\odot\,\mathrm{kpc}^{-2}\,\mathrm{yr}^{-1}$ for stars of 5–100 $M_\odot$. Merger-induced torques tend to drive gas toward the center, where high concentrations of it often dominate the dynamics. These induced concentrations may explain the high central phase-space densities observed in E's.

Globular-cluster formation appears to be a natural by-product of induced starbursts (e.g., NGC 4038/39, 7727, 3921, 7252, 5128, 1275). There is growing evidence that the globulars may form preferentially in high-pressure regions from Giant Molecular Clouds shocked by the surrounding starburst-heated gas. Even at $z \approx 0$, gas-rich DD mergers can apparently about double the number of globular clusters.

Galactic winds are radial 10^2–$10^3\,\mathrm{km\,s^{-1}}$ outflows of gas heated mostly by starburst supernovae. These winds appear "mass loaded" by factors of 3–6. In M82, the Fe-rich wind is estimated to eject $\sim 10^8\, M_\odot$ of gas into the halo over 20–30 Myr. Such winds presumably played a major role in the chemical evolution of early-type galaxies, but details remain unclear.

Nuclear activity often accompanies merger-induced gas inflows. Observationally, assessing the relative contributions from active galactic nuclei and central starbursts remains a challenging task. About 80% of all host galaxies of low-z QSOs appear to be interacting and 30% appear to be merging. There is evidence that the QSO activity tends to peak shortly before the nuclei of two galaxies merge, and in at least the case of OX 169 the variable Hβ emission-line components suggest the presence of *two* separate engines (Stockton & Farnham 1991).

6. Interactions and Mergers in Clusters

Clusters are complex environments for galaxy evolution. Two properties of cluster galaxies have led to major questions. First, is the morphology–density relation (Oemler 1974; Dressler 1980) a result of birth or evolution? And second, is the Butcher-Oemler (1978, 1984) effect mainly due to ram-pressure stripping or interactions and mergers? Many astronomers have doubted whether galaxies in clusters can merge because of their high mean relative velocities. If not, why are there so many E + S0 galaxies in clusters?

Yet, there has long been observational evidence for interactions and mergers in clusters, and recently this evidence has grown stronger.

For example, the Hercules Cluster abounds in pairs of interacting disks. The Coma Cluster harbors the well-known DD merger NGC 4676 (TT; Barnes 1998) in its outskirts. And in the Virgo Cluster, systems like NGC 4438/35 show that even $\sim 10^3$ km s^{-1} collisions can strongly affect member galaxies (Combes et al. 1988; Kenney et al. 1995). Clearly, strong tidal interactions and mergers occur in clusters to the present time.

Mergers must have played a major role in shaping the E and S0 galaxies of the Virgo Cluster, especially the most massive ones (as judged by their having Messier numbers). Many of these galaxies feature oddly rotating subsystems (M86, NGC 4365, NGC 4550), ripples (M85, M89), and bimodal globular-cluster populations (M49, M87), all signatures of past disk mergers. This confirms predictions of N-body simulations which have long suggested that "the upper end of the mass spectrum is most strongly affected by the merging process occurring predominantly during the expansion and subsequent collapse of the cluster" (Roos & Aarseth 1982).

There is also growing evidence for interactions and mergers in clusters at $z \approx 0.2$–0.5. Even before the advent of *HST*, groundbased observations suggested that many blue galaxies in Butcher-Oemler clusters are distorted disks or feature excess companions (Lavery & Henry 1988–1994). Observations with *HST* show that a fair fraction of these galaxies are interacting or merging while a majority appear to be disturbed gas-rich disks (Dressler et al. 1994; Couch et al. 1994; Barger et al. 1996). Perhaps the most interesting recent result is that at $z \approx 0.4$ the E population appears to be fully in place, while there are significantly fewer S0 galaxies and more late-type spirals and Irregulars than at $z \approx 0$ (Oemler et al. 1997).

I believe that a viable scenario for the formation of E and S0 galaxies in clusters is as follows. Major DD mergers occurred relatively early because the dynamical friction is strongest for equal-mass galaxies. These mergers formed the bulk of the ellipticals. Minor mergers took longer on average because of their lesser dynamical friction (deceleration being approximately proportional to mass). They—and perhaps also multiple interactions—slowly transform spirals into S0 galaxies. Presumably, it is the evolving substructure of clusters that allows mergers to occur to the present time.

Finally, field galaxies experienced the same processes, but on a longer time scale because of their lower spatial density. This is why we still see ellipticals forming (NGC 3921, NGC 7252, ULIR galaxies) and why there are fewer S0 galaxies and more spirals in the field than in clusters. If this cluster- and field-evolution scenario is correct in its essence, Hubble's morphological sequence may rank galaxies mainly by the number and mass ratio of mergers in their past history.

I gratefully acknowledge support through NSF Grant AST-95 29263.

References

Appleton, P.N., & Struck-Marcell, C., 1996, *Fundamentals Cosmic Phys.* **16**, 111
Arnaboldi, M., et al., 1997, *A. J.* **113**, 585
Ashman, K.M., & Zepf, S.E., 1992, *Ap. J.* **384**, 50
Barger, A.J., et al., 1996, *M. N. R. A. S.* **279**, 1
Barnes, J.E., 1988, *Ap. J.* **331**, 699
Barnes, J.E., 1992, *Ap. J.* **393**, 484
Barnes, J.E., 1998, this volume
Barnes, J.E., & Hernquist, L.E., 1992, *Ann. Rev. Astron. Astroph.* **30**, 705
Bender, R., 1996, *New Light on Galaxy Evolution* (Kluwer, Dordrecht), p. 181
Bertola, F., Buson, L.M., & Zeilinger, W.W., 1992, *Ap. J.* **401**, L79
Butcher, H., & Oemler, A., 1978, *Ap. J.* **219**, 18
Butcher, H., & Oemler, A., 1984, *Ap. J.* **285**, 426
Combes, F., Dupraz, C., Casoli, F., & Pagani, L., 1988, *A. & A.* **203**, L9
Couch, W.J., Ellis, R.S., Sharples, R.M., & Smail, I., 1994, *Ap. J.* **430**, 121
Davies, R.L., 1996, *New Light on Galaxy Evolution* (Kluwer, Dordrecht), p. 37
Dressler, A., 1980, *Ap. J.* **236**, 351
Dressler, A., Oemler, A., Butcher, H.R., & Gunn, J.E., 1994, *Ap. J.* **430**, 107
Dubinski, J., Mihos, J.C., & Hernquist, L., 1996, *Ap. J.* **462**, 576
Duc, P.-A., & Mirabel, I.F., 1994, *A. & A.* **289**, 83
Faber, S.M., et al., 1995, *Stellar Populations* (Kluwer, Dordrecht), p. 249
Gardiner, L.T., & Noguchi, M., 1996, *M. N. R. A. S.* **278**, 191
Geisler, D., Lee, M.G., & Kim, E., 1996, *A. J.* **111**, 1529
González, J.J., 1993, Ph.D. thesis, UC Santa Cruz
Hernquist, L., & Barnes, J.E., 1991, *Nature* **354**, 210
Hernquist, L., & Spergel, D.N., 1992, *Ap. J.* **399**, L117
Hibbard, J.E., et al., 1994, *A. J.* **107**, 67
Hibbard, J.E., & Mihos, J.C., 1995, *A. J.* **110**, 140
Hibbard, J.E., & van Gorkom, J.H., 1996, *A. J.* **111**, 655

Hubble, E. 1936, *The Realm of the Nebulae* (Yale Univ. Press, New Haven), Chap. 2
Hunsberger, S.D., Charlton, J.C., & Zaritsky, D. 1996, *Ap. J.* **462**, 50
Ibata, R.A., Gilmore, G., & Irwin, M.J., 1995, *M. N. R. A. S.* **277**, 781
Jore, K.P., Broeils, A.H., & Haynes, M.P., 1996, *A. J.* **112**, 438
Kauffmann, G., 1996, *M. N. R. A. S.* **281**, 487
Kenney, J.D.P., Rubin, V.C., Planesas, P., & Young, J.S., 1995, *Ap. J.* **438**, 135
Kennicutt, R.C., Schweizer, F., & Barnes, J.E., 1998, *Galaxies: Interactions and Induced Star Formation*, eds. D. Friedli, L. Martinet, & D. Pfenniger (Springer, Berlin)
Kuijken, K. 1993, *P. A. S. P.* **105**, 1016
Lake, G., & Dressler, A., 1986, *Ap. J.* **310**, 605
Lavery, R.J., & Henry, J.P. 1988, *Ap. J.* **330**, 596
Lavery, R.J., & Henry, J.P., 1994, *Ap. J.* **426**, 524
Lin, D.N.C., Jones, B.F., & Klemola, A.R., 1995, *Ap. J.* **439**, 652
Malin, D., & Hadley, B., 1997, *The Second Stromlo Symposium: The Nature of Elliptical Galaxies* (A.S.P., San Francisco), p. 460
Mihos, J.C., Dubinski, J., & Hernquist, L., 1998, *Ap. J.* **494**, in press
Miller, B.W., Whitmore, B.C., Schweizer, F., & Fall, S.M., 1997, *A. J.* **114**, in press
Mirabel, I.F., Lutz, D., & Maza, J. 1991, *A. & A.* **243**, 367
Morganti, R., et al., 1997, *A. J.* **113**, 937
O'Connell, R.W., 1980, *Ap. J.* **236**, 430
O'Connell, R.W., 1994, *The Nuclei of Normal Galaxies* (Kluwer, Dordrecht), p. 255
Oemler, A., 1974, *Ap. J.* **194**, 1
Oemler, A., Dressler, A., & Butcher, H.R., 1997, *Ap. J.* **474**, 561; and private commun.
Pfenniger, D., 1998, this volume
Plana, H., & Boulesteix, J., 1996, *Fresh Views of Elliptical Galaxies* (A.S.P., San Francisco), p. 113
Prada, F., Gutiérrez, C.M., Peletier, R.F., & McKeith, C.D., 1996, *Ap. J.* **463**, L9
Prieur, J.-L., 1990, *Dynamics and Interactions of Galaxies* (Springer, Berlin), p. 72
Reshetnikov, V.P., & Combes, F. 1994, *A. & A.* **291**, 57
Richter, O.-G., Sackett, P.D., & Sparke, L.S., 1994, *A. J.* **107**, 99
Roos, N., & Aarseth, S.J., 1982, *A. & A.* **114**, 41
Rubin, V.C., Graham, J.A., & Kenney, J.D.P. 1992, *Ap. J.* **394**, L9
Schechter, P.L., & Dressler, A., 1987, *A. J.* **94**, 563
Schiminovich, D., et al., 1994, *Ap. J.* **423**, L101
Schiminovich, D., et al., 1995, *Ap. J.* **444**, L77
Schweizer, F., 1982, *Ap. J.* **252**, 455
Schweizer, F., 1995, *Stellar Populations* (Kluwer, Dordrecht), p. 275
Schweizer, F., 1996, *A. J.* **111**, 109
Schweizer, F., Miller, B.W., Whitmore, B.C., & Fall, S.M., 1996, *A. J.* **112**, 1839
Schweizer, F., & Seitzer, P., 1988, *Ap. J.* **328**, 88
Schweizer, F., & Seitzer, P., 1992, *A. J.* **104**, 1039
Searle, L., & Zinn, R. 1978, *Ap. J.* **225**, 357
Stockton, A., & Farnham, T., 1991, *Ap. J.* **371**, 525
Thakar, A.R., Ryden, B.S., Jore, K.P., & Broeils, A.H., 1997, *Ap. J.* **479**, 702
Toomre, A., 1977, *Evolution of Galaxies & Stellar Populations* (Yale, New Haven), p.401
Toomre, A., & Toomre, J., 1972, *Ap. J.* **178**, 623 (TT)
Tóth, G., & Ostriker, J.P., 1992, *Ap. J.* **389**, 5
van Gorkom, J.H., et al., 1989, *A. J.* **97**, 708
van Gorkom, J.H., & Schiminovich, D., 1997, *The Second Stromlo Symposium: The Nature of Elliptical Galaxies* (A.S.P., San Francisco), p. 310
Walker, I.R., Mihos, J.C., & Hernquist, L., 1996, *Ap. J.* **460**, 121
Whitmore, B.C., et al., 1995, *Ap. J.* **454**, L73
Wielen, R. (ed.) 1990, *Dynamics and Interactions of Galaxies* (Springer Verlag, Berlin)
Zaritsky, D., & Rix, H.-W., 1997, *Ap. J.* **477**, 118
Zwicky, F., 1956, *Ergebnisse d. exakten Naturw.* **29**, 344

A REVIEW OF HIGH-REDSHIFT MERGER OBSERVATIONS

R.G. ABRAHAM[1]
Royal Greenwich Observatory
Madingley Road, Cambridge
CB3 OEZ United Kingdom

Abstract. Evolution in the merger rate as a function of redshift is *in principle* the key observable testing hierarchical models for the formation and evolution of galaxies. However, *in practice*, direct measurement of this quantity has proven difficult. In this opening review I outline the current best estimates for the merger rate as a function of cosmic epoch, focusing mostly upon recent advances made possible by deep ground-based redshift surveys and morphological studies undertaken with *HST*. I argue that a marriage of these techniques, in an attempt to determine the space density of mergers amongst the abundant morphologically peculiar population at high redshifts, is probably the most promising currently-available avenue for determining the prevalence of mergers at high redshifts. However, resolved kinematical studies, which seem set to become available in the next few years, are probably the best hope for a definitive determination of the space density of mergers at high redshifts.

1. Observational Techniques:

Three observational techniques have been used to probe changes in the merger rate as a function of cosmic epoch. These are (a) studies of angular and physical correlation functions, (b) pair counts, and (c) morphological studies. The advantages (+) and disadvantages (−) of each of these approaches for studying high-redshift mergers can be crudely summarized as follows.

[1]Current address: Institute of Astronomy, Cambridge University, Madingley Road, Cambridge CB3 OHA, UK.

1.1. CORRELATION FUNCTIONS

+ A close connection to large-scale structure work via the clustering statistics $w(\theta)$ and $\xi(r)$. Since mergers can be identified as the end-products of large-scale clustering, changes in the correlation function at small radius provide a conceptual link between the large-scale and small-scale regimes being probed with the same statistics.
+ Biases in measuring correlation functions are fairly well-understood.

– Measurement of correlation functions is best suited to large samples.
– These statistics are hard to measure at small radii, particularly when galactic components (such as giant HII regions) become difficult to distinguish from merging companions. This is often the case when probing distant galaxies in the rest-frame ultraviolet, *eg.* in the Hubble Deep Field, where knots of star-formation cannot easily be distinguished from companions (Colley et al. 1996).
– *Not really measuring merger rate (see below).*

1.2. PAIR COUNTS APPROACH

+ Simple statistics to measure (described in detail in the next section).
+ Conceptually the pair-counts approach is an integration over the two-point correlation function, with better signal-to-noise properties at small radii, so the biases are also fairly well-understood.

– Like correlation functions, the statistic becomes ambiguous when merging companions become indistinguishable from galactic components.
– *Not really measuring merger rate (see below).*

The most important criticism common to both the correlation function and pair counts approaches is that neither provides a direct measure of the merger rate. The fundamental difficulty is *the uncertain physical timescale over which merging occurs*, given physical proximity between galaxies. Both statistics probe the fuel "reservoir" available for close gravitational interaction, but what we really seek is an understanding of the rate at which the merger "engine" operates in converting these galaxies into the by-products of mergers. To understand this, one must observe the mergers in progress. So the third approach must necessarily be a morphological one:

1.3. MORPHOLOGICAL APPROACH

+ Direct observation of *mergers in progress.*

– Poorly-understood biases (*eg.* morphological K-corrections).

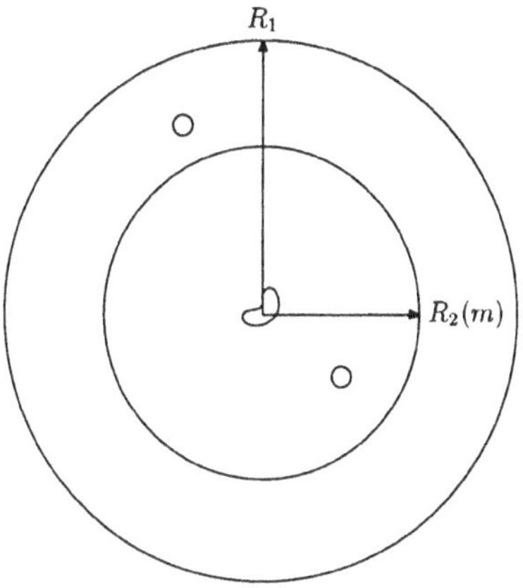

Figure 1. Cartoon showing the search criteria for the "mixed" redshift-photometric analysis of pair-density evolution undertaken by Yee & Ellingson (1995). The search radius R_1 is fixed by the physical scale being probed, while radius R_2 varies with the companion magnitude.

Because of space limitations, and because they are well-reviewed elsewhere, this review ignores studies of the correlation function (eg. Neuschaefer et al. 1997), and touches only rather superficially on recent studies on pair-counts in §2, in order to focus mostly on summarizing very recent progress made on morphological studies at high-redshift with *HST* in §3.

2. Evolution in Density of Pairs

Because existing redshift surveys are not yet deep enough to allow a direct analysis of pair-density evolution to be undertaken, pair count studies can be broadly grouped into two categories: (1) pure photometric analyses with no redshift information, and (2) photometric searches around galaxies with known redshifts. A flavor for the methodology adopted in the latter "mixed" category of redshift-photometric surveys is sketched out in Fig. 1, which illustrates the technique adopted by Yee & Ellingson (1995). Searches are conducted within an aperture projected onto the sky, defined by a projected *physical* radius of $R_1 = 20h^{-1}$ kpc of a galaxy of known redshift. But because low-redshift galaxies will have a large search radius that is likely to be littered with faint unrelated background galaxies, additional restrictions must be placed on the search radius based on the apparent magnitude of the putative companion galaxy. Each companion must lie within a radius $R_2(m) = D_{\rm nm}(m)/3$ where $D_{\rm nm}(m)$ is the expected nearest neighbor distance for galaxies brighter than companion magnitude m ($D_{\rm nm}(m) = \frac{1}{2\sigma(m)^{1/2}}$ (Rose 1977), where $\sigma(m)$ is the surface density of galaxies brighter than m.) A correction can then be made for the expected mean number of unrelated galaxies in the search area, assuming Poisson statistics: $\overline{N} = P_{\rm o}\left[1 + 2\ln\frac{R_1}{R_2(m_l)}\right]$

Results from all published surveys of pair-density evolution (from both classes of survey) are summarized below, in the form of power-law evolution in the density of pairs $(1+z)^n$:

2.1. PURE PHOTOMETRIC ANALYSES

1. Zepf & Koo (1989) $\Rightarrow (1+z)^{4.0\pm2.0}$
2. Burkey et al. (1994) $\Rightarrow (1+z)^{3.5\pm0.5}$
3. Carlberg, Pritchet, & Infante (1994) $\Rightarrow (1+z)^{3.5\pm1.0}$
4. Woods, Fahlman, & Richer (1995) $\Rightarrow$ No evolution!

2.2. SEARCHES AROUND GALAXIES WITH KNOWN REDSHIFTS

1. Yee & Ellingson (1995) $\Rightarrow (1+z)^{4.0\pm1.5}$
2. Patton et al. (1997) $\Rightarrow (1+z)^{2.8\pm0.9}$

The recent work by Patton et al. is based on the largest redshift sample to date: 545 field galaxies with a mean redshift of $z = 0.3$. Clearly these published surveys do not yet probe very far out in redshift space, but it is curious that with the exception of the pure photometric study by Woods and collaborators, all pair-count work to date (but not all correlation function work, eg. Neuschaefer et al. 1997) is roughly consistent with the $(1+z)^{2.7\pm0.5}$ increase in co-moving luminosity density from the Canada-France Redshift survey (Lilly et al. 1996). It is perhaps worth re-stating that the pair fraction increase is *not* the same thing as the merger rate increase: conversion between these is sensitive to assumptions made with regard to merger timescales. No particular consensus exists amongst the various authors regarding the appropriate conversion between pair density and merger rate. For example, given pair evolution with the form $(1+z)^n$, the merger rate is variously assumed to take the form $(1+z)^{n-1}$ by Burkey et al., $(1+z)^{n+1}$ by Carlberg et al., and $(1+z)^n$ by Yee & Ellingson. It is interesting that from the distribution of the projected distances of companions as a function of redshift, Patton et al. conclude that mergers are likely to occur over timescales of 150 Myr – 400 Myr.

3. High-Redshift Morphology

Convincing observations of high-redshift galaxy morphology have only become possible recently, with WF/PC2 observations on *HST*, and the resulting flood of imaging data has resulted in a great increase in our understanding of the field galaxy population at high redshifts. Selected highlights (with apologies to many colleagues whose work cannot be included here due to space limitations) from recent *HST*-based imaging work, relevant to high-redshift mergers, are summarized below. I then go on to describe what seem to me a number of "key issues" with regard to understanding the connection between these imaging observations and high-redshift merger scenarios.

3.1. *HST* FIELD SURVEYS WITH MORPHOLOGICAL INFORMATION

Early work from the Medium Deep Survey (MDS) reported an excess of faint peculiar systems (Griffiths et al. 1994), but the excess become much more convincing with the extension of this work to include number counts as a function of magnitude (Glazebrook et al. 1995; Driver et al. 1995; Abraham et al. 1996). Worries with regard to bulk misclassification of peculiar galaxies have been eased by the incorporation of objective machine-based classifications, which are now routinely being used to supplement (and in some cases replace) visual morphological classifications from a number of different surveys (Abraham et al. 1996; Odewahan et al. 1996; Naim et al. 1997; Brinchmann et al. 1997). The incorporation of redshift information, probing the regime out to roughly $z < 1$, has generally confirmed the photometric work, and in turn pushed much further in terms of our understanding of the contributions of the morphological classes to the star-formation history of the Universe (Cowie, Hu, & Songaila 1995; Schade et al. 1996; Pascarelle et al. 1996; Brinchmann et al. 1997). Imaging follow-up observations of Lyman-limit systems have in turn yielded views of systems at $z > 2$ (Giavalisco et al. 1996). Because of their unprecedented depth, and intensive spectroscopic and photometric redshift follow-ups (Cohen et al. 1996; Lanzetta et al. 1996; Mobasher et al. 1996; Lowenthal et al. 1997; Phillips et al. 1997; Sawicki et al. 1997), Hubble Deep Field observation of faint galaxy morphology are the likely to yield the greatest insights into evolving distribution of morphological types at high redshifts for some time to come (Abraham et al. 1996; van den Bergh et al. 1996; Bouwens et al. 1997; Guzman et al. 1997). Eagerly anticipated are the results from high-resolution infrared observations in the HDF with NICMOS, and forthcoming observations of the HDF South. Deeper observations (and, perhaps more importantly, greater sky coverage) will perhaps have to await completion of the *HST* Advanced Camera, and hopefully *NGST*.

While recent work (Fig. 2) has left little doubt that much of the high-redshift Universe is morphologically peculiar (about 1/3 of all galaxies by $I_{300W} \sim 24$ mag), the nature of these systems is currently unknown. In the context of the present meeting, the bottom-line question seems to be: what fraction of morphologically peculiar galaxies are mergers in progress? Unfortunately this question cannot be answered at the present time. Part of the problem is simply one of subjective and confusing taxonomy used to describe the morphologies of faint galaxies. For example, are "Morphologically Peculiar" (Griffiths et al. 1994), "Chainlike" (Cowie, Hu, & Songaila

1995), "Blue Nucleated" (Schade et al. 1996), "Irregular/Peculiar/Merger" (Glazebrook et al. 1995; Abraham et al. 1996a,b; Brinchmann et al 1997), and "Tadpole" (van den Bergh et al. 1996) galaxies similar objects simply denoted by different names, or a true reflection in the diversity of the morphologies of galaxies in the distant Universe? At a more mundane level, is bandshifting of the rest-frame of observation (the so-called "morphological K-correction") playing an important role, by fooling us into mistaking systems with normal UV morphology for new classes of galaxy? I have argued (Abraham et al. 1996a,b) that problems with both taxonomy and morphological K-corrections can be circumvented by abandoning conventional galaxy classification schemes in favor of objective classifications, based on measurement of simple structural parameters (such central concentration and asymmetry), and by calibrating such measurements against simulations of the appearance of galaxies at a range of redshifts. The best example of this approach is recent work by Brinchmann et al. (1997). In this work, *HST* imaging and ground-based spectroscopic data for ~ 300 galaxies with good completeness to $I < 22$ mag are used demonstrate that the peculiar galaxy excess is already in place at redshifts where morphological K-corrections have a minimal effect. Simple and objective approaches to quantifying galaxy morphology seems to me the safest course until the observational biases are better understood, but of course one pays a price by "smoothing over" much of the interesting diversity (denoted with correspondingly charming nomenclature) seen in galaxy forms on deep images.

What fraction of this diverse set of peculiar galaxies are actually mergers in progress? The problems inherent in answering this question are illustrated in Color plate 2 (p. xx), which shows candidate Lyman limit systems in the Hubble Deep Field with $I_{814} < 25$ mag. Obviously most are morphologically peculiar, but how many readers can honestly claim that these resemble the appearance of local merging systems? The difficulties inherent in identifying mergers amongst the distant morphologically peculiar population are made clear by simulations recently published by Hibbard and Vacca (1997), who use *HST* WF/PC camera ultraviolet data to predict the appearance of the high-redshift counterparts to local merging starburst systems. Because the usual signatures of mergers (tidal tails, distorted disks) are no longer visible at high redshifts, merging starbursts seem to provide at least qualitatively reasonable counterparts to many faint peculiar galaxies (Fig. 3). While very suggestive, such simulations need to be interpreted with some caution, because they do not explicitly account for evolutionary effects which are known to be important at high redshifts. For example, in the simulations shown in Fig. 3 the smooth components of the galaxy are invisible largely because their stellar populations are evolved in the local reference images and are subject to strong K-corrections. However at

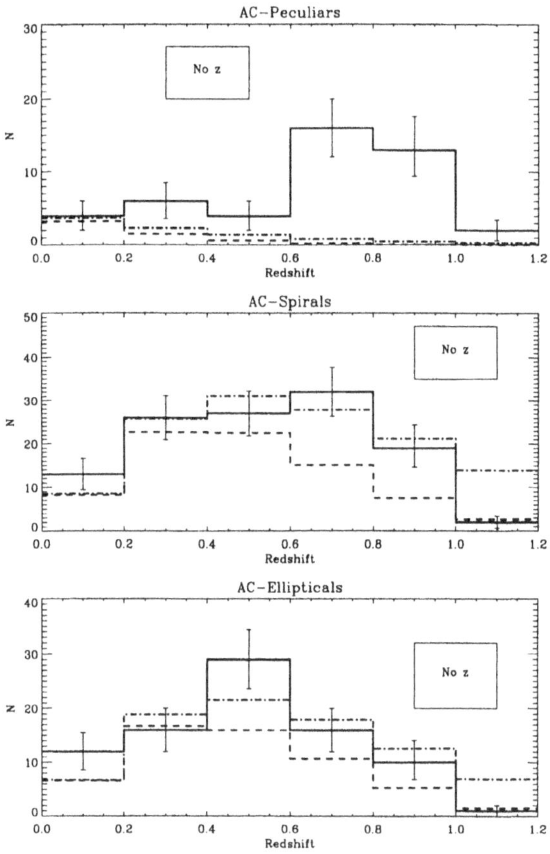

Figure 2. Morphologically segregated redshift distributions (taken from Brinchmann et al. 1997) for irregular/peculiar/merging galaxies (top), spirals (middle), and ellipticals (bottom). The sample consists of $\sim$ 300 galaxies from the CFRS (Lilly et al. 1995) and LDSS (Ellis et al. 1996) redshift surveys. Classifications have been based upon measurements of central concentration and asymmetry on deep *HST* WF/PC2 I_{814}-band images, calibrated by simulations for the effects of bandshifting of the rest-frame of observation. (Although over the redshift range probed such effects are small.) Dashed and dot-dashed lines are the predictions of no-evolution, and mild evolution models. Note the marked excess in the number of irregular/peculiar/merging systems at high redshifts.

$z \sim 2$ *all* stellar populations are relatively young. More detailed simulations incorporating explicitly the effects of evolution would be valuable.

Clearly the best way forward will be to incorporate dynamical information to determine directly which peculiar galaxies show distinct kinematical sub-components. Unfortunately these observations are not currently feasible, although they may soon become possible with adaptive optics and the new generation of 8m-class telescopes. In the meantime, a promising ap-

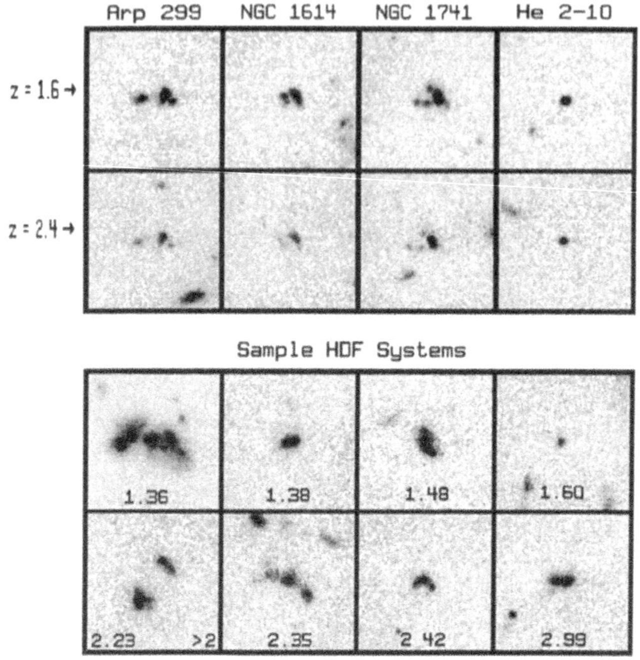

Figure 3. (Top montage:) Simulations (based on ultraviolet *HST* WF/PC observations) taken from Hibbard & Vacca (1997), showing the predicted appearance of local well-known merging starburst galaxies as seen at high-redshifts in the Hubble Deep Field. (Bottom montage:) Morphologically similar galaxies in the Hubble Deep Field.

proach to quantifying the fraction of mergers amongst the distant peculiar galaxy population may be to measure statistics which are relatively insensitive to image distortions resulting from bandshifting and surface-brightness biases, but which track probable merger activity. One such statistic is the "Lee Ratio", a measure of image bimodality. This statistic been applied to images of galaxies in the CFRS survey (Fig. 4) and to HDF galaxies, with the result that around $\sim$ 40% of faint peculiar systems are significantly bimodal, with an $\sim (1+z)^3$ increase in the merger rate (Le Févre et al., in preparation).

Another promising approach to quantifying the fraction of high-redshift mergers is to better understand the nature of the star-formation activity in distant peculiar galaxies, in order to test if their star-formation histories are consistent with merging. One overlooked method for accomplishing this is modeling of the *internal* pixel-by-pixel colors of these galaxies. This approach splits the galaxy up into components under the assumption that morphology can be used to identify stellar populations. In prin-

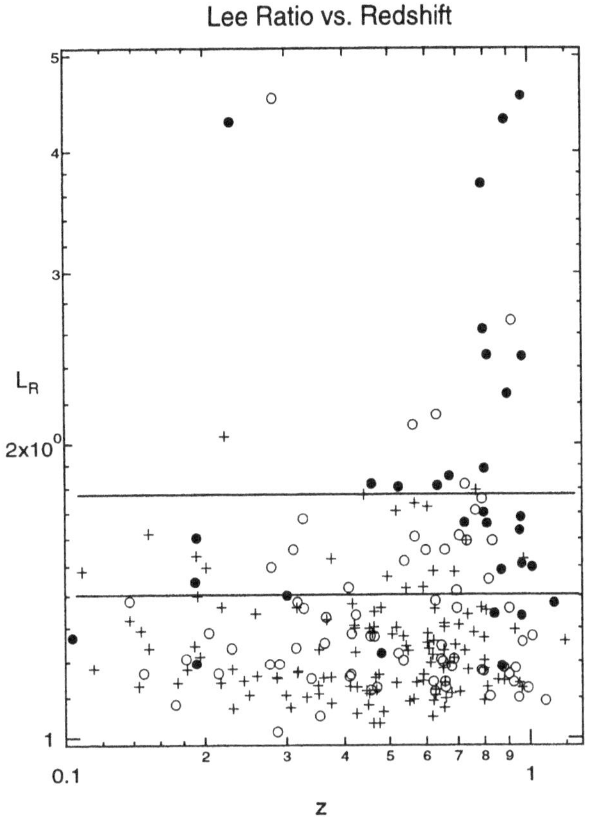

Figure 4. Lee ratio L_R bimodality index for *HST* WF/PC2 images of galaxies in the Canada-France Redshift survey, taken from Le Févre et al. (in preparation). Plot symbols correspond to visual classifications of the galaxies as "major mergers" (solid circles), "minor mergers" (open circles), and undisturbed galaxies (crosses). There is a marked increase in the number of merger candidates at $z > 0.5$. Note also the fairly good agreement between classification as a major merger and high Lee ratio.

ciple, this breaks much of the degeneracy inherent in population synthesis modeling which treat galaxies as point sources, and puts constraints not only on merger activity, but also on the dust content, relative ages of disk and bulge, and general star-formation history. As an example, Color plate 3 (p. xxi) shows two "chain galaxies" in the Hubble Deep Field. The resolved color analysis indicates that these systems are showing well-organized star-formation activity – a strong constraint on possible merging scenarios. The "knots" in these galaxies are unlikely to be individual galaxies; star-formation seems to be triggered linearly along the body of the galaxy, starting from a initial "seed" starburst, with individual knots that are about as luminous as "super star clusters" that are the putative progenitors of globular clusters. If such systems are mergers in progress then a mechanism for igniting the chain reaction via mergers must be invoked. Work is currently in progress (Abraham et al. 1998) to apply these ideas to all peculiar galaxies in the HDF in order to determine which galaxies are the strongest merger candidates. Until kinematical studies become available with resolution sufficient to detect merging subsystems at high redshifts, color-based studies may be the best way to detect distinct physical subcomponents in morphologically peculiar galaxies.

4. Conclusion

The most recent pair-count analyses seem to suggest that the fraction of physical pairs grows as $\sim (1+z)^3$. However, the conversion between pair fraction and merger rate is uncertain, and the pair count work so far published is limited to fairly low redshifts ($\langle z \rangle \sim 0.3$). At higher redshifts morphological work with *HST* indicates that by $I = 24$ mag something over 30% of all galaxies are morphologically peculiar. Simulations and follow-up spectroscopic work suggest this excess in morphologically peculiar systems is a physical effect, and not merely the result of "morphological K-corrections". The fraction of mergers amongst these morphologically peculiar galaxies is unknown, because obviously merging local systems, such as nearby major starburst galaxies, no longer appear like conventional mergers at high redshifts. A preliminary analysis of image bimodality (a robust parameter that in principle flags major mergers even at high redshift) amongst $I < 22$ mag peculiar systems with $z < 1$ suggests that around 40% the morphologically peculiar galaxies are strongly bimodal, and thus probably merging. This is consistent with a merger rate increasing like $\sim (1+z)^3$. Internal color analyses of morphologically peculiar systems is an interesting next step in probing high-redshift mergers, leading ultimately to kinematical investigations in the next few years.

5. Acknowledgments

I thank my collaborators Richard Ellis, Nial Tanvir, Sidney van den Bergh, Karl Glazebrook, Andy Fabian, and Basilio Santiago for their contributions to our high-redshift morphology projects. I also thank Jarle Brinchmann and Olivier Le Févre for permission to show figures in advance of publication, and Bill Vacca for useful discussions.

References

Abraham, R. G., Tanvir, N. R., Santiago, B. X., Ellis, R. S., Glazebrook, K., & van den Bergh, S. (1996) MNRAS, 279, 47

Abraham, R. G., van den Bergh, S., Glazebrook, K., Ellis, R. S., Santiago, B. X., Surma, P., & Griffiths, R. E. (1996) ApJ, 107, 1

Burkey, J. M., Keel, W. C., Windhorst, R. A., & Franklin, B. E. (1994) ApJ, 429, 13

Carlberg, R. G., Pritchet, C. J., & Infante, L. (1994) ApJ, 435, 540

Clements, D. L. & Couch, W. J. (1996) MNRAS, 280, 43

Cohen, J. G., Cowie, L. L., Hogg, D. W., Songaila, A., Blandford, R., Hu, E. M., & Shopbell, P. (1996) ApJ, 471L, 5

Colley, W. N., Rhoads, J. E., & Ostriker, J. P., Spergel, D. N. (1996) ApJ, 473L, 63

Cowie, L. L., Hu, E. M., & Songaila, A. (1995) AJ, 110, 1576

Cowie, L. L., Hu, E. M., & Songaila, A. (1995) Nature, 377, 603

Driver, S. P., Windhorst, R. A., Phillipps, S., & Bristow, P. D. (1996) ApJ, 461, 525

Driver, S. P., Windhorst, R. A., Ostrander, E. J., Keel, W. C., Griffiths, R. E., & Ratnatunga, K. U. (1995) ApJ, 449, 23

Driver, S. P., Windhorst, R. A., & Griffiths, R. E. (1995) ApJ, 453, 48

Ellis, R. S., Colless, M., Broadhurst, T., Heyl, J., & Glazebrook, K. (1996) MNRAS, 280, 235

Giavalisco, M., Steidel, C. C., & Macchetto, F. D. (1996) ApJ, 470, 189

Giavalisco, M., Livio, M., Bohlin, R. C., Macchetto, F. D., & Stecher, T. P. (1996) AJ, 112, 369

Glazebrook, K., Ellis, R., Santiago, B., & Griffiths, R.(1995) MNRAS, 275, 19

Griffiths, R. E., et al. (1994) ApJ, 435L, 19

Guzman, R., Gallego, J., Koo, D. C., Phillips, A. C., Lowenthal, J. D., Faber, S. M.; Illingworth, G. D., & Vogt, N. P. (1997) ApJ, 489, 559

Hibbard, J. E., & Vacca, W. D. (1997) AJ, 114, 1741

Lanzetta, K. M., Yahil, A., & Fernandez-Soto, A. (1996) Nature, 386, 759

Lilly, S. J., Le Févre, O., Crampton, D., Hammer, F., & Tresse, L. (1995) ApJ, 455, 501

Lilly, S. J., Le Févre, O., Hammer, F., Crampton, D. (1996), ApJ, 460L, 1

Lowenthal, J. D., Koo, D.C., Guzman, R., Gallego, J., Phillips, A. C., Faber, S. M., Vogt, N. P., Illingworth, G. D., Gronwall, C. (1997) ApJ 481, 673

Mobasher, B., Rowan-Robinson, M., Georgakakis, A., & Eaton, N. (1996) MNRAS, 282, 7

Naim, A., Ratnatunga, K. U., & Griffiths, R. E. (1997) ApJ, 111, 357

Naim, A., Ratnatunga, K. U., & Griffiths, R. E. (1997) ApJ, 476, 510

Neuschaefer, L. W., Im, M., Ratnatunga, K. U., Griffiths, R. E., & Casertano, S. (1997) ApJ, 480, 59

Odewahn, S. C., Windhorst, R. A., Driver, S. P., & Keel, W. C. (1996) ApJ, 472, 13

Owens, E. A., Griffiths, R. E., & Ratnatunga, K. U. (1996) MNRAS, 281, 153

Pascarelle, S. M., Windhorst, R. A., Driver, S. P., Ostrander, E. J., & Keel, W. C. (1996) ApJ, 456, 21

Patton, D. R., Pritchet, C. J., Yee, H. K. C., Ellingson, E., & Carlberg, R. G. (1997) ApJ, 475, 29

Phillips, A. C., Guzman, R., Gallego, J., Koo, D. C., Lowenthal, J. D., Vogt, N. P., Faber, S. M., & Illingworth, Garth D. (1997) ApJ, 489, 543

Rose, J. A. (1977) ApJ, 211, 311

Sawicki, M. J., Lin, H., & Yee, H. K. C. (1997) AJ, 113, 1

Schade, David, Lilly, S. J., Crampton, David, Hammer, F., Le Fevre, O., & Tresse, L. (1995) ApJ, 451, 1

van den Bergh, S., Abraham, R. G., Ellis, R. S., Tanvir, N. R., Santiago, B. X., & Glazebrook, K. G. (1996) AJ, 112, 359

Windhorst, R. A., Gordon, J. M., Pascarelle, S. M., Schmidtke, P. C., Keel, W. C., Burkey, J. M., & Dunlop, J. S. (1994) ApJ, 435, 577

Woods, D., Fahlman, G. G., & Richer, H. B. (1995) ApJ, 454, 32
Yee, H. K. C., & Ellingson, E. (1995) ApJ, 445, 37
Zepf, S. E., & Koo, D. C. (1989) ApJ, 337, 34

WHAT CAN WE LEARN FROM THE LOCAL GROUP ABOUT THE ROLE OF INTERACTIONS IN GALAXY FORMATION

K.C. FREEMAN

Mount Stromlo and Siding Spring Observatories
The Australian National University
Canberra, AUSTRALIA

1. Introduction

This talk is mainly about the halos and bulges of Local Group galaxies, in particular those of the Milky Way and M31, and what they can tell us about the role of interactions in galaxy formation. Bulges are potentially of particular interest for galaxy formation because they are widely suspected to be the seeds for galaxy formation. Sites of active star formation at high redshift are often regarded as spheroids in the process of formation. In this talk, I will say little about disks because their formation is conceptually understood, although many details remain uncertain. To form such flat systems, a fairly undisturbed dissipative process is needed. This process occurred fairly early: the disks appear to be in place by a redshift $z = 1$ (Ellis 1997), with a distribution of scalelengths that is similar to the distribution at $z = 0$.

2. The Milky Way

The Milky Way has two non-disk luminous components: the metal-poor stellar halo and the bulge. Interactions and accretion of small satellite systems may be a very important element in the formation of the halo. On the other hand, the bulge may tell us little about the interaction history of the Milky Way.

2.1. THE METAL-POOR HALO

This component, which includes the metal-poor globular clusters and field stars, has a mass of only about 1×10^9 $M_\odot$. Its stars have a heavy element abundance [Fe/H] < -1. It is kinematically hot and slowly rotating, and is probably unrelated to the bulge: the kinematics of the halo stars ([Fe/H]$<$ -1) and the more metal-rich bulge stars are very different in the region where they overlap (*eg.* Morrison & Harding 1993). This slowly rotating (maybe even retrograde: Majewski 1992) subsystem in an otherwise rapidly rotating galaxy requires explanation. It is now fairly widely believed that the metal-poor halo comes mainly from the accretion of small metal-poor dwarf galaxies during and after disk formation, as suggested by Searle and Zinn (1978). The evidence includes:

- the weak dependence of the kinematics of halo stars on metallicity (*eg.* Beers & Sommer-Larsen 1995). A stronger dependence would be expected in a more monolithic dissipative picture.
- the anisotropy of the velocity ellipsoid for halo stars changes with radius, from radial anisotropy in the inner halo to tangential anisotropy in the outer halo (Sommer-Larsen et al. 1994). This can be readily understood from the effects of dynamical friction and tidal disruption on small dwarf galaxies (*eg.* Quinn & Goodman 1986).
- the presence of moving stellar groups (*eg.* Eggen 1979) and stellar streams (Majewski 1994, Lynden-Bell & Lynden-Bell 1995) in the halo. These groups and streams presumably represent the debris of accreted and disrupted satellites.
- on-going accretion: the apparently disrupting Sgr dwarf (Ibata et al. 1994) is a very direct example. The young metal-poor stars discovered by Preston et al. (1994) are another example. These stars are typically older than about 3 Gyr, have abundances [Fe/H] < -1, a velocity dispersion of about 90 km s^{-1} and rotation of about 130 km s^{-1} (*ie.* intermediate between the halo and the disk). These stars contribute about 10% of the local halo density, and could come from the accretion of Carina-like dwarf spheroidal systems with intermediate-age populations. The total associated accreted mass is about 10^8 $M_\odot$.

In summary, the process of halo-building by accretion is a continuing process in the Milky Way, not something that happened only long ago.

2.2. THE BULGE OF THE MILKY WAY

Although the origin of bulges is an interesting and important problem, and may in some cases be relevant to the formation of the parent galaxies, it is clear that bulges are not an essential element of galaxy formation, because many disk galaxies do not have bulges. In the Milky Way, the boxy bulge is barlike. Long ago, de Vaucouleurs (1964) classified our Galaxy as SAB(rs)bc from a range of morphological arguments, and Gerhard (1997) has summarized the detailed recent evidence for the galactic bulge/bar. In brief, the evidence includes:

- the distribution and kinematics of bulge stars (the *COBE*/DIRBE light distribution, clump stars, OH/IR stars, planetary nebulae ...).
- the gas distribution and kinematics in the inner Galaxy.
- the high optical depth for microlensing towards the galactic bulge.

N-body simulations of self-gravitating disks strongly suggest that most of the small boxy/peanut bulges, like the bulge of the Milky Way, are bars arising from planar and vertical instabilities of disks (*eg.* Combes et al. 1990; Pfenniger & Friedli 1991). Some of these structures may themselves be triggered by interactions (*eg.* Noguchi 1987). Until now, observational verification of the bar-like nature of boxy bulges has been difficult, because boxy bulges are seen most clearly in edge-on spirals, and then it is not so clear whether they are barred. Recently Kuijken & Merrifield (1995) have devised a kinematical test of the bar-like nature of near-edge-on boxy/peanut bulges. This test, which depends on the properties of the two principal orbit families in the gravitational field of a rotating bar, is particularly effective and direct for galaxies with extended emission lines in the region of the bulge. Bureau (1997) has applied the Kuijken-Merrifield test to 19 edge-on boxy/peanut bulges and a non-boxy control sample. He found that 15 of the 19 have extended emission lines, so could show the Kuijken-Merrifield effect. Of these 15, 11 show the effect clearly, 3 are very dusty so the effect may be masked, and one galaxy is disturbed by interaction. None of the 7 galaxies in the non-boxy control sample shows the effect. One can conclude from his work that most boxy/peanut bulges are indeed bar-like.

If these boxy bar/bulges arise from instabilities of an equilibrium disk or a forming disk, then they are just consequences of disk formation and dynamics, and are not themselves the seeds of galaxy formation. If this is true, then they are probably not very relevant to understanding interactions in galaxy formation. (The small boxy bulges are certainly interesting in other contexts. I have skipped over some striking properties of these small bulges, such as their exponential structure (not $r^{1/4}$, Courteau et al. 1996) and the similar exponential scale heights ($\simeq$ 300 pc) of the Milky Way's disk and bulge.)

3. The Bulge of M31

M31 has a *real* $r^{1/4}$ bulge, from a radius of 200 pc out to 20 kpc (Pritchet & van den Bergh 1994). Its bulge is flattened, with an axial ratio of about 0.6 (E4).

The chemical properties of the outer bulge or halo of M31 are very different from the halo of the Milky Way. The mean abundance [Fe/H] $\simeq -0.6$, much higher than for the Milky Way. The outer bulge shows a large spread in chemical abundance, from about [Fe/H] $= -2$ to -0.2 (*eg.* Durrell et al. 1994, Couture et al. 1995, Rich et al. 1996, Holland et al. 1996), with little or no abundance gradient out to a radius of 40 kpc.

I suspect that the prominent, more metal-rich bulge dominates the metal-poor population at all radii, as it does in the giant ellipticals. Is it just a semantic point, whether one calls this "outer bulge" a bulge or a halo? I think not, and most of the rest of this talk will be used to argue why not.

4. The Structure and Chemical Properties of Large Bulges

These large bulges, as in M31, M104 (the Sombrero galaxy) and NGC 7814, follow the $r^{1/4}$ surface brightness distribution. An $r^{1/4}$ light distribution is usually associated with a fairly violent merger or aggregation history (*eg.* van Albada and van Gorkom 1977, Barnes 1988). Chemically, the bulges of spirals show a (Mg/Fe) - absolute magnitude relation in the same sense as for ellipticals: the brighter bulges show a more marked overabundance of Mg relative to Fe (Jablonka et al. 1996). The usual interpretation of this effect is that, after the first major burst of star formation, supernova-driven winds in the more luminous systems remove the remaining gas quickly, and so reduce the subsequent Fe-enrichment by the slower SN of type I. If this is all correct, then it implies that the formation of large bulges occurred early and quickly, much as for giant ellipticals.

5. The Rotation of Bulges and Ellipticals

Bulges lie close to the oblate isotropic rotator curve in the $(V/\sigma - \epsilon)$ plane while, from the kinematics of their *inner* regions ($r \lesssim r_e$), the giant ellipticals mostly lie well below the oblate curve. This suggested that the giant ellipticals have lower specific angular momentum J/M relative to the spirals (see Fall, 1983). The apparent difference in specific angular momentum was a long-standing puzzle, because cosmological simulations give similar distributions of J/M in high density and low density regions of the universe, which are usually associated with ellipticals and spirals respectively. So what has happened to the angular momentum of the giant ellipticals?

6. The Outer Regions of Giant Ellipticals

Planetary nebulae can be used to study the kinematics of the outer regions of giant ellipticals ($r \gtrsim 20$ kpc), well beyond the radius accessible to integrated light spectroscopy. Three giant ellipticals have so far been studied: Cen A (Hui et al. 1995), NGC 1399 (the cD galaxy in the Fornax cluster; Arnaboldi et al. 1994) and NGC 1316 (a large late-merger system in the Fornax cluster; Arnaboldi et al. 1998). In all of these systems, the planetary nebulae studies show that the outer regions are relatively rapidly rotating. When this outer angular momentum is included, the specific angular momenta of the giant ellipticals does indeed appear to be similar to that of the spirals, as predicted by the simulations. Most of the angular momentum of these giant ellipticals resides in their outer regions.

We might expect some morphological evidence of rapid rotation in the outer regions of giant ellipticals. This is seen. The giant elliptical M87 in the Virgo cluster appears as an almost round E0 to E1 system on sky survey images. However, a very deep image of M87, prepared by D. Malin, shows that the outer regions are much more elliptical, about E4. A similar effect was seen by Porter et al. (1991) in a photometric study of brightest cluster ellipticals. They found that the mean ellipticity of these galaxies increases strongly with radius, as we would expect if their outer regions were rotating more rapidly.

Why does the angular momentum of the giant ellipticals reside in their outer regions? This kind of angular momentum segregation is seen in simulations of hierarchical galaxy formation (*eg.* Quinn et al. 1988), and in the outcome of major and multiple mergers (*eg.* Weil & Hernquist 1996). The angular momentum is transported outwards by torques generated while the system is out of equilibrium. If we accept this segregation of angular momentum as indicating a history of hierarchical aggregation or multiple mergers, then the internal angular momentum distribution in large bulges (like those of M31 and M104) gives a pointer to their formation history.

At this time, data on the kinematics of the outer regions of bulges is available only for M104, from unpublished planetary nebulae observations (Freeman et al. 1998). Even for this system the data reach out to only moderately large radii ($\simeq$ 12 kpc). From previous work (Kormendy & Illingworth 1982, Jarvis and Freeman 1986), it was already known that the inner regions of M104 lie close to the oblate isotropic rotator curve in the $(V/\sigma - \epsilon)$ plane, unlike the giant ellipticals. The velocity dispersion is roughly constant with radius, at about 200 km s^{-1}. The mean stellar rotational velocity of the bulge is about 100 km s^{-1} near the equatorial plane and remains constant with radius in the outer regions. Comparing the kinematics of M104 with those of the giant ellipticals NGC 1399 and

NGC 1316, we note that the ellipticals have slow rotation in the inner regions (unlike M104) but the rotation in their outer regions is again about half of the velocity dispersion, as in M104.

If angular momentum redistribution has occurred in M104, then it was clearly not as extreme as in the giant ellipticals, despite the very well established structural $r^{1/4}$ law. It would be interesting to know the properties of the angular momentum distribution in the outer bulge of M31: this would not be difficult to measure.

Why do the inner regions of large bulges rotate rapidly, while the inner regions of giant ellipticals do not? What is the essential difference in their formation histories? (We note again that the *total* specific angular momenta for large spirals and large ellipticals appear to be fairly similar).

One possibility is that bulge formation went on in the presence of a substantial envelope of high angular momentum gas, which later dissipated to form the disk. Some of this gas may have funneled into the inner bulge, through the torques that redistribute angular momentum, and so produced a more rapidly rotating inner bulge. This envelope of gas is presumably absent in the formation of the diskless giant ellipticals.

7. Summary

- the halo of the Milky Way continues to be built by the accretion of dwarf galaxies.
- small boxy bulges like that of the Milky Way are mostly barlike, probably grow from the disk, and probably tell us little about the interaction history of their parent galaxies.
- the angular momentum distribution within large bulges, out to large radii, is a useful clue to their merger/aggregation history. M31 is important in this context because the kinematics of its outer bulge are relatively easy to study.
- The difference between outer stellar halos (as in the Milky Way) and outer bulges (as in M31 and M104) may have to do with the answer to the following question:

> did the stars that now inhabit the outer regions form *together with* the aggregation of the dark corona, so that the redistribution of angular momentum also affected the stellar system
>
> OR
>
> did the stellar bulge/halo form *later* by accretion, so was not affected by this angular momentum redistribution?

References

Arnaboldi, M. et al. 1994. ESO Messenger, June 1994, p 40.
Arnaboldi, M. et al. 1998. In preparation.
Barnes, J. 1988. Astrophys.J., 331, 699.
Beers, T. & Sommer-Larsen, J. 1995. Astrophys.J.Suppl., 95, 175.
Bureau, M. 1997. Poster paper at this symposium.
Combes, F. et al. 1990. Astron.Astrophys., 233, 82.
Courteau, S. et al. 1996. Astrophys.J., 457, L73.
Couture, J. et al. 1995. Astron.J., 109, 2050.
de Vaucouleurs, G. 1964. In "The Galaxy and the Magellanic Clouds", ed F. Kerr & A. Rodgers (Sydney: Aust. Acad. Sci), p 88.
Durrell, P. et al. 1994. Astron.J., 108, 2114.
Ellis, R.S. 1997. Talk presented at IAU Symposium 183.
Eggen, O.J. 1979. Astrophys.J., 229, 158.
Fall, S.M. 1983. In "Internal Kinematics and Dynamics of Galaxies", ed E. Athanassoula (Dordrecht: Reidel), p 391.
Freeman, K.C. et al. 1998. In preparation.
Gerhard, O. 1997. Talk presented at IAU Symposium 184.
Holland, S. et al. 1996. Astron.J., 112, 1035.
Hui, X. et al. 1995. Astrophys.J., 449, 592.
Ibata, R. et al. 1994. Nature, 370, 194.
Jablonka, P. et al. 1996. Astron.J., 112, 1415.
Jarvis, B. & Freeman, K.C. 1986. Astrophys.J., 295, 324.
Kormendy, J. & Illingworth, G. 1982. Astrophys.J., 256, 460.
Kuijken, K. & Merrifield, M. 1995. Astrophys.J., 443, 13.
Lynden-Bell, D. & R. 1995. Mon.Not.R.Astron.Soc., 275, 429.
Majewski, S. 1992. Astrophys.J.Suppl., 78, 87.
Majewski, S. 1994. Astrophys.J., 431, L17.
Morrison, H.L & Harding, P.X. 1993. Publ.Astron.Soc.Pac., 105, 977.
Noguchi, M. 1987. Mon.Not.R.Astron.Soc., 228, 635.
Pfenniger, D. & Friedli, D. 1991. Astron.Astrophys., 252, 75.
Porter, A. et al. 1991. Astron.J., 101, 1561.
Preston, G. et al. 1994. Astron.J., 108, 538.
Pritchet, C. & van den Bergh, S. 1994. Astron.J., 107, 1730.
Quinn, P. & Goodman, J. 1986. Astrophys.J., 309, 472.
Quinn, P. & Zurek, W. 1988. Astrophys.J., 331, 1.
Rich, R.M. et al. 1996. In "Formation of the Galactic Halo ... Inside and Out", ed H. Morrison and A. Sarajedini (San Francisco: ASP), p 544.
Searle, L. & Zinn, R. 1978. Astrophys.J., , 225, 357.
Sommer-Larsen, J. et al. 1994. Mon.Not.R.Astron.Soc., 271, 743.
van Albada, T.S. & van Gorkom, J. 1977. Astron.Astrophys., 54, 121.
Weil, M. & Hernquist, L. 1996. Astrophys.J., 460, 101.

THE MAGELLANIC STREAM AND THE MAGELLANIC CLOUD SYSTEM

M. FUJIMOTO
Uedayama 4-901, Tempaku, Nagoya 468, Japan

T. SAWA
Department of Physics and Astronomy
Aichi University of Education, Kariya 470, Japan

AND

Y. KUMAI
Faculty of Commerce
Kumamoto Gakuen University, Kumamoto 862, Japan

1. Introduction

A tidal model has been introduced to the triple system of the Galaxy, Large and Small Magellanic Clouds (the LMC and SMC hereafter) and successfully reproduced the Magellanic Stream (Murai and Fujimoto 1980; Lin and Lynden-Bell 1982; Gardiner et al. 1994; Gardiner and Noguchi 1995; Lin et al. 1995), a narrow band of diffuse atomic hydrogen gas emerging from the SMC region, passing by the South Galactic Pole along an overhead great circle spanning over 100° (Wannier and Wrixon 1972; Mathewson et al. 1974). The LMC and SMC have a hydrogen bridge and common envelope (Hindman 1964; McGee and Milton 1966) and, therefore, we can consider that they have been in a binary state for the Hubble time, revolving together around the Galaxy with a halo whose mass is larger than $10^{12} M_{\odot}$ if the flat rotation curve extends up to more than 100 kpc. The strong gravitational force due to this heavy halo attracts the Magellanic Stream and produces the high negative radial velocities (Murai and Fujimoto 1980).

The tidal model seems to be realistic when we view some characteristics of the SMC with two-peaked radial velocities separating each other with $\sim$40 km s^{-1}: The SMC is elongated and partially splitting over more than 15 kpc along the line of sight. (These famous phenomena have been found by many investigators. See their names referred to in a summary paper

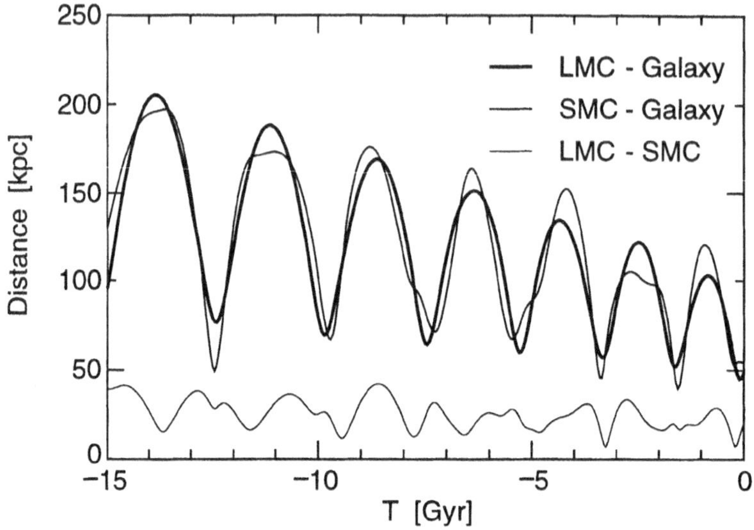

Figure 1. Time variation of the Galaxy-Magellanic Clouds distance and the LMC-SMC separation. The LMC and SMC have remained in a binary state for the past 15 Gyr.

by Mathewson and Ford 1983). The dynamics can be explained uniquely in terms of the tidal interaction that the SMC approached the LMC so closely a few 10^8yr ago and then the far-less-massive SMC suffered a deep tidal damage.

Since we support this tidal model for the Magellanic Stream, we are introducing in this paper more data that favor the past gravitational interaction between the LMC and SMC in a binary state.

2. The Magellanic Stream and the Binary Orbits of the LMC and SMC

The radial velocities of the LMC and SMC have been already measured, whereas their transverse velocities (proper motions) could not be measured until only three years ago (Section 3). As Murai and Fujimoto (1980) applied, we search for binary orbits of the LMC and SMC backward in time by assuming various values for their transverse velocities. For the following model for the Galaxy-LMC-SMC system, we take again $2 \times 10^{10} M_\odot$ and $2 \times 10^9 M_\odot$ for the LMC and SMC masses, and 220 km s^{-1} for the velocity amplitude of the flat rotation curve of the Galaxy.

Fig. 1 shows a typical result for the binary orbits of the LMC and SMC which have endured in the past 15 Gyr. On the basis of such binary orbits, we apply particle simulations to reproduce the geometrical as well as

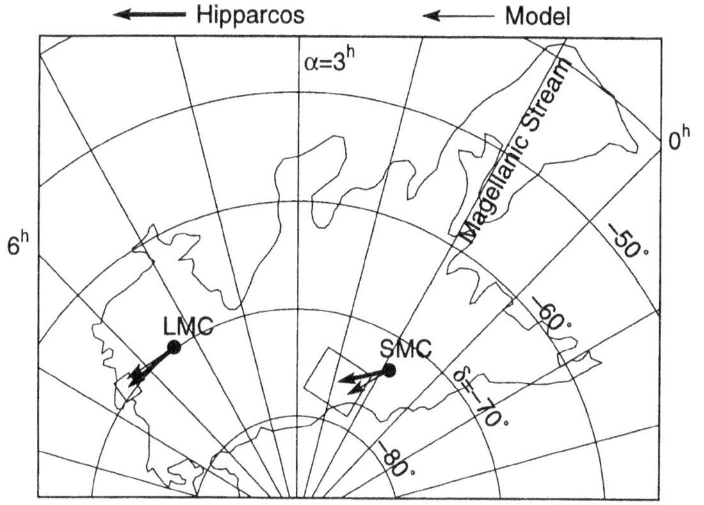

Figure 2. Observed and model proper motions of the LMC and SMC. The former data are due to Kroupa & Bastian (1997). The model proper motions are shown to fall well within the error boxes of observations.

dynamical structures which resemble quantitatively the Magellanic Stream. As discussed so far in many papers (see references in § 1), the Magellanic Stream can be modeled as a tidal debris torn off from the less-massive SMC and elongated due to the gravitational force of the Galaxy. We note that two diametrically-opposite structures are pulled out usually from the tidally perturbed object: One is heading, and the other is trailing. The present case for the SMC, the heading gaseous debris flowed into the LMC region and forms the common envelope of gas enclosing the LMC and SMC.

3. Observations Supporting the Tidal Model

3.1. PROPER MOTIONS OF THE LMC AND SMC

Jones et al. (1994) and Kroupa et al. (1994) have first obtained reliable data about the proper motion of the LMC by measuring extremely slight changes of stellar positions relative to the far background galaxies. The magnitude and direction of the proper motion are consistent with those predicted by Lin and Lynden-Bell (1982) and Gardiner et al. (1993): The LMC moves with ~300 km s^{-1}, trailing the Magellanic Stream counterclockwise as seen towards the galactic center from the present position of the Sun (Murai and Fujimoto 1980).

Recently Kroupa and Bastian (1997) used the *Hipparcos* data and obtained the proper motions of both the LMC and SMC: $\mu_\alpha = 1.94 \pm 0.29$ mas

y^{-1}, $\mu_\delta = -0.14\pm0.36$ mas y^{-1} for the LMC, and $\mu_\alpha = 1.23\pm0.84$ mas y^{-1}, $\mu_\delta = -1.21 \pm 0.75$ mas y^{-1} for the SMC. We recognize in Fig. 2 that these data are nearly coincident with our model proper motions. Furthermore, as Kroupa and Bastian (1997) pointed out, the observed proper motions of the LMC and SMC are qualitatively the same in magnitude and direction, implying that these two dwarf irregulars are in a binary state and revolve around the Galaxy with similar high velocities, as leading the Magellanic Stream. If we adopt the above mean values of μ_α and μ_δ for the LMC, the last pericenter distance of the LMC is 48 kpc, and its orbital plane is approximately perpendicular to the line joining the Galactic center and the present position of the Sun, but the northern part is inclined toward us by 7°.

3.2. TIDAL DAMAGE OF THE SMC

When we examine in details the separation between the LMC and SMC in Fig. 1, we find a close approach of the SMC to the LMC 2×10^8 yr ago with impact parameter of 5 kpc or so, comparable to their sizes. Murai and Fujimoto (1980) have found that such dynamics occur in quite high probability in their model construction and concluded that the LMC and SMC collided and/or grazed recently in the near past. As mentioned already in the Section 1, the less-massive SMC is actually damaged due probably to this close encounter (Mathewson et al. 1986)

3.3. SPORADICALLY ENHANCED CLUSTER FORMATION

It is now widely known that massive star-clusters are easily generated in environments where interstellar gas clouds are in large-scale unorganized motion, such as in heavily interacting and merging gas-rich galaxies (Ashman and Zepf 1992, Holtzman et al. 1992, Whitmore et al. 1993, Schweizer et al. 1996). Fujimoto and Noguchi (1990) first pointed out that greatly disordered motion of interstellar gas produces young massive clusters in the LMC, and Kumai et al. (1993) found that galaxies associated with young massive clusters generally have interstellar gas in large-scale turbulent motion with velocity amplitude more than 50 km s^{-1}, based on the study of nearby galaxies including the Magellanic Clouds (see also Fujimoto and Kumai 1991).

According to the idea that disordered motion of interstellar gas (through high-velocity collisions between gas clouds) forms massive clusters, the tidal model predicts sporadic cluster formation history of the LMC and the SMC since serious disturbance is caused in their interstellar gas at their every close encounter. For example, Fig. 1 reveals that the LMC and SMC made close encounters 2×10^8 yr and $3 \sim 4 \times 10^9$ yr ago, with the impact param-

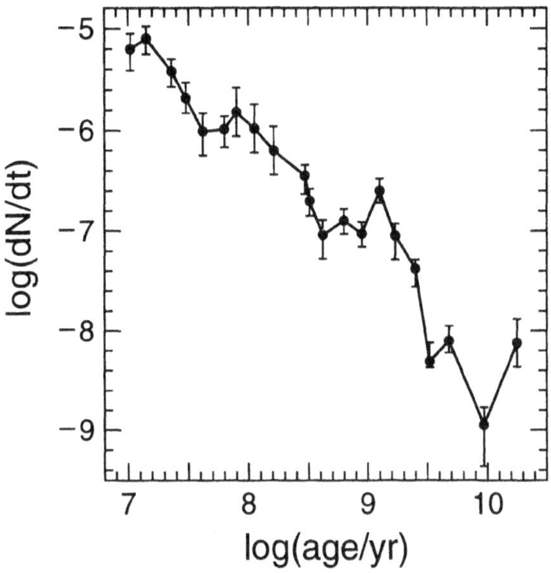

Figure 3. Age distribution of the LMC clusters, reproduced from Girardi et al. (1995). Salpeter initial mass function (with exponent = 1.35) is assumed for stars in a cluster, and correction is made for incompleteness due to magnitude bias. The error bars indicate the statistical uncertainties of cluster counts. The overall distribution is not smooth but has distinct humps at ages $\sim 10^8$ yr and $\sim 10^9$ yr, consistent with the prediction of the tidal model.

eters of $5 \sim 7$ kpc and relative velocity of more than 100 km s^{-1}. Then, the Magellanic Clouds gas were violently disturbed and we expect that numerous star clusters were formed associated with these events.

With such an expectation, we refer to a paper by Girardi et al. (1995) in which age distribution of more than six hundred LMC star cluster is obtained (for the SMC, unfortunately, there are no available data sample which contains significant number of clusters of known ages). Actually in Fig. 3, we recognize two humps in the age distribution, which approximately correspond to the above-mentioned epochs of close encounter of the Magellanic Clouds. Since the number of data clusters in Girardi et al. (1995) are large enough, this characteristic feature can be regarded as statistically significant, and this coincidence lends a strong support to the reality of the tidal model; the binary orbit chosen as in Fig. 1 where the LMC and SMC have revolved together around the Galaxy for the past, at least, $3 \sim 4$ Gyr, are likely the case.

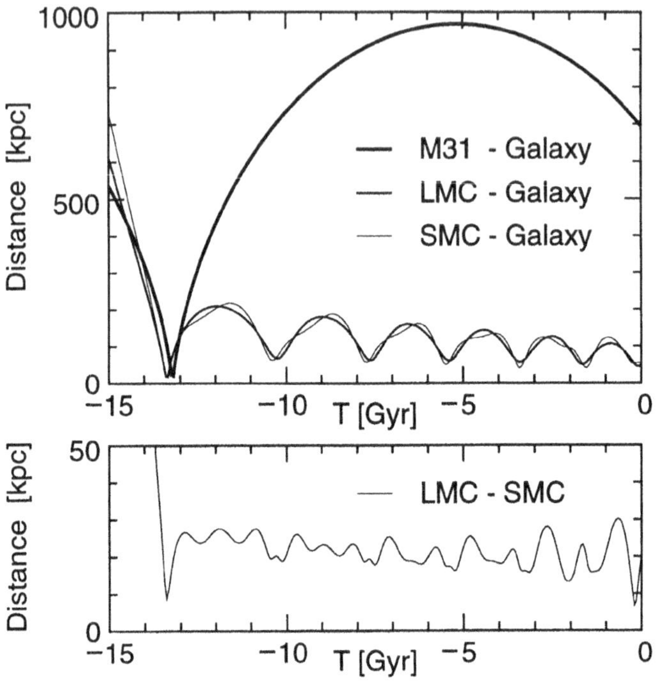

Figure 4. Time variation of the Galaxy-M31 distance and the Galaxy-Magellanic Clouds separation. The LMC-SMC separation is given in the lower diagram.

4. The Andromeda Galaxy and The Triple System of the Galaxy, LMC and SMC

When we explore the origin of the LMC and SMC, and their early dynamical relations to the Galaxy, it is inevitable to ask the gravitational influence from the Andromeda galaxy (M31) which must have been very close to the Galaxy in the expanding early universe. In a similar way to the timing arguments made by Kahn and Woltjer (1960), we trace back in time the motions of the M31 and the Galaxy, and, in this Section, the LMC and SMC as well.

M31 approaches the center of the Galaxy with a velocity of 117 km s^{-1} at present from the direction $l = 121°$, $b = -22°$ (Tully 1988). We assume the sum of the masses of the Galaxy and M31 to be $4 \times 10^{12} M_{\odot}$ with the flat rotation curves respectively of 220 km s^{-1} and 250 km s^{-1}, both truncated at 150 kpc from the center. The upper diagram of Fig. 4 shows the evolutions of the distances between M31 and the Galaxy, and that between the Galaxy and the LMC-SMC binary. The lower diagram shows the LMC-SMC separation. We find that the binary state of the LMC and

SMC are stable, and disrupted only when M31 and the Galaxy overlapped completely ~13 Gyr ago. When they are closely located by about 300 kpc apart or when their halo outer fringes touch each other, the binary structure of the LMC and SMC seem to be unaffected by the M31's gravity.

We have also traced back in time the motion of the four bodies (M31, the Galaxy, LMC and SMC) in some other cases that the unknown transverse motion of M31 is assumed. The orbits of the Galaxy and M31 are not radial but elliptical, avoiding their complete overlapping in early phase. It is obvious that M31's tidal force on the Galaxy is reduced compared with the previous case and therefore the binary structure of the LMC and SMC is more stable.

The LMC and SMC seem thus to have been formed together attracting each other, and bound gravitationally to the Galaxy, suggesting it unlikely that the Magellanic Clouds migrate via the M31 region initially and then to the Galaxy (Shuter 1992).

5. Conclusions

A tidal model for the Magellanic Stream is reexamined by postulating that the LMC and SMC have been in a binary state for the Hubble time and the high negative radial velocity of the Magellanic Stream is due to the strong gravitational attraction of the massive halo of the Galaxy.

All the data that we have discussed, the proper motions of the LMC and SMC, and the sporadic enhancement in formation history of the LMC clusters, are not only consistent with the tidal model but also strongly support the dynamics of the LMC and SMC in binary structure. These phenomena are rather difficult to understand, if we do not take into account the occasional close encounters between the LMC and SMC rotating around each other with time-variable radius.

The introduction of M31 to our tidal model offers a new key to explore the dynamical origin of satellite galaxies. Although more data are necessary, we could say primarily that the satellite dwarf irregulars, at least the LMC and SMC, were formed not isolated but closely related to their nearest-neighboring massive galaxies.

References

Ashman, K.M. & Zepf, S. 1992, *ApJ*, **384**, 50
Fujimoto, M., & Kumai, Y. 1991, *Ann. Phys. Coll., Suppl.* **3** vol. 16, 75
Fujimoto, M., & Kumai, Y. 1997, *AJ*, **113**, 249
Fujimoto, M., & Noguchi, F. 1990, *PASJ*, **142**, 505
Gardiner, L.T., Sawa, T., & Fujimoto, M. 1994, *MNRAS*, **266**, 567
Gardiner, L.T., & Noguchi, F. 1995, *MNRAS*, **278**, 191
Girardi, L., Chiosi, C., Bertelli,G., & Bressan, A. 1995, *A&A*, **298**, 87

Hindman, J.V. 1964, *Nature*, **202**, 377
Holtzman, J.A., & WFPC-team 1992, *AJ*, **103**, 691
Jones, B.F., Klemola, A.R., & Lin, D.N.C. 1994, *AJ*, **107**, 1333
Kahn, F.D., & Woltjer, L. 1959, *ApJ*, **130**, 705
Kroupa, P., & Bastian, U. 1997, *NewA*, **2**, 77
Kroupa, P., Röser, S., & Bastian, U. 1994, *MNRAS*, **266**, 412
Kumai, Y., Basu, B., & Fujimoto, M. 1993, *ApJ*, **404**, 576
Lin, D.N.C., & Lynden-Bell, D. 1982, *MNRAS*, **198**, 707
Lin, D.N.C., Jones, B.F., & Klemola, A.R. 1995, *ApJ*, **439**, 652
Mathewson, D.S., Cleary, M.N., & Murray, J.D. 1974, *ApJ*, **190**, 291
Mathewson, D.S., & Ford, V.L. 1983, in *IAU Symp.* **108**, Structure and Evolution of the Magellanic Clouds, ed. S. van den Bergh & K.S. de Boer (Dordrecht:Reidel), 125
Mathewson, D.S., Ford, V.L., & Visvanathan, N. 1986, *ApJ*, **301**, 664
McGee, R.X., & Milton, J.A. 1966, *Aust. J. Phys.*, **19**, 343
Murai, T., & Fujimoto, M. 1980, *PASJ*, **32**, 581
Olszewski,E.W., Schommer, R.A., Suntzeff, N.B., & Harris, H.C. 1991, *AJ*, **101**, 515
Schweizer, F., Miller, B.W., Whitmore, B.C., & Fall, S.M. 1996, *AJ*, **112**, 1839
Shuter, W.L.H. 1992, *ApJ*, **386**, 101
Tully, R.B. 1988, in Nearby Galaxies Catalog (Cambridge University Press), pp.9
Wannier, P., & Wrixon, G.T. 1972, *ApJ*, **173**, 119
Whitmore, B.C., Schweizer, F., Leithrer, C., Borne, K., & Roberts, C. 1993, *AJ*, **106**, 1354

THE NATURE AND FATE OF THE SAGITTARIUS DWARF GALAXY

R.A. IBATA
European Southern Observatory
Karl Schwarzschild Straße 2,
D-85748 Garching bei München, Germany

The Sagittarius dwarf galaxy, located at a distance of $\sim$ 15 kpc behind the Galactic bulge, is in the process of being tidally disrupted and assimilated into the Milky Way. This unique event allows the physics of galactic merging to be probed in unprecedented detail, and may shed light on the hypothesized population of primordial galaxies that merged to form the Milky Way.

Numerical simulations, constrained by the latest kinematic and structural data, show that the luminous component of this dwarf galaxy must reside within a substantial dark matter halo for the dwarf to have survived the Galactic tides long enough to be seen at the present time. The minimum mass of the dark halo is then $\sim 5 \times 10^8\ M_\odot$, which implies a global mass to light ratio of $\sim$ 100. It is found that the eventual fate of all the plausible models is to disrupt into a stream of particles that follows closely the original orbit of the dwarf through the Galactic halo, remaining stable for many Gyr. However, the timescale for the complete disruption of the Sagittarius dwarf is model-dependent and remains presently unknown.

1. Introduction

The Sagittarius dwarf spheroidal (Ibata, Gilmore & Irwin 1994, 1995), discovered during the course of a spectroscopic study of the bulge of the Milky Way (Ibata & Gilmore 1995a, 1995b), is the closest satellite companion galaxy to the Milky Way. This dwarf galaxy contains a mix of stellar populations ranging from relatively old stars – many RR Lyrae stars are observed (Mateo et al. 1995, Alard 1996) and at least one of its four globular clusters is as old as the oldest Galactic halo clusters (Richer et al. 1996; Chaboyer, Demarque and Sarajedini, 1996) – to intermediate age stars – several Carbon stars have been identified (Ibata et al. 1994, 1995). The dom-

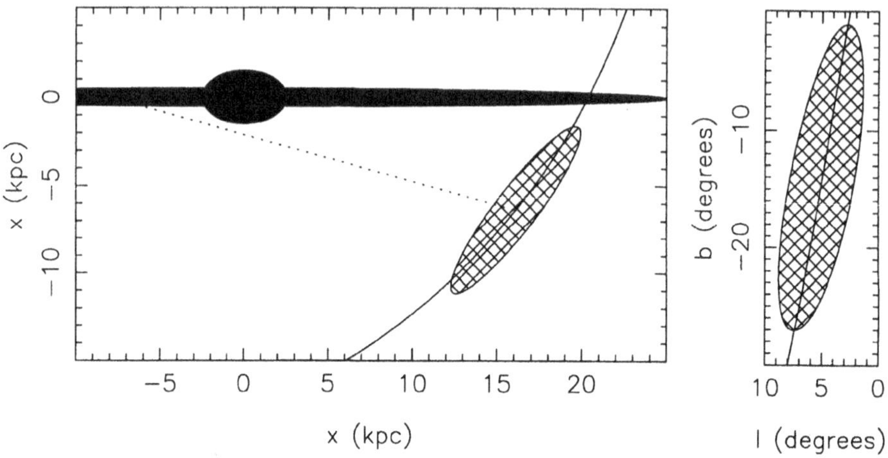

Figure 1. Schematic diagrams showing the position of the Sagittarius dwarf in relation to the Galaxy. In both diagrams, the orbital path of the dwarf is indicated with a solid line. The left hand panel is in the x–z plane of the Galaxy. The line of sight from the Sun to the center of the dwarf is marked with a dotted line. The right hand panel shows the projection of the Sagittarius dwarf on the sky.

inant population however, is 10 to 14 Gyr old (Fahlman et al. 1996) and has mean abundances between [Fe/H] = −0.8 and [Fe/H] = −1.2 (Whitelock, Catchpole & Irwin 1996). The full abundance range observed covers $\gtrsim$ 1dex around this mean.

The derived geometrical picture is shown schematically in Fig. 1. The Sagittarius dwarf is a prolate body with axis ratios $\sim$ 3:1:1, oriented approximately perpendicular to the plane of the Galaxy from Galactic latitude $b = -4°$ to $b = -32°$, with its longest dimension aligned $\gtrsim$ 10 kpc along the coordinate line $\ell = 5°$. Its center is located $\sim$ 25 kpc from the Sun and 16 ± 2 kpc from the Galactic center, near its point of closest approach to the Galaxy. Such proximity to the Galactic center is expected to induce huge tidal stresses in the dwarf, which will lead to its eventual destruction, with its stars and globular clusters eventually becoming members of the Galactic halo. Thus, the Sagittarius dwarf is perhaps a prime, and it is certainly the closest, example of a small galactic building block, as envisaged in the currently popular picture of hierarchical structure formation. Depending on its orbit and mass, the Sagittarius dwarf may also significantly influence the Milky Way. The proximity allows us to study this interacting pair of galaxies in great detail; one may obtain very precise kinematics (including proper motions), accurate abundance measurements, and very deep photometry, all with a precision unobtainable in any other system; the observational constraints are reviewed in Ibata et al. (1997).

2. The orbit of the Sagittarius dwarf

Determining the orbit of the dwarf is the most important first task in understanding the interaction. Several numerical studies (Oh et al. 1995, Piatek & Prior 1995, Johnston et al. 1995, Velazquez & White 1995) have shown that dwarf satellite galaxies become elongated in the tidal field of their massive companion. The elongation, it transpires, always points parallel to the plane of the dwarf's orbit, and approximately along the direction of motion. Since the Sagittarius dwarf is almost directly behind the Galactic center, as viewed from Earth, the projected elongation must be aligned with the proper motion vector (and the projection of the orbit) to very good approximation. From these constraints, the pole of its orbit is deduced to be at ($\ell = 94°, b = 11°$) (Lynden-Bell & Lynden-Bell 1996). The component of the proper motion of the central regions of the Sagittarius dwarf along the direction of the deduced orbit has been measured, (Irwin et al. 1996, Ibata et al. 1997), indicating that it is moving northwards with a transverse velocity of $250 \pm 90\,\mathrm{km\,s^{-1}}$.

Unfortunately, this proper motion measurement has too large an uncertainty to provide a useful determination of the orbit. To get around this problem, one can simulate N-body models of the dwarf on a grid of orbits and compare the present-day structure of the models to the accurately determined radial velocity gradient (which has been obtained over a large portion of the major axis of the dwarf from $b = -10°$ to $b = -30°$). Such numerical experiments also allow the structural evolution of the Sagittarius dwarf to be studied. Since 'King models' (King 1962) fit the structure of present day dwarf spheroidal galaxies well (Irwin & Hatzidimitriou 1995), it is natural to use these models to represent the structure of the initial proto-dwarf galaxy. A grid of 21 structural models was set up, sampling the plausible parameter space of concentration, central velocity dispersion and central density. The evolution of the models was calculated using an N-body tree-code algorithm (Richardson 1993), which was altered to include the forces due to the assumed fixed Galaxy potential and due to dynamical friction, as approximated by the Chadrasekhar formula (Binney & Tremaine 1987). The models, comprising either 4000 or 8000 particles, were evolved for 12 Gyr, a time equal to the age of the dominant stellar population in the Sagittarius dwarf (Fahlman et al. 1996). This work is described in detail in Ibata & Lewis (1997).

Each of the 21 King models was simulated on one or more of the orbits shown in Fig. 2. The orbits are obtained by integrating the path of a massless tracer particle backwards in time under the influence of the fixed Galactic potential described in Johnston et al. (1995). Though the initial radial velocity of the tracer particle is identical in all four orbits, the proper

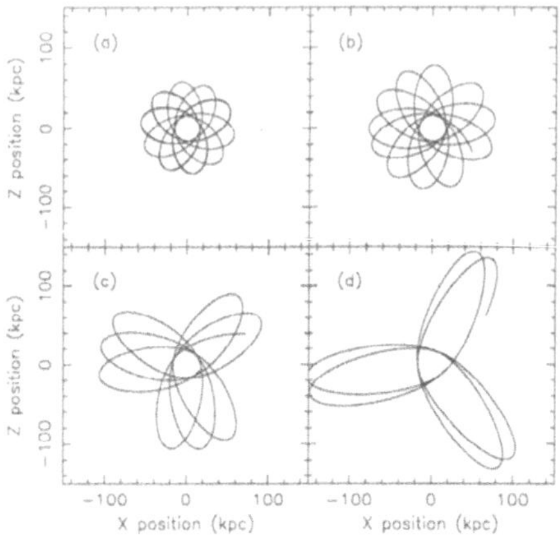

Figure 2. The four orbits on which the King models were launched. In the text, these orbits are referred to by the panel label in this plot.

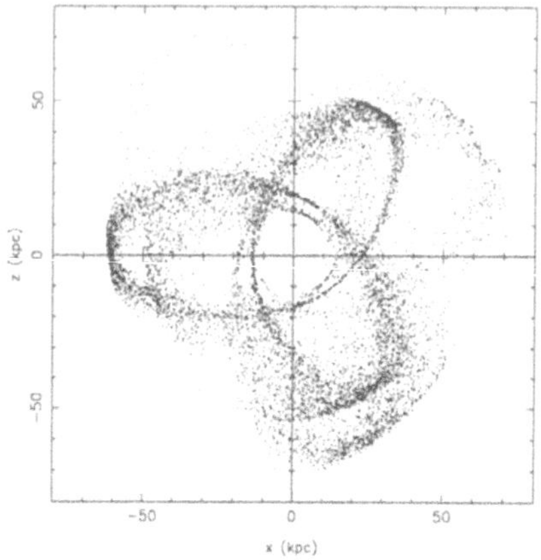

Figure 3. The end-point structure of the dwarf galaxy model whose initial structure was set up to match the present day observations of the Sagittarius dwarf.

motion increases from $250\,\mathrm{km\,s^{-1}}$ to $390\,\mathrm{km\,s^{-1}}$ from orbits 'a' to 'd'.

It transpires that all the models placed initially on the 'a' orbit give rise to a radial velocity gradient that is in good agreement with the observations, while those models on the longer period 'b', 'c' and 'd' orbits give progressively worse fits to the radial velocity data. This is illustrated in Fig. 4, which compares the observed radial velocity gradient along the major axis of the dwarf with the King models that managed to survive the interaction with the Milky Way. Thus the orbit of the Sagittarius dwarf is well approximated by the 'a' orbit, which has a short period $T \sim 0.7\,\mathrm{Gyr}$, and implies that there have been many collisions with the Galactic disk in the past.

3. Fate of the dwarf galaxy models

Models set up with an initial configuration to fit the observations of the present state of the Sagittarius dwarf turn out to be very fragile. In particular, setting the velocity dispersion to the observed value of $11.4\,\mathrm{km\,s^{-1}}$, the minor axis half-mass radius to the observed value of 0.55 kpc, and placing the model on the 'a' orbit, results in a model that becomes completely unbound after only 5.3 Gyr. This confirms, using constraints from a more complete data set, the findings of Velasquez & White (1995) and Johnston et al. (1995) that models of the Sagittarius dwarf, where light traces mass, are fragile. If so, one may deduce that the Sagittarius dwarf will be completely disrupted within $\sim 5\,\mathrm{Gyr}$, providing a source of new stars and

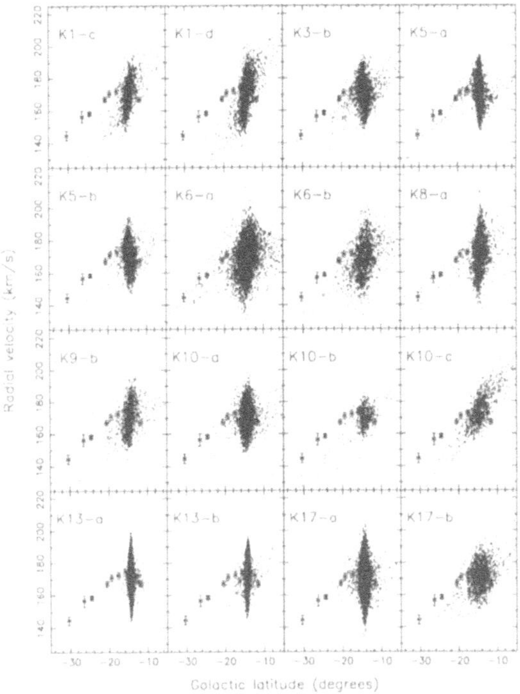

Figure 4. The radial velocity – Galactic latitude structure of the models that survive the tidal interaction with the Milky Way is compared to the observed radial velocity in fields along the major axis of the Sagittarius dwarf (points with error-bars). Each panel is marked with the model label from Ibata & Lewis (1997), where the letter following the hyphen identifies one of the four orbits of Fig. 2.

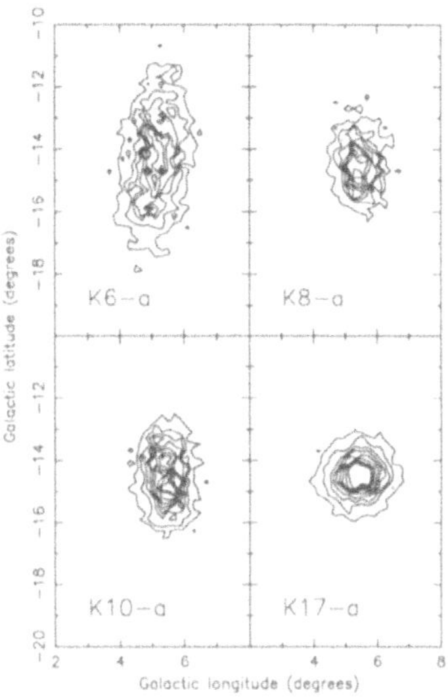

Figure 5. The final structure of the models that survived the Galactic tides and gave a good representation of the observed radial velocity gradient and radial velocity dispersion of the Sagittarius dwarf. However, all of the models have a *much* narrower half-mass radius than the observed half-brightness radius.

globular clusters to the Galactic halo. At the end of the integration, at 12 Gyr, the structure of the galaxy remnant is shown in Fig. 3. No bound concentration of particles remains; instead, streams of particles populate the orbital path of the former dwarf galaxy.

A search of parameter space with the extensive grid of models has revealed several configurations that do retain a bound core at the end of the simulation, and fit the radial velocity profile. Of these, the four models displayed in Fig. 5 also fit the observed radial velocity dispersion. However, no model manages to survive and be consistent with the observed minor axis width of the Sagittarius dwarf: we find that all models that have initially large half mass radii are rapidly destroyed by the Galactic tides. The half mass radius, it transpires, becomes smaller as disruption proceeds. So large initial half mass radii are required to match the observed half-brightness radius of $R_{\rm HB} = 550\,{\rm pc}$. Yet any model with such a large initial half-mass radius becomes completely unbound within a few orbital periods.

This problem was investigated by Ibata et al. (1997), who concluded that a self-consistent solution to the present existence of the dwarf can only be found if the requirement that light traces mass is relaxed. The tidal disruption of the Sagittarius dwarf can be impeded if the stellar component of the dwarf galaxy is enveloped in a halo of dark matter, which has a mass profile such that dark matter density is a factor of 2–3 larger than the mean Galactic density interior to its peri-galacticon distance. To be consistent with the observed low velocity dispersion of the stellar component embedded therein, the core radius of the dark halo would have to extend out to the photometric edge of the system. The deduced mass to light ratio would then be $M/L \sim 100$. With this model, the escape velocity from the center of the Sagittarius dwarf is substantial, $v_{\rm e} = 90\,{\rm km\,s^{-1}}$, which helps to explain the observed wide abundance range.

4. Survival of primordial galaxy fragments

All of the dwarf galaxy models give rise to streams of tidally disrupted material that follow the orbital path of the remnant quite closely. The disrupted fraction always exceeded 15%, so one may expect to find a sizeable fraction of the stellar component of the Sagittarius dwarf spheroidal, including perhaps one or more globular clusters, stretching along a ring around the sky. It will be very fruitful to detect such material, as its kinematics could provide a very sensitive test of the Galactic potential gradients.

One of the most significant consequences of the failure of the mass-traces-light models to reproduce the observations is that the merging fragments that made the Galaxy probably had a radially increasing mass to light ratio, so that the stars were more centrally concentrated and the dark matter more extended. Detailed numerical simulations are required to explore this question further, but one may expect such a structure to initially lose almost exclusively dark matter, with stars being lost only in the last stages of disruption. If dynamical friction causes significant orbital decay of the merging clumps, the luminous matter would be deposited more centrally than the dark matter in the global potential well, naturally giving rise to a radially increasing mass to light ratio in the halos of large galaxies.

5. Interaction with the Milky Way

The models of the Sagittarius dwarf galaxy that are designed to reduce tidal disruption have sufficient mass to affect the structure of the Milky Way. Indeed, in this case, the tidal forces on the Milky Way due to the Sagittarius dwarf will be substantially larger than those due to the Large Magellanic

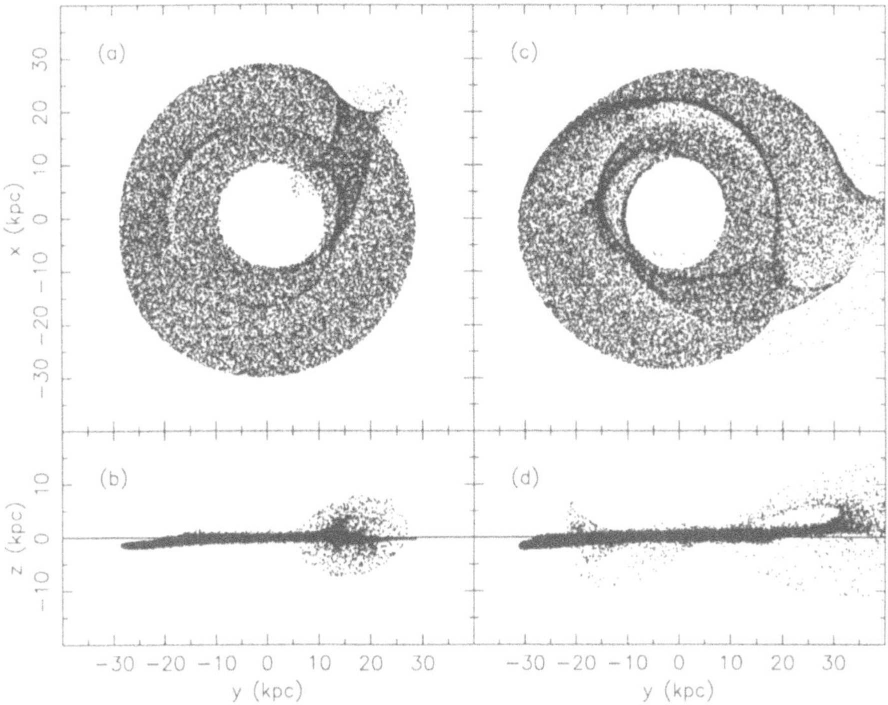

Figure 6. The structure of the Galactic H I disk is shown at 150 Myr and 0.7 Gyr after the impact in, respectively, the left- and right-hand panels. The mass of the dwarf in this simulation is $M = 10^9\ M_\odot$. The coordinate system is inertial, chosen such that the Galactic center lies at the origin and the *present* position of the Sun is at (-8,0,0). The Galactic plane is viewed from above, so the sense of Galactic rotation is clockwise in the upper panels.

cloud, which has previously been invoked as a possible perturber of the H I disk (Weinberg 1995). To investigate this possibility, hydrodynamical calculations were undertaken to simulate the collisional interaction between the Sagittarius dwarf and the Galactic outer H I disk (Ibata & Razoumov 1997). It is found that a significant distortion of the Galactic H I disk will be induced by the collision if the mass of the dwarf exceeds $\sim 10^9\ M_\odot$. Though the precise details of the interaction are compromised in these simulations by the lack of a live Galactic halo, it is found that for model masses $\gtrsim 5 \times 10^9\ M_\odot$, prominent spiral arms and a substantial lopsidedness in the outer disk are produced. Furthermore, a noticeable warp-like structure is induced in the disk (see Fig. 6). Thus the Sagittarius dwarf may have significantly affected the star formation history and structure of the outer Galaxy.

6. Conclusions

The Galaxy and its dwarf companion in Sagittarius provide an ideal laboratory in which we may probe the complex processes that take place during the merging of galaxies. Due to the accuracy of the data now being obtained, it is possible to model this interaction in great detail. Analysis of the kinematic data provide severe constraints on, and in fact almost completely exclude, standard numerical models of the tidal disruption of dwarf satellites near large galaxies, since according to these models Sgr should have been destroyed long ago. The implications of this conclusion for topical models of the growth of large galaxies by repeated mergers is considerable.

This review has discussed mainly the dynamical constraints. However, much interesting work remains to be done in understanding the chemical evolution of the dwarf (and the outer regions of the Galaxy) and how this is affected by the interaction. From this, one may gain insights into the chemical evolution of galaxies during a merger event, and in turn about dominant evolutionary processes occurring during hierarchical formation of 'normal'-sized galaxy units.

References

Alard, C. 1996, *Ap.J.* **458**, L17.
Binney, J. & Tremaine, S. 1987 *Galactic Dynamics* (Princeton University Press, Princeton).
Chaboyer, B., Demarque, P. & Sarajedini, A. 1996, *Ap.J.* **459**, 558.
Fahlman, G., Mandushev, G., Richer, H., Thompson, I., & Sivaramakrishnan, A. 1996, *Ap.J.* **459**, L65.
Ibata, R., Gilmore, G., & Irwin, M. 1994, *Nature* **370**, 194.
Ibata, R. & Gilmore, G. 1995, *MNRAS* **275**, 591.
Ibata, R. & Gilmore, G. 1995, *MNRAS* **275**, 605.
Ibata, R., Gilmore, G., & Irwin, M. 1995, *MNRAS* **277**, 781.
Ibata, R., Wyse, R., Gilmore, G., Irwin, M. & Suntzeff, N. 1997, *A.J.* **113**, 634.
Ibata, R. & Razoumov, A. 1997, submitted
Ibata, R. & Lewis, G. 1997, submitted
Irwin, M. & Hatzidimitriou , D., 1995, *MNRAS* **277**, 1354.
Irwin, M., Ibata, R., Gilmore, G., Suntzeff, N. & Wyse, R. 1996, in *Formation of the Galactic Halo*, ed A. Sarajedini (A.S.P., San Francisco).
Johnston, K. V., Spergel, D. N., & Hernquist, L. 1995, *Ap.J.* **451**, 598.
King, I. 1966, *A.J.* **71**, 64.
Lynden-Bell, D. & Lynden-Bell R., 1995, *MNRAS* **275**, 429.
Mateo M., Udalski A., Szymanski M., Kaluzny J., Kubiak M. & Kreminski W. 1995, *A.J.* **109**, 588.
Oh, K.S., Lin, D.N. & Aarseth, S.J. 1995, *Ap.J.* **442**, 142.
Piatek, S. & Pryor, C. 1995, *A.J.* **109**, 1071.
Richardson, D. 1993, Ph.D. Thesis, Cambridge.
Richer, H., Harris, W., Fahlman, G., Bell, R., Bond, H., Hesser, J., Holland, S., Pryor, C., Stetson, P., Vandenberg, D. & van den Bergh, S. 1996, *Ap.J.* **463**, 602.
Velasquez, H. & White, S. 1995, *MNRAS* **275**, L23.
Weinberg, D. 1995, *Ap.J.* **455**, L31.

A SEARCH FOR MOVING GROUPS IN THE GALACTIC HALO

LUIS A. AGUILAR
Inst. de Astronomía, UNAM
Apdo. Postal 877, 22800 Ensenada, BC, México

AND

RONNIE HOOGERWERF
Sterrewacht Leiden
P.O. Box 9513, 2300 RA Leiden, The Netherlands

Abstract. A new method to search for moving groups, specifically tailored for the *HIPPARCOS* database, is used to search for moving groups among the high velocity stars in this database. Although the sampled volume is small, the high quality of the astrometry makes this search interesting. No moving groups are detected, but limits in the velocity and number of members of possible non-detected groups are given.

1. Introduction

The idea that the galactic halo has been formed to a large extent by the accretion and tidal disruption of satellite systems has been gaining strength. Back in 1976, D. Lynden–Bell pointed out that the Magellanic clouds, several dwarf systems, and some distant globular clusters which are satellites of our Galaxy, lie near two great circles in the celestial sphere (Lynden–Bell 1976). More recently, Lynden–Bell and Lynden–Bell (1995), have analyzed this and other alignments, and identified streams associated to the Magellanic clouds, Fornax, and one possibly associated with the recently discovered dwarf system in Sagittarius (Ibata et al. 1994); additional streams related to outer globular clusters are shown.

Tantalizing evidence for the existence of structure is beginning to appear in surveys of halo stars too. Dionidis and Beers (1989), investigated the clustering properties of 4,400 field horizontal branch stars in the halo, and found an excess 2–point correlation at angular separations of less than $10'$, which at the typical depth of the sample corresponds to linear separations of less than 25 pc. Côté et al. (1993) conducted a radial velocity survey of 879 field stars within a 1° field toward $\ell = 277°$ and $b = 9°$; they found a clump of 18 stars within the 74–76 km/s radial velocity bin, where their expected number is only 5. Arnold and Gilmore (1992) have studied a color selected sample of 44 stars in 2 high galactic latitude fields; they found 4 stars at a distance of 30 kpc with a velocity dispersion of less than 12 km/s, and a systematic velocity with respect to the Galactic rest frame of 70 km/s, which they suggest may be the remnants of a disrupted halo cluster. Majewski et al. (1996) have found evidence for clumping in velocity in a study of halo stars toward the North Galactic Pole, in particular, they identified a retrograde rotating, halo moving group. They suggest that discrepancies in the halo kinematics derived from surveys conducted along different line of sights may be the result of a clumpy halo velocity distribution.

One argument used in the past against the accretion scenario is the apparent fragility of galactic disks when subject to accretion events (Tóth and Ostriker 1992). However, Velázquez and White (1997) have recently shown that the inclusion of a dynamically active halo can be an effective shield to the galactic disk.

An important consequence of this accretion scenario is that the phase space portrait of the halo, far from being a smooth distribution, should consist of a patchy aggregation of tidally disrupted systems that have been phase mixed all over the sky, but which retain kinematic memory of their existence as a coherent entity. Johnston et al. (1996) have studied the formation of these features within the context of satellite disruption within the halo of our Galaxy and proposed a method to identify them in sky surveys.

The challenges to discover these moving groups in the halo are enormous due to the distances involved and the fact that they can span large angles in the sky. The availability of an astrometric database of unprecedent accuracy (*HIPPARCOS*, ESA 1997) and the plans for follow ups (e.g. *GAIA*, Lindegren and Perryman 1996), usher a new era of opportunities to search for such moving groups.

2. The Method

The *Hipparcos* database provides positions with a median precision of 0.77 mas, parallaxes with a median precision of 0.97 mas, and proper motions with a median precision of 0.88 mas/yr, for 117,955 stars (ESA 1997). These are stars in the solar neighborhood brighter than $V \sim 12$, although completeness is achieved for stars brighter than $V \sim 7$–9 (field dependent) only. This information determines three spatial coordinates and the linear velocity on the plane of the sky; the radial velocity being the only infor-

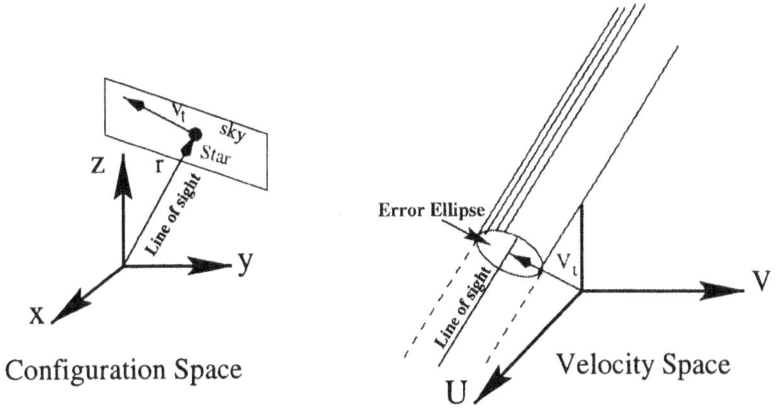

Figure 1. The determination of the spatial position and tangential velocity of a star (left), constrains the corresponding spatial velocity vector to be inside a cylindrical region in velocity space, parallel to the line of sight, offset from the origin by the magnitude of the tangential velocity, and of cross section given by the error ellipse of the latter velocity (right).

mation missing for the full phase space position. This information restricts the star to a line in velocity space, parallel to the line of sight and tangent to the measured tangential velocity vector, as any spatial velocity vector whose tip is on this line will project on exactly the same velocity on the plane of the sky for that particular direction (Fig. 1). In reality, the star's velocity vector is not confined to such a line, but to a "cylinder" of probability whose cross section is given by the uncertainties and correlation in the components of the tangential velocity vector.

If a group of stars is part of a moving group, their true spatial velocity vectors should lie within a neighborhood whose radius is set by the velocity dispersion of the group. As such, their corresponding cylinders in velocity space should all intersect within this neighborhood. As the probability of a chance intersection drops as the number of cylinders increases, a moving group should stand out against the statistical background if it is numerous enough and/or it is sufficiently constrained in velocity space.

This method has been used quite successfully, in combination with the more traditional convergent point method (Jones 1971), to find new members for OB associations. Results on the identification of new members, down to spectral type F, and refinement of the old membership lists for various associations in 21 fields in the sky, can be found in de Buijne et al. (1997), Hoogerwerf et al. (1997) and de Zeeuw et al. (1997).

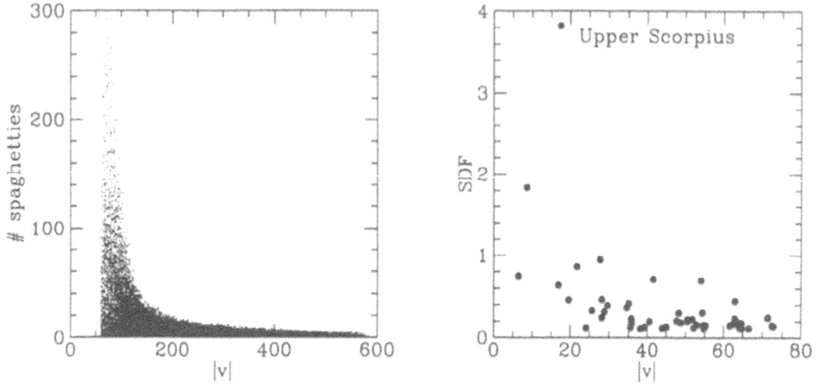

Figure 2. Left: The Magnitude of the velocity of the center is plotted against the number of cylinders (spaghetti) that intersect it, for each sphere searched in velocity space for our halo sample. Right: A similar diagram in which the number of intersections is replaced by the cylinder density function (SDF), directly proportional to the former, for the sample used by de Buijne et al. (1997) to identify the Upper Scorpious Association. The Upper Scorpious association is the point at $|v| \sim 17\,\mathrm{km/s}$ and $SDF \sim 3.8$.

3. The Sample

The vast majority of stars in the *HIPPARCOS* database are disk stars which produce a high background of intersections in velocity space. For this reason, it is necessary to restrict our study to a sample biased toward halo objects. Our sample consists of all stars with astrometry in the *Hipparcos* catalogue whose tangential velocities are larger than 60 km/s, and whose corresponding uncertainties are less than 10 km/s. This results in a list of 3,493 stars with no particular preferred direction in the sky and which lie roughly within 300 pc of the sun. These stars span spectral types from A to M and of all luminosity classes.

4. Results

We have searched in a cube of size 600 km/s in velocity space, excluding the region within 60 km/s of the sun. On this region we laid down a Cartesian grid of 10 km/s spacing; for each grid point, we computed the number of cylinders that pass within a 10 km/s sphere centered at this point. The results are shown in Fig. 2.

The left side of Fig. 2 plots the velocity of the center of the spheres against the number of cylinders that intersect them. At all velocities, the number of intersections distribution of spheres is smooth and tapers off toward large number of intersections; the upper envelope decreasing as the velocity of the spheres increases. There are no obvious spheres anomalously above the envelope in this diagram. Such an anomalous sphere would signal the existence of a suspicious intersection, difficult to explain as a chance occurrence. Contrast this situation with the right side of Fig. 2, which shows a similar diagram but for the 1,426 stars sample within the field $336° < \ell < 4°$, $6° < b < 33°$, in which the Upper Scorpious Association is quite conspicuous (de Buijne et al. 1997).

From Fig. 2 we learn that there are no obvious moving groups within our halo sample. We have examined in detail some points on the envelope of this figure, but in each case, the intersection could be explained as a spurious one. All what we can say at present, is that there are no moving groups in our sample, whose velocity with respect to the sun and number of member stars, places them above the upper envelope that is clearly defined by the left panel of Fig. 2.

This is a no detection result, but we can delineate in Fig. 2 the boundary of the region, within which we can exclude the presence of moving groups, in the *Hipparcos* subsample that we have studied. The sample used is quite small in terms of sampled volume, but the high precision of its information makes this a worthwhile search; particularly in the light of the success of the present method at identifying and refining membership lists, for OB associations in the solar neighborhood.

References

Arnold, R. & Gilmore, G., (1992), *MNRAS*, **257**, 225.

de Bruijne, J.H.J., Hoogerwerf, R., Brown, A.G.A., Aguilar, L.A., de Zeeuw, P.T., (1997) in *Proc. of the ESA Symposium: Hipparcos–Venice 97*, ESA SP–402, 575.

de Zeeuw, P.T., Brown, A.G.A., de Bruijne, J.H.J., Hoogerwerf, R., Lub, J., Le Poole, R.S., Blaauw, A., (1997) in *Proc. of the ESA Symposium: Hipparcos–Venice 97*, ESA SP–402, 575.

Côté, P., Welch, D.L., Fischer, P. & Irwin, M.J., (1993), *ApJ*, **406**, L59.

Dionidis, S.P. & Beers, T.C., (1989), *ApJ*, **340**, L57.

ESA, (1997) *The Hipparcos and Tycho Catalogues*, SP–1200.

Hoogerwerf, R., de Bruijne, J.H.J., Brown, A.G.A., Lub, J., Blaauw, A., de Zeeuw, P.T., (1997) in *Proc. of the ESA Symposium: Hipparcos–Venice 97*, ESA SP–402, 575.

Ibata, R., Gilmore, G.F. & Irwin, M.J., (1994), *Nature*, *370*, 194.

Johnston, K.V., Hernquist, L. & Bolte, M., (1996), *ApJ*, **465**, 278.

Jones, D.H.P., (1971), *MNRAS*, **152**, 231.

Lindegren, L., Perryman, M.A.C., (1996), *A& A SS*, **116**, 579.

Lynden–Bell, D., (1976), *MNRAS*, **174**, 695.

Lynden–Bell, D. & Lynden–Bell, R.M., (1995), *MNRAS*, **275**, 429.

Majewski, S.R., Hawley, S.L. & Munn, J.A., (1996), in *Formation of the Galactic Halo — Inside and Out*, ed. H.L. Morrison & A. Sarajedini. ASP Conf. Series **92**, 119.

Toomre, A. & Toomre, J., (1972), *ApJ*, **178**, 623.

Tóth, G. & Ostriker, J.P., (1992), *ApJ*, **389**, 5.

Velázquez, H. & White, S.D.M., (1997), *MNRAS*, *Submitted.*

CONSTRAINTS ON INTERACTIONS AND MERGERS FROM DSPH GALAXIES AND GLOBULAR CLUSTERS

E.K. GREBEL
Astron. Institut, Univ. Würzburg, Germany
Lick Observatory, UC Santa Cruz, USA

Observations at high redshifts are revealing numerous interactions and ongoing mergers. Our own Milky Way is currently merging with the Sagittarius dwarf spheroidal (dSph) galaxy. Past mergers with dwarf galaxies may have contributed significantly to the Galactic halo and possibly to the thick disk. The properties of Local Group dSphs and halo globular clusters impose constraints on the merger history of the Milky Way.

The apparent alignment of dwarf satellites of the Milky Way and 'young' halo globular clusters along two polar great circles suggests streams of dwarf satellites that are gradually accreted and contribute the 'young' globulars (Majewski 1994, *ApJ*, **431**, L17). However, Sagittarius contributes both intermediate-age and old globulars. Absolute space motions for 30 Galactic globular clusters imply that several 'young' globulars are not associated with the proposed streams. Moreover we find clusters on prograde and retrograde orbits to cover similar age ranges and metallicities.

The large differences in star formation histories and metallicity (Grebel 1997, *Reviews in Modern Astronomy*, **10**, 29) seem to contradict that dSphs are remnants of two massive galaxies that merged with the Milky Way (Gerola et al. 1983, *ApJ*, **268**, L75).

Absolute space motions are lacking for most dwarf galaxies, but the direction of motion is also indicated by tidal tails if galaxies suffer tidal interactions. Extra-tidal stars were detected around all Milky Way dSphs (Irwin & Hatzidimitriou 1995, *MNRAS*, **277**, 1354). We show that the position angles of most dSphs (exception: Leo II) are aligned with the proposed great circles.

We find that Andromeda's satellites all seem aligned along a single polar great circle. Dwarf galaxies observed at the present epoch may preferably constitute galaxies on stable polar orbits – survivors of a previously much more numerous dwarf galaxy population.

TOTAL MASS AND LUMINOSITY OF STAR FORMING REGIONS IN THE LMC: PREDICTED INFRARED FLUXES

V. MISSOULIS, A. DAPERGOLAS AND E. KONTIZAS
Astronomical Institute, National Observatory of Athens
M. KONTIZAS
Department of Physics, National University of Athens
AND
S. OLIVER
Astrophysics Group, Blackett Laboratory, Department of Physics, Imperial College, London

We have investigated 4 Shapley constellations (1996, *A&A*, **308**, 40) to define the associated stellar complexes and their properties. From observed star counts and recent spectral classification we derive an age estimate and the upper mass limit of individual stars in these regions.

Theoretical isochrones and the star counts allow us to find the total mass and luminosity in the optical band of each complex. For the stellar complex in Shapley I there are *IRAS* observations (1991, *A&A*, **245**, 635) providing 12–100μm flux data for this region. From our observations in the optical bands, theoretical isochrones and the available infrared observations we have derived the ratios of $L_{\rm IR}/L_{\rm B} = 80$ and $M/L_{\rm IR} = 0.03$ with errors of a factor of 2. The first ratio is much larger than in the majority of luminous infrared galaxies (IRG) and the second ratio is typical for an IRG. The temperature of the dust in Shapley I should be smaller than in an IRG since the stellar population in an IRG is younger than the age of 10^8 yr for Shapley I. Nevertheless the star formation in an IRG is taking place only in the central regions whereas in a star forming region it is taking place in the whole volume, but there is also a large number of old stars. So the ratio $M/L_{\rm IR}$ in both cases is affected equivalently by stars which do not heat the dust. However, the ratio $L_{\rm IR}/L_{\rm B}$ is affected strongly by stars which do not heat the dust in an IRG, but is affected only slightly in a star forming region.

Acknowledgements: One of us (V.Missoulis) is grateful to the LOC for financial support to attend the General Assembly.

THE MAGELLANIC STREAM REVISITED

- The Mass of the Galaxy in the Range of 100 kpc -

T. MURAI
Department of Physics, Nagoya University
Chikusa-ku, Nagoya, 464-01 Japan

There is still a controversy on the mass of the Galaxy in the deep halo, some advocate a conservative view that the rotation velocity ultimately decays in accordance with the Keplerian law at the distance of 50 kpc, while others have come to consider that the rotation curve of the Milky Way, essentially stays flat or is still increasing at the distance of the Magellanic Clouds and the Magellanic Stream.

On the basis of accurate observed data of the spatial location of the LMC, SMC and the Magellanic Stream and the distribution of their radial velocities, it is clarified that the halo of the Galaxy has dark matter, resulting in a flat rotation curve with the terminal velocity of the order of 250 km/s. It is to be noticed that the tidal interaction of the LMC and the SMC around the Galaxy has produced a number of characteristics – a series of bursts of star formation, kinematic peculiarities within both Clouds, collision-induced imprints, etc. (Westerlund 1997). All of these characteristics have been revealed by observations, and could be interpreted as a result of at least two close encounters of the LMC with the SMC, which can occur only in the deep gravitational potential of dark matter as shown by a tidal simulation of Murai & Fujimoto (1980, 1986).

References

Murai, T., & Fujimoto, M. 1980, *PASJ*, **32**, 581
Murai, T., & Fujimoto, M. 1986, *Ap. Space Sci.*, **119**, 169
Westerlund, B.E. 1997, *The Magellanic Clouds*, Cambridge University Press, Cambridge.

ARE THE BULGE C-STARS RELATED TO THE SAGITTARIUS DWARF GALAXY?

YUEN KEONG NG
Padova Astronomical Observatory
Vicolo dell'Osservatorio 5, I-35122 Padova, ITALY

When the Sagittarius dwarf galaxy (SDG) crosses the galactic plane star formation might occur in its tidal tail. Carbon stars, due to their brightness, are ideal tracers for such an event. Ng & Schultheis (1997) suggested that the Bulge carbon stars (Azzopardi et al. 1991; ALRW) might be at SDG's distance. In the galactic Bulge these stars are completely different from carbon stars found in the LMC, SMC and dwarf spheroidal galaxies. If one keeps these stars in the Bulge and tries to explain their properties through a binary evolution scenario, one has to explain why they are not found up to the red giant branch tip: there is a $1\overset{m}{.}5$ discrepancy. Ng (1997) argues that part of the mystery concerning the luminosity of the ALRW stars would be solved if they are related to the SDG. It is demonstrated that the carbon star sequence is in that case similar to the observed SMC sequence. The carbon stars are in that case not metal-rich as previously thought, but they have a metallicity comparable to the LMC, with an age between $0.1-1$ Gyr. A significant fraction of the carbon stars still have luminosities fainter than the lower LMC limit of $M_{\rm bol} \simeq -3\overset{m}{.}5$. A similar trend is present among carbon stars found in other dwarf spheroidals. A scenario through binary evolution is assumed for these stars, because TP-AGB models cannot explain their origin through a single-star evolution scenario, even if they form immediately after entering the TP-AGB phase.
Acknowledgements for financial support received from the ASI.

References

Azzopardi M., Lequeux J., Robeirot E., & Westerlund B.E. 1991, *A&AS*, **88**, 265 (ALRW)
Ng Y.K. 1997, *A&A*, in press
Ng Y.K., & Schultheis M. 1997, *A&AS*, **123**, 115

COMPRESSIVE EFFECTS ON THE GALACTIC GLOBULAR CLUSTERS BY GRAVITATIONAL DISK-SHOCKING

MASAAKI SHIMADA
Department of Astronomy, Kyoto University
Kyoto 606-01, Japan

What is our problem? Observed figures of the Galactic globular clusters are somewhat elliptic in many cases. For not a few clusters (e.g. 23 of 72), the observed ellipticities (Shawl & White 1987) are ≥ 0.1 within $\sim 2r_{\rm h}$. By a series of numerical experiments, we study whether or not tidal force from the Galactic disk can produce the observed ellipticities in such an inner region (Shimada 1996)

Method and Model: We integrate the equations of motion of 10^5 cluster stars. The clusters mean force field is calculated by the SCF method (Hernquist & Ostriker 1992), and the tidal force field is given from a plane parallel disk model. We use three kinds of model orbits of the cluster: 1) single passage through the disk, 2) oscillation around $Z = 0$, and 3) stationary at $Z = 0$. Models 2 and 3 correspond to circular motions around the Galactic center. We draw density contour maps of projected images of the clusters and measure their ellipticities, ϵ.

Results and Concluding Remarks: Although the clusters exhibit $\epsilon > 0.1$ around $r \sim 2r_{\rm h}$ in some cases, $\epsilon < 0.1$ holds at almost all times in an innermost region $r \sim r_{\rm h}$. Therefore, we can conclude that the tidal force from the Galactic disk is not the principal cause of the observed ellipticities of the Galactic globular clusters. The principal cause may be rotation of the clusters (Einsel & Spurzem 1997)

References

Shawl, S.J. and White, R.E. 1987, *Ap J*, **317**, 246
Shimada, M. 1996, *Prog. Theor. Phys.*, **96**, 537
Hernquist, L. and Ostriker, J.P. 1992, *Ap J*, **386**, 375
Einsel, Ch. and Spurzem, R. 1997, *astro-ph/9704284*

AN HI SEARCH FOR M81 GROUP DWARF GALAXIES

W. VAN DRIEL
Unité Scientifique Nançay, Observatoire de Paris, France
R.C. KRAAN-KORTEWEG
Astronomy Department, Universidad de Guanajuato, Mexico
B. BINGGELI
Astronomical Institute, University of Basel, Switzerland
AND
W.K. HUCHTMEIER
Max-Planck-Institut für Radioastronomie, Bonn, Germany

The M81 group has great advantages for dwarf galaxy studies: it has about three times the dwarf content of the Local Group but is at only about a quarter of the Virgo cluster distance. We searched for H I in 23 optically selected dwarf members and possible members of the M81 group with the Nançay decimetric radio telescope in the velocity range of −529 to 1826 $\mathrm{km\,s^{-1}}$ with considerably better sensitivity (2–4 mJy rms.) than previous surveys. Half the objects observed are irregular dwarfs, expected to be H I-rich and detectable at Nançay if at the distance of the M81 group (4 Mpc).

Only three objects (Kar 1N, UGC 4998 and UGC 5658) were detected, and their high radial velocities (between 600 and 1150 $\mathrm{km\,s^{-1}}$) show them to lie *behind* the M81 group. The unexpectedly low detection rate is likely due to confusion with strong foreground (Galactic) as well as local (M81 group) H I lines: The mean redshift of the M81 group is $\langle V \rangle = 95$ $\mathrm{km\,s^{-1}}$, while Galactic H I emission typically dominates in the range $-150 \leq V \leq 115$ $\mathrm{km\,s^{-1}}$, and the local extended H I envelope in the immediate vicinity of M81 dominates at $-280 \leq V \leq 355$ $\mathrm{km\,s^{-1}}$.

H I emission of low velocity H I-rich members of the M81 group may thus still remain hidden among strong Galactic and M81 group H I lines – optical redshifts are required to determine group memberships (van Driel et al. 1998, *A&AS*, **127**, 1).

THE HIGH-VELOCITY CLOUDS: GALACTIC OR INTERGALACTIC?

HUGO VAN WOERDEN AND ULRICH J. SCHWARZ
Kapteyn Institute, Postbus 800, Groningen, The Netherlands
REYNIER F. PELETIER
Kapteyn Institute, Postbus 800, Groningen, The Netherlands; Instituto Astrofisica Canarias, La Laguna, Tenerife, Spain; Dept. of Physics, Univ. of Durham, Durham DH1 3LE, UK
BART P. WAKKER
Dept. Astronomy, Univ. Wisconsin, Madison WI 53706, USA
AND
PETER M. W. KALBERLA
Radio-astr. Institut, Univ. Bonn, 53121 Bonn, Germany

Nature and origin of the high-velocity clouds (HVCs) remain enigmatic after thirty years (Wakker & van Woerden 1997, *ARA&A*, **35**, 217), owing to lack of distance information. Hypotheses range from supernova shells at 100 pc to intergalactic clouds at 1 Mpc. On statistical grounds, Blitz et al. (1996, *BAAS*, **28**, 1349) claim that the HVCs are "remnants of Local Group formation, best explained as members of the Local Group of galaxies". Reliable distances must come from the presence or absence of absorption at the HVC's velocity in spectra of stars at different distances. For Complex A, MgII absorption is seen in *HST* spectra of the Seyfert galaxy Mrk 106, but not in the star PG0859+593 at 4 kpc (Wakker et al. 1996, *ApJ*, **473**, 834). La Palma spectra of the RR Lyr star AD UMa at 11 kpc distance show CaII absorption by Complex A at both K and H, which is lacking at 4 kpc. These absorptions are not confused with stellar metal lines. Our distance bracket $4 < d < 11$ kpc places Complex A in the Galactic Halo, at $2.5 < z < 7.5$ kpc above the plane; a distance similar to Local Group galaxies is excluded. The HI mass implied lies between 0.15 and 1.2 times 10^6 $M_\odot$. Our result precludes local origins for this HVC. It allows an origin in a Galactic Fountain, or in interaction of infalling intergalactic material (from the Magellanic System or the Local Group) with the Galactic Halo.

A NEAR INFRARED SURVEY OF THE LMC

TAKEHIKO WADA AND MUNETAKA UENO
Dept. of Earth Science & Astronomy, Univ. of Tokyo,Japan

AND

TOSHIKAZU EBISUZAKI AND YOSUKE OHNO
Computational Science Lab., The Institute of Physical and Chemical Research, Japan

We surveyed the central $3° \times 6°$ region of the Large Magellanic Cloud (LMC) with angular resolution of $10''.0$ at the J(1.25μm), H(1.65μm) and K$'$(2.15μm) bands. The observations were performed from Oct to Nov, 1994 at the Siding Spring Observatory. We used a 25 cm/F3.5 Newtonian telescope equipped with a 512x512 PtSi-Camera (Ueno et al. 1992). The plate scale was $4.''6 \times 6.''0$/pixel and the total field of view was $40.'2 \times 52.'3$. The system was attached onto the Automated Patrol Telescope of the University of New South Wales. The limiting magnitudes of the survey were 13.6, 11.9, and 10.0 magnitude (3σ) at the J, H, and K$'$ bands, respectively. The positions and the J, H, and K$'$ magnitudes were derived for 1599 point sources whose S/N ratio were more than four at the K$'$-band. The dominant components of the sources are red super giants (RSG), luminous M-type giants and AGB stars in the LMC. The distributions of RSG, luminous giants and CO line flux are spatially different from each other. This supports the idea that the position of the active star forming region has changed during the past 10^8 years. The detected sources are cross-identified with the *IRAS* Point Source Catalog. Sixty-two of 680 *IRAS* sources in the region spatially coincide with the detected sources and the spectrum energy distributions from 1.25μm to 100μm were derived for these 62 sources. Half of them are well described by a black body spectrum, while the others show a flat or redder spectrum. These latter sources are supposed to be stars with dust envelopes.

References

Ueno, M., Ito, M., Kasaba, Y., & Sato, S. 1992, *SPIE*, **1762**, 423

N-BODY SIMULATIONS OF THE MAGELLANIC SYSTEM INCLUDING GAS DYNAMICS AND STAR FORMATION

A.M. YOSHIZAWA AND M. NOGUCHI
Astronomical Institute, Tohoku University
Aoba, Sendai 980-77, Japan

The system of the Magellanic Clouds is considered to be dynamically interacting among themselves and with our Galaxy. This interaction is thought to be the cause of many complicated features seen in the Magellanic Clouds and the Magellanic Stream (see Westerlund 1990, *A&AR*, **2**, 27). In order to better understand the formation and evolution of the Magellanic System, we carry out realistic N-body simulations of the tidal distortion of the Small Magellanic Cloud (SMC) due to our Galaxy and the Large Magellanic Cloud (LMC).

The present work is based on the tidal model (e.g., Murai & Fujimoto 1980, *PASJ*, **32**, 581; Gardiner & Noguchi 1996, *MNRAS*, **278**, 191). A novel point of this work is that *it takes into account the motion of interstellar gas clouds and the star formation process.* We succeeded in reproducing a gas stream with almost no stars as a result of the close encounter between the SMC and the LMC plus our Galaxy about 1.5 Gyr ago. We also reproduced the spatial distribution of gas particles around the Inter-Cloud Region and the star formation property in the center and the vicinity of the SMC. These features have been caused by the latest encounter with the LMC about 0.2 Gyr ago. The major achievement of this work is to provide a possibility of explaining observed structural, kinematic, and star formation properties of the SMC, and other related tidal features in the Magellanic System, *without resorting to non-gravitational processes such as ram pressure due to the hot halo gas of our Galaxy or collisions of the SMC gas with the high velocity clouds.* A detailed comparison of the model results with observational data would enable a deeper understanding of the gas dynamical processes and star formation activity in this nearest interacting galaxy system. A full paper will be published elsewhere.

TIDAL DWARF GALAXIES

P.-A. DUC
ESO, Karl-Schwarzschild Strasse 2
D-85748 Garching bei München, Germany
<*pduc@eso.org*>

AND

I.F. MIRABEL
CEA, SAp, C.E. Saclay
91191 Gif/Yvette Cedex, France, & Instituto de Astronomía y Física del Espacio, Argentina
<*mirabel@discovery.saclay.cea.fr*>

Abstract. We review the observational evidences for tidal dwarf galaxies, a class of small galaxies formed out of the tidal debris of collisions between massive galaxies. Tidal dwarfs are found far from the interacting parent galaxies, associated with massive clouds of atomic hydrogen located at the tip of long tidal tails. These newly formed galaxies are among the best cases for the study of galaxy formation in the nearby Universe.

1. Introduction

Most studies of interacting/merging systems focus on their central regions, which are dramatically affected by the collision. In particular the gas often looses angular momentum and sinks into the galactic cores, triggering nuclear starbursts and the formation of young star clusters. On the other hand tidal forces may pull out from the outer regions into intergalactic space stars and interstellar gas shaping rings, bridges, plumes and tails. The amount of matter lost during that outflow can be as large as one third of the mass in the pre-encounter disks. In interacting systems the bulk of the atomic hydrogen is in fact located outside the galaxy bodies (see review by F. Combes in this volume and examples below).

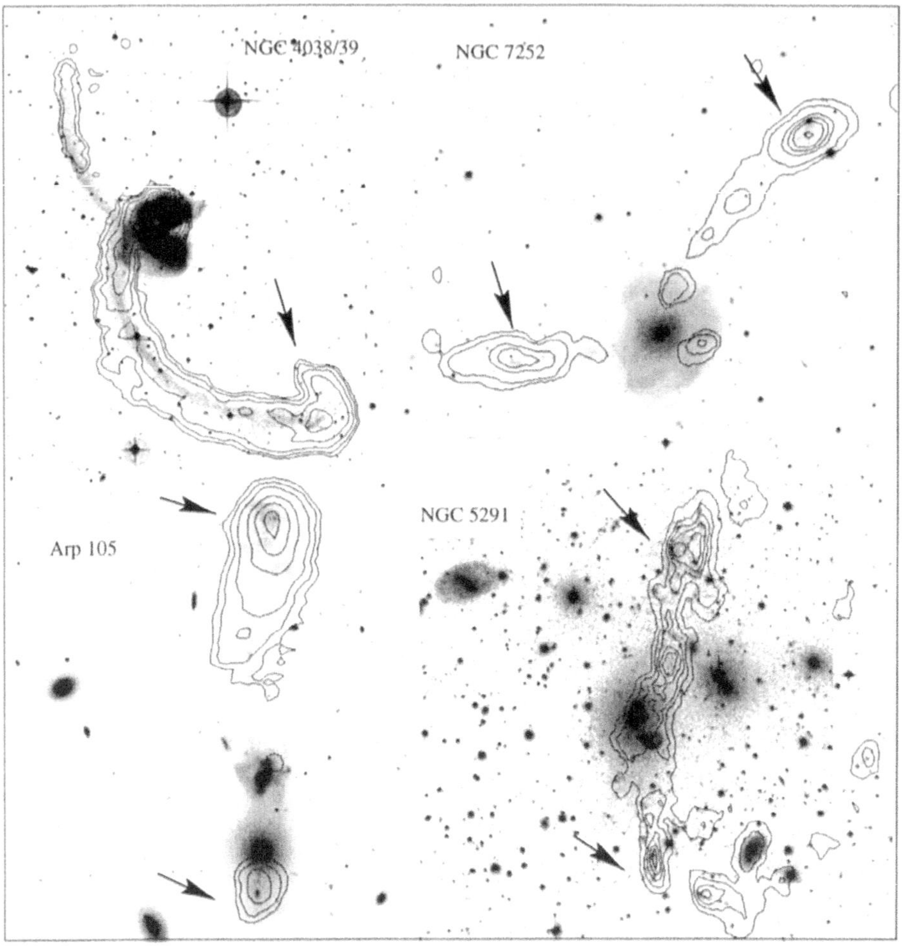

Figure 1. Tidal dwarf galaxies in four interacting systems. They are seen at the end of optical tidal tails or HI plumes (black contours) and indicated by arrows. The optical images are clock-wise from the Digital Sky Survey, Duc (1995), Duc & Mirabel (1997) and Duc & Mirabel (1994). The HI maps are from Hibbard et al. (1997), Hibbard et al. (1994), Malphrus et al. (1997) and Duc et al. (1997)

Nevertheless until recently only few studies have been devoted to the properties of tidal features (e.g. Wallin 1990; Schombert et al. 1990). Tails are basically used as tracers of past minor/major mergers. Their shapes are useful to constrain the parameters in numerical models of the collision, including the mass of dark matter in the parent galaxies (Dubinski et al. 1996). The interest in these collisional debris was revived when it was discovered that they host active star forming regions (Schweizer 1978; Mirabel et al. 1992) and are actually a nursery of small galaxies, the so-called "tidal

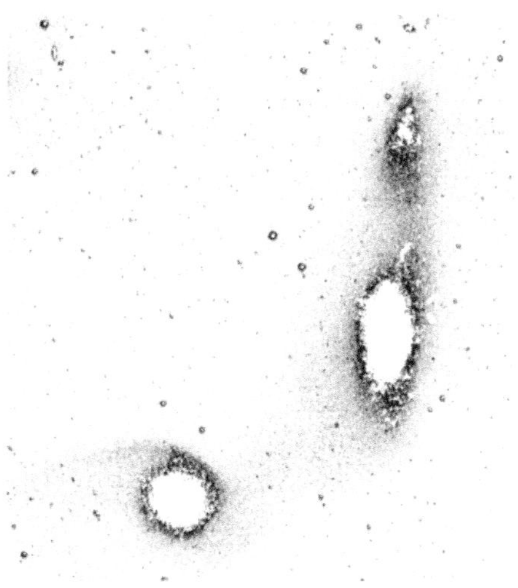

Figure 2. Optical image of the disk-disk system NGC 2992/93 with the Hα emission superimposed in white. Note how star formation as traced by the ionized gas is enhanced simultaneously in the galaxy body and at the tip of the tidal tail.

dwarf galaxies" (hereafter TDGs). New optical and HI observations have shown that TDGs actually form a class of "recycled" objects with some characteristics similar to the more classical dwarf irregulars (dIrrs) and blue compact dwarf galaxies (BCDGs). Here we review prototype interacting systems that exhibit the formation of tidal dwarfs, detail the general properties of TDGs and give some hints for their origin and future evolution.

2. Case studies

We have carried out multi-wavelength observations of several interacting systems in the nearby Universe. Optical imaging and spectroscopy were obtained with the CFHT at Mauna Kea and the ESO/NTT at la Silla Observatory; near-infrared imaging at the ESO/MPI 2.2m, and HI at the VLA. HI data come either from our own observations, or were kindly provided to us by other groups.

Fig. 1 and Fig. 2 present different examples of interactions: spiral-spiral collisions (NGC 4038/39, "The Antennae"; NGC 2992/93), complete merger between spirals (NGC 7252) and encounters involving early-type galaxies (Arp 105, "The Guitar galaxy"; NGC 5291). Long tidal tails are clearly seen emanating from the parent galaxies. At their tip, at distances up to 100 kpc from the nuclei, small irregular objects are found with absolute magnitudes typical of dwarf galaxies. They host blue compact clumps

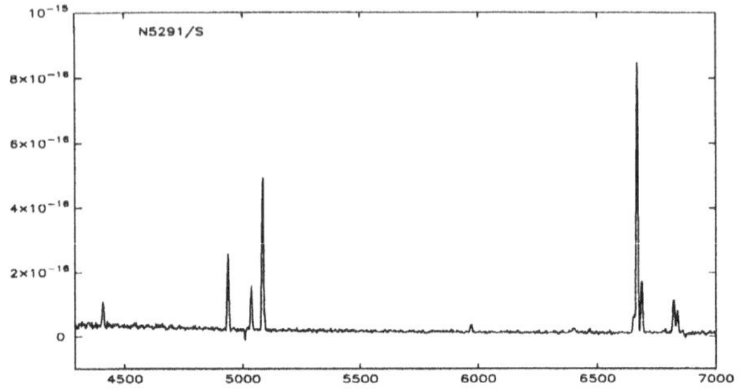

Figure 3. Optical spectrum of one of the tidal dwarfs near NGC 5291 showing emission lines typical of HII regions. From Duc & Mirabel (1997).

that also show up in maps of the ionized gas (see Fig. 2).

The spectra of the optical condensations exhibit emission lines, typical of HII regions ionized by massive OB stars younger than 10 Myrs (e.g. Fig. 3). Given the time scale for the formation of clumps in tails – typically 1 Gyr –, the stars at the tip of the antennae must have been born in situ. Therefore interactions not only trigger star formation in the main body of the parent galaxies but also far from the nuclei in the remote tidal features.

The contours of the HI column density are superimposed on the optical images in Fig. 1. It is clear that the central regions of the parent galaxies contain little atomic gas, whereas the optical tails, and especially the tidal dwarfs at their tip, are associated with HI clouds as massive as $6\times10^9\ M_\odot$. Similar HI distributions were observed by Hibbard & van Gorkom (1996) in several other interacting systems. On the other hand the molecular gas tends to concentrate in the central regions fueling the nuclear starburst of interacting galaxies (e.g. Young & Scoville 1991). Such a spatial segregation of the different gas components is clearly seen in Arp 105 (Duc et al. 1997; Fig. 1).

3. Properties of tidal dwarf galaxies

Table 1 summarizes the statistics for the main properties of the 20 TDGs sofar studied.

3.1. STELLAR POPULATIONS AND GAS CONTENT

Tidal dwarf galaxies may be made of two stellar components: young stars, formed from the recent collapse of expelled HI clouds, and possibly an older star population coming from the disk of their parent galaxies. Using our aperture photometry and spectroscopic data, we could estimate the relative proportion of both populations and conclude that TDGs actually

TABLE 1. Integrated properties of tidal dwarf galaxies

Properties	Units	Mean	Min	Max
Absolute Blue Magnitude	mag	-14.8	-12.1	-18.8
$B-V$ color index	mag	0.3	0.0	0.7
Star Formation Rate	$\log(M_\odot/\text{year})$	-1.1	-3.6	0.3
HI Mass	$10^9\ M_\odot$	1.6	0.2	6.0
O Abundance	12+log(O/H)	8.5	8.3	8.6

split into two categories 1) extremely young objects, most probably forming their first generation of stars (e.g.: the dwarfs around NGC 5291, Duc & Mirabel 1997), with high star formation rates equivalent to those observed in blue compact dwarf galaxies, and 2) galaxies dominated by an old stellar population originally from the disk of their progenitors, and that look like dwarf irregulars (e.g. the galaxy North of NGC 2992, Fig. 2). Both type of galaxies separate on an optical/near infrared color-color diagram, as shown in Fig. 4. The optical $B-V$ colors of the "young" TDGs and BCDGs are similar. The near-infrared $V-K$ color index of TDGs appears though to be redder on average, which could be due to a difference in metallicity between both classes of objects. The colors of the "old" TDGs are similar to these of the outer parts of their parent galaxies.

In all these objects, the equivalent width of the optical Balmer lines indicates that the current star–forming episode is younger than 10 Myrs. Important HI reservoirs – between $5\times10^8\ M_\odot$ and $5\times10^9\ M_\odot$ – can sustain the star formation for several Gyrs.

The molecular gas content of tidal dwarfs is still largely unknown. Smith & Higdon (1994) failed to detect any CO emission in the tails of a few interacting systems. CO was also reported to be very weak in blue compact dwarf galaxies despite their high star formation rate. This was interpreted as a metallicity effect. However TDGs seem to be quite metal rich systems in which CO should be easier to detect.

3.2. METALLICITY

Fig. 5 shows the oxygen abundance vs. absolute magnitude of a sample of TDGs and nearby dIrrs. The abundances have been estimated in the ionized gas from the [OIII]/Hβ line ratio. Their uncertainties are discussed in Duc & Mirabel (1997). Clearly TDGs are more metal rich than classical dwarfs of the same luminosity. They have metallicities $Z_\odot/3$ on average, a value that is typical of the outer regions of spirals. They do not follow the correlation found for field dwarf and giant galaxies between luminosity

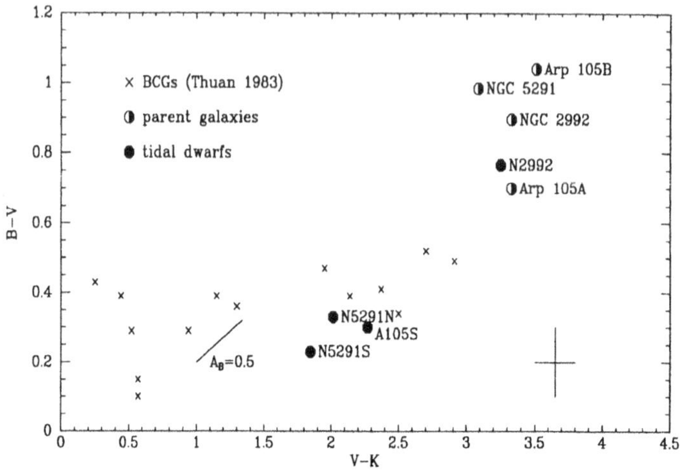

Figure 4. Color-color diagram of tidal dwarfs. For reference, the colors of the outer regions of the parent galaxies are also indicated and a sample of blue compact dwarfs has been added. The cross at the lower right corner indicates typical error bars.

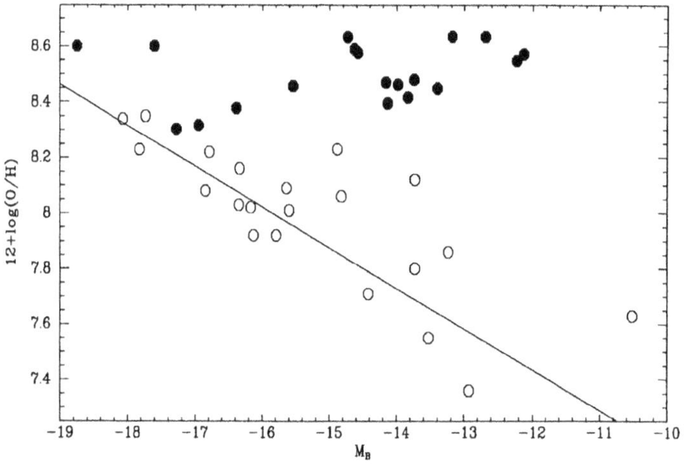

Figure 5. Oxygen abundance vs absolute blue magnitude for our sample of tidal dwarfs (black points) and a sample of isolated dwarf galaxies (open points; from Richer & McCall 1995).

(hence mass) and metallicity. Being "recycled" objects, formed from pre-enriched material, tidal dwarfs got as an heritage from their parents this relative high metal content.

3.3. DYNAMICS

Little is still known about the internal dynamics of tidal dwarfs. First indications are that the most massive TDGs may be gravitationally bound (Hibbard et al. 1997; Malphrus et al. 1997). Hints for rotation were found in the HI cloud associated with a TDG in Arp 105 (see Fig. 6). Furthermore, strong velocity gradients of the ionized gas have been found in several objects, which suggests rotation too (Fig. 7). Therefore, some objects in tidal tails may already be dynamically independent. Further 3D kinematical studies would be necessary to verify this assertion and estimate their dark matter content, predicted to be low in numerical simulations (Barnes & Hernquist 1992).

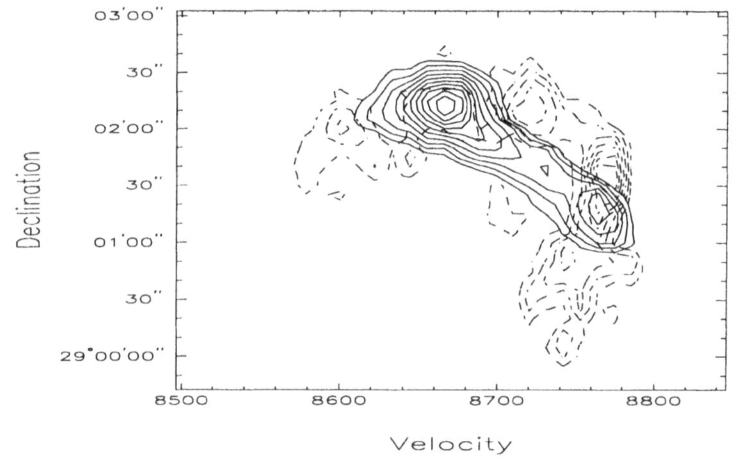

Figure 6. Position-Velocity diagram of the HI northern tidal tail in Arp 105. Two components were identified: the expanding HI tidal tail (dashed contours) and a kinematically decoupled, possibly rotating, component (continuous contours) associated with the formation of a tidal dwarf at the tip of the tail; adapted from Duc et al. (1997).

4. Origin and evolution of tidal dwarf galaxies

4.1. INGREDIENTS FOR THE FORMATION OF TDGS

Obviously star-forming tidal dwarf galaxies only form after an encounter between gas-rich galaxies. The nature of the collision is a priori also important. One should differentiate disk-disk collisions from collisions involving early type galaxies. In the first case, tidal features are made out of expelled gas and stars. In the second case, stellar tails are difficult to form because of the pressure-supported nature of the stellar dynamics of ellipticals, whereas the HI component, mostly of external origin and often supported by rotation, is easier to disrupt. Pure HI tails will then be formed. Finally there are

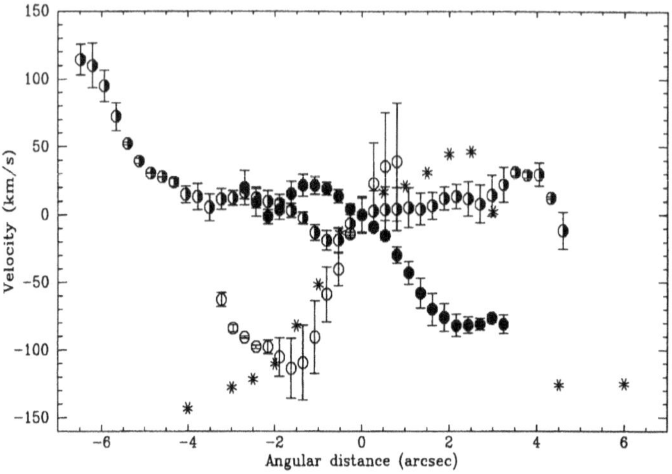

Figure 7. Velocity profiles of the ionized gas for several tidal dwarfs; adapted from Duc & Mirabel (1997).

instances where HI and stellar tails have different positions and extensions. The initial properties of the resulting TDGs will depend on their location. This explains why two categories of TDGs have been identified.

4.2. MODELS FOR THE FORMATION OF TDGS

Models for the formation of TDGs put forward two mechanisms: a local dynamical instability in the old stellar populations of tidal tails, followed by accretion of gas (Barnes & Hernquist 1992) or the collapse of a supermassive cloud triggering precipitous star formation activity (Elmegreen et al. 1993). Because of the variety of the stellar and gas content of TDGs one could argue that both mechanisms play a role: the stellar scenario would apply for TDGs born in HI+stellar tails; the gas scenario for TDGs born in pure HI tails. However it seems that even for TDGs belonging to the first category, the HI masses and resulting potential well are large enough to trap the old tidal star population. As a consequence, despite the diversity of initial environments, all TDGs may have been formed by the same mechanism.

What causes the collapse of tidally expelled gas clouds is still unclear. Other environmental effects such as ram pressure by the intergalactic medium may play a role. Strong asymmetries of the HI distribution in some interacting systems suggest some kind of compression at the interface with the IGM. Hawarden & Chaytor (1996) find in NGC 5291 a possible expansion of the x-ray emission along the HI tail and its associated HII regions. This hypothesis should be further investigated and modelized. In x-ray clusters, besides tidal forces, ram pressure could even have taken part in the pulling out of the HI clouds from spirals known to be HI deficient. Instances of enhanced star formation in stripped clouds have been presented by J. Kenney (this volume).

4.3. SURVIVAL OF TDGS

Do tidal dwarf galaxies contribute significantly to the overall population of dwarf galaxies? The answer to this fundamental question relies on the knowledge of the frequency of tidal interactions between galaxies, which is still controversial, and on the life-time of TDGs. The latter is limited by the hostile environment for TDGs located in the vicinity of giant parent galaxies. They may fall back on their progenitors in time scales of 1 Gyr, as put forward by Hibbard & Mihos (1995), or be tidally disrupted. It is therefore expected that only the most massive TDGs that are far away from their progenitors will survive in timescale of Gyrs. This limits the number of galaxies produced to one or two per colliding system. In this context, it is not surprising that no luminous star–forming TDGs are found at the base of the tidal tails too close from the parent galaxies.

From an observational point of view, the census of TDGs is not an easy task. TDGs should obviously be searched in the environment of interacting galaxies and a priori in high density regions such as groups and clusters. Instances of small star-forming entities were discovered in the arms of the Stephan's Quintet by Ohyama et al. (1997). Hunsberger et al. (1996) claim from the analysis of photometric data that half of the dwarf galaxies in Hickson compact groups could be of tidal origin. However, one should note that once the stellar/gaseous "bridge" between the parent and child galaxies has dissipated, it is difficult to re-establish a link between the two. Our study has shown that a good genetic fingerprint of TDGs is their high metallicity. In this respect, several studies have put forward trends for dwarf galaxies in clusters to be be more metallic than field dwarfs (Bothun et al. 1985; Vilchez 1995). Since the collision rate is enhanced in denser environments, it is tempting to argue that a significant fraction of dwarfs in clusters could be recycled objects. A bimodal star formation history is also a strong signature for tidal dwarfs. Evolutionary Synthesis Models simulating a burst of star formation on top of the underlying component of old galaxies reproduce well the TDG star formation history and will give constraints for their future evolution (Fritze - v. Alvensleben & Duc 1997).

4.4. TDGS AS LABORATORIES

Tidal dwarf galaxies are of particular interest to studies of galaxy formation in the nearby Universe. They are recently formed galaxies from the collapse of massive gas clouds. Contrary to BCDGs, TDGs born in pure HI tails are not contaminated by old stellar populations, allowing a better study of the parameters that rule star formation in galaxies.

References

Barnes, J. E. and Hernquist, L.: 1992, *Nature* **360**, 715
Bothun, G. D., Mould, J. R., Wirth, A., and Caldwell, N.: 1985, *AJ* **90**, 697
Dubinski, J., Mihos, C., and Hernquist, L.: 1996, *ApJ,* **462**, 576
Duc, P.-A and Mirabel, I. F.: 1994, *A&A,* **289**, 83
Duc, P.-A.: 1995, *Ph.D. thesis*, Université Paris VI
Duc, P.-A., Brinks, E., Wink, J. E., and Mirabel, I. F.: 1997, *A&A,* **326**, 537
Duc, P.-A. and Mirabel, I. F.: 1997, submitted to A&A
Elmegreen, B. G., Kaufman, M., and Thomasson, M.: 1993, *ApJ* **412**, 90
Fritze-v.Alvensleben, U. and Duc, P.-A.: 1997, in *IAU JD2-050P*
Hawarden, T.G. and Chaytor, D.H. *BAAS*,**189**, 120.17
Hibbard, J., van der Hulst, J., and Barnes, J.: 1997, *in preparation*
Hibbard, J. E., Guhathakurta, P., van Gorkom, J. H., and Schweizer, F.: 1994, *AJ* **107**, 67
Hibbard, J. E. and Mihos, J. C.: 1995, *AJ* **110**, 140
Hibbard, J. E. and van Gorkom, J. H.: 1996, *AJ* **111**, 655
Hunsberger, S. D., Charlton, J. C., and Zaritsky, D.: 1996, *ApJ* **462**, 50
Malphrus, B., Simpson, C., Gottesman, S., and Hawarden, T. G.: 1997, *AJ* **114**, 1427
Mirabel, I. F., Dottori, H., and Lutz, D.: 1992, *A&A* **256**, L19
Ohyama, Y., Nishiura, S., Murayama, T. and Taniguchi, Y., 1997, *preprint*
Richer, M. G. and McCall, M. L.: 1995, *ApJ* **445**, 642
Schombert, J. M. and Wallin, J. F. and Struck-Marcell, C.: 1990, *ApJ* **99**, 497
Schweizer, F.: 1978, in E. Berkhuijsen and R. Wielebinski (eds.), *Structure and Properties of Nearby Galaxies*, p. 279, Dordrecht, D. Reidel Publishing Co.
Smith, B. J. and Higdon, J. L.: 1994, *AJ* **108**, 837
Thuan, T. X.: 1983, *ApJ* **268**, 667
Vilchez, J. M.: 1995, *AJ* **110**, 1090
Wallin, J. F.: 1990, *AJ* **100**, 1477
Young, I. S. and Scoville, N. Z.: 1991, *ARA&A* **29**, 581

GALAXY INTERACTIONS: THE HI SIGNATURE

R. SANCISI
Kapteyn Astronomical Institute
University of Groningen
The Netherlands

Abstract. HI observations are an excellent tool for investigating tidal interactions. Ongoing major and minor interactions which can lead to traumatic mergers or to accretion and the triggering of star formation, show distinct HI signatures. Interactions and mergers in the recent past can also be recognized in the HI structure and kinematics. Recent 21cm line surveys of large samples of galaxies indicate that at least one out of every four galaxies shows signs of a present interaction or merger/accretion events in the recent past.

1. Introduction

This review focuses on the role of neutral hydrogen in galaxy interactions. Neutral hydrogen serves, in the first place, as a very good tracer of tidal disruption. This is simply because it is generally more extended than the stars in the outer parts of galaxies and is, therefore, the component most affected by the interaction. Indeed HI tails and bridges or other peculiar features in the HI distribution and kinematics around images of optically undisturbed galaxies are sometimes the only evidence that the galaxies are really interacting.

The HI gas plays a fundamental role in the formation of disks and in their evolution. In the course of an interaction the HI complexes can be either directly accreted or displaced by the tidal forces and fall back onto the interacting systems later. In the latter case a reservoir of fresh material is created which, at a later stage, contributes to the process of disk building and star formation.

It is also interesting to note that the extended HI structures of tidal origin observed in the neighborhood of interacting systems provide large cross sections, of order 100 kpc diameter, as needed for systems capable of explaining the absorption lines in the spectra of QSOs.

In this study a distinction is made between two types of interactions: a major and a minor one. The "major" one involves systems of comparable masses and usually produces large tidal effects. It may lead to the destruction of disks and to mergers and elliptical galaxies as end products. The "minor" interaction takes place between a main galaxy and one or more satellites or companions of small mass (mass ratio usually less than 0.1). It leads to gas accretion and the building up of disks, and may cause localized star formation and starbursts. Attention is also drawn to a number of systems which despite their isolation show the peculiar HI properties typical of interacting systems. They may, therefore, have had some recent encounter and at present be in an advanced stage of accretion.

A brief review of the morphology and main properties of interacting systems is presented. A few examples are illustrated (Fig. 1) and briefly discussed. At the end, the question of the frequency of galaxy interactions and infall is addressed.

2. Major Interactions

There are several cases of multiple systems with two or more members which show heavily perturbed HI images: in addition to the gas which is seen associated with the individual galaxies, there are cloud complexes, usually long tails, bridges or ring-like structures, in the regions near and around them. A list of well-known, representative cases is given in Table 1. For many of these systems it is the peculiar gas picture (Fig. 1) that unmistakably points at the ongoing strong tidal interaction.

The galaxy pairs in the table (see also the sample of Chengalur et al. 1995b) present a characteristic HI picture with long tails and bridges. M51 and its companion are a classic example. The VLA observations of Rots et al. (1990) show a highly disturbed picture. The most striking feature is a 90 kpc long HI tail connected loosely to the outer disk of M51, which has no optical counterpart.

A number of optically disturbed systems, characterized by bridges and long tails, were proposed by Toomre (1977) as a possible sequence of ongoing galaxy mergers. Four of these and one system at a slightly earlier stage have been recently imaged in HI with the VLA (Hibbard 1995, Hibbard and van Gorkom 1996). These observations seem to indicate some trends along the merging sequence. In the early stages, large amounts of HI are still present within the galaxy disks. In the final stages there is little or no HI within the remnant bodies. Tidal material is seen falling back towards the remnant, as beautifully illustrated by the observations of NGC 7252

TABLE 1. Multiple Systems

M81/M82/NGC3077	Yun et al. 1994
NGC3165/3166/3169	Haynes 1981
NGC3623/3627/3628 (Arp 317, Leo Triplet)	Haynes et al. 1979
NGC4631(=Arp 281)/4656/4627	Rand 1994
NGC7448/7463/7464/7465	Haynes 1981
Stephan's Quintet	Shostak et al. 1984
Galaxy-Magellanic Clouds	Mathewson et al. 1974 (and refs.)
M96 Group (Leo ring)	Schneider et al. 1989
NGC5194(M51)/5195 (Arp 85)	Rots et al. 1990
NGC520 (Arp 157)	Hibbard and van Gorkom 1996
NGC678/680	Haynes 1981
NGC1510/1512	Hawarden et al. 1979
NGC3226/3227 (Arp 94)	Mundell et al. 1995
NGC3690/IC694 (Arp 299)	Stanford and Wood 1989
NGC3921 (Arp 224, MKN 430)	Hibbard and van Gorkom 1996
NGC4038/4039 (Arp 244, The "Antennae")	van der Hulst 1979
Arp 295	Hibbard and van Gorkom 1996
NGC4485/4490 (Arp 269)	Viallefond et al. 1980
NGC4676 (Arp 242, "The Mice")	Hibbard and van Gorkom 1996
NGC4725/4747(=Arp 159)	Wevers et al. 1984
NGC7252 (Arp 226)	Hibbard et al. 1994
IIZw70-71	Balkowski et al. 1978
UGC6922/6956	Appelton 1983
HI 1225+01	Chengalur et al. 1995a

(Hibbard et al. 1994) shown here in Fig. 1. The HI is almost completely concentrated in tidal tails which are often more extended than their optical counterparts, while massive concentrations of molecules are found in the central region.

One of the key questions about the merger of two disk systems is whether the end product is always an elliptical galaxy or whether it is possible, depending on the kind of impact, for a disk to survive or re-form. There are indications from the observations (cf. van Gorkom and Schiminovich 1997, Stanford and Wood 1989, see also NGC 3310 below) and from numerical work (Barnes 1996) that disks can be quite robust, and can survive or form again under certain encounter-merger conditions.

Another class of objects that may be the result of collisions and interactions as described here are the ring galaxies and the associated remarkable HI structures such as Arp 143 (Appleton et al. 1987).

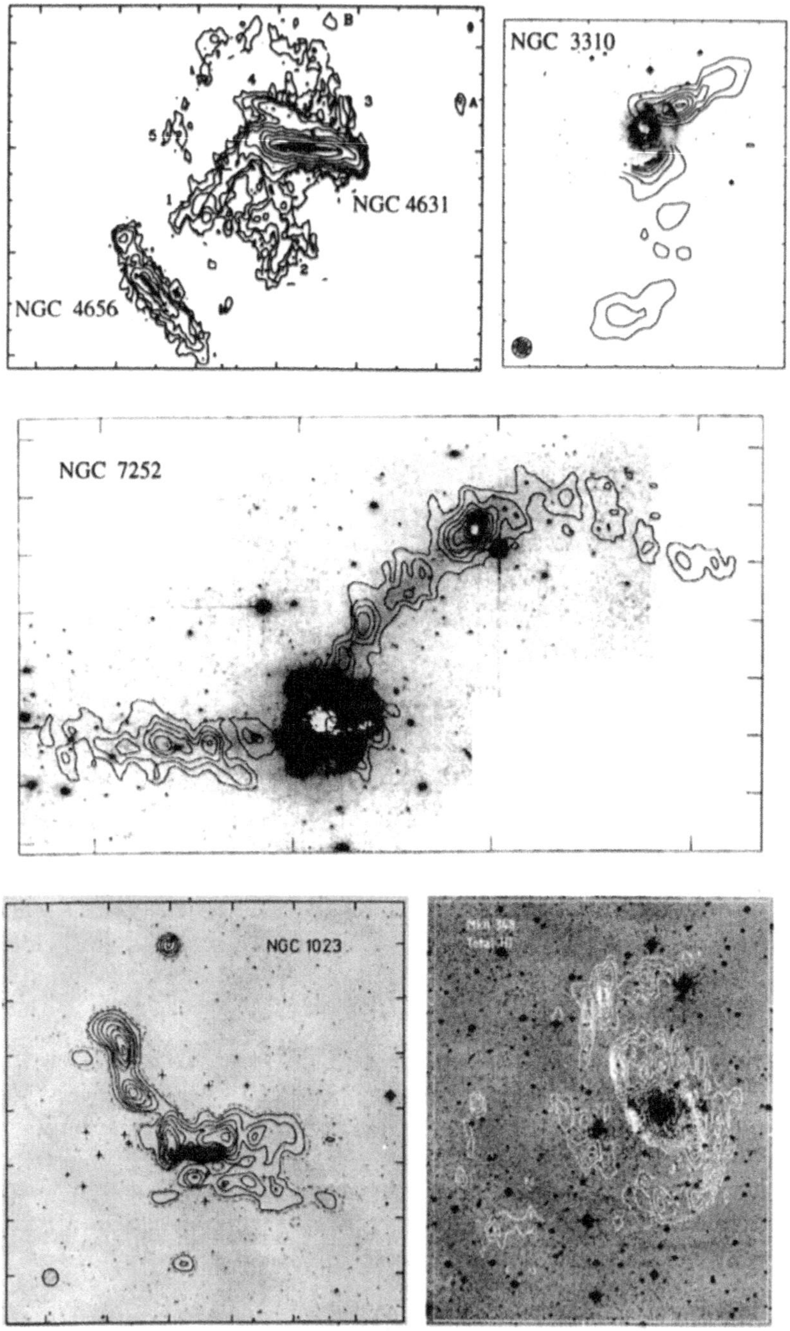
NGC 4631
NGC 4656
NGC 3310
NGC 7252
NGC 1023

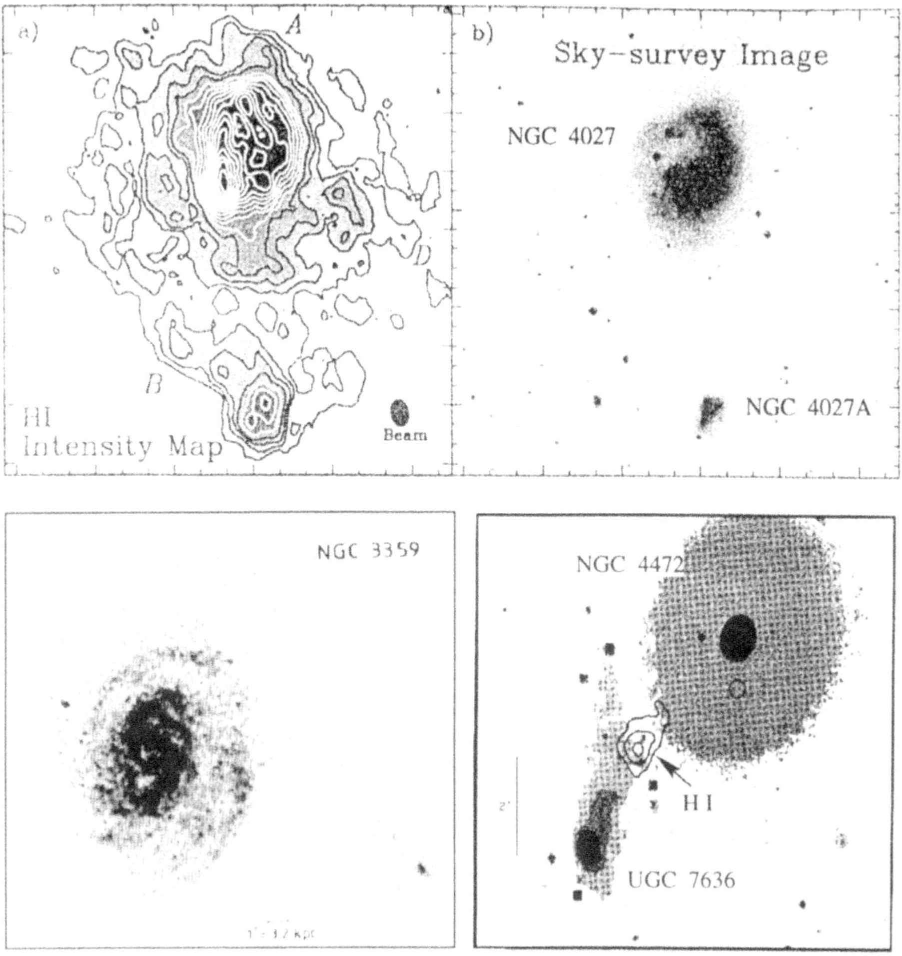

Figure 1. HI maps (see facing page) superposed on the optical pictures of the interacting system NGC 4631/4656/4627 (Rand 1994) and of the "merger" galaxies NGC 3310 (Kregel and Sancisi 1998), NGC 7252 (Hibbard 1995), NGC 1023 (Sancisi et al. 1984) and Mkn 348. The HI map of NGC 3310 shown here is incomplete. It shows only the two extended tails and not the main body HI. The HI map of Mkn 348 was obtained with the WSRT (Sancisi, unpublished). The lowest contour and the contour step are 6 10^{19} cm^{-2}, the beam FWHM is $25'' \times 48''$.
This page shows, as representative examples of galaxies interacting with dwarf companions, the spirals NGC 4027 (Phookun et al. 1992) and NGC 3359 (Kamphuis and Sancisi 1993) and the elliptical NGC 4472 (McNamara et al. 1994).

3. Minor Interactions. Accretions

A large number of galaxies have dwarf companions and when mapped in HI show clear indications of present tidal interactions. The best examples known are listed in Table 2, and a few are illustrated in Fig. 1. The prototype is NGC3359. The HI map of this galaxy clearly shows a small companion with a long tail pointing back to NGC 3359 and almost connecting with its extended HI layer (Kamphuis and Sancisi, 1993). The companion is a hydrogen-rich object with a very faint optical counterpart. Its total HI mass is 10^8 $M_\odot$, only 2 percent of the HI mass and 0.1 percent of the total mass of NGC 3359. Its head-tail structure indicates that it is probably being tidally disrupted.

TABLE 2. Galaxies with dwarf companions and/or peculiar structures

NGC262 (Mkn348)	Heckman et al. 1982, Simkin et al. 1987
NGC628	Kamphuis and Briggs 1992
NGC1023 (Arp 135)	Sancisi et al. 1984
NGC1961 (Arp 184)	Shostak et al. 1982
NGC2146	Fisher and Tully 1976, Briggs 1988
NGC2782 (Arp 215)	Smith 1994
NGC2865	Schiminovich et al. 1995
NGC3067	Carilli and van Gorkom 1992
NGC3310 (Arp 217)	Mulder et al. 1995, Kregel and Sancisi 1998
NGC3359	Kamphuis and Sancisi 1993
NGC3656 (Arp 155)	Balcells and Sancisi 1996
NGC4027 (Arp 22)	Phookun et al. 1992
NGC4254	Phookun et al. 1993
NGC4472 (Arp 134)	McNamara et al. 1994
NGC4565	Rupen 1991
NGC4694	Cayatte et al. 1990, van Driel and van Woerden 1989
NGC4826	Braun et al. 1994
NGC5128 (Cen A)	Schiminovich et al. 1994
NGC5635	Saglia and Sancisi 1988
NGC5457 (M101, Arp 26)	van der Hulst and Sancisi 1988, Kamphuis 1993

Very similar situations are found in NGC 4565-4565A (Rupen, 1991) and in NGC 4027-4027A (Phookun et al 1992). The HI masses of the companions are usually less than 10 percent of the HI masses of the main galaxies and their systemic velocities are generally close to those of the larger systems. The picture which emerges from the observations is that of the capture of a gas-rich dwarf by a massive system followed by tidal disruption and accretion of the dwarf, but only minor damage to the main galaxy.

The specific cases just mentioned probably represent early stages of the interaction-accretion process. More advanced cases have also been observed. The giant nearby spiral galaxy M101 may be in such a stage. An HI complex of about 2×10^8 $M_\odot$ moving with velocities of up to 150 km/s with respect to the disk and the corresponding large hole in the HI layer are interpreted as being due to a collision with a dwarf companion which has gone through the HI layer of M101 creating a large cavity (van der Hulst and Sancisi 1988, Kamphuis 1993). The high velocity gas will eventually rain back down onto the M101 disk.

Like M101, there are more systems which do not have any obvious bright companions and yet when mapped in HI display peculiar features which are reminiscent of those seen associated with interacting systems. They may represent cases of past interactions and be at present in an advanced merger stage in which the victim is no longer visible. Interesting examples are those of NGC 1023, NGC 3310, NGC 628 and Mkn 348. NGC 1023 is an SO galaxy surrounded by a clumpy and irregular HI structure of total mass 1.5×10^9 $M_\odot$ (Sancisi et al. 1984) reminiscent of the tails and bridges found in the interacting multiple systems (see Fig. 1). NGC 3310 is a peculiar (Arp 217) Sbc-type starburst galaxy. Mulder et al. (1995) have shown the presence of extended HI, which has a well developed two-tail structure (Fig. 1, Kregel and Sancisi 1998). This must be an advanced merger of two galaxies, probably of similar masses, which has either preserved the old disk of one of the progenitors or, perhaps more likely, has led to the formation of a new disk. Although the optical images of both NGC 1023 and NGC 3310, as of the several other Arp objects in the table, already show some peculiarities, it is their HI structure and kinematics that fully reveal the ongoing, possibly "major" mergers. In contrast to these, NGC 628 and Mkn 348 are different and very interesting because they have a very clean, regular optical image. Only the HI betrays a possible recent accretion. For NGC 628 this is indicated by the presence in its outer parts of two high-velocity HI complexes, which are symmetrically placed with respect to the galaxy center. These complexes have sizes of tens of kiloparsecs, HI masses of 10^8 $M_\odot$, and maximum velocity excesses of 100 km/s (Kamphuis and Briggs, 1992). For Mkn 348 a probable past interaction and gas accretion is suggested by the presence of an enormous HI envelope and a large tail-like extension to the east (see Fig. 1, cf. also Simkin et al. 1987).

It is important to note that in all these cases of spiral galaxies, a careful study of the structure and kinematics of the HI is necessary to distinguish between configurations that can be considered "normal" and configurations that are definitely "peculiar" and point at a recent interaction and infall. There are distinct and recognizable signatures in the HI that make the difference possible to determine, but it is not always easy to draw the line between effects due to the internal metabolism of the galaxy and those due to the environment. For example it is interesting to consider the asymmetries affecting spiral galaxies, which seem to occur quite frequently (Richter and Sancisi, 1994). The asymmetry is seen in the HI structure and, even more often and more strikingly, in the HI kinematics (Swaters et al., 1998). Should the origin of such asymmetries be attributed to past interactions and accretion events? This is not at all obvious and there may be other explanations. It should be noted that asymmetries are also found in the optical and that for them a past interaction is the favored interpretation (see paper by Zaritsky in this volume).

Neutral hydrogen found in elliptical galaxies with shells, like NGC 5128 (Cen A) (Schiminovich et al. 1994) and NGC 2865 (Schiminovich et al. 1995) (see also Schiminovich, this Symposium), and also near ellipticals with dwarf companions, like NGC 4472 (McNamara et al. 1994) shown here in Fig. 1 and NGC 3656 (Balcells and Sancisi, 1996), clearly indicates that the accretion phenomenon illustrated above for spirals is probably playing an important role in all types of galaxies, including ellipticals.

4. Conclusions

There is clear evidence from HI observations that a number of galaxies are undergoing strong tidal interactions. There is a wide range of interactions, from those occurring in groups or between galaxies of comparable masses, which may lead to mergers, to the less disruptive ones between a galaxy and one or more small companions which contribute new material to the disk and may produce starburst phenomena.

There are also systems which are not seen presently interacting with other galaxies, but possess peculiar HI morphologies and kinematics similar to those found in the tidal cases. These may represent more advanced stages of accretion-merger processes with the victims no longer visible. The interactions with small companions and these probable cases of recent mergers form circumstantial evidence that even in the present epoch there is episodic infall of gas onto galaxies.

How often do interactions, strong or mild, within multiple systems or with small companions take place? and how important are they for the formation of elliptical galaxies or for the building up of disks, the triggering of star bursts and for galaxy evolution in general? In the past years, a large number of galaxies have been mapped in HI with the Westerbork Synthesis Radio Telescope and with the VLA. A first estimate made on the basis of about hundred galaxies led to the conclusion (Sancisi 1993) that almost half of the systems observed showed signs of either present or recent interactions. The incompleteness and the various biases of such a spurious sample, however, make such an estimate highly uncertain.

A very recent HI survey carried out by Verheijen (1997) for a magnitude and volume limited sample of galaxies from the Ursa Major cluster provides more solid statistical evidence on the frequency of tidal interactions and of accretion phenomena. This cluster differs from Virgo or Coma type clusters. It has a very low velocity dispersion and a long crossing time, comparable to the Hubble time. It has no central concentration and no X-ray emitting gas and the sample is dominated by late type systems. It can be considered, therefore, representative for a galaxy population in the field. Out of the 40 galaxies mapped in HI, about 10 show clear signs of interactions with small companions or have peculiar structures. About half of the sample galaxies show asymmetries in their kinematics or in the HI density distribution and at least 30% are warped.

The available evidence from HI observations indicates, therefore, that at least 25% of field galaxies are undergoing now or have undergone in the recent past some kind of tidal interaction. If one is willing to accept the lopsided kinematics and structure as evidence of past interactions and mergers, as proposed in optical studies (see Zaritsky, this Symposium), then the conclusion would be that more than 50% of present day galaxies have been through one or more merger events in a recent past. If lumps of gas with HI masses of order 10^{8-9} $M_{\odot}$ (as indicated by the 21 cm observations) are accreted at a rate of 1 per 10^{8-9} yr, the mean accretion rate for the gas would be at least 0.25 or, perhaps, even more than 0.5 $M_{\odot}$/yr. If dark matter were also included the total mass accretion rate could easily be a factor 10 higher.

I am grateful to M. Franx, T. van der Hulst, W. Lane and M. Verheijen for valuable comments and to H. Hoekstra for helping with the figure.

References

Appleton, P.N., 1983, *MNRAS*, **203**, 533.
Appleton, P.N., Ghigo, F.D., van Gorkom, J.H., Schombert, J.M., Struck-Marcell, C., 1987, *Nature*, **330**, 140.
Balcells, M., Sancisi, R., 1996, *A.J.*, **111**, 1053.
Balkowski, C., Chamaraux, P., Weliachew, L., 1978, *A&A*, **69**, 263.
Barnes, J.E., 1996, in New Light on Galaxy Evolution, eds. R. Bender, R.L. Davies, Kluwer Acad. Publ., p. 191.
Braun, R., Walterbos, R.A.M., Kennicutt, R.C., Tacconi, L.J., 1994, *Ap.J.*, **420**, 558.
Briggs, F.H., 1988, in QSO Absorption Lines, eds J.C. Blades, D. Turnshek, C. Norman, Cambridge University Press, p. 275.
Carilli, C.L., van Gorkom, J.H., 1992, *Ap.J.*, **399**, 373.
Cayatte, V., van Gorkom, J.H., Balkowski, C., Kotanyi, C., 1990, *A.J.*, **100**, 604.
Chengalur, J.N., Giovanelli, R., Haynes, M.P., 1995a, *A.J.*, **109**, 2415.
Chengalur, J.N., Salpeter, E.E., Terzian, Y., 1995b, *A.J.*, **110**, 167.
van Driel, W., van Woerden, H., 1989, *A&A*, **225**, 317.
Fisher, J.R., Tully, R.B., 1976, *A&A*, **53**, 397.
van Gorkom, J.H., Schiminovich, D., 1997, in The Nature of Elliptical Galaxies, ASP Conference Series No. 116, eds. M. Arnaboldi, G.S. Da Costa & P. Saha, p. 310.
Hawarden, T.G., van Woerden, H., Mebold, U., Goss, W.M., Peterson, B.A., 1979, *A&A*, **76**, 230.

Haynes, M.P., 1981, *A.J.*, **86**, 1126.
Haynes, M.P., Giovanelli, R., Roberts, M.S., 1979, **Ap.J.**, **229**, 83.
Heckman, T.M., Sancisi, R., Sullivan, III, W.T., Balick, B., 1982, *MNRAS*, **199**, 425.
Hibbard, J.E., 1995, Ph.D. Thesis, Columbia University.
Hibbard, J.E., van Gorkom, J.H., 1996, *A.J.*, **111**, 655.
Hibbard, J.E., Guhathakurta, P., van Gorkom, J.H., Schweizer, F., 1994, *A.J.*, **107**, 67.
van der Hulst, J.M., 1979, *A&A*, **71**, 131.
van der Hulst, J.M., Sancisi, R., 1988, *A.J.*, **95**, 1354.
Kamphuis, J., 1993, PhD Thesis, University of Groningen.
Kamphuis. J., Briggs, F., 1992, *A&A*, **253**, 335.
Kamphuis, J., Sancisi, R., 1993, in Panchromatic View of Galaxies, eds. G. Hensler, Ch. Theis, J.S. Gallagher, Editions Frontieres, p. 317.
Kregel, M., Sancisi, R. 1998, in preparation.
Mathewson, D.S., Cleary, M.N., Murray, J.D., 1974, *Ap.J.*, **190**, 291.
McNamara, B.R., Sancisi, R., Henning, P.A., Junor, W., 1994, *A.J.*, **108**, 844.
Mulder, P.S., van Driel, W., Braine, J., 1995, *A&A*, **300**, 687.
Mundell, C.G., Pedlar, A., Axon, D.J., Meaburn, J., Unger, S.W., 1995, *MNRAS*, **277**, 641.
Phookun, B., Mundy, L.G., Teuben, P.J., Wainscoat, R.J., 1992, *Ap.J.*, **400**, 516.
Phookun, B., Vogel, S.N., Mundy, L.G., 1993, *Ap.J.*, **418**, 113.
Rand, R.J., 1994, *A&A*, **285**, 833.
Richter, O.-G., Sancisi, R., 1994, *A&A*, **290**, L9.
Rots, A.H., Bosma, A., van der Hulst, J.M., Athanassoula, E., Crane, P.C., 1990, *AJ*, **100**, 387.
Rupen, M.P., 1991, *A.J.*, **102**, 48.
Saglia, R.P., Sancisi, R., 1988, *A&A*, **203**, 28.
Sancisi, R., 1993, in Physics of Nearby Galaxies: Nature or Nurture?, eds. T.X. Thuan, C. Balkowski, J.T.T. Van, Editions Frontieres, p. 31.
Sancisi, R., van Woerden, H., Davies, R.D., Hart, L., 1984, *MNRAS*, **210**, 497.
Schiminovich, D., van Gorkom, J.H., van der Hulst, J.M., Kasov, S., 1994, *Ap.J.*, **423**, L101.
Schiminovich, D., van Gorkom, J.H., van der Hulst, J.M., Malin, D.F., 1995, *Ap.J.*, **444**, L77.
Shostak, G.S., Hummel, E., Shaver, P.A., van der Hulst, J.M., van der Kruit, P.C., 1982, *A&A*, **115**, 293.
Shostak, G.S., Sullivan, W.T., Allen, R.J., 1984, *A&A*, **139**, 15.
Schneider, S.E., Skrutskie, M.F., Hacking, P.B., Young, J.S., Dickman, R.L., Claussen, M.J., Salpeter, E.E., Houck, J.R., Terzian, Y., Lewis, B.M., Shure, M.A., 1989, *A.J.*, **97**, 666.
Simkin, S.M., van Gorkom, J.H., Hibbard, J.E., Su, H-J., 1987, *Sci.*, **235**, 1367.
Smith, B.J., 1994, *A.J.*, **107**, 1695.
Stanford, S.A., Wood, D.O.S., 1989, *ApJ*, **346**, 712.
Swaters, R.A., Schoenmakers, R.H.M., Sancisi, R., van Albada, T.S., 1998, in preparation.
Toomre, A., 1977, in The Evolution of Galaxies and Stellar Populations, eds. B.M. Tinsley and R.B. Larson, (New Haven: Yale Univ.), p. 401.
Verheijen, M.A.W., 1997, PhD Thesis, University of Groningen.
Viallefond, F., Allen, R.J., de Boer, J.A., 1980, *A&A*, **82**, 207.
Wevers, B.M.H.R., Appleton, P.N., Davies, R.D., Hart, L., 1984, *A&A*, **140**, 125.
Yun, M.S., Ho, P.T.P., Lo, K.Y., 1994, *Nature*, **372**, 530.

TIDAL INTERACTIONS IN M81 GROUP

M.S. YUN
National Radio Astronomy Observatory
P.O Box 0, Socorro, NM 87801 (USA)

1. Introduction

At a distance of 3.3 Mpc, M81 group is one of the nearest groups of galaxies and thus an ideal place to study various galactic phenomena. Observations at all wavelengths can be made with high sensitivity and excellent spatial resolution ($1'' = 15$ pc). The three core members M81, M82, and NGC 3077 are known to be closely interacting from the 21cm HI emission studies (van der Hulst 1979, Yun et al. 1993, 1994), and all three galaxies show evidence for an AGN or starburst activity, possibly induced by the on-going tidal interactions. A more distant member NGC 2976 also shows evidence for tidal interaction within the group (Appleton et al., 1981). Therefore M81 group is an excellent test ground for studying tidal interaction and tidally induced activities.

The central six main members of the group are listed in Table 1. Their optical appearances suggest little evidence for any on-going tidal interactions (see Fig. 1). The aperture synthesis maps of their 21cm HI emission,

TABLE 1. M81 Group Members

name	RA (B1950)	Dec (B1950)	$v_{\rm hel}$ (km/s)	log $L_{\rm B}$ ($L_\odot$)	log $L_{\rm FIR}$ ($L_\odot$)	D_{25} (′)	$d_{\rm p}$(M81) (kpc)
M81	09:51:27.3	69:18:08	−34	10.28	8.33	27	-
M82	09:51:43.5	69:55:01	+203	9.69	10.24	11	33
NGC 3077	09:59:19.9	68:58:30	+14	9.05	8.43	5.4	42
NGC 2976	09:43:11.5	68:08:45	+3	9.11	8.40	5.9	74
HO I	09:36:00.9	71:24:55	+136	8.03	-	3.6	135
IC 2574	10:24:41.3	68:40:18	+47	9.10	7.86	13	164

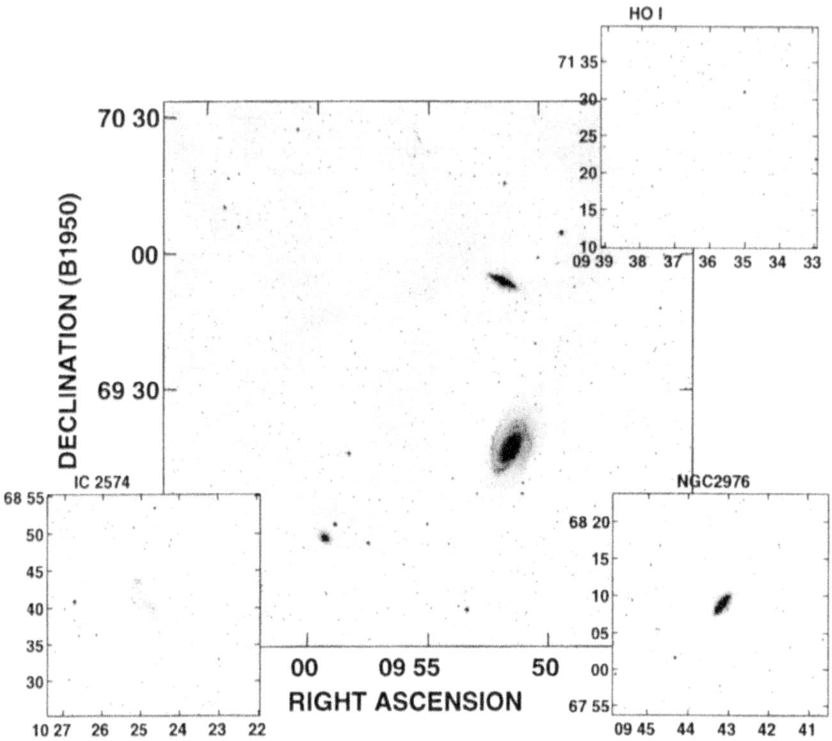

Figure 1. Optical images of the M81 Group galaxies.

however, show clear evidence for on-going tidal disruptions in M81, M82, NGC 3077, and NGC 2976 (see Fig. 2 and Appleton et al. 1981). The HI distribution in the more distant members HO I and IC 2574 appears undisturbed (D. Westpfahl, private communication). It is somewhat peculiar that the most luminous member M81 also has the most blueshifted velocity within the group (Table 1), but this probably reflects its orbital motion with respect to the two nearby companions M82 and NGC 3077. The mean group line-of-sight velocity, 55 km s^{-1}, is nearly identical to the mean of the three outer members, 62 km s^{-1}.

In the remainder of this paper, I will focus only on the tidal interactions involving the inner most three members (M81, M82, & NGC 3077) because the HI data suggest that their interactions are the most current. The HI data also suggest that NGC 2976 has passed through the group center in the recent past, but the tidal remnants associated with this older interaction are likely largely erased by the more recent ones. First, the 21cm HI observations of the three galaxies are summarized in §2. Numerical model-

TABLE 2. Summary of the HI Observations

name	$N_{\text{HI,peak}}$ (10^{21}cm^{-2})	M_{HI} ($10^9 M_\odot$) VLA	ADS[†]	$\langle V_{\text{HI}} \rangle$ (km/s)	ΔV (km/s)
M81	10.6 ± 0.2	2.81 ± 0.56	2.19 ± 0.22	−35	30~45
M82	10.3 ± 0.2	0.80 ± 0.16	0.72 ± 0.07	+230	40~140
NGC 3077	10.7 ± 0.2	0.69 ± 0.14	1.00 ± 0.10	+0	30~40
Concentration I	7.8 ± 0.2	0.31 ± 0.06	0.20 ± 0.05	+50~+150	45~100
Concentration II	5.1 ± 0.2	0.26 ± 0.05	-	+120	25~30
S. Tidal Bridge	2.4 ± 0.2	0.26 ± 0.05	-	−200~+0	10~30
N. Tidal Bridge	1.6 ± 0.2	0.20 ± 0.04	-	+0~+120	20~30

[†] Single dish measurements by Appleton, Davies, & Stephenson (1981).

ing of the tidal interactions and its implications are discussed in §3 and §4. Finally, the summary is given in §5.

2. HI Observations

The 21cm HI emission map of the central 2° region of the M81 group at 1′ resolution (Fig. 2) is produced using a mosaic of 12 separate pointings at the Very Large Array[1] (VLA; see Yun 1992 and Yun et al. 1994 for more detailed description of the observations). A total of $5.6 \times 10^9 M_\odot$ of atomic hydrogen is detected in this group, mostly concentrated on the three individual galaxies. The two bright HI knots found east of M81 disk ("Concentrations I & II") have the HI masses comparable to a dwarf galaxy, and the Concentration I may be associated with HO IX, a clump of stars previously suggested to be a satellite companion of M81. The gas kinematics are consistent with simple disk rotations or tidal perturbations superposed on a disk rotation in the three galaxies. The derived properties of the individual HI features are summarized in Table 2.

By far the most remarkable features seen in HI are the tidal bridges connecting all three galaxies. The bridge connecting M81 and NGC 3077 ("South Tidal Bridge") was previously mapped by van der Hulst (1979), and this feature appears to loop around NGC 3077 and pointing toward M82 ("North Tidal Bridge"). Both of these tidal features are massive - each $2 \times 10^8 M_\odot$ in HI mass and ≥ 30 kpc in length. The HI complex associated with NGC 3077 is displaced by 4 kpc to the southeast of the stellar galaxy, and a bright loop of HI connects the HI peak with the stellar peak. The HI

[1]The National Radio Astronomy Observatory is a facility of the National Science Foundation operated under cooperative agreement by Associated Universities, Inc.

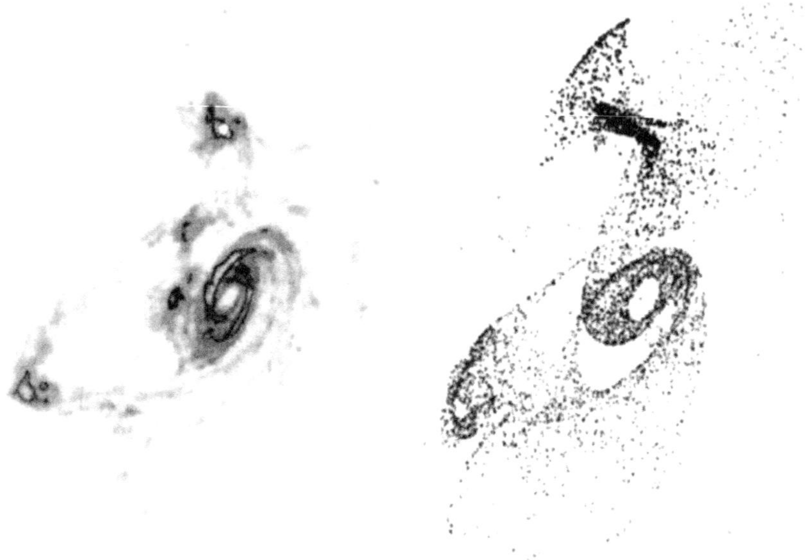

Figure 2. Comparison of the 21cm HI emission (left) and the numerical simulation (left).

disk of M82 is severely disrupted, contributing to the HI bridge connecting with M81 (see Yun et al. 1993 for a detailed study). These tidal features are completely absent in optical light (Fig. 1), and one would not have guessed that these three galaxies are strongly interacting based on the optical image alone. This in turn suggests that the frequency of tidal interactions among galaxies may far exceed any estimate based on optical images alone.

3. Numerical Modeling

Disk galaxies are highly coherent objects in phase space, and therefore numerical simulation of tidally interacting disk galaxies can be a relatively straightforward task as long as gravity is the dominant force, even in cases where tidal features display complex morphology and kinematics. The pioneering work by Toomre & Toomre (1972) has convincingly demonstrated that the filamentary features associated with close pairs of galaxies are indeed the tidal debris resulting from galaxy-galaxy collisions. They have further suggested that such tidal disruptions may "stoke the furnace" and result in enhanced starburst or AGN activities. Recent reviews on numerical modeling of galaxy interaction and mergers are found in Barnes (1998) and also in the contributions by J. Barnes and C. Mihos in this volume.

Modeling an interacting pair of galaxies is an extension of a simple

TABLE 3. Input parameters for the modeling.

	M81	M82	NGC 3077
RA (1950)	$09^h51^m30^s$	$09^h51^m44^s$	$09^h59^m22^s$
Dec. (1950)	$+69°18'.3$	$+69°54'.9$	$+68°58'.5$
Relative Velocity (km s^{-1})	0	+230	+36
Total Mass ($10^{10} M_\odot$)	20	1-4	1-4
Disk Radius (kpc)	36	17	17
Disk Inclination	59°	82°	45°
Disk P.A.	150°	28°	225°
$\alpha^\dagger$ (kpc)	0.3	0.1	0.1

† softening parameter in Plummer potential.

2-body interaction, which is a problem with a well known solution. The modeling a system of *three* interacting galaxies is a far more difficult problem since gravitational interaction involving three bodies has no simple analytical solution. Therefore, the problem of modeling the tidal interactions among M81, M82, and NGC 3077 is first approached as a problem of understanding two separate pair-wise interactions. This may be a reasonable approximation since M81 is much more massive than its two companions. The modeling is further simplified by performing only the "restricted 3-body" calculation (tracing the motions of test particles within a time dependent potential, as in Toomre & Toomre 1972) rather than attempting full N-body calculations with self-consistent galaxy models (e.g. Barnes & Hernquist 1996). Some of the consequences of using these simplifications are discussed below.

The code, originally developed by N. Brouillet (L'université Bordeaux), computes the motions of a disk of massless test particles that are initially in circular orbit about its central (Plummer) potential. The initial parameter searches roughly determine the M81-M82 and M81-NGC3077 orbits separately, and then a full three galaxy interaction and the resulting responses of the disk particles are realized by combining the pair-wise solutions. The input parameters for the model calculations are listed in Table 3, and the best model solution obtained is shown in comparison with the HI data in Fig. 2. The numerical simulation not only matches the observed HI distribution well, but it also matches the velocity field also (not shown here) even though no effort was made to do so. The best model parameters are listed in Table 4.

The details of the tidal interactions are shown in Fig. 3 by displaying the particles representing three galaxies separately along with the traces

TABLE 4. Best 3-body numerical model parameters.

	M81	M82	NGC 3077
Mass ($10^{10} M_\odot$)	20	4	2
Orbit Eccentricity	–	1.5	0.7
Perigal. Distance (kpc)	–	25	16
Time since pericenter	–	2.2×10^8 yrs	2.8×10^8 yrs
Orbit Inclinations, i	–	10°	135°
Perigal. Argument, ω	–	235°	270°
Orbit Longitude, λ	–	265°	175°

of the companion orbits. The tidal disruption of the M82 disk is done entirely by M81 alone while the disruption of both M81 and NGC 3077 disks require both of their two companions. The key difference between this simulation from the previous simulation of M81 group (Cottrell 1977, van der Hulst 1977, Killian 1978, Brouillet et al. 1991) is the inclusion of an HI disk surrounding NGC 3077. All attempts to produce the North Tidal Bridge (between M82 and NGC 3077) as the tidally disrupted outer disk of M81 failed, and NGC 3077 was given its own disk in order to reproduce this feature (nearly $10^9 M_\odot$ of HI is associated with NGC 3077, further justifying that NGC 3077 initially possessed its own disk, which is now nearly unrecognizable). The original HI disk of NGC 3077 is first disrupted by M81 and is then stretched out further like a streamer during the subsequent passage by M82 (see a GIF movie of this interaction at **http://www.aoc.nrao.edu/~myun/movie.gif**). The apparent smooth connection of the North Tidal Bridge (NTB) with the South Tidal Bridge (STB – tidally disrupted outer disk of M81) near NGC 3077 seems fortuitous, but it is not at all an accident since tidal tails tend to point back and finger their perturbers (M82 for the NTB and NGC 3077 for the STB).

4. Discussions

The most interesting physical parameters derived from this simulation are the time scales of interactions. The time since the nearest approach for M82 and NGC 3077 are 2.2×10^8 and 2.8×10^8 years, respectively. These agree very well with the estimated ages of on-going starbursts in M82 and NGC 3077 (see Rieke et al. 1980), and such a good agreement is a strong support for the idea that the tidal disruptions are indeed responsible for the "stoking of the furnaces". The actual inflow of the gas into the nuclear region is likely driven by a hydrodynamic process not included in the present

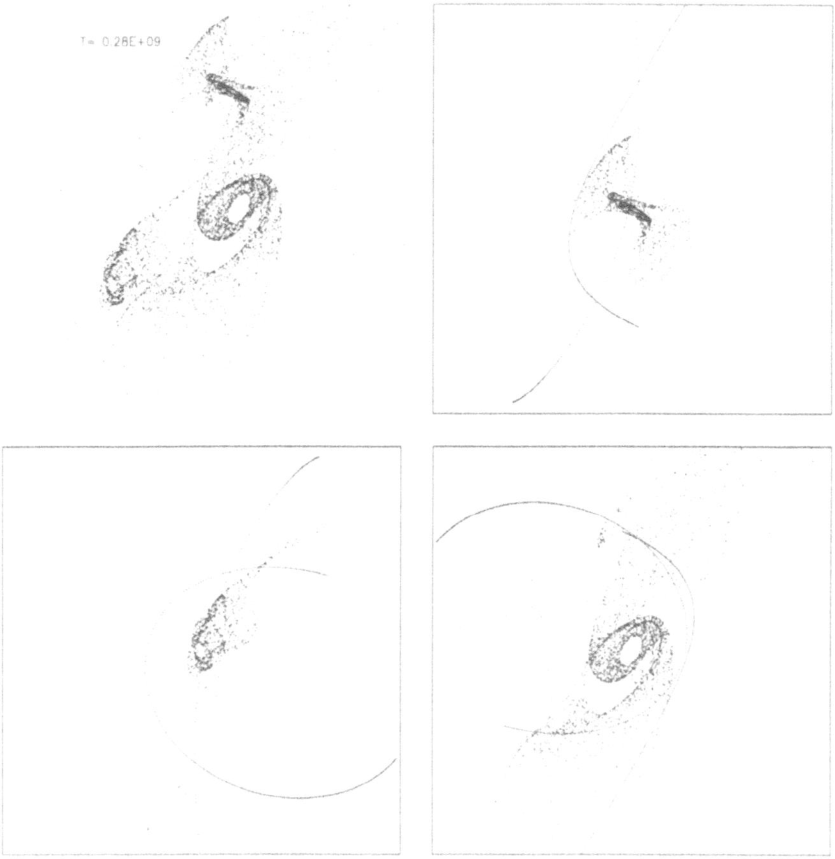

Figure 3. Tidal disruptions of the individual galaxies are shown separately to illustrate the course and nature of tidal interactions and the resulting tidal features.

model, but such gas inflow has been seen in other simulations such as by Mihos & Hernquist (1996) and Barnes & Hernquist (1996).

The details of the inner disk of M81 (and the others) are not well reproduced by this simulation, but this is the result of using a Plummer potential rather than a self-consistent galaxy model with full N-body calculation. If self-gravity and hydrodynamical effects are included in describing the response of the HI disks, the tidal features reproduced by the simulation should have sharper appearances as seen in the HI map. The full N-body calculations using self-consistent model galaxies, on the other hand, have demonstrated that dynamical dissipation of orbital energy and angular momentum can significantly alter the course of tidal interaction, often leading to mergers of galaxies (see Barnes 1998 and references therein). If the dissipation of orbital energy and angular momentum also occurs in this system, then the parabolic orbits describing the observed interactions are indeed only the approximations of the motions of individual galaxies past their initial collisions. A full N-body simulation of this system has been attempted previously using the orbits similar to derived here, but this simulation resulted in the mergers of the companions onto M81, probably because the dynamical dissipation associated with the truncated isothermal model halo used was too efficient (Yun 1992). The evolution of outer tidal features and dynamical decay of the orbits depend critically on the physical nature of the halo, and I speculate that simulating *non-merging* interacting systems using self-consistent N-body model will first require a better understanding of the nature of the galactic halos.

5. Summary

A tidal interaction scenario involving M81, M82, and NGC 3077 was derived using a restricted 3-body calculation. This scenario can successfully account for the complex system of tidal features seen in the 21cm HI emission map, despite the multitude of simplifying assumptions incorporated into the modeling. The derived time scale for tidal interactions agree very well with the estimated ages of starbursts seen in M82 and NGC 3077, lending further support that these starburst are the direct results of the tidal disruptions suffered within last 3×10^8 years.

References

Appleton, P.N., Davies, R.D., Stephenson, R.J. 1981, MNRAS, 195, 327
Barnes, J.E. 1988, ApJ, 331, 669
Barnes, J.E. 1998, in *Galaxies: Interactions and Induced Star Formation*, eds. D. Friedli, L. Martinet, & D. Pfenninger (Springer-Verlag)
Barnes, J.E., Hernquist, L. 1996, ApJ, 471, 115
Brouillet et al. 1991, A&A, 242, 35
Cottrell, G.A. 1977, MNRAS, 178, 577
Killian, D.J. 1978, PhD thesis, University of Florida
Mihos, J.C., Hernquist, L. 1996, ApJ, 464, 641
Rieke et al. 1980, ApJ, 238,24
Toomre, A. & Toomre, J. 1972, ApJ, 178, 623
van der Hulst, J.M. 1977, PhD thesis, University of Groningen
van der Hulst, J.M. 1979, A&A, 75, 97
Yun, M.S. 1992, PhD thesis, Harvard University
Yun, M.S., Ho., P.T.P., Lo, K.Y. 1993, ApJ, 411, L17
Yun, M.S., Ho., P.T.P., Lo, K.Y. 1994, Nature, 372,530

EXTENDED GAS IN INTERACTING SYSTEMS

F. COMBES
Observatoire de Paris, DEMIRM
61 Av. de l'Observatoire, F-75 014 Paris, France

Abstract. HI observations have revealed large gaseous extensions in interacting and merging systems. The interstellar gas is obviously dragged out in tidal tails during an encounter, and the percentage of HI in the tails increases with the merging stage. However, the opposite is true for the molecular gas, which is observed highly concentrated towards the nuclei of interacting galaxies, amounting to a significant fraction of the dynamical mass. Statistically, there appears to be more gas *observed* in interacting galaxies than in normal, isolated ones. As N-body simulations show, the gas is driven inwards in the interaction process by the strong gravity torques, before being consumed through star formation in the triggered starbursts. We review here all observations that could bring more knowledge about the state of the gas in the outer parts of galaxies, and about accretion processes. The link with the observations of the Lyα absorbers at low and high redshifts is discussed.

1. Fate of the gas in interacting systems

One of the main striking features about interacting and merging galaxies, is the presence of large tidal tails of matter dragged out of the galaxies; recent VLA maps have revealed huge HI extensions with respect to the optical systems, as though most of the neutral gas was splashed all around. Yun et al. (1993) have found large quantities of HI all around the M81/M82/NGC 3077 system, and Hibbard (1995) shows in his thesis an evolving sequence of interacting/merging galaxies, where the HI extensions are conspicuous. More precisely, the percentage of the total HI found in the tails/extensions is increasing with the merging stage, from 20% in the M81 system, to 80% in the merger remnant NGC 7252. Of course, this can be explained by the tidal potential in the frame of the target growing as r^2 as a function of the target radius, and because matter in the outer parts is less bound to the galaxies. But this gives a wrong idea of the fate of the gas in interacting systems. In fact, with all probability, the gas dragged out remains bound to the system, and will rain back onto the merger remnant, after some billion years. Already, Hibbard (1995) shows that the gas at the bottom of the tails in NGC 7252 is infalling. This progressive infall will take place through phase wrapping, and shells and loops will form.

What is seen in the molecular phase is just the contrary: apparently large H_2 concentrations pile up at the galaxy nuclei in interacting systems. Up to 50% of the dynamical mass could be under the form of molecular hydrogen in merging systems (Scoville et al. 1991). Some ultraluminous infrared galaxies, which are also mergers and starbursts possess 10^{10} $M_\odot$ of H_2 gas, about 10 times more than in their spiral precursors. There exist some hints that there is more gas in interacting galaxies (Braine & Combes 1993). Although the H_2/CO conversion ratio is not well known for these peculiar objects, there is evidence for denser gas in these systems, through the HCN/CO ratio (Solomon et al. 1997). In summary, the observations suggest that the HI gas is dragged outwards, while the H_2 gas is driven inwards, to be consumed in star formation. In fact these two tracers (HI and CO) shed light on two aspects of the same gas component.

The global result is that most of the dissipative component loses angular momentum, and falls inwards towards the nucleus of the merger, although part of it is heated in shocks and pass in the coronal phase (seen in X-rays).

2. Extension of gas in normal galaxies

2.1. SPIRAL GALAXIES

It is well known that the HI gas sometimes extends much farther than the stars in spiral galactic disks, and they are precious tools to probe the rotation curve and the presence of dark matter. However, large extensions such as in NGC 628, where $R_{\rm HI} = 5\ R_{25}$ for instance, are quite exceptional (Kamphuis & Briggs 1992); only about 10% have $R_{\rm HI}$ larger than 2.5 R_{25} (Huchtmeier & Seiradakis 1985). In a sample of about 100 galaxies, chosen to search precisely for extended HI disks, Broeils (1992) did not find many extended gaseous disks. In fact, only regular galaxies were selected, to be able to exploit the rotation curves, and therefore no strongly interacting galaxies are included. The HI diameter, defined at a surface density of 1 $M_\odot$ pc^{-2}, is about twice the optical diameter D_{25} (Broeils & van Woerden 1994). The HI-to-optical-diameter ratio does not depend on morphological type or luminosity (fig. 1), but there is a strong correlation indicating that $M_{\rm HI} \propto D_{\rm HI}^2$, which means that HI surface density, averaged over the whole HI disk, is constant from galaxy to galaxy, and all over the Hubble sequence

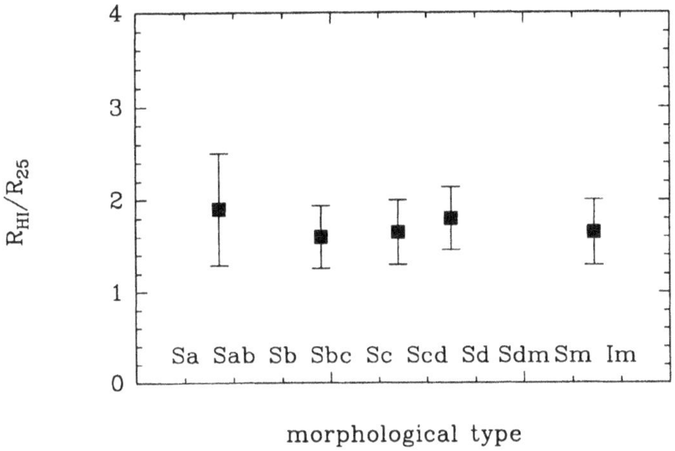

Figure 1. Ratio of HI radius (defined by a surface density of 1 $M_{\odot}$ pc^{-2}) to optical radius R_{25} as a function of morphological type, from Broeils & Rhee (1997). The error bars are 1 σ dispersion.

(Broeils & Rhee 1997). Besides, there is a strong dependence of the ratio M_{HI}/M_{tot} with type, which confirms that the percentage of gas is larger in small late-type galaxies, and decreases all over the Hubble sequence (which is also a mass sequence).

2.2. DWARF IRREGULAR GALAXIES

Since dwarf irregular galaxies are particularly rich in HI gas, it was first thought that there might exist big HI envelopes around these objects, or even a large number of isolated gas clouds still waiting to form their first stars. It turned out however that big gas extensions, such as in DDO 154 are very uncommon (e.g. Hoffman et al. 1996), and that the HI-to-optical-radius ratio is very similar in dwarfs and in spirals (cf. fig. 2).

2.3. THE IONIZATION EDGE OF THE HI DISK

HI observations have shown that the neutral gas radial distribution is much smoother than that of stars (exponential decrease), or molecular gas as traced by CO (following more or less the blue luminosity). The HI surface density is statistically falling as the dark matter surface density, determined from rotation curves (Bosma 1981, Freeman 1993). This implies an HI column density falling as $1/r$ at large radii (because the gas layer is flaring linearly with radius, the spatial density in fact is going as $1/r^2$). It would

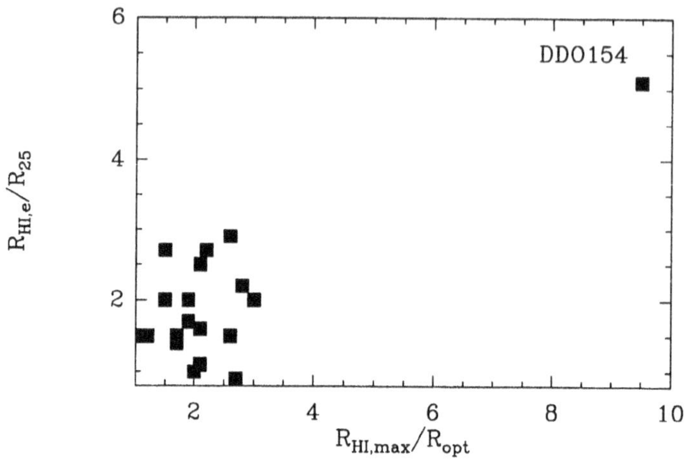

Figure 2. The ratio of maximum HI radius to optical radius, and the ratio of HI radius at 1/e of the peak flux to the optical isophotal radius, for all resolved dwarfs, studied by Hoffman et al. (1996). Only DDO 154 is outstanding in its HI size.

be interesting to observe the HI until the edges of the neutral disk, to trace its physical condition and also the distribution of the dark matter at large radii, but this is a hard task, due to sensitivity problems. It has been done in a few cases only, in NGC 3198 (van Gorkom 1991) and M33 & NGC 3344 (Corbelli et al. 1989). The results show a sudden decrease from a column density of $N_{HI} = 2 \cdot 10^{19}$ cm^{-2} to $2 \cdot 10^{18}$ cm^{-2} in a very short length-scale of 2-4 kpc (of course limited by the spatial resolution of the observations). This sharp edge has been interpreted in terms of the ionization front of the gas disk (Corbelli & Salpeter 1993): the data can be reproduced if the extragalactic background, essentially from the quasar UV light, provides an ionization rate of $\xi \sim 2 \cdot 10^{-14}$ s^{-1}, a value corresponding to the study of low-redshift Lyman-α absorption lines (Madau 1992).

The conclusion is that the gas surface density is likely to continue to decrease as $1/r$ at large distances, but it is difficult to see it. It is even possible that the gas density in the outer parts is highly underestimated, if it is clumpy down to very small scales, and under a cold molecular phase (Pfenniger et al. 1994, Pfenniger & Combes 1994). Large HI extensions, that are only seen when galaxies are interacting, and molecular gas concentrations, could then reveal the presence of the clumps when they are stirred up by tidal interactions, driven inwards and concentrated.

2.4. HOW TO TRACE GAS AT LARGE RADII?

Gas at large radii is difficult to trace, because of its low average column density, its low temperature, low metallicity Z and dust content (Z is decreasing exponentially with radius, e.g. Smartt & Rolleston 1997). Even at any radii, the usual tracers are far from perfect: the HI line can be optically thick (e.g. Burton 1992), CO emission can be absent by lack of excitation for example (Adler et al 1991). Nelson et al. (1996) claim to have detected cold dust at large radii in nearby galaxies through its 100μ emission; but the very cold dust could be more easily traced at 1.3mm. Recent continuum maps at this wavelength (which is in the Rayleigh-Jeans domain), are still limited by sensitivity, but they reveal interesting radial distributions. In galaxies where the interstellar medium (ISM) is dominated by the molecular component, like NGC 891, the 1.3mm flux radial distribution is superposable to the CO one (Guélin et al. 1993), suggesting that the CO emission is directly proportional to metallicity (like dust emission since $S_{1.3mm}$ is proportional to the dust temperature, the column density of the gas, and the metallicity). On the contrary, in galaxies where the ISM is dominated by the atomic hydrogen, like NGC 4565, the 1.3mm continuum flux follows more the HI emission, which is decreasing more slowly with radius than the CO one (Neininger et al. 1996), but still the dust emission is falling more rapidly than the HI, because of its Z dependency. In spite of sensitivity and metallicity limitations, dust emission could be one of the best way to trace extended cold gas (e.g. Combes & Pfenniger 1997).

2.5. STABILITY OF EXTENDED GASEOUS DISKS

If extended gaseous disks exist, their gravitational stability raises interesting problems. Since the HI gas at large radii is usually observed lopsided (Richter & Sancisi 1994), or with multiple spiral arms and large-scale instabilities, it must possess a minimum of self-gravity (Pfenniger et al. 1994). As for small scale instabilities, as well as the vertical equilibrium, constraints can be put on the flattening of the dark matter haloes, which determines its volume density in the plane, as shown in fig. 3 (cf. Olling 1995, 1996, Becquaert & Combes 1997).

3. Intergalactic medium

3.1. GALAXY CLUSTERS

Through frequent tidal interactions or harassment (Moore et al. 1996), and ram-pressure stripping (Cayatte et al. 1990), the extended gaseous disks are truncated in rich clusters. Globally, the HI surface density is conserved for spirals in clusters, only their HI-to-optical-radius ratio is smaller (Cayatte et

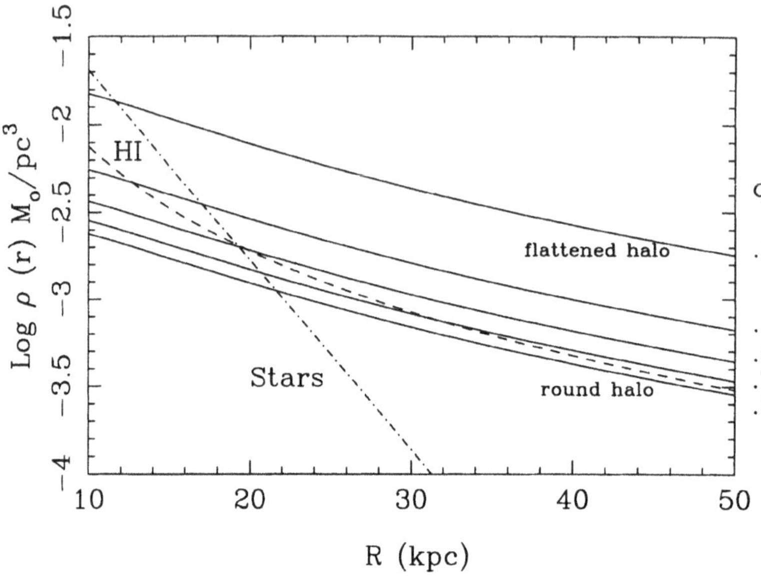

Figure 3. Volume density as a function of radius for the various mass components in a typical galaxy like the Milky Way: exponential stellar disk (dot-dash), flaring HI layer (dash), and isothermal spheroidal halo of flattening q (full lines).

al. 1994), and since their CO content is also normal (Casoli et al. 1996), their central gaseous disk appears not too much perturbed, except for a small number of anemics. The gas coming from the outer parts of galaxies must be now under the form of X-ray emitting hot gas, at the virial temperature of the cluster, and enriched in heavy elements from the galaxies (Renzini 1997).

3.2. LYMAN-α ABSORBERS

The intergalactic medium is traced through Lyman-α absorption in front of quasars, especially well sampled at $z = 2$ from the ground. Space observations have shown that the absorbers are still present until $z = 0$, although with some evolution in number density. The distribution of HI column densities derived from these probes is a continuous power-law (slope $\sim$ -1.6) up to damped Ly-α values of 10^{20}-10^{21} cm^{-2}, comparable to the outer parts of galaxy disks. Between $z = 2$ and $z = 3$, most of the baryons are in the Lyman-α forest with $N_{\rm HI}$ between 10^{14} and 10^{16} cm^{-2}, but the medium is strongly photoionized, with the neutral fraction around 10^{-6}-10^{-5} (the total column density is thus much higher). Typical sizes are 200kpc, certainly in filamentary structures. Recent ionized helium detections (Jacobsen et al. 1994, Davidsen et al. 1996) reveal that the most diffuse regions of the intergalactic medium are filled with ionized gas, in other words the Gunn-Peterson effect is detected in HeII.

What is less well known is the high end of the density spectrum, and whether small clumps of dense molecular gas exist in the intergalactic medium. In the frame of the fractal model or hierarchy of gas structures proposed for the ISM, physical mechanisms can explain the dynamical equilibrium and the non star formation (Pfenniger & Combes 1994). These structures, down to the smallest H_2 clumps, could be formed at high redshift ($z > 100$), as soon as the cooling time is much smaller than the Hubble time. The clumps will form before re-heating, and represent an alternative model to solve the cooling catastrophe.

In any case, galaxies are not isolated objects, they are connected to the gas reservoirs of the intergalactic medium, which contains most of the baryons at $z = 2$. This should be taken into account when considering galaxy interactions. If the external gas has a flattened geometry, a big disk or ring can form in a merging sequence (as observed in NGC 520 for instance). The interaction produces compression, higher densities and higher neutral fraction (which goes as the square of the density) and the gas will become visible in the HI line.

4. Conclusion

There are various pieces of evidence for large gas reservoirs in the outer parts of galaxies. This gas is driven inwards during galaxy interactions, through the associated gravity torques. It can help to explain mergers and the triggered starbursts, permanent gas accretion in warps, polar rings, etc. If the medium is clumpy and fractal, the gas mass could be underestimated by factors 2-10, and this can have important cosmological consequences.

References

Adler D.S., Allen R.J., Lo K.Y.: 1991, ApJ 382, 475
Becquaert J-F., Combes F.: 1997, A&A 345, 41
Bosma A.: 1981, AJ 86, 1971
Braine J., Combes F.: 1993, A&A 269, 7
Broeils A.: 1992, PhD Thesis, Groningen
Broeils A., van Woerden H.: 1994, A&AS 107, 129
Broeils A., Rhee M.H.: 1997, A&A 324, 877
Burton W.B.: 1992, in "The Galactic Interstellar Medium", Saas-Fee Advanced Course 21, ed. D. Pfenniger & P. Bartholdi, p. 1
Casoli F., Dickey J., Kazes I. et al.: 1996, A&A 309, 43
Cayatte V., Balkowski C., van Gorkom J.H., Kotanyi C.: 1990, AJ 100, 604
Cayatte V., Kotanyi C., Balkowski C., van Gorkom J.H.: 1994, AJ 107, 1003
Combes F., Pfenniger D.: 1997, A&A 327, 453

Corbelli E., Salpeter E.E.: 1993, ApJ 419, 94 & 104
Corbelli E., Schneider S.E., Salpeter E.E.: 1989, AJ 97, 390
Davidsen A.F., Kriss G.A., Zheng W.: 1996, Nature 380, 47
Freeman K.C.: 1993, in "Physics of nearby galaxies, Nature or Nurture?" ed. T.X. Thuan, C. Balkowski, Van J.T.T., Editions Frontières, Gif-sur-Yvette, p. 201
Guélin M., Zylka R., Mezger P.G. et al.: 1993, A&A 279, L37
Hibbard J.E.: 1995, PhD thesis, Columbia University
Hoffman G.L., Salpeter E.E., Farhat B. et al.: 1996, ApJS 105, 269
Huchtmeier W.K., Seiradakis J.H.: 1985, A&A 143, 216
Jacobsen P. et al.: 1994, Nature 370, 35
Kamphuis J., Briggs F.: 1992, A&A 253, 335
Madau P. 1992, ApJ 389, L1
Moore B., Katz N., Lake G.: 1996, ApJ 457, 455
Nelson A.E., Zaritsky D., Cutri R.M.: 1996, AAS, 189, 6706
Neininger N., Guélin M., Garcia-Burillo S., Zylka R., Wielebinski R.: 1996, A&A 310, 725
Olling R.P.: 1995, AJ 110, 591
Olling R.P.: 1996, AJ 112, 457
Pfenniger D., Combes F., Martinet L.: 1994, A&A 285, 79
Pfenniger D., Combes F.: 1994, A&A 285, 94
Renzini A.: 1997, ApJ 488, 35
Richter O-G., Sancisi R.: 1994, A&A 290, L9
Smartt S.J., Rolleston W.R.: 1997, ApJ 481, L47
Scoville, N.Z., Sargent, A.I., Sanders, D.B., Soifer, B.T.: 1991 ApJ, 366, L5
Solomon P.M., Downes D., Radford S.J.E., Barrett J.W.: 1997, ApJ 478, 144
van Gorkom J.H.: 1991, in "Atoms, Ions and Molecules", ed. A.D. Haschik, ASP Conf. Ser. 16, 1
Yun M.S., Ho, P.T.P., Lo K.Y.: 1993, ApJ, 411, L17

COLLISIONAL RING GALAXIES

P.N. APPLETON
Department of Physics and Astronomy
Iowa State University, Ames, IA 50011, USA

1. Introduction

Ring galaxies are believed to represent a special case of a collision between two galaxies, in which one of the galaxies impacts and passes through the center of another disk system (e.g. Lynds & Toomre 1976). Although rare, this kind of low orbital–angular–momentum collision leads to a recognizable structure, namely a luminous blue star–forming ring (Appleton & Marston 1997), which should be easily identifiable even at moderate redshift. Indeed, Lavery et al. (1996) have used this fact, and their relative rarity at low-redshift, to conclude that rings (and therefore presumably all collisions) are over-represented in deep *HST* fields.

At low redshift, ring galaxies have been recognized as remarkable laboratories for the study of massive star formation (Fosbury & Hawarden 1977, Higdon 1995, Higdon & Wallin 1997; see also Appleton & Struck-Marcell 1996 for a recent review). In the collisional picture, the expanding near–circular density–wave, which is driven into the disk of the "target" galaxy by the gravitational perturbation of the "intruder", creates massive stars along the crest of the wave which should all have approximately the same age. If it can be shown that the massive stars are created almost simultaneously along the wave, then observed differences in the properties of the knots which make up the ring must be due to intrinsic variations in their properties. In addition, as the wave moves radially outwards, models predict that it should leave a trail of massive stars and star clusters in its wake (Appleton & Struck Marcell 1987a). The action of the wave may be to map stellar evolution radially across the face of the galaxy (see Marcum, Appleton, & Higdon 1992). Given an understanding of the basic dynamics of ring galaxies, they are a nice environment for testing stellar evolutionary models of some of the most massive star formation regions found in nature.

2. Ring Galaxies as Laboratories for Studying Massive Star Formation

It was recognized early-on that the central perturbation caused by a head-on collision would drive density waves through the disk of a galaxy and that the subsequent star formation triggered by the passage of the wave might be a powerful test of star formation and stellar evolution in collisional systems (Appleton & Struck-Marcell 1987a, Struck-Marcell & Appleton 1987). Observations have confirmed many of the predictions of the earlier models in the sense that there is now compelling evidence that many ring galaxies do indeed owe there morphology to a propagating wave of star formation. Perhaps the most compelling evidence for this comes from thickness of the blue star forming ring in the Cartwheel ring galaxy (Borne et al. 1995, Struck et al. 1996). The ring at its narrowest is approximately 0.5-1kpc in width, which corresponds to a timescale of 8-16 million years if the ring is expanding at 60 km/s (Higdon 1996). The overall expansion time of the ring is at least 100 million year, and so the ring must represent the coherent triggering of massive O/B stars around the ring. From the careful modeling of the stellar evolution of knots in 10 ring galaxies, Bransford et al. (1997) has found a very similar story, in which the ages of most knots in ring galaxies are between 10 and 80 Myrs, and in all cases are significantly shorter than the ring expansion timescales derived from model-fitting the kinematics of the rings. These ground-based results also show that rings are mildly metal-poor and show little variation in metallicity from knot to know within a single galaxy.

In addition to the simultaneous triggering of the star birth in the wave, as it expands outwards, another prediction of the models seems to be consistent with observations, namely the expectation that stellar evolution will be mapped across the face of the galaxy as the wave expands. The idea here is that the youngest stars are born in the ring as it expands outwards compressing new ISM material. The models show that newly formed stars created in the wave are left behind by the wave, creating an age gradient in the radial direction. We would expect to find the oldest stars and star clusters in the inner regions of the galaxy, and the youngest stars would be found at the current position of the wave (the present position of the ring). Such a picture would naturally lead to radial color gradients and these are observed in many ring galaxies (Marcum, Appleton, & Higdon 1992, Appleton & Marston 1997). A radial distribution of dust could also cause a radial reddening-gradient and so mapping the dust in these galaxies is a priority (see below).

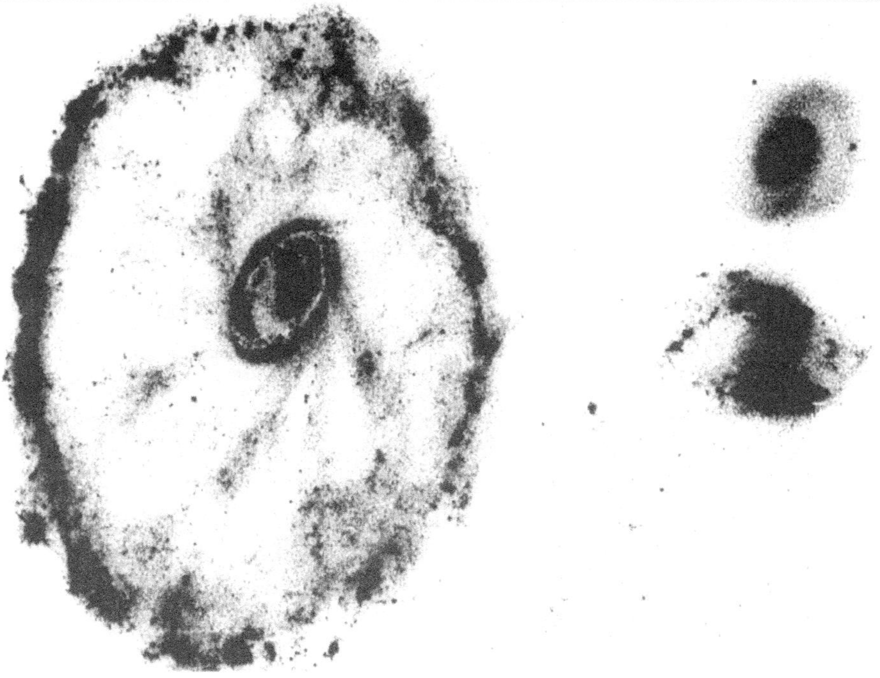

Figure 1. *HST* WFPC2 F450W Image of the Cartwheel Ring Galaxy.

3. *HST* Observations of the Cartwheel Ring Galaxy: Possible Formation of a Thick-disk Pre-Globular Cluster Population

Recent observations of the Cartwheel ring galaxy with *HST* (Borne et al. 1995, Struck et al. 1996, Appleton et al. 1997) have revealed a huge variety of structure in this classical ring galaxy, both in the outer and inner rings, and in the spokes between them. Fig. 1 shows a B-band *HST* image of the Cartwheel ring. The image show many star clusters and associations in the outer ring, as well as compact sources between the rings. The galaxy shows a dramatic difference in color between the massive star forming knots in the outer ring, and the inner regions of the galaxy. The range of color variation is from $-0.7 < B - I < 2.4$. In a recent analysis of the colors and luminosity of star clusters in the Cartwheel, Jacobs (1996), and Appleton et al. (1997) have concluded that the clusters found between the two rings are consistent with the colors and luminosity of a faded population of star clusters deposited about 100 million years ago, when we might have expected the expanding ring–wave to have passed through that region of the galaxy. These “inter-ring” knots are mainly unresolved at the resolution of the Wide-Field camera on WFPC2 (at an assumed distance of 120 Mpc for the Cartwheel, this corresponds to $\sim$ 50pc) and many are close to the completion limit of the observations ($B > 25.5$ mag). A number are also slightly resolved (at the 50 pc level) and may be regions of faint secondary star formation.

We are currently testing various hypothesis about the origin of the inter-ring clusters using survival analysis to investigate the shape of the luminosity function for these fainter clusters as they approach the limit. Preliminary results suggest that these clusters represent *faded remnants of the most luminous star clusters and associations seen in the outer ring* at the present time. Although these inter-ring clusters are still quite young (if the above hypothesis is correct), some of them may survive long into the future and may eventually become globular clusters with metallicities presumably in the range of those seen in other rings (1/2 to 1/5 solar in oxygen and nitrogen). Any metallicity gradient present in the original target disk would also be mimicked by the thick-disk population of aging clusters. It is interesting that a possible "thick-disk" globular clusters population has been seen in the Sombrero galaxy (= NGC 4594) (Forbes, Grillmar, & Smith 1997). Perhaps near-central collisions between gas-rich progenitors might be one mechanism for creating such a population.

4. Molecules and Dust in Ring Galaxies

The recent discovery by Gao et al. (1997), that the ring galaxy Arp 118 (NGC 1144) is extremely bright in the ^{12}CO line (it is twice as bright in the CO line as Arp 220) suggests that violent collisions are capable of converting significant amounts of neutral hydrogen into molecules. In a conventional galaxy interaction or merger, it is believed that gas is funneled into the center of the galaxies by the action of tides. However, when two disks collide head-on, it is not obvious what causes an enhancement in the ratio of H_2 to HI, and yet such an enhancement is observed (Horellou et al. 1995). We have recently obtained HI observations of Arp 118 with the VLA, and have discovered rather weak HI emission, but very strong absorption lines against two sources of radio emission in NGC 1144. The thesis observations of C. McCain (with K. Freeman at Stromlo-private communication) suggests a collision between two gas rich disks in which large-scale shocks are present. Understanding the mechanisms that can convert large amounts of neutral hydrogen into molecules in this process will be a major challenge of models of the hydrodynamics. The recent models of Struck (1997) attempt to investigate the different thermal phases of the ISM during such violent collisions, and such work is producing interesting results, including the formation of new gas disks around the companion.

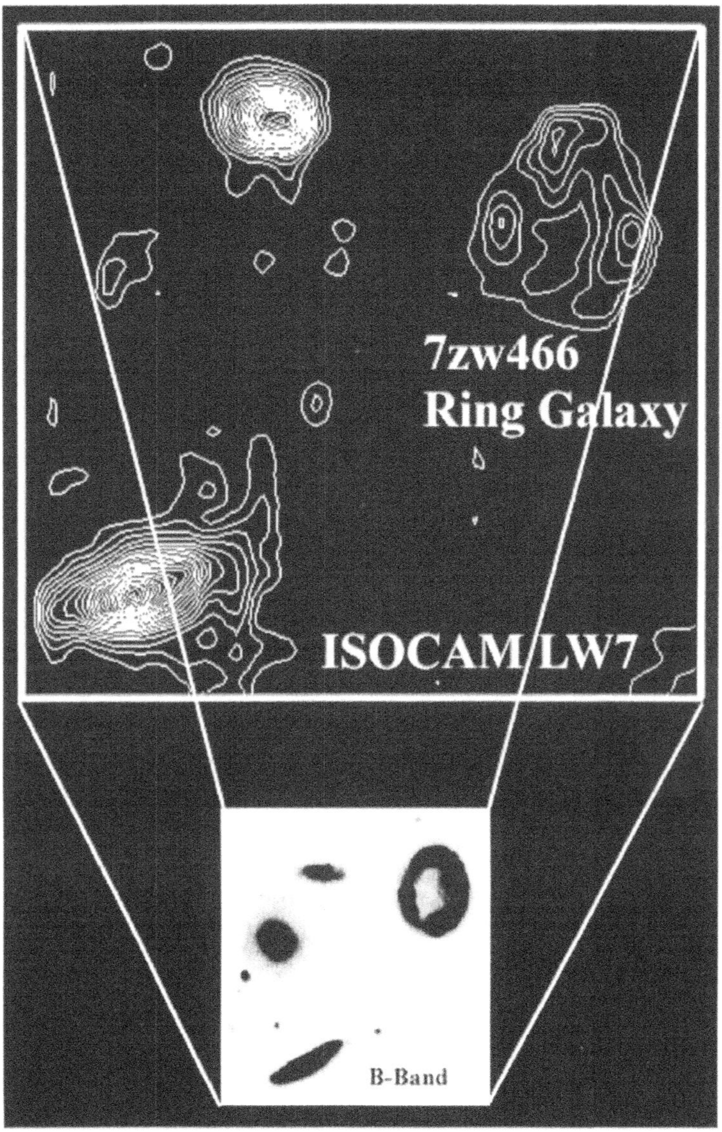

Figure 2. *ISO* $\lambda 12\mu m$ contour map of the VII Zw 466 Group (Optical Inset). Note that the elliptical galaxy is not detected. The brightest emission is from the northern background galaxy and the edge-on disk.

It has been know for some time that ring galaxies emit about half of their bolometric luminosity in the Far-IR (Appleton & Struck 1987, Wakamatsu & Nishida 1988). Until recently little was known about the distribution of dust in ring galaxies. For example, optical spectra of VI Zw 466 (Bransford et al. 1997) show that the Balmer decrement in this galaxy (usually taken as a indication of the optical extinction) is quite similar from one knot to the next around the ring, suggesting uniform extinction. However *ISO* observations (see Fig. 2) by Appleton et al. (1997) reveal a major asymmetry in the dust emission at Mid-IR wavelengths, showing strong emission along the north-western quadrant of the ring where the λ3cm radio continuum emission is also stronger. The *ISO* emission pinpoints the regions of warm dust in the galaxy, perhaps indicating regions where young stars are more dust-enshrouded.

The Cartwheel at *ISO* wavelengths (Charmandaris et al. 1997) shows a rather different picture. At $\lambda 15\mu$m, where thermal emission from warm dust dominates, only one small intensely powerful cluster of HII regions of the ring (the region called Knot A by Higdon 1995) is detected. This seems surprising because much of the southern quadrant of the Cartwheel's outer ring shows very powerful Hα emission (Higdon 1995). Could the HII regions in the Cartwheel be creating small fountains which have lifted the grains to a larger scale-height that in VIIZw466? The lower metallicity of the Cartwheel's outer ring may also be a factor, although oddly most of the star–forming outer ring is detected by *ISO* at $\lambda 7\mu$m, where PAHs are believed to be the main source of emission, and grains have clearly been formed. Another interesting result from the *ISO* observations of the Cartwheel is the detection of strong $\lambda 15\mu$m emission from the center of the galaxy (Charmandaris et al.) in the regions of the second ring and nucleus. Since little or no signs of star formation is found at optical wavelengths, the source of the central emission remains a mystery. It may originate in a dust-enshrouded starburst, in grains heated by infalling gas from the HI plume (see Struck et al. 1996) or from a stronger than normal UV radiation field from a post starburst population.

5. Conclusions

Recent work on ring galaxies has led to the following conclusions:

a) ring galaxies have sub-solar metallicities in the range 1/2 to 1/5 solar in [O/H] and [N/H] ratios (Bransford, Appleton, & Marston 1997). There is a suggestion of an increase in the mean nitrogen abundance for rings of larger linear size, but the oxygen abundances show no trend with ring diameter. The increase in nitrogen abundance with ring diameter probably reflects the original nitrogen abundance of the target galaxy before the collision. There is no clear decrease in metallicity with ring diameter, as might be expected if the rings travel down metallicity gradients present in the target galaxy. (A glaring exception is the Cartwheel which has a very metal-poor outer ring (Fosbury & Hawarden 1977).)

b) the colors of the knots in ring galaxies (after correcting for the effects of reddening) are consistent with very young star formation regions with typical ages in the range 20-80 Myrs (see BAM). These ages are significantly shorter than the expansion ages of the rings of 200-300 million years, providing strong evidence that the star formation in the ring waves are created by a single coherent event, the expanding wave, which triggers star formation through compression of the disk.

c) by a comparison with models of starburst populations it is possible to show that most ring knots are totally dominated by the young stellar population, and show only a small contribution from the underlying stellar density wave. Although Lynds & Toomre (1976) were probably correct in their suggestion that the rings are produced by a crowding of disk material due to the collisional perturbation, the rings are defined more by the *gas response and young stellar population, than by the underlying old population.*

d) *HST* observations of the Cartwheel ring (Borne et al. 1996, Struck et al. 1996, Jacobs 1997, Appleton et al. 1997) show that the ring is resolved into hundreds of star forming regions which show a large scatter in color consistent with clusters of young stars 10-20 million years old. The smoothed-out light distribution is consistent with a radial expanding starburst population superimposed on a $1/r^2$ disk of red light. Approximately 50 star clusters are seen between the inner and outer rings in the Cartwheel, and these clusters have the colors and magnitudes consistent with faded versions of the brightest clusters seen currently in the outer ring. This can be taken as evidence that the outer starburst ring has expanded to its present size by moving through the inner disk and depositing star clusters in its wake. We note that some of the star clusters seen between the rings may also be regions of faint secondary star formation formed in the "spokes" which connect the inner to outer rings. New narrow-band imaging of the Cartwheel by J. Higdon (ATNF) may help to shed light on this topic.

e) The star clusters seen between the inner and outer rings of the Cartwheel have absolute magnitudes on the range -13 to -10 and $B-I$ colors of 0.8-1.2 mag. (assuming $D = 120$ Mpc). If they are evolving star clusters formed by the radial passage of the density wave, then they may hint at a way of forming a thick disk population of globular clusters. Such star clusters, if they could survive for another few 10^9 yrs, would have the luminosity and colors of globular clusters, although they would mimic any metallicity gradient present in the original galaxy. Collisions of this kind (which drive radial expanding waves) may help to explain the possible thick-disk globular clusters seen in galaxies like the Sombrero galaxy.

f) *ISO* observations of two ring galaxies (Cartwheel—Charmandaris et al.

1997, VIIZw466—Appleton et al. 1997) show emission from PAHs, and probably larger grains in thermal equilibrium with the radiation field from the young stars. In the Cartwheel, the shorter-wavelength (PAH) emission follows the distribution of young stars better than the longer-wavelength (thermal) IR, suggesting that the smaller PAH grains are more sensitive to the uv field than the larger grains.
g) The *ISO* observations of the inner regions of the Cartwheel reveal a relatively powerful 15μm source which seems coincident with the inner ring and bulge (Charmandaris et al. 1997). Since only very weak HII regions are seen from this regions we speculate that the emission comes from either an obscured starburst, grains heated by infalling material from a tidal stream, or grains heated by a post-starburst population.

References

Appleton, P. N., Jacobs, M., Struck, C., Borne, K. & Lucas, R., 1997, ApJ, (preprint)
Appleton, P. N. & Marston, A. P., 1997, *AJ*, **113**, 201
Appleton, P. N., & Struck, C. 1996, *Fund. of Cos. Phys*, **16**, 111.
Appleton, P. N. & Struck-Marcell, C., 1987a, *ApJ*, **318**, 103
Appleton, P. N. & Struck-Marcell, C., 1987b, *ApJ*, **312**, 566
Borne, K. D., R. A. Lucas, P. Appleton, C. Struck, A. B. Schultz, & L. Spight 1995, *in Science with the Hubble Space Telescope II*, eds. P. Benevenui, F. P. Machetto, & E. J. Schreier (Washington: U.S. Gov. Printing Office), p. 239
Bransford, M. A., Appleton, P. N. & Marston, A. P., 1997, *AJ*, (preprint)
Charmandaris, V.., Appleton, P. N. & Mirabel, I. F., 1997, *ApJ*, (preprint)
Fosbury, R. A. E. & Hawarden, T. G., 1977,*MNRAS*, **178**, 473
Forbes, D. A., Grillmar, C. J. & Smith, R. C. 1997, *AJ*, **113**, 1648
Gao, Y., Solomon, P. M., Downes, D. & Radford, S. J. E. *ApJL*, **481**, L35
Higdon, J. L, 1995, *ApJ*, **455**, 524
Higdon, J. L, 1996, *ApJ*, **467**, 241
Higdon, J. L. & Wallin, J. F. 1997, *ApJ*, **474**, 686
Horellou, C., Casoli, F., Combes, F. & Dupraz, C. 1995, *A&A*, **298**, 743
Jacobs, M. 1996, MS Thesis, Iowa State University (Ames, Iowa)
Lavery, R.J., Seitzer, P., Suntzeff, N.B., Walker, A.R. & Da Costa, G.S. 1996, *ApJL*, **467**, L1.
Lynds, R. & Toomre, A. 1976, *ApJ*, **209**, 382
Marcum, P. M., Appleton, P. N. & Higdon, J. L. 1992, *ApJ*, **399**, 57
Marston, A. P. & Appleton, P. N., 1995,*AJ*, **109**, 1002
Struck, C., 1997, *ApJ*, (In Press)
Struck, C., Appleton, P.N., Borne, K.D. & Lucas, R.A. 1996, *AJ*, **112**, 1868
Struck-Marcell, C. & Appleton, P. N., 1987, *ApJ*, **323**, 480
Wakamatsu,K.-I. & Nishida, M. T. 1987, *ApJL*, **315**, L23

NUMERICAL SIMULATIONS OF M51

M. ANTONIOLETTI AND A.H. NELSON
Department of Physics and Astronomy, University of Wales College Cardiff, PO Box 913, Cardiff, CF2 3YB

No conference on the interaction of galaxies would be complete without a contribution on M51 (NGC5194) and its companion NGC5195. Much observational and theoretical work has been carried out to try to understand this interacting pair, and to elucidate the morphological features and weak AGN which are thought to be the result of the interaction.

On the theoretical front Toomre and Toomre (1972) carried out restricted two body calculations, using mass points for 5194 and 5195 together with test particle discs. This simple calculation, involving only a few hundred particles, was surprisingly successful, generating the outer tidal tail and bridge arms, and establishing a first approximation to the orbit of 5195 around 5194. In this orbit 5195 travels in a plane highly inclined to the disc plane of 5194, going from the foreground to around behind 5194, and moving to the north as it makes one disc plane crossing 200 Myears before the time of the contemporary image.

Later, Hernquist & Barnes (Hernquist 1990) used a fully self-gravitating model, involving a bulge/halo/disc model for 5194, and distributed mass for 5195. This calculation utilized the very effective tree-code gravity method, which allows the calculation of the complex interaction to be carried out with sufficient accuracy, and confirmed the basic orbit of Toomre and Toomre. The self-gravity and higher resolution obtained from using a much greater number of particles also enabled Hernquist & Barnes to obtain well formed inner spiral structure, as well as the outer tail and bridge. However they pointed out that a better fit could be obtained to the pitch angle of the inner spiral arms, and to the degree of development of the outer southern tail, if the calculation was continued to a later time for the contemporary image than that suggested by Toomre & Toomre. On the other hand this yields a line of sight relative velocity between 5194 and 5195 which is smaller than is observed.

Recently Byrd & Salo (1995) have used a model involving gas discs in both 5194 and 5195. Their calculation involved a Finite Difference technique for the gravity on two moving polar grids centred on the two galaxies, together with inelastic collisions for the gas particles. The greater responsiveness of the dissipative gas particles to the strongly non-axisymmetric particles yields strong spiral arms in the gas with distinct bends (similar to those observed in M51). They also proposed that the contemporary image is much later than that proposed by Toomre & Toomre, viz. 400 Myears after the first disc plane crossing. This means that 5195 (still in a highly inclined orbit) crosses the plane for a second time, and consequently at the beginning of the calculation starts in the background of M51. By this device they obtain sufficient time for the arms, bends and southern tail to develop, but also obtained the observed relative velocity between the two galaxies.

Here we report calculations using a halo/bulge/stellar disc/gas disc model for both 5194 and 5195, and using tree-code gravity and Smoothed Particle Hydrodynamics. We utilized the orbit of Hernquist & Barnes, confirming that a later time produces a better representation of the morphological features, but still having a problem with the relative velocity. The contemporary stellar and gas images are shown in figures 1 and 2, with figures 3 and 4 showing close-ups of M51. In addition we obtain some features not seen in previous calculations, but present in the observations :-

- In addition to broad stellar arms, a central bar appears in the stellar component. This has the degree of ellipticity and orientation of the major axis (in the rotational sense just behind 5195's orbit) as is observed in IR observations of the old stellar component.
- Strong narrow gas arms, with the observed distinct bends, and resolved shocks in the inner side of the arms, in agreement with CO observations.
- A nuclear gas ring - although with a radius half the size of the one observed.
- Warping on the outer disc and tail leading to apparent counter-rotation as viewed along the line of sight.

These features, we believe, are not highly sensitive to minor modifications of the orbit, and are likely to persist if we use the modified orbit of Byrd & Salo. They demonstrate conclusively that the major morphological features of M51, even at its centre, are dictated by the dynamics of the interaction with 5195.

References

Byrd, G., & Salo, H. 1995. Astro. Lett & Comm. 31 193
Hernquist, L. 1990. Heidelberg Conference on Dynamics & Interactions of Galaxies 108
Toomre, A., & Toomre, J. 1972. Ap.J. 178 623

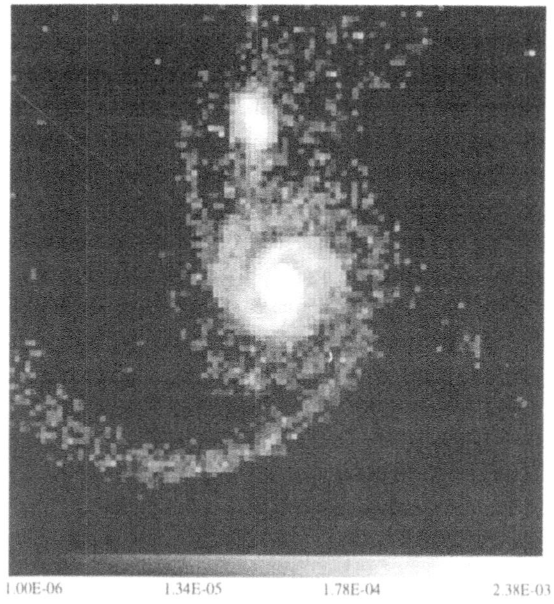

Figure 1. Stellar image of the 5194/5195 pair.

Figure 2. Gas image of the 5194/5195 pair.

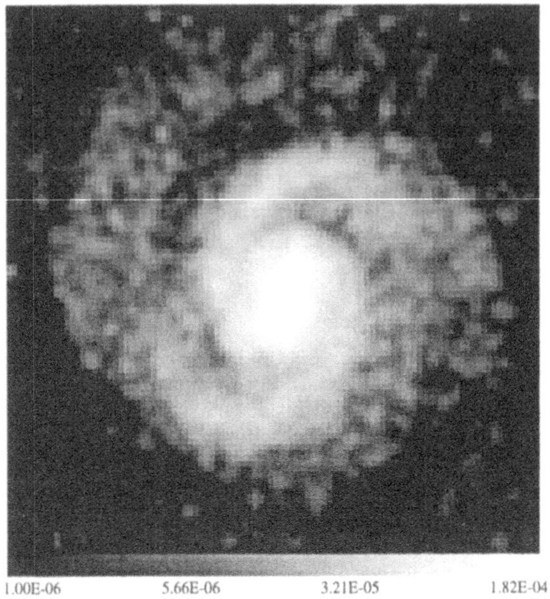

Figure 3. Close up of the Stellar image of M51.

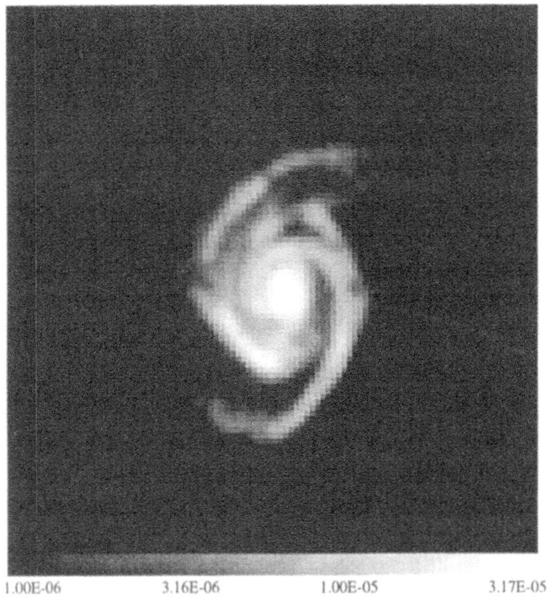

Figure 4. Close up of the Gas image of M51.

GALAXY INTERACTIONS IN THE LOCAL VOLUME

I.D. KARACHENTSEV AND D.I. MAKAROV
Special Astrophysical Obs., Russian Academy of Sciences
N.Arkhyz, Stavropolsky kraj, KChR, 357147, Russia

1. Introduction

Imaging of a large number of faint galaxies with the *Hubble Space Telescope* reveals a significant excess of interacting and peculiar objects among them (Griffiths et al., 1994, Burkey et al., 1994). Being expected in general, this result needs, however, to be expressed quantitatively. To estimate evolutionary changing of the relative number of interacting galaxies, $\delta_{\rm int}(z)$, from their redshift z, one must determine reliably the local value, $\delta_{\rm int}(0)$. But due to a subjective manner of definition of interacting galaxy system, the published values of $\delta_{\rm int}(0)$ differ from one author to another several times. The most appropriate sample to estimate $\delta_{\rm int}(0)$ would be quite a representative sample of very nearby galaxies, which is restricted by their distance or radial velocity. Below we consider data on such a fair sample, named the Local Volume (=LV), which is the most complete one at the present moment.

2. Basic properties of the Local Volume sample

Kraan-Korteweg & Tammann (1979) compiled a catalog of nearby galaxies having radial velocities corrected for the solar motion within 500 km/s. After avoiding of probable Virgo cluster members, their sample consisted of 179 galaxies. Later the list of LV galaxies was updated essentially in three different ways: a) accidental revealing of new dwarf galaxies with radial velocities below 500 km/s during optical and HI surveys of the known catalogs; b) HI observations of new galaxies which have been found in a strong extinction zone along the Milky Way; c) special searches for and subsequent HI surveys of dwarf galaxies in the vicinity of nearby groups.

As a result, at the present moment the population of the LV increased up to 277 galaxies, including 255 galaxies with $V_0 < 500$ km/s, and 22

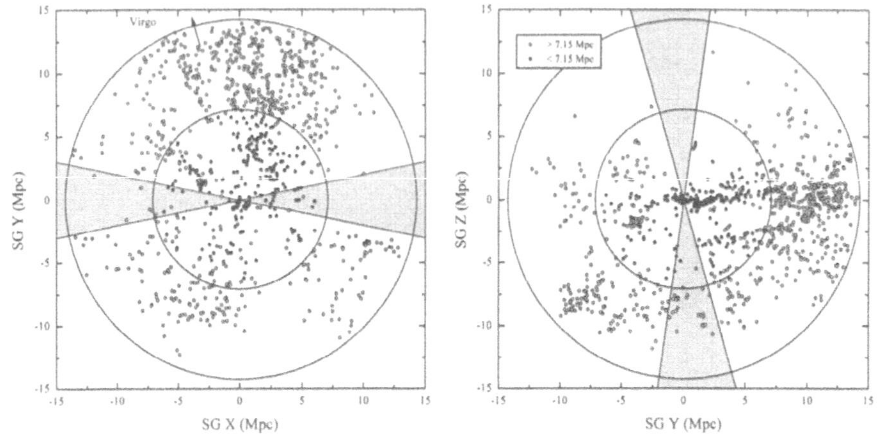

Figure 1. Distribution of nearby galaxies in the Supergalactic Cartesian coordinates. Filled circles are galaxies with a distance $D < 7.15$ Mpc, open circles are more distant ones with $V_0 < 1000$ km/s. The shaded cone indicates an absorption zone. The arrow shows a direction toward the Virgo cluster.

spheroidal dwarf systems, which have no radial velocities but are probable companions of the known nearby galaxies.

We should emphasize that besides its numerical growth the LV sample is enriched strikingly by data on galaxy distances. As it is known, due to coherent and random non-Hubble motions the radial velocity of a nearby galaxy is rather unreliable indicator of its distance. During the 1990s in the Special Astrophysical Observatory there was undertaken a project of CCD imaging of more then 100 nearby galaxies with the goal of measuring their distances via the brightest stars. Together with data obtained at other observatories the number of the LV galaxies having direct (photometric) distance estimates has reached 158. Among the rest of the LV galaxies 35 are with distances adopted according to their membership in the known groups, and only 84 have kinematic distances, $D = V_0/H$, derived under the local Hubble parameter $H = 70$ km/s/Mpc (Karachentsev & Makarov, 1996). Note that in the initial sample of Kraan-Korteweg & Tammann (1979) the photometric distance estimates were known for 15 galaxies only. Therefore, the present 3-D structure of the LV looks much more detailed than decade ago.

Fig. 1 shows the LV galaxy distribution (filled circles) in projection onto the Local Supercluster plane and also edge-on. More distant galaxies with radial velocities $V_0 < 1000$ km/s are indicated with open circles. The region of strong extinction in the Milky Way is shaded.

Distribution of the LV galaxies by their angular diameters and distances

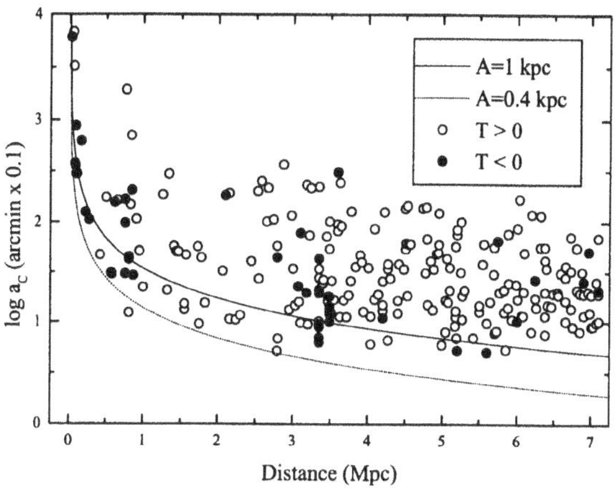

Figure 2. Distribution of the LV galaxies vs. their distance and angular diameter, corrected for inclination and absorption. The lines correspond to the linear diameter of 1 Kpc and 0.4 Kpc. The early type galaxies are indicated by filled circles.

is shown in Fig. 2. Two lines in it correspond to the values of linear diameter: 1.0 Kpc and 0.4 Kpc. As one can see from the diagram, at distances $D > 4$ Mpc there is a deficit of galaxies with small linear diameters. The data of Fig. 2 allow us to judge the completeness of our sample. A simple analysis yields the completeness level to be about 70%. Taking into account this correction, the mean number density of the Local Volume is 0.25 gal/Mpc^3.

3. The nearby interacting galaxies

The most systematic searches for interacting galaxies were undertaken by Vorontsov-Velyaminov (1959, 1977), who published the Atlas of 850 interacting galaxies, and also by Arp (1966), whose Atlas contains images of 338 peculiar and interacting objects. Among 603 isolated pairs of galaxies in the northern sky 314 interacting system were noted in his Catalogue by Karachentsev (1987). Vorontsov-Velyaminov's survey was based on an inspection of the Palomar Observatory Sky Survey prints, whereas Arp used high resolution photographs obtained on the 200″ telescope.

In Table 1 we collected all cases when a galaxy from the Local Volume is present in the Atlases of Vorontsov-Velyaminov (=VV) or Arp. First three columns contain the galaxy common name and its number in both Atlases, the forth one indicates the value of "Tidal Index" described below, and the last column gives brief remarks concerning interaction features. Basing on these data and taking into account the incompleteness of both the surveys in the southern sky, we conclude that the local value of the relative abundance of interacting galaxies is $\delta_{\rm int}(0) = (5–6)\%$. However, we should note that the cases when a galaxy is classified as interacting by both the authors seen to be too infrequent. This may be caused by prevalence of a subjective factor when classifying signs of interaction even for the nearest galaxies. In many cases the irregular patchy shape of dwarf galaxies gives an erroneous impression of an interacting system. In fact, judging by remarks in Table 1, only two galaxies, M 101 and its companion NGC 5474, may be considered as actually interacting system with tidal distortions of their spiral structures. Therefore, when the single objects of irregular shape are excluded, the estimate of $\delta_{\rm int}(0)$ drops to ~1%.

TABLE 1. Nearby interacting and peculiar galaxies.

Name	VV	Arp	Θ	Remarks
M 32	—	168	5.97	Compact
N 1313	436	—	−1.58	Irregular
N 1569	—	210	0.36	MERGER? double nuclei
N 2537	138	6	−1.58	Irregular,HSB
Holmb.II	—	268	0.61	Irregular
DDO 53	499	—	0.56	Irregular
M 81	—	—	1.01	MERGER, Arp's loop
M 82	—	337	2.30	MERGER?
N 3077	—	—	2.27	MERGER, "Garland"
VII Zw403	574	—	−0.08	Irregular
N 3738	—	234	−0.40	Irregular
N 4190	104	—	0.19	Irregular
P 42134	42	211	−1.04	Irregular
GR 8	558	—	−1.12	Irregular
N 5128	—	153	0.38	MERGER!
N 5238	828	—	−0.58	Irregular
U 8638	133	—	−1.15	Irregular
M 101	344	26	0.70	Mutually interacting
N 5474	344	26	1.91	Mutually interacting
N 5477	561	—	1.48	Irregular
Milky W.	—	—	2.71	MERGER,polar Magell. ring
N 6946	—	29	−1.35	Irregular arm

Beside $\delta_{\rm int}(0)$, the local relative number of merging galaxies, $\delta_{\rm merg}(0)$, is of evolutionary interest too. We did not find any estimate of this quantity in literature. Based on large scale photographs and CCD frames available almost for all the LV galaxies, we may rank among mergers the six nearby galaxies indicated in Table 1. Among them there is a classic merger example,

NGC 5128 = Centaurus A, and also our Galaxy with its polar Magellanic ring. Therefore, in a first approximation one may adopt $\delta_{\rm merg}(0) = (2\text{–}3)\%$ as the local abundance of mergers. Note, that signs of merging, as well as interaction, are seen more frequently among the most luminous galaxies. For double galaxies this effect was described by Karachentsev (1987).

4. Tidal index as a measure of interaction

For objective description of interaction we use a quantitative approach, where each LV galaxy i has a tidal index

$$\Theta_i = \max\{\log(M_k/D_{ik}^3)\} + C, \qquad k = 1, 2 \ldots N. \tag{1}$$

Here M_k and D_{ik}, are mass and 3D separation of a neighboring galaxy. In such a way we find for each galaxy its "Main Disturber" (=MD), which produces the highest tidal action on the galaxy. Calculating the Θ index we also took into account more distant galaxies with radial velocities up to 1000 km/s (see Fig. 1). The value of the arbitrary constant C was taken from the following condition. A galaxy, which is interacting with its MD, may be described with a cyclic Keplerian period

$$t_{ik} = D_{ik}^{2/3} \cdot G^{-1/2} \cdot (M_i + M_k)^{-1/2}, \tag{2}$$

where G is the gravity constant. We determine the value of C from the requirement that Θ is zero when the Keplerian period equals cosmological time, $1/H$. In the units of $\{M_\odot, \text{Mpc}\}$ this gives $C = -11.75$. Using the idea of critical density, $\rho_c = 3H^2/8\pi G$, one can express the condition $\Theta = 0$ in another way: $\rho_k(\Theta = 0) = \rho_c$, where ρ_k means a density excess, caused by the MD with a mass M_k at a distance D_{ik}. Therefore, we have a reason to consider a galaxy with tidal index $\Theta < 0$ a rather isolated one, which is causatively unrelated to its neighbors. As is seen from Table 1, the majority of irregular galaxies which have been classified as interacting by Vorontsov-Velyaminov and Arp are actually isolated objects.

The total mass of a galaxy was determined as $M = \kappa V_{\rm m}^2 \cdot A_{25}/2G$, where $V_{\rm m}$ is the galaxy rotation curve amplitude derived from its HI line width after correction for inclination and turbulent velocity in the manner described in PGC-ROM (Paturel et al., 1996). The dimensionless constant κ is adopted to be 2.5 in order to take into consideration that the typical rotation curve of a galaxy extends 2.5 times further than its standard optical radius (Broeils, 1992). In the case of a galaxy with the unknown HI linewidth or with the inclination angle $i < 40°$ its mass estimate was made via luminosity L taking into account its morphological type: $(M/M_\odot) = \kappa \cdot (8 - 0.4 \cdot T) \cdot (L/L_\odot)$. This kind of mass estimate was applied to 46% of the LV sample. The general list of the galaxies in the Local Volume with indication of their distances, masses, tidal indices, and the Main Disturber names was published in Karachentsev & Makarov (1997). The updated version of the list is accessible via e-mail from `dim@sao.ru`.

5. The tidal index and other properties of the LV sample

Fig. 3 reproduces a distribution of the number of the LV galaxies vs. their tidal index. The histogram of $N(\Theta)$ has an asymmetric shape with a maximum at $\Theta \simeq -1$. More than half of the galaxies (55%) are situated in the region of $\Theta < 0$, i.e. they may be considered as rather isolated objects. For them the Θ-parameter has rather a sense of isolation index than tidal. The galaxies of early types, $T < 0$, are indicated on the histogram by a gray area. On the average they have higher values of the Θ-index, which demonstrates the existence in the LV of the known effect of segregation of elliptical and spiral galaxies depending on the density of their environment.

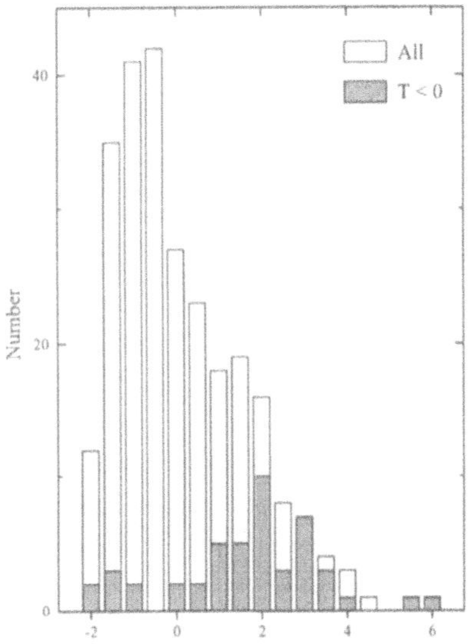

Figure 3. Distribution of the LV galaxies on their Tidal Index, Θ. The early type objects are shaded.

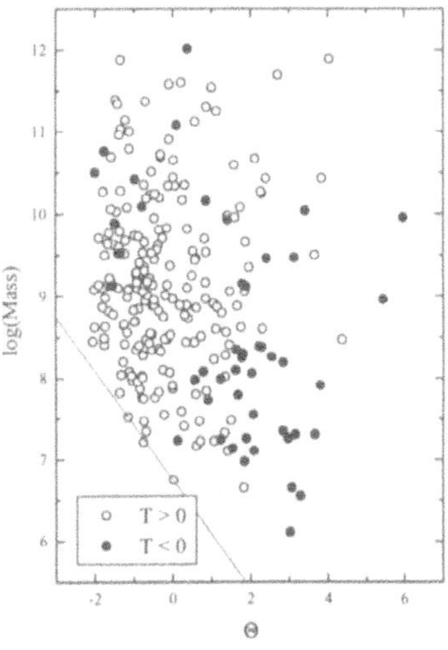

Figure 4. Masses of the LV galaxies versus their Tidal Indices. Early and late galaxy types are indicated by filled and open circles. An empty lower left corner is caused by the Milky Way "shadow".

The distribution of the LV galaxies vs. their masses and tidal indexes (Fig. 4) shows that these values are practically uncorrelated. The dotted line in the lower left corner indicates a zone of observational selection, where isolated objects of low mass are absent due to the Milky Way influence. With correction of this feature the distribution of $\{M, \Theta\}$ does not find any noticeable segregation of the galaxies by their masses depending on the number density of neighbors. However, it attracts attention that on the diagram the galaxies of early and late types (filled and open circles) have essentially different distribution. All dwarf elliptical and spheroidal galaxies with $\log M < 9$ have $\Theta > 0$, i.e. they occur only in dense regions around massive MDs. The observed absence of isolated objects among spheroidal dwarfs may be considered as an argument in favor of their origin from HI rich irregular dwarf systems with tidal evolution effects.

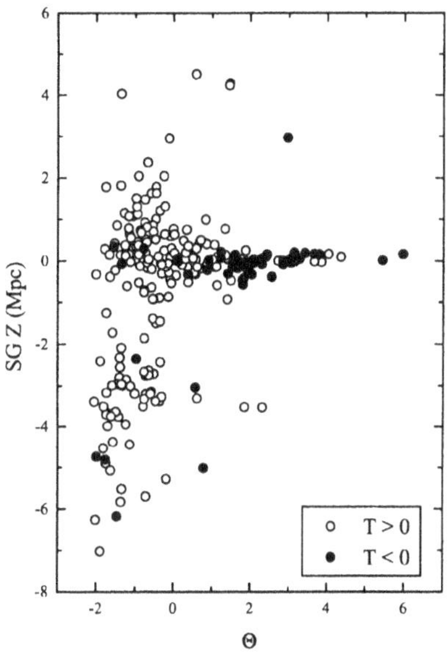

Figure 5. Map of Tidal Index of the LV galaxies versus their distance from the Supergalactic plane. E+Sph and S+Irr types are shown with different symbols.

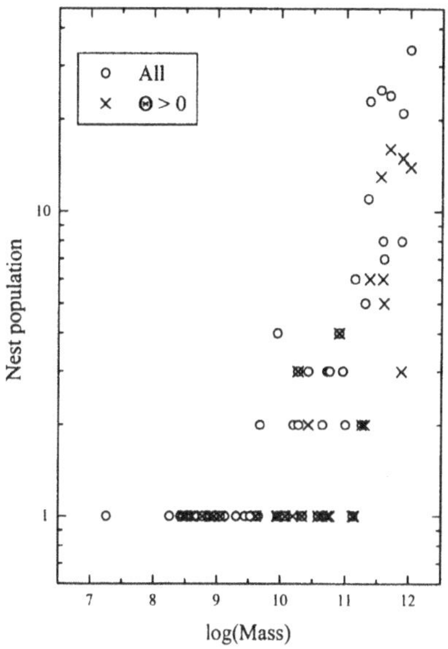

Figure 6. Number of galaxies in a nest around Main Disturber versus its mass. Circles are the total nest population, crosses are the same with $\Theta > 0$.

Fig. 5 presents the distribution of the LV galaxies vs. their tidal indices and distances from the Local Supercluster plane. As it is seen from the data, the early type galaxies (filled circles) are concentrated toward the LS plane stronger than spiral and irregular ones, which confirms again the existence of the effect of morphological segregation within the Local Volume. One can also note another feature. Among the galaxies having a high (positive) tidal index more than 90% of them are situated in a narrow layer, $|SGZ| < 0.5$ Mpc, while isolated objects spread throughout the Local Volume. Such a feature may indicate the presence of two subsystems in the LV: "disc" and "bulge", where conditions for galaxy interaction are very different.

A relative number of galaxies, which are the Main Disturbers for one or several neighbors, make up 24%. This value is defined by the properties of the 3D structure of the LV, as well by the galaxy mass function. Being ordered according to their MDs, the LV galaxies form "nests" with a population k from 1 to 34 members. Obviously, the nest population must be higher the more massive its MD is. Such a relation is actually seen in Fig. 6, where the nests (open circles) have a tendency to follow the line $k \sim M^{1/2}$. In particular, the seven most massive galaxies control 51% of the total LV population within their zones of gravitational influence.

However, we should remark that the majority of nest members around MDs have tidal indices $\Theta < 0$, i.e. their crossing time with respect to the MD exceeds the cosmological time, $1/H$. If one takes into consideration as real companions of MDs only the galaxies with $\Theta > 0$ (crosses in Fig. 6), the typical nest population drops about 2 times.

In conclusion we note that the Local Volume contains some objects, not distinguished by signs of peculiarity or interaction, which nevertheless need special explanations. As an example we mention the case of NGC 404. This compact, bright lenticular galaxy has the corrected radial velocity V_0 = +195 km/s or kinematic distance of 2.8 Mpc. Within ~2 Mpc around it there are no other galaxies. NGC 404 has a rather low tidal index, $\Theta = -0.78$, and its MD is M 31. Judging by its HI flux, NGC 404 is moderately rich in neutral gas, and also several dust clouds are seen near its nucleus. The origin of such a very isolated compact galaxy seems a puzzle. Probably, the galaxy is the final stage of consecutive merging of members of the former group. Such kind of nearby unusual objects deserve the closer attention of observers as well as theorists.

Acknowledgements

This work is supported by INTAS-RFBR grant No 95-IN-RU-1390, and RFBR grant No 97-02-27101.

References

Arp H. (1966) Atlas of Peculiar Galaxies, *Asroph. J. Suppl.*, **14**, p.1

Broeils A. (1992) *Dark and Visible Matter in Spiral Galaxies*, Dissertation, Groningen

Burkey J.M., Keel W.C., Windhorst R.A., Franklin B.E. (1994) *Asrophys.J.*, **429**, p.413

Griffiths R.E., Casertano S., Ratnatunga K.U. et al. (1994) *Astrophys.J.*, **435**, L19

Karachentsev I.D. (1987) *Binary galaxies*, Moscow, Nauka

Karachentsev I., Makarov D. (1996), *Astron.J.*, **111**, p.794

Karachentsev I., Makarov D. (1997), *Astrophysical Letters & Communications*, accepted

Kraan-Korteweg R.C., Tammann G.A. (1979), *Astron.Nachr*, **300**, p.181

Paturel G., Bottinelli L., Di Nella H. et al. (1996) Principal Galaxy Catalogue, PGC-ROM, Observatoire de Lyon

Vorontsov-Velyaminov B. (1959) Atlas and Catalogue of Interacting Galaxies, Part I, Moscow Univ.

Vorontsov-Velyaminov B. (1977), *Astron. Asrtophys. Suppl.*, **28**, p.1

LOPSIDED GALAXIES AND THE SATELLITE ACCRETION RATE

DENNIS ZARITSKY
UCO/Lick Observatory and Univ. of Calif. Santa Cruz
Santa Cruz, CA, USA

AND

HANS-WALTER RIX
Steward Observatory
Univ. of Arizona, Tucson, AZ, USA

1. Introduction

Although current observations and theoretical models indicate that galaxy mergers and interactions are catalysts in the process of galaxy evolution, we have only a limited quantitative understanding of some basic aspects of the process. For example, the rate at which galaxies merge is poorly constrained. We can simplify the problem by considering only disk galaxies, which because of the fragility of their disks (cf. Tóth and Ostriker 1992) have presumably not suffered a major merger. Even so, these galaxies have almost certainly experienced the infall of small companion galaxies at some time. The Milky Way is currently experiencing the accretion of the Sagittarius dwarf (Ibata, Gilmore, & Irwin 1994) and will eventually accrete the Magellanic Clouds (Tremaine 1976). To understand how galaxies evolve, we need to have quantitative knowledge of the accretion rate as a function of mass for all types of galaxies. Here we consider only the accretion of companion galaxies ($\sim$ 10% by mass) onto large spiral galaxies.

How can we measure the accretion rate of small companion galaxies onto large spirals? First, we identify a signature of the accretion event. Second, we measure the frequency of the signature among a carefully chosen set of galaxies. Third, we derive the lifetime of such a signature after the accretion event. By combining the frequency and duration of the signature, we then calculate the accretion rate. In this paper, we describe such a calculation done by attributing disk lopsidedness to the relatively recent accretion of a moderate mass satellite. Because lopsidedness is not necessarily uniquely associated with accretion, such an estimate will result in an upper limit on the accretion rate. Future work, which combines several signatures and more detailed dynamical modeling of the various processes that may create the signature, will lead to improved measurements. The results and techniques described here are presented in a series of papers (Rix & Zaritsky 1995; Zaritsky 1995; Zaritsky & Rix 1997).

2. Lopsidedness in Spiral Galaxy Disks

The candidate signature for lopsidedness must be a relatively simple quantity that is easy to measure and robust to minor variations in galaxy type, inclination, luminosity, and surface brightness. This leads one to low order distortions. The most global of these is an $m = 1$ distortion, or lopsidedness. The ubiquity of lopsidedness was first noted by Baldwin, Lynden-Bell, and Sancisi (1980) and then studied in HI line profiles by Richter & Sancisi (1994). We will focus on lopsidedness in the stellar component. The advantage of using the stellar component is that the stars trace lopsidedness at smaller radii, where lopsidedness is more quickly dynamically erased and so provides a finer chronometer of the interaction.

The measurement of lopsidedness in disk galaxies is done photometrically. We decompose the light in circular annuli into a Fourier series in position angle. For the purpose of this presentation, we only discuss the amplitudes of the $m = 0$ and 1 terms, although higher order terms are interesting for other reasons. We quantify lopsidedness by referring to the average of the ratio of the amplitude of the $m = 1$ term, A_1, to the amplitude of the $m = 0$ term, A_0, between 1.5 and 2.5 disk scale-lengths. This quantity is referred to as $\langle A_1 \rangle$.

We observed a sample of nearly face-on disk galaxies, brighter than $m_B = 13.5$, with relative velocities $V < 5500$ km s^{-1}, and with kinematic inclinations less than 32°. The kinematic inclination is determined by inverting the Tully-Fisher relation and comparing to the observed HI line-width. Our data consist of I(0.8μm) and K$'$(2.2μm) images, although not both for all of the galaxies. The exposure times are short (15 min in I and 30 on-object in K$'$), so obtaining the data is straightforward. The measurement of lopsidedness is consistent between the two bands (Zaritsky & Rix 1997). The final sample consists of 60 galaxies.

The distribution of $\langle A_1 \rangle$ is nearly uniform between 0 and 0.25, with a few galaxies beyond 0.25 (cf. Fig. 1). We have chosen somewhat arbitrarily $\langle A_1 \rangle \geq 0.2$ to denote galaxies that are "significantly" lopsided. We will adopt this as the definition for the rest of the discussion; however, more detailed treatments of this issue must consider the distribution of $\langle A_1 \rangle$

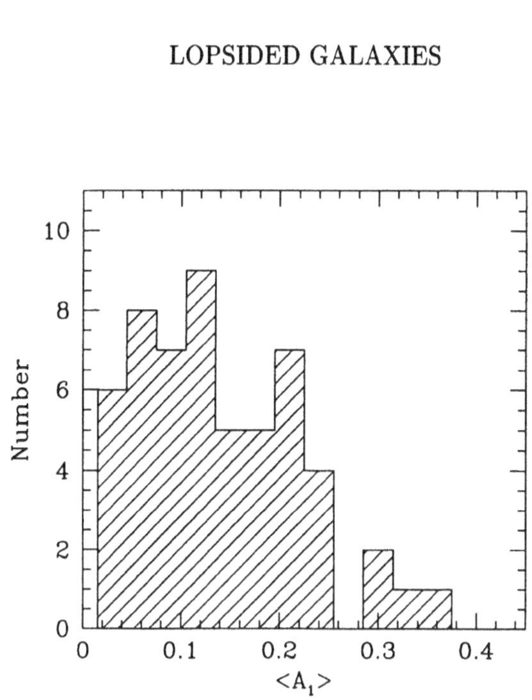

Figure 1. The distribution of the lopsidedness parameter, $\langle A_1 \rangle$. This distribution is drawn from the combined I and K′ data.

rather than simply modeling the number of galaxies beyond the threshold.

Nearly one-third of the spirals in our sample of 60 have $\langle A_1 \rangle > 0.2$. If we can associate such large lopsidedness with accretion events then we can begin to measure the accretion rate.

3. Accretion and Lopsidedness

Is the level of lopsidedness observed consistent with the infall of a moderate (10% by mass) satellite? Simulations of such an infall event were performed by Walker, Mihos, and Hernquist for the infall of a satellite onto a disk galaxy. In this simulation the satellite has 10% of the mass of the disk of the parent (1% of the total mass of the parent). Such a system is comparable to the LMC/MW system. In the simulation the parent has $\langle A_1 \rangle$ (measured in the same manner as for the data) that is 0.25±0.03, 0.22±0.02, and 0.18±0.01 at times 1, 1.125 and 2.5 Gyr after the beginning of the simulation (at which time the satellite is at a radius of 6 disk scale lengths on an orbit inclined 30° to the disk plane).

This single simulation demonstrates that the accretion of a small companion is a plausible mechanism for generating lopsidedness of the observed magnitude. However, a variety of simulation parameters (involving different mass ratios, orbital angular momentum, and orbital inclination) must

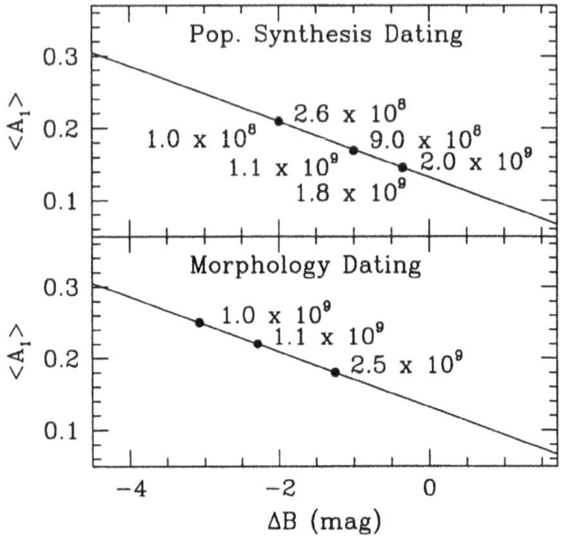

Figure 2. Comparison of the estimated chronology using the population synthesis models (see §4) and the dynamical simulation described in this section. For the population synthesis ages above the line come from the 0.5 Gyr burst models and the ages below come from the 0.1 Gyr burst model.

be considered to derive the expected distribution of lopsidedness. In addition, it is likely that other phenomenon will also generate lopsidedness. For example, close interactions between galaxies may generate lopsidedness of a similar magnitude as the infall of a small companion (Weinberg 1994). Others have examined dynamical instabilities as the source of lopsidedness (cf. Weinberg 1994; Sellwood & Valluri 1997). Currently, the incidence of lopsidedness can only be taken as a possible indication of an accretion event.

The simulations also allow us to measure the lifetime of significant distortions, and therefore obtain the third piece of information required to measure the accretion rate. However, further exploration of parameter space is also critical here. In the one simulation available so far, the lopsidedness parameter remains above 0.2 for about 1 Gyr (Fig. 2). Combining the fraction of spirals that exhibit such lopsidedness (~ 0.3) with the lifetime (~ 1 Gyr) implies that an accretion event happens at most once every three Gyr for the typical field spiral galaxy. However, as we discuss next, this result may be biased high.

4. Luminosity Enhancement as a Chronometer and Possible Bias

Interactions among massive galaxies are known to enhance star formation activity (Larson & Tinsley 1987; Lonsdale, Persson, & Mathews 1984; Kennicutt et al. 1987; Lavery & Henry 1988). Therefore, it is plausible to conjecture that the accretion of a small satellite might do the same. If so, then our *magnitude limited* sample of galaxies would be biased in favor of systems that have recently accreted a satellite and hence brightened. We discuss that possible bias below, but first discuss how a triggered starburst can be used to determine the time since the interaction.

We have used Bruzual & Charlot (1993) stellar population models to place limits on how the galaxy colors will change during 10% burst (*e.g.*, 10% of the stars in the galaxy were born during the interaction). In the models presented here, we assume that the underlying galaxy light is dominated by an old population of stars (models with a base of younger stars show less pronounced changes in both the color and $M_{\rm B}$). From such models we can predict the change in both the $B-R$ color and in $M_{\rm B}$ as a function of time since the triggered burst (Fig. 3).

From the dynamical simulation we can determine the relationship between the time since the accretion event and lopsidedness, and from the stellar synthesis models we can determine the relationship between the time since the burst and lopsidedness. If we can connect the accretion event with the burst (*e.g.*, the burst happens 100 Myr after the accretion) then we can synchronize the two chronometers and use the two to measure the time elapsed since the accretion event and to check the results. The inferred ages from each method are shown in Fig. 2 — assuming that the starburst begins at the same time as the dynamical simulation.

We have searched for signs of a connection between recent star formation and lopsidedness by defining a "blue" color as the galaxy's deviation from the B-band Tully-Fisher relation. There is a marginally significant correlation, in the expected sense, between this color and the lopsidedness measurement. If a line is fit to this relationship, and the effective search volumes for normal and lopsided galaxies are calculated, then we find that we have effectively searched for lopsided galaxies in a volume that is four times larger than that searched for normal galaxies. This difference in search volumes causes a corresponding decrease in our estimate of accretion rate. Hence, it is evident that either one must obtain volume limited samples, or have an excellent understand of the degree of brightening of a galaxy during an interaction.

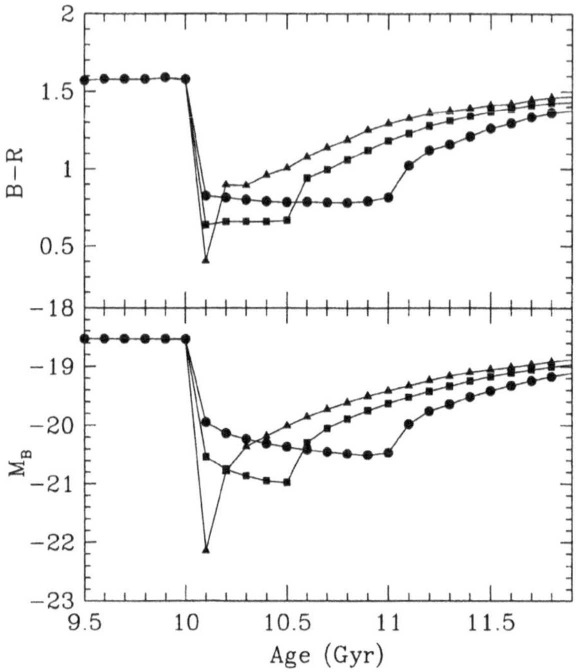

Figure 3. The effect of a 10% starburst on the color and luminosity of the galaxy. Three models are plotted: a 1.0 Gyr burst model (filled circles), a 0.5 Gyr burst model (filled squares), and a 0.1 Gyr burst model (filled triangles).

5. Directions Toward Progress

This work described here is a first attempt to use lopsidedness to measure the accretion rate. There is potential in the method, even though there are clear avenues for progress. First, we should obtain a volume limited sample to eliminate the issue of brightening due to the interaction. Second, we should obtain a sample of early type galaxies to determine the Hubble-type lopsidedness connection. Ongoing work by Rudnick and Rix (1997, *in prep.*) show that the fraction of highly lopsided galaxies is nearly the same in early type spirals ($\sim 20\%$). Third, we should examine a larger suite of simulations to determine the variations in lopsidedness created by different accretion events. This is being done by Velazquez, Dubinski, and Zaritsky. Fourth, we need to understand alternative phenomena that generate $m = 1$ distortions.

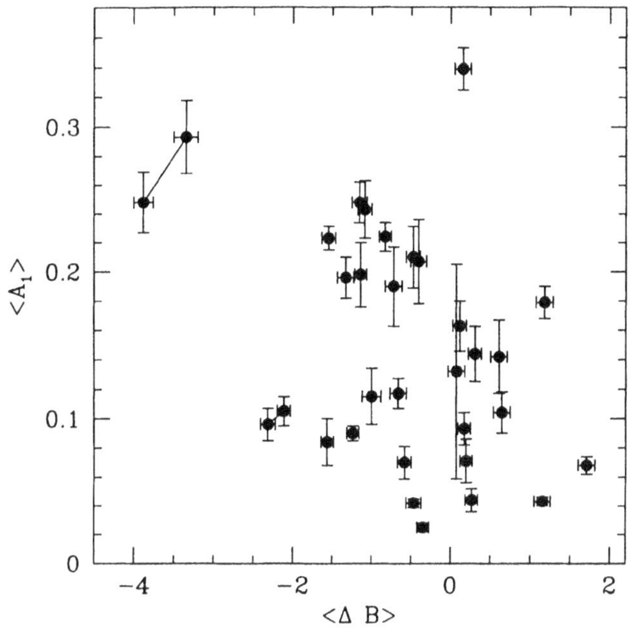

Figure 4. Color vs. Lopsidedness. We plot the mass-weighted color, $\langle \Delta B \rangle$, vs. the lopsidedness parameter, $\langle A_1 \rangle$. The two pairs of points connected by a line indicate the different measurements obtained for the same galaxies in I and K′ and with slightly different adopted parameters.

6. Summary

The accretion rate can be constrained by identifying a signature of accretion and measuring the frequency and lifetime of that signature. We have used disk lopsidedness as that signature. We estimate that the accretion rate is between 0.07 and 0.25 satellites per Gyr for satellites that are about 10% of the mass of the spiral disk.

References

Baldwin, J., Lynden-Bell, D., & Sancisi, R., (1980), *MNRAS*, **193**313
Bruzual, A. G., & Charlot, S. (1993), *ApJ*, **405**, 538
Ibata, R.A., Gilmore, G., & Irwin, M.J. (1993), *Nature*, **370**, 194
Kennicutt, R.C. et al. (1987), *AJ*, **93**, 1011
Larson, R.B., & Tinsley, B.M. (1978), *ApJ*, **370**, 194
Lavery, R.J., & Henry, J.P. (1988), *ApJ*, **330**, 596
Lonsdale, C., Persson, S.E., & Mathews, K. (1984), *ApJL*, **287**, 95L
Richter, O., & Sancisi, R (1994), *AA*, **290**, 9
Rix, H.-W., & Zaritsky, D. (1995), *ApJ*, **447**, 82
Sellwood, J.A., & Valluri, M. (1997), *MNRAS*, **287**, 124

Tóth, G., & Ostriker, J.P. (1992), *ApJ*, **389**, 5
Walker, I.R., Mihos, J.C., & Hernquist, L.(1996), *ApJ*, **460**, 121
Weinberg, M. (1994), *ApJ*, **421**, 481
Zaritsky, D., & Rix, H.-W. (1997), *Ap.J.*, **448**, L17
Zaritsky, D., & Rix, H.-W. (1997), *Ap.J.*, **477**, 118

"E+A" GALAXIES: ENVIRONMENT AND EVOLUTION

ANN I. ZABLUDOFF
UCO/Lick Observatory and University of California, Santa Cruz, CA, USA

1. Introduction

One important approach to the study of galaxy evolution is to identify those galaxies whose spectral and/or morphological characteristics suggest that they are in transition. For example, "E+A" galaxies,[1] which have strong Balmer absorption lines and no significant [OII] emission, are generally interpreted as post-starburst galaxies in which the star formation ceased within the last $\sim$ Gyr (Fig. 1). This transition between a star forming and non-star forming state is a critical link in any galaxy evolution model in which a blue, star forming disk galaxy evolves into a S0 or elliptical. Another possible evolutionary track is that the star formation in an "E+A" resumes at some later time, if enough gas remains in the galaxy after its starburst ends. Given this ambiguity, it is important to investigate (1) the environment's role in "E+A" evolution, (2) the stellar and gas morphologies of "E+A"s, (3) the likely progenitors of "E+A"s, and (4) how common the "E+A" phase is in the evolution of galaxies.

This proceeding summarizes recent results from several inter-related projects designed to address these questions. These projects focus on a sample of 21 nearby "E+A" galaxies ($0.05 < z < 0.15$; Zabludoff et al. 1996) drawn from the Las Campanas Redshift Survey (Shectman et al. 1996). These studies include VLA and *HST* observations, in addition to comparisons of these data with galaxy-galaxy interaction simulations and stellar population synthesis models. My collaborators are D. Zaritsky (UCO/Lick),

[1]The term "E+A" is a bit of a misnomer. The Mg, Fe, and Ca lines observed in the spectra of these galaxies are consistent with the stellar populations of ellipticals or "E"s. The additional "A" designation arose from the galaxies' strong Balmer absorption lines, which are characteristic of A stars. Because the morphologies of "E+A"s now appear to range from spheroidals to disks, a more apt, and exclusively spectroscopic, designation is "K+A" (cf. Franx 1993). Nevertheless, we use "E+A" throughout this paper for historical reasons.

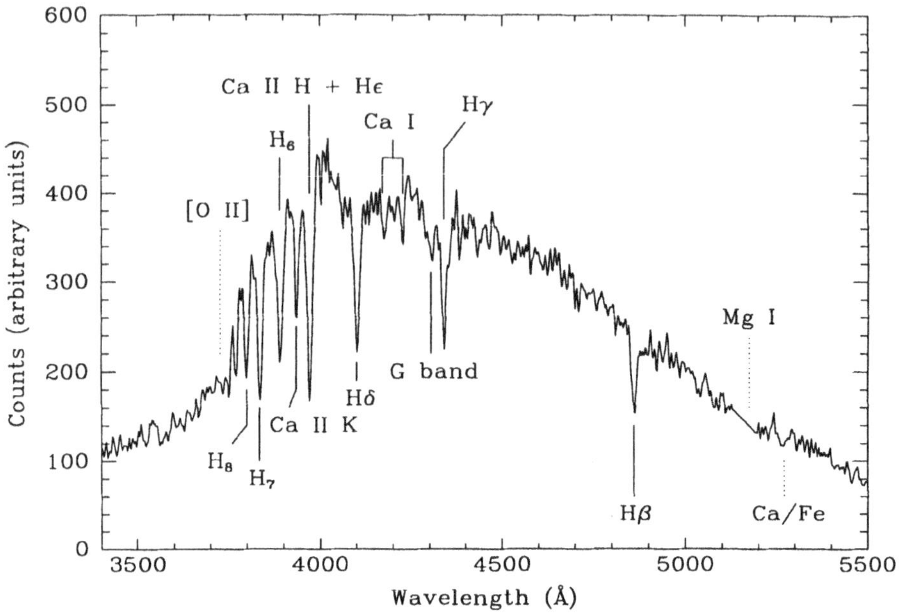

Figure 1. Identification of lines in the rest frame spectrum of an "E+A" galaxy, which is dominated by a young "A" stellar component. The residual sky line at 5577 Å has been excised. Note the absence of [O II] emission.

J. van Gorkom (Columbia), C. Mihos (Case Western), I. Smail (Durham), G. Bruzual (CIDA), S. Charlot (IAP), M. Franx (Leiden), and R. Bernstein (OCIW).

2. Environment and "E+A" Evolution

The role of environment in the evolution of "E+A" galaxies, specifically in producing the initial burst of star formation, in ending it, and in allowing it to resume, is unknown. In past work, the detection of "E+A"s almost exclusively in distant clusters led to speculation that these galaxies represent an evolutionary sequence unique to or most efficient in cluster environments. The existence of such a cluster-dependent evolutionary sequence would suggest that the cluster environment, in the form of the intracluster medium, galaxy harassment, or the global potential, is responsible for the recent star formation history of "E+A"s and, by extension, for the Butcher-Oemler effect (Butcher & Oemler 1978) in clusters. In contrast to this line of reasoning, Schweizer (1982, 1996) and others find several nearby "E+A"s that appear to lie outside the hot, dense environments of clusters and that have highly disturbed morphologies consistent with the products

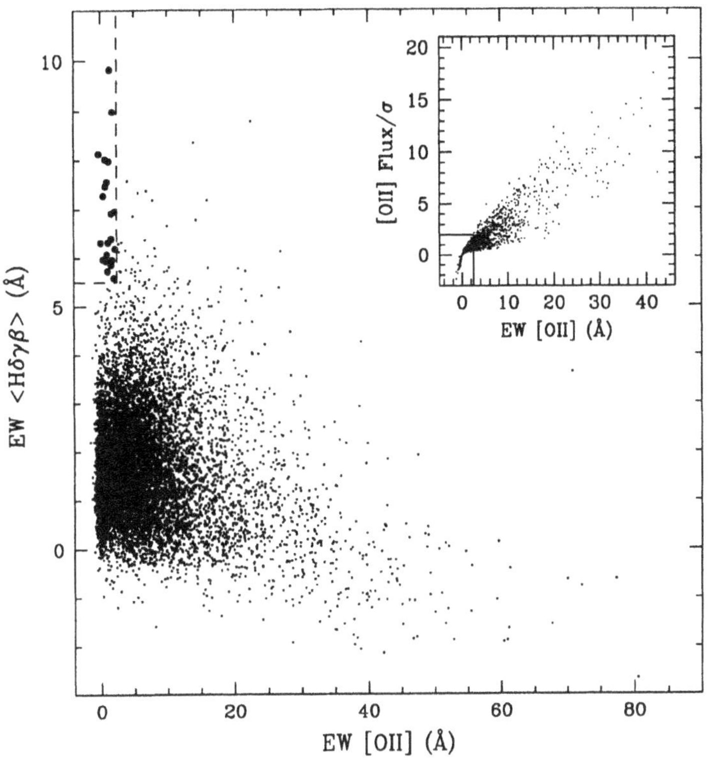

Figure 2. Plot of average Balmer line absorption $\langle H \rangle$ *vs.* [O II] line emission EW[O II] for the 11113 LCRS galaxies with $S/N > 8$ and $0.05 < z < 0.13$. The dashed line encloses the region, $\langle H \rangle > 5.5$ Å and EW[O II] < 2.5 Å, from which the sample of 21 "E+A" galaxies (large points) is drawn. The inset shows that EW[O II] cut excludes galaxies with a more than 2σ detection.

of galaxy-galaxy mergers. To isolate at least one mechanism that governs the evolution of "E+A" galaxies requires a statistical inventory of the environments in which "E+A"s form. The Las Campanas Redshift Survey (LCRS), which includes high signal-to-noise spectra for $\sim$ 11000 galaxies with $0.05 < z < 0.15$, is the ideal sample with which to characterize the environments of "E+A"s.

To identify "E+A" galaxies in the LCRS having properties consistent with those of known "E+A"s, we plot the distribution of Balmer absorption line and [OII] emission line strength for the LCRS galaxies (Fig. 2; Zabludoff et al. 1996). The "E+A"s are selected to have the strongest Balmer absorption lines (the average of the equivalent widths of Hβ, γ, δ is > 5.5 Å) and weakest [O II] emission-line equivalent widths (< 2.5 Å, which corresponds to a detection of [O II] of less than 2σ significance) of any of the galaxies in the survey. We test whether these 21 "E+A"s lie in rich clusters in several ways, including calculating the local galaxy density around each "E+A" and also checking whether the "E+A" is a member of a rich cluster in the LCRS group catalog (Tucker 1994). Surprisingly, a large fraction ($\sim 75\%$) of nearby "E+A"s lie in the field, well outside of clusters and rich groups of galaxies. We conclude that interactions with the cluster environment are not essential for "E+A" formation and therefore that the presence of these galaxies in distant clusters does not provide strong evidence for the effects of cluster environment on galaxy evolution.

If one mechanism is responsible for "E+A" formation, then the observations that "E+A"s exist in the field and that at least five of the 21 in our sample have clear tidal features argue that galaxy-galaxy interactions and mergers are that mechanism. The most likely environments for such mergers are poor groups of galaxies, which have lower velocity dispersions than clusters and higher galaxy densities than the field. Groups are correlated with rich clusters and, in hierarchical models, fall into clusters in greater numbers at intermediate redshifts than they do today (cf. Bower 1991; Lacey & Cole 1993; Kauffmann 1994). When combined with the strong evolution observed in the field population (cf. Broadhurst et al. 1988; Lilly et al. 1995), our work suggests that the Butcher-Oemler effect may reflect the typical evolution of galaxies in groups and in the field rather than the influence of clusters on the star formation history of galaxies.

3. Stellar and HI Morphologies of "E+A"s

Is the transition between galaxy types implied by the post-star formation spectrum of an "E+A" seen in its morphology? The two *HST* images that we have obtained to date suggest a morphological transition. One "E+A" has an E type morphology, but has extended tidal tails. The other "E+A" is a barred S0. If star formation does not resume in these galaxies, they will look like early types after their blue stars die. One interesting unanswered question is why galaxies of such different morphologies have such similar spectra.

There is substantial evidence that galaxy-galaxy interactions increase star formation rates. While the effects of such interactions are consistent with the starburst history of "E+A"s, the mechanism by which the star formation stops is still a mystery. The HI morphologies of "E+A"s can provide some clues.

In the first LCRS "E+A" for which we have obtained VLA data (Fig. 3), there are clear HI tidal tails similar to those of the Antennae (Hibbard & Mihos 1995). These tails support the galaxy-galaxy interaction picture for "E+A" formation. Perhaps even more interesting is the distribution of the

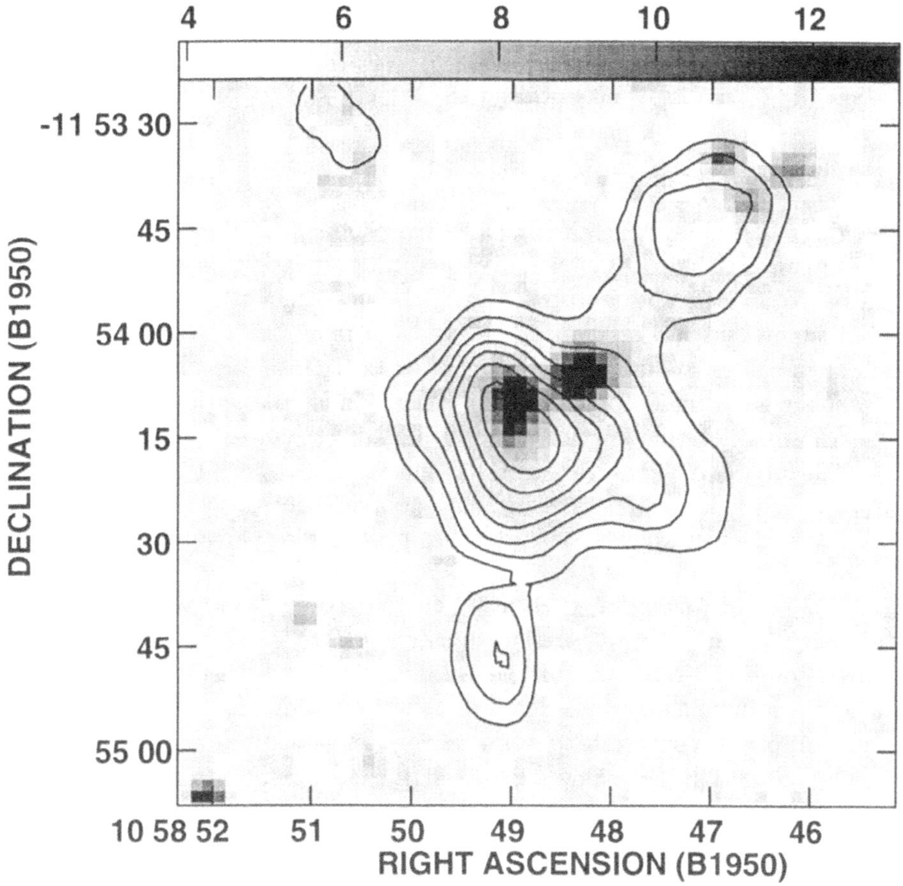

Figure 3. The HI distribution superposed on the Scanned Digitized Sky Survey SERC b_J image of one of the bluest "E+A"s in the sample. The central HI gas has a mass of $\sim 5 \times 10^9 M_\odot$, consistent with that in late type galaxies, but it is extended over more than $50h^{-1}$ kpc.

gas not in the tails. The gas mass is comparable to that in disk galaxies ($\sim 5 \times 10^9\ M_\odot$), but it is extended over $50h^{-1}$ kpc. Thus, the lack of star formation in this "E+A" is not due to an absence of gas, but perhaps to the low density of that gas. It is possible that this gas will someday fall back into the galaxy and generate new star formation.

The rarefied HI gas in this "E+A" suggests that the subsequent evolution of such galaxies could be affected by environment. Extended gas is easier to strip by ram pressure than that in galactic disks. Therefore, if this "E+A" formed in a subcluster, instead of the field, it is likely that the effects of the intracluster medium would preclude subsequent star formation.

While derived from only one VLA observation to date, this speculation may illuminate one source of the difference between the galaxy morphologies in clusters and in the field. In this spirit, our current observational program is to compare the gas distributions in cluster and field "E+A"s.

4. "E+A" Progenitors

As discussed above, the clear tidal features in some "E+A"s indicate that these galaxies have evolved morphologically, in addition to spectroscopically. To identify the morphologies of the most likely "E+A" progenitors, we assume that two progenitors merge to form an "E+A" and derive limits on their gas-to-stellar masses from the strength of the "E+A" starburst (as inferred from a comparison of the Balmer absorption lines and the 4000Å break strengths with stellar population synthesis models; Bruzual & Charlot 1995). If the gas-to-stellar masses of the "E+A" progenitors are consistent with those of gas-rich, disk galaxies, and a particular "E+A" is an S0 or E, then we can conclude that a morphological transformation has occurred.

For most of the 21 LCRS "E+A"s, the (HI+H_2)-to-stellar mass ratios of a pair of Sa-c spirals provide sufficient gas to generate burst strengths corresponding to 10-30% of the total stellar mass in the "E+A". Note that we assume a standard Scalo IMF and a star formation efficiency (*i.e.*, fraction of gas converted to stars) of 50% or less. However, for the bluest three "E+A"s in the sample (*e.g.*, Fig. 1), the burst strengths of $\sim$ 50% cannot be reproduced without relaxing some of these assumptions. For example, either both merging progenitors are late Sd disks, or at least one is a low surface brightness, Malin I type galaxy, or the star formation efficiency of the resulting burst is an extraordinary 100% (in contrast with the $\sim$ 50% efficiencies of the brightest *IRAS* ultraluminous galaxies). Although based on stellar population synthesis models that are still incomplete, these results support the picture that "E+A"s are a phase of galaxy evolution in which blue, star-forming disk galaxies are transformed via galaxy-galaxy encounters into early type S0 and E galaxies.

5. How Common is the "E+A" Phase?

Are "E+A"s rare objects or do they represent a short-lived phase in the evolution of many galaxies? From a comparison of the 21 "E+A" spectra with stellar population synthesis models, we estimate that the duration of the "E+A" phase is < 0.8 Gyr. The fraction of galaxies that are "E+A"s in the nearby universe is $21/11113 = 0.002$. Therefore, at least 4% of galaxies could have passed through an "E+A" phase within a Hubble time, a fraction which would constitute a significant number of the early types in the field. We plan to improve this estimate by comparing the *HST* images, HI maps, and follow-up long-slit spectra with simulations of galaxy-galaxy interactions (cf. Mihos & Hernquist 1994). The internal kinematics and morphological features on small scales should better constrain the time elapsed since the starburst ended and thus the duration of the "E+A" phase in galaxies.

References

Bower, R. (1991), *MNRAS*, **248**, 332
Broadhurst, T.J., Ellis, R.S. & Shanks, T. (1988), *MNRAS*, **235**, 827
Bruzual, A. G., & Charlot, S. (1993), *ApJ*, **405**, 538
Butcher, H.R. & Oemler, A. (1978), *ApJ*, **219**, 18
Franx, M. (1993), *ApJL*, **407**, L5
Hibbard, J.E., & Mihos, J.C. (1995), *AJ*, **110**, 140
Kauffmann, G. (1995), *MNRAS*, **274**, 153
Lacey, C. & Cole, S. (1993), *MNRAS*, **262**, 627
Lilly, S.J., Tresse, L., Hammer, F., Crampton, D., & Le Fevre O. (1995), *ApJ*, **455**, 108
Mihos, J.C., & Hernquist, L. (1994), *ApJ*,**431**, 9
Schweizer, F. (1982), *ApJ*, **252**, 455
Schweizer, F. (1996), *AJ*, **111**, 109
Shectman, S.A., Landy, S.D., Oemler, A., Tucker, D.L., Lin, H., Kirshner, R.P., & Schechter, P.L. (1996), *ApJ*, **470**, 172.
Tucker, D.L., (1994), Ph.D. thesis, Yale University.
Zabludoff, A.I., Zaritsky, D., Lin, H., Tucker, D., Hashimoto, Y., Shectman, S.A., Oemler, A., & Kirshner, R.P. (1996), *ApJ*, **466**, 104.

RADIO CONTINUUM OBSERVATIONS OF NGC 1961: INTERACTION WITH THE INTERGALACTIC MEDIUM OR THE REMNANT OF A MERGER?

U. LISENFELD
Universidad de Granada, Spain

AND

P. ALEXANDER AND G. POOLEY
MRAO, Cavendish Laboratory, Cambridge, UK

We present new radio continuum images of the supermassive, peculiar galaxy NGC 1961 at 1.5, 4.9, 8.4 and 15 GHz. NGC 1961 (Arp 184) is a very massive (dynamical mass $> 10^{12}\,M_{\odot}$) Sb galaxy with a peculiar and asymmetric optical appearance. Furthermore, it exhibits an unusual, asymmetric HI distribution with an extensive wing of gas extending 30 kpc to the north-west and a sharp edge to the south-east (Shostak et al. 1982).

Our observations allow us to separate the thermal and the nonthermal radio emission and to determine the nonthermal spectral index distribution. This spectral index distribution in the galactic disk is unusual: at the maxima of the radio emission the synchrotron spectrum is very steep, indicating aged cosmic ray electrons. Away from the maxima the spectrum is much flatter. We discuss various possibilities to explain this peculiar behavior (for more details see Lisenfeld et al. 1998) and conclude that the most likely cause are variations of the star formation rate (SFR) in the past. The steep spectra of the synchrotron emission at the maxima indicate that a strong decline of the SFR has taken place at these sites. The extended radio emission is a sign of recent cosmic ray acceleration, probably by recent star formation. We suggest that a violent event in the past, most likely a collision with an intergalactic gas cloud or a merger, has caused the various unusual features of the galaxy.

References

Lisenfeld, U., Alexander, P., Pooley, G.G., & Wilding, T. 1998, *MNRAS*, submitted
Shostak, G.S., Hummel, E., Shaver, P.A., van der Hulst, & J.M., van der Kruit, P.C. 1982, *A&A*, **115**, 293

DYNAMICAL PROPERTIES OF TIDALLY-INDUCED GALACTIC BARS

T. MIWA AND M. NOGUCHI
Astronomical Institute, Tohoku University
Aoba, Sendai 980-77, Japan

We have simulated a series of N-body models of tidal encounters between a disk galaxy and a perturbing galaxy to investigate the dynamical properties of tidally-induced galactic bars (Noguchi 1987, *MNRAS*, **228**, 635). We have also calculated N-body models on isolated galaxies with bar-unstable disks (Ostriker & Peebles 1973, *ApJ*, **186**, 467) to make a comparison between galactic bars of different origins (Miwa & Noguchi 1997, submitted).

It is found from our simulations that the tidally-induced bars sometimes roate quite slowly and have *inner Lindblad resonances (ILR) near the bar ends*, whereas the spontaneously-formed bars have no ILR and end near corotation owing to their large pattern speed. Since the difference in resonance structures affects the kinematics of the interstellar gas, these peculiar tidal bars may give us a new way of creating the vast morphological and kinematic variety observed in real barred galaxies, which may not be explained solely by the spontaneous bars.

The slow rotation of tidal bars is caused by two major factors. First, the smallness of the disk mass fraction in some models leads to a small pattern speed. Second, the angular momentum transfer from the inner disk to the perturber at perigalactic passage serves to reduce the pattern speed. "Massive disk models" in which the disk is stabilized by large random motions in the disk stars are quite sensitive to the second process.

This complicated behavior of tidal bars can be understood naturally by recognizing *two regimes of tidal bar formation*. When the tidal perturbation is weak, it works only as a trigger of bar formation and the bar properties are determined largely by the internal structure of the target galaxy. On the other hand, a sufficiently strong tidal perturbation washes out the intrinsic property of the target galaxy and imposes the bar as a common characteristic determined by the parameters of the tidal encounter.

STAR FORMATION IN COLLISIONS BETWEEN TWO GAS-RICH DISK GALAXIES

CURTIS STRUCK
Iowa State University
Dept. of Physics and Astronomy
Ames, IA 50011 USA

We present the results of a modest grid of three-dimensional (SPH) simulations of nearly head-on collisions between two model galaxies, each consisting of a rigid halo and a gas-rich disk component. The companion to primary mass ratio used is typically 0.2 - 0.35. Simple models of radiative cooling and heating from young star activity are also included in the simulations. Star formation is assumed to occur when the local gas density exceeds a threshold value. In these collisions the primary gas disk remains largely intact, though a gas bridge is splashed out between the two galaxies. Star formation occurs in asymmetric ring-like waves in the primary. The bridge is predicted to have essentially no ongoing star formation. These results agree with observations of the Cartwheel and VII Zw 466 ring galaxies.

The companion gas disk is usually disrupted by the impact, but reforms via accretion from the bridge. Star formation is often delayed in the companion until the gas disk has reformed. Interestingly, star formation can also be reduced or delayed in the central regions of the primary, apparently as a result of accretion heating. For example, in models with a diskless companion star formation is strongest in the first ring wave, while in some models with a companion gas disk the third ring, which forms at the end of the accretion phase, is strongest. These results suggest the general possibility that heating due to high rates of accretion is able to delay starbursts in collisional systems. For further details see Struck (1997, and Struck et al. 1996, for applications to the inner ring of the Cartwheel).

References

Struck, C. 1997, *ApJS*, **113**, in press
Struck, C., Appleton, P.N., Borne, K.D., & Lucas, R.A. 1996, *AJ*, **112**, 1868

TOWARDS AN INTERACTION MODEL OF M81, M82 AND NGC 3077

R.C. THOMSON, S. LAINE AND A. TURNBULL
Department of Physical Sciences, University of Hertfordshire, Hatfield, Herts AL10 9AB, UK

1. Introduction

Thomson (1992) suggested that the Cen A group of galaxies may be the result of one interaction. Here we report full N-body simulations made to study this proposal in the M81-M82-NGC 3077 group.

2. Numerical Techniques and Initial Conditions

The mass of the companion was 0.1 of the main galaxy mass. The initial orbits of the companion were either prograde or retrograde, and mildly hyperbolic. The simulations were made with a tree-SPH code.

3. Results and Discussion

The companion always loses a large fraction of its mass near the closest approach. However, the satellite morphology is not as strongly distorted as in the test particle simulations of Thomson (1992). All our experiments resulted in a merger.

4. Conclusion

The interaction scenario, where M82 is a shredded disk remnant and NGC 3077 the bulge of the companion, is not reproduced in a mildly hyperbolic full N-body simulation.

References

Thomson, R.C. 1992, *MNRAS*, **257**, 689

AN HI LINE SURVEY OF POLAR RING GALAXIES

W. VAN DRIEL
Unité Scientifique Nançay, Observatoire de Paris, France
M. ARNABOLDI
Osservatorio Astronomico di Capodimonte, Naples, Italy
F. COMBES
DEMIRM, Observatoire de Paris, France
AND
L.S. SPARKE
Astronomy Dept., University of Wisconsin-Madison, U.S.A.

We have observed a total of 74 polar ring galaxies (PRGs), PRG candidates and related objects at 3 radio observatories: Green Bank (Richter et al. 1994, *AJ*, **107**, 1), Effelsberg (Huchtmeier 1997, *A&A*, **319**, 401) and Nançay (van Driel et al. 1997, *A&AS*, in press; van Driel et al. 1998, *A&A*, in prep.). Our main aim is to identify systems suitable for H I synthesis mapping, crucial for an understanding of their dynamical state. Most objects were selected using one or more of the following 3 criteria: known redshift (<8000 km/s), or blue magnitude brighter than 15.5 mag, or kinematically confirmed PRG or good PRG candidate. A total of 62 objects were detected, 39 of which with more than one telescope.

A large range in gas-richness is found among the objects observed. The kinematically confirmed PRGs are all gas-rich, with an average $M_{\rm HI}/L_{\rm B}$ ratio of 1.1 $M_\odot/L_{\odot,\rm B}$ (typical for gas-rich late-type galaxies), while the PRG candidates and related objects show a much larger range; some are as gas-rich as confirmed PRGs, while a lot of them have as little H I as detected normal S0 galaxies (about 0.15 $M_\odot/L_{\odot,\rm B}$) – radio synthesis imaging of PRGs shows the H I to be associated with the polar ring and not with the equatorial S0 disk. The Tully-Fisher (TF) relation, made without correcting profile widths for inclination (a correct procedure for true PRGs with their edge-on rings), shows the largest deviations from the average relation for non-confirmed PRG candidates and related objects.

DYNAMICS OF MERGERS & REMNANTS

JOSHUA E. BARNES
Institute for Astronomy, University of Hawai'i
2680 Woodlawn Drive, Honolulu, Hawai'i, 96822, USA

Abstract. This review focus on some issues which seem relevant to recent discussions: (1) how halo structure influences tail length, (2) the fate of power-law density cusps, (3) the results of unequal-mass disk galaxy mergers, and (4) the behavior of hot and cold gas in merging disk galaxies.

1. Halos & Tail Length

The tails of interacting disk galaxies were lucidly explained by Toomre & Toomre (1972). Briefly, tails develop when tidal forces tear galactic disks apart; material on the side of the disk furthest from the companion galaxy, suddenly free of the gravitational pull which had kept it in a circular orbit, escapes along a nearly linear trajectory and so produces an ever-lengthening tail. Self-consistent simulations of galaxies possessing modest dark halos have elaborated but not fundamentally modified this basic picture (Negroponte & White 1982, Barnes 1988). A recent study claims that the long tidal tails observed in many interacting systems can't escape from dark halos if these halos have more than about ten times the luminous mass (Dubinski, Mihos, & Hernquist 1996). But this claim is model-dependent, as shown by the experiments described below.

In these experiments, each galaxy models had three components: a bulge, a disk, and a halo. The bulges were Hernquist (1990) models with mass[1] $M_{\rm b} = 0.0625$ and scale radius 0.04168; beyond a radius of 4.0 the bulge density tapered away smoothly. The disks were exponentials, with mass $M_{\rm d} = 0.1875$, radial scale length $\alpha^{-1} = 1/12$, and vertical scale height $z_0 = 0.005$. The halos were based on the Navarro, Frenk, & White (1996; hereafter NFW) model, which has a logarithmically divergent mass.

[1]Unless otherwise noted, results are quoted in arbitrary units with $G = 1$.

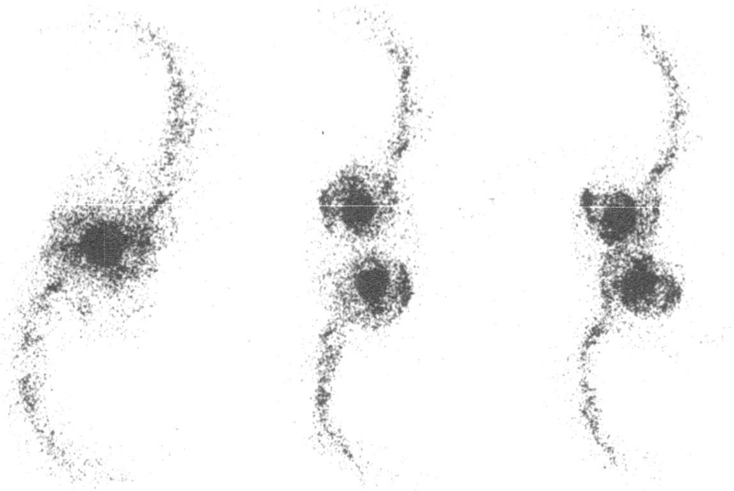

Figure 1. Tails produced by parabolic encounters with halo-to-luminous mass ratios of 5:1 (left), 10:1 (middle), and 20:1 (right).

To obtain finite total masses the NFW halo profiles were tapered:

$$\rho(r) = \begin{cases} \dfrac{M_{\rm a}}{\log(2) - 0.5} \dfrac{1}{4\pi r(a+r)^2} & \text{if } r < b \ , \\ \rho_{\rm b}(b/r)^2 \exp[-\gamma((r/b)^2 - 1)] & \text{otherwise} \ , \end{cases} \tag{1}$$

where $M_{\rm a}$ is the mass within the halo scale radius a and the parameters $\rho_{\rm b}$ and γ were chosen so that both $\rho(r)$ and its first derivative are continuous at $r = b$. All the models used halo mass and length scales $M_{\rm a} = 0.2$ and $a = 0.2$; the taper radius b was used to adjust the total halo mass $M_{\rm h}$ as follows:

b	0.55285	2.8099	34.95
$M_{\rm h}$	1.25	2.50	5.00

(2)

Combining these halos with the standard bulge and disk yielded composite models with luminous-to-dark ratios of 1:5, 1:10, and 1:20, respectively. All these models had identical, fairly flat rotation curves out to at least six disk scale lengths.

All calculations started with the two galaxies only 5 length units apart; thus the low-mass halos were initially well-separated while the others significantly overlapped. The potential energy of each initial configuration was used to assign the galaxies relative speeds consistent with asymptotically parabolic orbits; the directions of the initial velocities were set so as to obtain pericentric separations of about 0.2 length units in each case. As shown in Figure 1, all three experiments produced acceptable tails.

The success of this tail-making exercise is hardly unexpected. To make a proper tail, the velocity of the tail material must exceed the local escape velocity. The NFW model, with $\rho \propto r^{-3}$ at large radii, has a *finite* escape velocity even though its total mass diverges; consequently it is possible to obtain proper tails with encounters involving arbitrarily massive halos. The models used by Dubinski et al., on the other hand, have potential wells which become significantly deeper as total halo mass increases. *Long tails constrain potential well depth, but not halo mass itself.*

2. Cusp Survival

Especially since the refurbishment of *HST* it's become clear that few if any early-type galaxies have constant-density cores; instead, the luminosity profiles of some galaxies are power-laws, while others show a break and a more gradual rise to the innermost point (eg. Faber et al. 1997). Core profile shape is correlated with luminosity; bright Es have breaks, while faint ones have power-laws. It is natural to ask what merging will do to this dichotomy. A simple argument outlined below suggests that merging preserves steep power-law cusps.

Pressure-supported systems may be approximately described by isotropic distribution functions of the form $f \simeq f(E)$; in such systems, most of the variance of the actual DF is due to its variance with binding energy E. Systems with central cusps have singular distribution functions with formally infinite phase-space densities; for example, a density profile $\rho \propto r^{-2}$ implies that the amount of mass with phase-space densities above f scales as $M(> f) \propto f^{-1/2}$. When systems merge, violent relaxation (Lynden-Bell 1967) spreads material over a range of binding energies; subsequently, phase mixing averages phase-space density on surfaces of constant E. But this violent relaxation is *incomplete*; numerical studies of mergers show that potential fluctuations die down before binding energies are completely randomized (e.g. White 1987). So phase mixing can only average over a limited range of fine-grained phase-space densities. The final coarse-grained DF, which describes the remnant at later times, is thus similar to the initial DF; in particular, the DFs of remnants produced by mergers of galaxies with cusps should still be singular after "the dust settles".

Merger simulations indicate that neglect of velocity anisotropy does not badly compromise this argument. The experiments reported here involved parabolic collisions between identical "gamma" models (Dehnen 1993; Tremaine et al. 1994), with inner density power-law slopes of $\gamma = 1$, 1.5, and 2; each model had total mass $M = 1$ and half-mass radius $r_{1/2} = 0.25$. These simulations used $N = 65536$ bodies, advanced with a leap-frog time-step of $\Delta t = 1/512$; the smoothing scale in the force calculation was

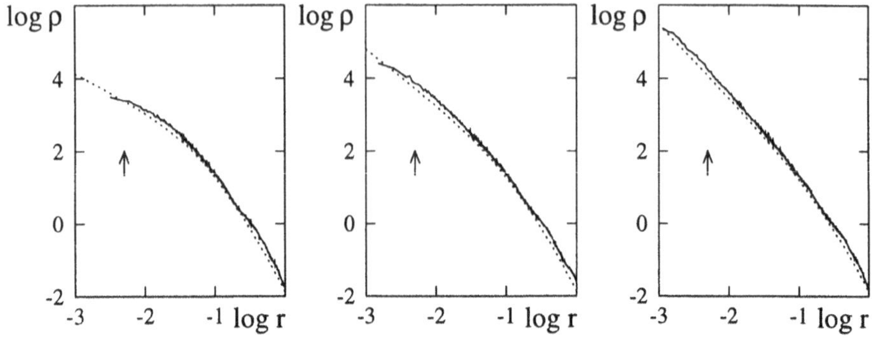

Figure 2. Spherically-averaged density profiles of merger remnants (solid lines) compared with initial models (dotted lines). Results are shown for initial models with $\gamma = 1$ (left), 1.5 (middle), and 2 (right). The arrow in each plot indicates the smoothing length.

0.005. The models were launched on asymptotically parabolic orbits which reached a pericentric separation of $r_{\rm p} = 0.25$ at time $t = 1$; by time $t = 8$ the systems had merged and largely relaxed. Figure 2 compares the density profile of each remnant with the gamma model of its progenitors. As in earlier studies (eg. White 1978; Villumsen 1983), remnant density profiles are closely related to those of the initial systems. Shown here more clearly than before is that steep power-law cusps survive merging.

Such results have implications for dark halos (e.g. NFW, Fukushige & Makino 1997) as well as for visible galaxies. One implication for galaxies is that *mergers can't transform the power-law profiles of faint Es into the broken profiles of bright Es unless violent relaxation is somehow prolonged*, perhaps by the effects of a pair of massive black holes (e.g. Makino & Ebisuzaki 1996, Quinlan & Hernquist 1997). If broken profiles are produced by inspiraling black holes, the central regions of bright Es should be effectively homogenized, with little velocity anisotropy and relatively weak color and metallicity gradients.

3. Disk Destruction

Galactic disks are fragile; while accretions of low-mass satellites do little harm mergers with comparable objects "scramble" disk galaxies into hot spheroids. Between these extremes, disks may be damaged but not altogether obliterated (eg. Walker, Mihos, & Hernquist 1996). Clustering models suggest that typical mergers involve objects with broadly distributed mass ratios; a ratio of 3:1 seems typical. What happens to a disk galaxy which merges with a companion one-third as massive?

A partial answer to this question emerges from a modest survey of

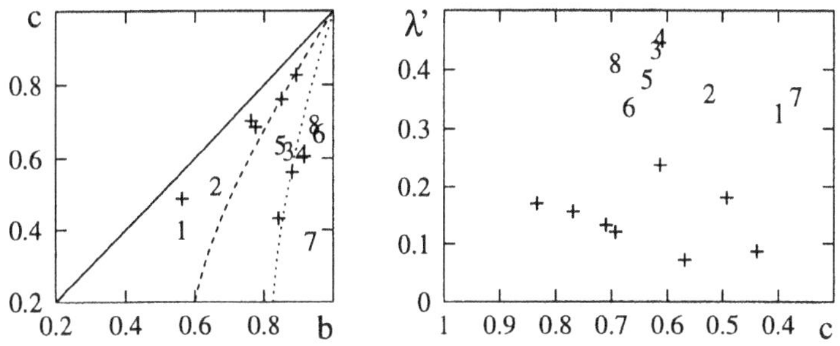

Figure 3. Remnants of 3:1 mergers, numbered 1 to 8, compared with remnants of 1:1 mergers, indicated by crosses. (Left) axial ratios: b and c are the intermediate and minor axis ratios, respectively. (Right) rotation *vs.* minor axis ratio: λ' is a dimensionless measure of angular momentum (Barnes 1992), equal to unity for a pure, co-rotating disk.

unequal-mass encounters (Barnes 1998). In these experiments, both galaxies were bulge/disk/halo systems; the larger galaxy had 3 times the mass of the smaller, and rotated $\sqrt[4]{3}$ times faster in accord with the Tully-Fisher relationship. The galaxies were launched on initially parabolic orbits and went through several passages before merging; remnants were evolved for several more dynamical times before being analyzed.

Figure 3 compares the shapes and kinematics of the 3:1-mass remnants with a companion sample of 1:1 remnants. The plot on the left shows axial ratios, computed from the moment of inertia of the more tightly-bound half of the luminous material in each remnant. As a group, the 3:1 remnants are fairly oblate, while the 1:1 remnants are more triaxial. The plot on the right shows angular momentum content plotted against minor axis ratio; the 1:1 remnants, which are flattened by velocity anisotropy, rotate at less than half the typical rate of the 3:1 remnants. On the whole, the 3:1 remnants are dynamically unlike their equal-mass counterparts.

What Hubble type would these objects be assigned? Of the eight cases reported here, numbers 3, 4, 5, 6, & 8 are fairly oblate and rapidly rotating. Moreover, many stellar orbits in these remnants are roughly circular; at any given energy, most of the minor-axis tube orbits have nearly the maximum possible angular momentum. Finally, several of these remnants actually look disk-like when viewed edge-on. Apparently, *some 3:1-mass merger remnants are more like S0 galaxies than ellipticals.*

Some early-type disk galaxies may thus owe their relatively hot disks and massive spheroids to unequal-mass mergers (Schweizer, this volume). If unequal-mass mergers are as common as cosmological models indicate, they could form a large fraction of S0 galaxies.

4. Hot & Cold Gas

Interstellar material, though only a modest part of the total mass, plays a profound role in disk galaxy mergers: it reveals the large-scale kinematics of tidal features (Hibbard et al. 1994), powers starbursts in luminous IR galaxies (Sanders & Mirabel 1996), and builds dense central regions in early-type merger remnants (Schweizer 1990, Kormendy & Sanders 1992). The complex behavior of interstellar gas is incompletely captured by the present generation of numerical experiments. Most simulations have in effect modeled the gas isothermally (Barnes & Hernquist 1996, and references therein); this rather drastic simplification tends to obscure the relationship between the thermal history and dynamical behavior of gas in interacting galaxies.

Color plate 4 (p. xxii) presents a simulation which gives some indication of the dynamics of a multi-phase medium in a merger of equal-mass disk galaxies. On the left are the stars, which evolved collisionlessly. In the middle is "hot" gas, assumed to be too tenuous to cool on dynamical timescales. On the right is "cool" gas with $T \simeq 10^4$ K, which serves as a proxy for molecular and atomic interstellar material.

The bulk of the hot component roughly followed the stellar distribution throughout the calculation. A small amount was heated above 10^6 K during the first interpenetrating passage of the two disks, but most of the gas warmed up only *after* this passage. When the tidally disturbed disks formed bars, the resulting noncircular motions caused shocks which heated the gas to several 10^5 K. The final encounter of the two disks heated the gas to the virial temperature, forming a pressure-supported atmosphere about as extended as the stellar component (Barnes & Hernquist 1996). In contrast, *ROSAT* observations of the Antennae (Read et al. 1995) and other starburst systems show actual outflows of gas at $\sim 10^6$ K; energy sources associated with ongoing starburst probably power these outflows.

Gas which cools only in the later stages of an encounter retains much of its initial angular momentum. Unless strong shocks develop during violent relaxation, this gas can't become radically segregated from collisionless stuff; if it dissipates after the remnant's potential settles down, it form a disk rotating in the same direction as the stellar remnant. Gradual return of gas from tidal tails (eg. Hibbard & van Gorkom 1996) can build up disks with radii of many kpc. Since the returning gas generally falls in at an angle to the principal planes of the remnant, such disks are likely to be warped.

Infall of material from tidal tails may have interesting consequences even before the galaxies actually merge. In the right-hand column, third frame from the top, a ring of gas is seen around the more face-on galaxy. A video shows that this ring formed from gas returning from the tidal tail extending to the left of this disk (Barnes & Hernquist 1998). This case was particularly favorable since the disk lay in the orbital plane, but the other disk in this simulation, though inclined by 71°, also developed a ring. The extended ring of molecular gas and star formation in the northern disk of Antennae (Color plate 5, p. xxiii) may likewise have resulted from material returning from the gas-rich southern tail; tests of this idea await more detailed analysis and dynamical modeling of this system.

Dissipation at early stages of galactic encounters has the effect of driving gas inward to pool within the central kpc of interacting disks (Icke 1985, Noguchi 1988) and at radii an order of magnitude smaller in merger remnants (Negroponte & White 1982, Barnes & Hernquist 1991). This material must lose *most* of its angular momentum to become so concentrated, and gravitational interaction with collisionless material seems to be the crucial brake on the rotation of the gas. What little angular momentum the gas retains may be poorly correlated with the angular momentum of the rest of the remnant, giving rise to kinematically decoupled central structures (Hernquist & Barnes 1991).

Still missing from these experiments is a proper treatment of interactions between various phases of the interstellar material. Simulations with star formation and stellar evolution are needed to incorporate "feedback" effects; one promising approach to this problem is described by Gerritsen & Icke (this volume). Hot gas can ionize neutral material as it falls back from the tidal tails; this may explain the lack of HI in the bodies of merger remnants like NGC 7252 (Hibbard et al. 1994). If the pressure of the hot gas is high enough, it may implode molecular clouds, triggering galaxy-wide starbursts (Jog, this volume); the extensive star formation in the Antennae could have been triggered in this fashion. Finally, the ram pressure of the hot gas may impart significant momentum to cooler interstellar material, thereby explaining the rather curious offsets between stellar and gaseous tails observed in some interacting systems (Schiminovich et al. 1995, Hibbard & van Gorkom 1996, Hibbard & Yun 1998).

Thus, the new data from multi-wavelength studies of systems like the Antennae offer a strong motivation for numerical simulations incorporating both hot and cool interstellar gas. Such simulations are needed to interpret the observations and to test theories of galactic transformation via violent interactions and mergers. Some questions which might be answered in this way include: What's going on in the "overlap" region of the Antennae; is this an interpenetrating encounter of two gas-rich systems? Did overpressure of hot gas trigger the *galaxy-wide* starbursts seen in this system? Will the resulting stars and star clusters spread throughout the remnant or concentrate in its central regions? Do systems like the Antennae give rise to ultraluminous IR galaxies like Arp 220? What powers the outflows of X-ray gas in the Antennae and in ULIR galaxies? Can such winds clean out the dusty central regions of ULIR galaxies, possibly exposing central AGNs? What happens to the outflowing gas; how much is returned to the intergalactic medium, and how much eventually falls back?

This work was supported by NASA through grant NAG 5-2836.

References

Barnes, J.E. 1988, *Ap.J.* **331**, 699.
Barnes, J.E. 1992, *Ap.J.* **393**, 484.
Barnes, J.E. 1998, in *Interactions and Induced Star Formation: Saas-Fee Advanced Course 26*, eds. D. Friedli, L. Martinet, D. Pfenniger (Berlin: Springer-Verlag), p. 275.
Barnes, J.E. & Hernquist, L. 1991, *Ap.J.* **370**, L65.
Barnes, J.E. & Hernquist, L. 1996, *Ap.J.* **471**, 115.
Barnes, J.E. & Hernquist, L. 1998, *Ap.J.* **495**, 187; see `http://www.ifa.hawaii.edu/~barnes/tog2.html`.
Dehnen, W. 1993, *MNRAS* **265**, 250.
Dubinski, J., Mihos, J.C., & Hernquist, L. 1996 *Ap.J.* **462**, 576
Faber, S.M. et al. 1997, *A.J.* **114**, 1771.
Fukushige, T. & Makino, J. 1997, *Ap.J.* **477**, L9.
Hernquist, L. 1990, *Ap.J.* **356**, 359.
Hernquist, L. & Barnes, J.E. 1991, *Nature* **354**, 210.
Hibbard, J.E. et al. 1994, *A.J.* **107**, 67.
Hibbard, J.E. & van Gorkom, J.H. 1996, *A.J.* **111**, 655.
Hibbard, J.E. & Yun, M.S. 1998, poster at 191st AAS meeting; see `http://www.cv.nrao.edu/~jhibbard/aas98/aas98.html`
Icke, V. 1985, *A.A.* **144**, 115.
Kormendy, J. & Sanders, D.B. 1992, *Ap.J.* **390**, L53.
Lynden-Bell, D. 1967, *MNRAS* **136**, 101.
Makino, J. & Ebisuzaki, T. 1996, *Ap.J.* **465**, 527.
Navarro, J.F., Frenk, C.S., & White, S.D.M. 1996, *Ap.J.* **462**, 563.
Negroponte, J. & White, S.D.M. 1982, *MNRAS* **205**, 1009.
Noguchi, M. 1988, *A.A.* **203**, 259.
Quinlan, G.D. & Hernquist, L. 1997, *New Astr.* **2**, 533.
Sanders, D.B. & Mirabel, I.F. 1996, *Ann. Rev. Astr. Ap.* **34**, 749.
Schiminovich et al. 1995, *Ap.J.* **444**, 77.
Schweizer, F. 1990, in *Dynamics and Interactions of Galaxies*, ed. R. Wielen (Berlin: Springer Verlag), p. 60.
Toomre, A. & Toomre, J. 1972, *Ap.J.* **178**, 623.
Tremaine, S. et al. 1994, *A.J.* **107**, 634.
Villumsen, J.V. 1983, *MNRAS 204*, 291.
Walker, I.R., Mihos, J.C., & Hernquist, L. 1996, *Ap.J.* **460**, 121.
White, S.D.M. 1978, *MNRAS* **184**, 185.
White, S.D.M. 1987, in *Structure and Dynamics of Elliptical Galaxiest*, ed. T. de Zeeuw (Dordrecht: Reidel), p. 339.

THE STRUCTURE OF MERGER REMNANTS OF COMPACT GROUPS OF GALAXIES: SOME PRELIMINARY RESULTS

E. ATHANASSOULA, CH. L. VOZIKIS
Observatoire de Marseille
2, Place Le Verrier, 13248 Marseille Cedex 04, France

Abstract. We present some preliminary results on the properties of multiple merger remnants. We discuss their axial ratios, their V/σ ratio as a function of ellipticity, and their circular velocity profiles.

1. Introduction

Since Toomre's pioneering suggestion that elliptical galaxies could be the result of a merger (Toomre & Toomre 1972) a number of numerical simulations have focused on the properties of merger remnants of galaxy pairs (see e.g. the reviews of Barnes & Hernquist 1992, or Barnes 1994 and references therein) in order to compare them with those of ellipticals. Although such comparisons show in general good agreement, they also put forward many points of disagreement, like the form of the radial density profile in the innermost parts, or the misalignment between the minor axis and the angular momentum vector. A variation to the original merging scenario is that elliptical galaxies are indeed merger remnants but not necessarily of pairs, but rather result from multiple mergers. In order to test this we are currently studying the properties of merger remnants of compact groups of five galaxies. Our work is complementary to that of Weil & Hernquist (1996), who have more particles per simulation and therefore can examine features we cannot. However, we examine a larger variety of initial conditions and thus can study their effect on the structure of the merger remnant.

The simulations we discuss here have been presented by Athanassoula, Makino & Bosma (1997). For the purposes of the present work we extended them in time to make sure that the remnants have reached equilibrium. In some of our simulations a common halo encompasses the whole group, while in others galaxies have individual halos. We have also used different values of halo-to-total mass, different density profiles for the distribution of mass in the group and in the common halo, and different initial kinematics of the group. Here we present only some preliminary results. Further results will be presented elsewhere (Vozikis & Athanassoula, 1998).

2. Axial ratios

For all remnants of our simulations we calculated the axial ratios b/a and c/a for the luminous and dark material separately. For the luminous material we find average values $\langle c/a \rangle = 0.7$ and $\langle b/a \rangle = 0.9$. These values depend neither on whether the halo was common or individual, nor on the central concentration of the initial configuration. On the other hand the spread around the mean value is much bigger for simulations starting with individual halos than for those starting with common halos. For the dark material we find average values of $\langle c/a \rangle = 0.85$ and $\langle b/a \rangle = 0.95$, i.e. the dark matter is distributed in a more spherical manner than the luminous material. Simulations with common halos lead to remnants whose halos are more spherical than do simulations starting with individual halos. Simulations with a higher halo-to-total mass ratio give remnants with more spherical halos than simulations starting with lower halo-to-total mass ratios.

3. V/σ as a function of ellipticity

Fig. 1 shows the position of the merger remnants evolving from simulations initially in virial equilibrium on the V/σ as a function of ellipticity diagram. The values displayed were obtained by viewing the remnant perpendicular to the axis of rotation and taking a mass weighted average for all the material within the half mass radius. The area on this diagram occupied by the luminous material of the remnants is roughly the same as that occupied by big ellipticals (e.g. Davies 1987), although in the observational sample there are some faster rotators than we have in our simulations. The dark matter part of the remnants coming from simulations with initially individual halos covers a similar part of the diagram, due to the fact that the two components in such simulations have similar evolutionary histories. On the other hand the dark matter of remnants from simulations with initially common halos clusters in less elongated and less anisotropic regions, due to the fact that the initial common halos are spherical and isotropic.

Taking into account the values of the triaxiality parameter T (Franx, Illingworth & de Zeeuw 1991), as well as the values of the angle between the angular momentum vector and the minor axis, we find the following general trend for the luminous part of the remnants. Remnants with relatively high values of V/σ (i.e. relatively fast rotators) are flatter, more oblate-like (i.e.

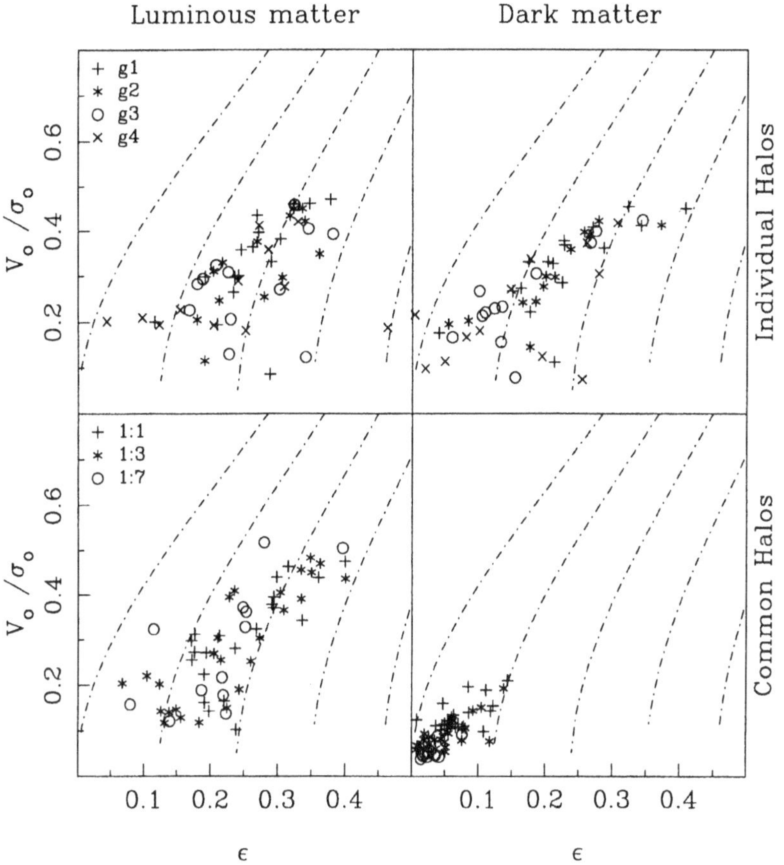

Figure 1. Location of the merger remnants from simulations initially in virial equilibrium on the v/σ as a function of eccentricity diagram. The two left panels correspond to the luminous matter and the two right ones to the dark matter. The two upper panels refer to simulations with initially individual halos and the two lower ones to simulations with initially common halos. Different symbols correspond to different types of galaxies in the initial group (cf. Athanassoula, Makino & Bosma (1997)), or different ratios of luminous to dark matter, as given in the figure. The dot-dashed lines correspond to the results for oblate spheroidals with values of the anisotropy parameter δ (from left to right) 0., 0.1, 0.2, 0.3 and 0.4.

have low values of the triaxiality parameter T) and show a good alignment between the angular momentum vector and the minor axis. On the other hand remnants which are slow rotators (i.e. have a relatively low values of V/σ) are more spherical-like, have higher values of the triaxiality parameter and do not present a good alignment between the angular momentum vector and the minor axis.

4. Circular velocity profiles

Assuming spherical symmetry, we can define the circular velocity profile from

$$V_c(r) = \sqrt{M(r)/r} \tag{1}$$

where $M(r)$ is the total mass within the radius r. We have calculated this profile for all the merger remnants coming from our simulations and find that it depends on the initial conditions of the simulations. Thus remnants of groups with initially individual halos have circular velocity profiles that decrease fast with radius. Similar profiles can be found for cases where the dark matter was initially very centrally concentrated, or where it was a relatively small contribution to the total mass. On the other hand merger remnants from simulations with initially common halos, a relatively high halo-to-total mass ratio and an initial density profile which is not too centrally concentrated give circular velocity profiles which are rather flat or present two maxima, one near the center, due to the luminous material and the other quite far out, due to the dark matter. If these distinctions carry over to the rotation curves of these remnants then, by comparing them with observed rotation curves, we could get some information on the initial distribution of dark matter in groups and its relative amount.

Acknowledgments We thank A. Bosma for useful discussions and J.-C. Lambert for his help with the administration of the simulations. We also thank the INSU/CNRS and the University of Aix-Marseille I for funds to develop the necessary computing facilities. CV gratefully acknowledges an EEC TMR fellowship, which made this work possible.

References

Athanassoula, E., Bosma, A., Lambert, J.-C., Makino, J. (1997) Performance and accuracy of a GRAPE-3 system for collisionless N-body simulations, *MNRAS*, (in press), astro-ph/9707079.

Athanassoula, E., Makino, J., Bosma, A. (1997) Evolution of compact groups of galaxies I. Merging rates, *MNRAS*, **286**, pp. 825-838.

Barnes, J. (1994) Interactions and mergers in galaxy formation, in *The formation and evolution of galaxies*, eds. C. Muñoz-Tuñón & F. Sánchez, Cambridge univ. press, pp. 399-453.

Barnes, J. Hernquist, L. (1992) Dynamics of interacting galaxies, *Ann. Rev. Astron. Astrophys.*, **30**, pp. 705-742.

Davies, R. (1987) The stellar kinematics of elliptical galaxies, in *Structure and dynamics of elliptical galaxies*, ed. Tim de Zeeuw, Reidel Pub., pp. 63-77.

Franx M., Illingworth G.D. and de Zeeuw T. (1991) The ordered nature of elliptical galaxies: implications for their intrinsic angular momenta and shapes, *Ap. J.*, **438**, 112

Toomre A., Toomre J. (1972) Galactic bridges and tails, *Ap. J.*, **178**, pp 623-666.

Vozikis, Ch. L., Athanassoula, E. (1998), in preparation.

COUNTERROTATION IN GALAXIES

F. BERTOLA AND E.M. CORSINI
Dipartimento di Astronomia, Università di Padova
vicolo dell'Osservatorio 5, I-35122 Padova, Italy

1. Introduction

The phenomenon of counterrotation is observed when two galaxy components have their angular momenta projected antiparallel onto the sky. It follows that if the two components rotate around the same axis, the counterrotation is intrinsic. On the contrary the counterrotation is only apparent if the rotation axes are misaligned and the line-of-sight lies in between the two vectors or their antivectors. In the case of intrinsic counterrotation the two components can be superimposed or radially separated.

As far as the two components are concerned, stars are observed to counterrotate with respect to other stars or gas. The counterrotation of gas versus gas has been also detected. Up to now, the number of galaxies exhibiting these phenomena are ~ 60, with morphological types ranging from ellipticals to S0's and to spirals. Previous reviews about counterrotation are those of Rubin (1994b) and Galletta (1996).

When a second event occurs in a galaxy, such as the acquisition of material from outside, it is likely that the resulting angular momentum of the acquired material is decoupled from the angular momentum of the preexisting galaxy. Counterrotation is therefore a general signature of material acquired from outside the main confines of the galaxy. Good examples of such cases are ellipticals with a dust lane or gaseous disk along the minor axis and polar ring galaxies, where the angular momenta are perpendicular. It should be noted that recently attempts have been made to explain special cases of stars versus stars counterrotation in disk galaxies as due to a self-induced phenomenon in non-axisymmetric potentials (Evans & Collett 1994, Wozniak & Pfenninger 1997).

In the following we discuss the phenomenon according to the morphological type and to the kind of counterrotation.

2. Elliptical Galaxies

2.1. GAS VS. STARS

Disks of ionized gas, which appear as dust lanes crossing the stellar body when seen on edge, have been detected in a large fraction ($\sim 50\%$) of ellipticals (e.g. Macchetto et al. 1996). Typical masses of dust and ionized gas in ellipticals are $M_{\rm dust} = 10^5 - 10^6 M_\odot$ and $M_{\rm HII} = 10^3 - 10^5 M_\odot$ (e.g. Bregman et al. 1992). The kinematical decoupling generally observed (Bertola et al. 1990) between the gaseous and the stellar components suggests the gas is settled or in the process of settling in the equilibrium configurations. In triaxial ellipticals they are the planes orthogonal to the shortest and to the longest axes.

Intrinsic counterrotation of gas and stars in ellipticals has been observed in Anon 1029−459 (Bertola et al. 1988), IC 2006 (Schweizer et al. 1989), NGC 3528 (Bertola et al. 1988), NGC 5354 (Bettoni et al. 1995), NGC 5898 (Bettoni 1984, Bertola & Bettoni 1988), and NGC 7097 (Caldwell et al. 1986).

2.2. STARS VS. STARS

Kinematically decoupled cores are observed in ellipticals when their stellar velocity field show a discontinuity between the rotation of the inner and the outer regions. Isolated cores have been detected in 17 ellipticals (Mehlert et al. 1997 and references therein) and in IC 4889 (Bertola et al. 1992). About half of them are stellar nuclear disks (Carollo et al. 1997).

Stellar counterrotation characterizes IC 1459 (Franx & Illingworth 1988), IC 4889 (Bertola et al. 1992), NGC 1439, NGC 1700 (Franx et al. 1989), NGC 4472 (Davies & Birkinshaw 1988), and NGC 3608 (Jedrzejewsky & Schecther 1988), NGC 4816 (Mehlert et al. 1997), NGC 5322 (Bender 1988), NGC 7796 (Bertin et al. 1996). At least in the first five galaxies the counterrotation is intrinsic as indicated by the lack of velocity gradient along the minor axis. Minor axis observations are not available for the remaining four galaxies.

Several formation scenario have been proposed. They are the dissipationless minor merging with a compact low-luminosity elliptical (Kormendy 1984, Balcells & Quinn 1990), the accretion of gas-rich dwarf companions and subsequent star formation as in dust-lane ellipticals (Franx & Illingworth 1988), the dissipational major merging between an elliptical and a disk galaxy or between two disk galaxies (Schweizer 1990, Hernquist & Barnes 1991), the hierarchical merging of dynamically hot, but still partially gaseous objects (Bender & Surma 1992), and the interaction involving an elliptical with an embedded disk (Hau & Thomson 1994).

2.3. GAS VS. GAS

In the E4 galaxy NGC 1052 two dimensional high resolution Hα spectroscopy has revealed in the central ± 2 kpc two apparently counterrotating ionized gas components (Plana & Boulesteix 1996). The two components are interpreted as produced by two distinct structures superimposed by a projection effect and settled onto the two orthogonal equilibrium planes of a triaxial elliptical galaxy. They have a mass of $M_{\mathrm{HII},1} \sim 10^3\ M_\odot$ and $M_{\mathrm{HII},2} \sim 10^5 M_\odot$ respectively.

The case of the E5 galaxy IC 4889 (Vega et al. 1997, these proceedings) is more complex. In addition to the above mentioned counterrotating core, two gas disks settled onto the allowed planes of a triaxial galaxy apparently counterrotate. Both cases indicate multiple acquisition events.

3. Disk Galaxies

3.1. GAS VS. STARS

A compilation of 36 S0 galaxies with ionized gas based on the samples of Bertola et al. (1992) and of Kuijken et al. (1996) lists 12 objects with gas counterrotating with respect to the stars. The large fraction of S0's with kinematical decoupling between stars and gas is consistent with the idea that the gas in lenticulars is acquired from outside (Bertola et al. 1992). Episodic infall of external gas, continuous infall of external gas and dissipational merging with a gas-rich dwarf companion all have been investigated by means of numerical simulation (Thakar & Ryden 1996, Thakar et al. 1997) as viable mechanisms for producing an overall gaseous counterrotation in disk galaxies. In the infall case the acquisition rate, the initial angular momentum and the gas clumpiness state are crucial parameters to successfully build up the counterrotating gas disk without dynamically heating the preexisting stellar disk. The same is true for the mass of the captured satellite in the merging case.

In the edge-on peanut S0 NGC 128 (Emsellem & Arsenault 1997) a tilted gaseous disk counterrotates with respect to the stars. It has been interpreted as due to acquired gas settled onto the so-called anomalous orbits in a tumbling triaxial potential, as in the case of the barred S0 NGC 2217 (Bettoni et al. 1990) and NGC 4684 (Bettoni et al. 1993).

In the early Sa NGC 3626 (Ciri et al. 1995) the neutral, ionized and molecular gas (Garcia-Burillo et al. 1997), amounting to $\sim 10^9 M_\odot$, counterrotates at all radii with respect to the stars. It should be noted that, in spite of the presence of two faint amorphous dust and possibly stellar spiral arms, the morphology of NGC 3626 resembles that of more typical S0 galaxies.

3.2. STARS VS. STARS

3.2.1. *Stellar Disk vs. Stellar Disk*

The S0 galaxy NGC 4550 has two cospatial counterrotating stellar disks, one of them corotating with the gaseous one (Rubin et al. 1992). The two disks have exponential surface brightness profiles with the same central surface brightnesses and scale lengths (Rix et al. 1992). In the early-type spirals NGC 4138 (Jore et al. 1996) and NGC 7217 (Merrifield & Kuijken 1994) about 20% - 30% of the stars in the disk are in retrograde orbits constituting a kinematically distinct component. Both disks are equally extended but have different levels of surface brightness. In NGC 7217 the gaseous disk rotates in the same direction as the primary (i.e. the more massive) stellar disk. The contrary is true for NGC 4138.

The presence of a counterrotating stellar disk is interpreted as the result of a subsequent stellar formation in an accreted gaseous disk. In the case of NGC 7217 where the ionized gas rotates in the same direction as the more massive disk, Merrifield and Kuijken (1994) suggested that the actual retrograde (i.e. the less massive) disk formed first. An alternative to the second event scenario has been proposed for NGC 4550 by Evans & Collett (1994). They stated that whenever oval disks become circular or triaxial halos in which they are embedded become axisymmetric, then box orbits are scattered equally into clockwise and counterclockwise rotating tube orbits. In this way two identical counterrotating stellar disks can be built. The presence of three gaseous rings in NGC 7217 has suggested to Athanassoula (1996), that in this galaxy the bar has decayed, causing the counterrotation of the less massive stellar disk.

Extended stellar counterrotation in disks of S0s and spirals seems to be a rare phenomenon according to the results of Kuijken et al. (1996). They measured carefully the line-of-sight velocity distribution along the major axes of 28 S0 galaxies, without finding any new case. Indeed they estimated that less than 10% of S0's show large-scale counterrotation with more than 5% of disk stars on retrograde orbits. If counterrotating stars formed from acquired gas, this result contrasts with the large fraction of S0's exhibiting counterrotating disks of ionized gas. However, most of these gas disks are too small to produce the large-scale counterrotation observed in systems like NGC 4138, NGC 4550 and NGC 7217.

The Sa galaxy NGC 3593 (Bertola et al. 1996) is composed of a small bulge, a first, radially more concentrated, stellar disk that contains about 20% of the total luminous mass and dominates the star kinematics in the inner ± 1 kpc, and a second counterrotating stellar disk, radially more extended, dominating the outer kinematics. The two disks have exponential luminosity profiles with different scale lengths and central surface bright-

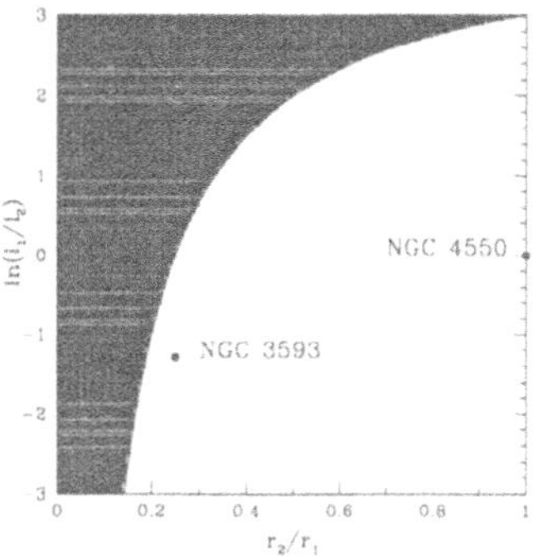

Figure 1. Detectability of two counterrotating stellar disks having exponential luminosity profiles of scale lengths r_1 and r_2 and central intensities I_1 and I_2. We assume to disentangle the two disks if the intensity of the smaller one (disk 2) is at least the 5% of that of the greater one (disk 1) at all radii lower than r_1 (non-hatched area). The dots correspond to the observed cases of NGC 4550 (Rubin et al. 1992, Rix et al. 1992) and NGC 3593 (Bertola et al. 1996).

nesses. The gaseous disk counterrotates with respect to the more extended disk. The existence of cases like NGC 3593, where two disks of different scale lengths and central surface brightnesses counterrotate, suggest that they can be numerous if not a general case. Fig. 1 shows the vast area to be explored in the plane central surface brightness vs. scale length within the limit of present day detectability.

Instabilities in disk galaxies with counterrotating stars and/or gas have been recently investigated by Sellwood & Merritt (1994) and Lovelace et al. (1997). Due to the results to their numerical simulations Comins et al. (1997) suggest that one-armed spiral features may characterize disk galaxies with counterrotation. Spirals with tightly-coiled narrow arms are the candidates for Howard et al. (1997).

3.2.2. *Stellar Bulge vs. Stellar Disk*

In the Sb NGC 7331 (Prada et al. 1996) the bulge counterrotates with respect to the disk. The galaxy appears morphologically smooth and undisturbed, but the analysis of the stellar LOSVD reveals the presence in the inner ± 1.4 kpc of two counterrotating stellar components. Near-infrared photometry shows that the radial surface brightness profile of the slow-rotating component follows that of the bulge, while the fast-rotating follows the disk. Also the Sb spiral NGC 2841 seems to have a counterrotating bulge (Prada 1997, private communication).

Did disk galaxies really start out as 'undressed spheroids' and the disks formed gradually over several several billion of years as suggest by Binney

& May (1986)? To answer this question we need to know how unique are the cases of NGC 2841 and NGC 7331. The absence of a velocity gradient in the central (±1 kpc) regions of the Sa NGC 4698 would suggest the presence of a counter or orthogonally rotating bulge (Bertola et al. 1997, in preparation).

3.2.3. *Stellar Counterrotation in Early-Type Barred Galaxies*

The stellar rotation curve along the major axis of the bar of eleven early-type barred galaxies seen at intermediate inclination show a 'waving pattern'. Nine are listed by Bettoni & Galletta (1997) while two, namely NGC 5005 and NGC 5728 were found by Prada (1997, private communication). The observed amplitude of the oscillations is lower than 30 km s^{-1} producing in the rotation curve a region of minimum, which sometimes can reach negative values.

The phenomenon, which was pointed out by Bettoni (1989), has been recently interpreted by Wozniak & Pfenninger (1997). They used self-consistent models of barred galaxies with reproduce the observed wavy pattern as due to the presence of retrograde orbits with a local concentration in the region of minimum velocity. The origin of such a stellar counterrotation is not necessarily external, as acquired gas can be trapped on this family of retrograde orbits and then eventually form stars. These counterrotating gas and stars could produce an embedded retrograde secondary bar. Friedli (1996) demonstrated that galaxies having two nested and counterrotating bars are stable and long-lived systems. He also predicted the peculiar signatures characterizing their kinematics, which is not yet observed.

3.3. GAS VS. GAS

NGC 4826, a nearby and relatively isolated Sab(s) spiral called 'Devil-Eye' or 'Black-Eye' contains two nested counterrotating gaseous disks (Braun et al. 1992). Radio and optical observations (Braun et al. 1994, Rubin 1994a) revealed an inner disk of about 1 kpc radius containing $\sim 10^7 M_\odot$ in H I and $\sim 10^8 M_\odot$ in H_2 and a counterrotating outer gas disk extending from 1.5 to 11 kpc and containing $\sim 10^8 M_\odot$ in H I. They are coplanar to the stellar disk. Stars corotate with the inner gas, but beyond the dust lane less than 5% of them ($\sim 10^8 M_\odot$) corotate with the outer gas (Rix et al. 1995). The kinematical features of NGC 4826 are interpreted (Rix et al. 1995) considering an original gas-poor galaxy with prograde gas which slowly acquires a comparable mass of external retrograde gas. The new counterrotating gas settles in the outer parts of the stellar disk, leaving undisturbed the galaxy morphology (Walterbos et al. 1994). Another case we interpret as constituted by two counterrotating gas components as in

NGC 4826, is the edge-on S0 NGC 5252 (Held et al. 1992)

The late-type galaxy NGC 253 (Anantharamaiah & Goss 1996) contains in the central $\sim$ 150 pc region three nested structures of ionized gas: a counterrotating inner disk, an orthogonal rotating ring and an outer ring rotating in the same sense as the galactic disk.

In the edge-on S0 NGC 7332 Fisher et al. (1994) detected a second gas component corotating with the stars in addition to the main counterrotating gaseous disk. Recently two dimensional high resolution Hα spectroscopy (Plana & Boulesteix 1996) has revealed that the two ionized gas counterrotating structures are apparently superimposed. They extend to about 4 kpc and contain $\sim 10^5 M_{\odot}$ in H II. The accretion of newly supplied counterrotating gas onto an existing corotating gas disk investigated by Lovelace & Chou (1996) could be a possible formation scenario for NGC 7332.

4. Conclusions

Counterrotation occurs in a wide variety of forms (gas vs. stars, stars vs. stars, gas vs. gas). It is present in galaxies of different morphological types, ranging from ellipticals to early-type disk galaxies.

The counterrotation of stellar vs. stellar disks is the type of counterrotation we expect to be prevailing, since it is the end result of gas vs. stars counterrotation. Therefore its frequency could be very high if the area described in Fig. 1 is carefully inspected, using the state of the art analysis of the shape and asymmetries of the line profiles of the absorption lines. This would indicate that acquisition and merging events are common phenomena in the history of galaxies.

We are grateful to D. Burstein for useful comments on the manuscript.

References

Anantharamaiah, K.R., Goss, W.M. 1996, ApJ, 466, L13
Athanassoula, E. 1996, In: R. Buta, D.A. Crocker, B.G. Elmegreen (eds.) ASP Conf. Ser. 91, "Barred Galaxies". ASP, San Francisco, p. 309
Balcells, M., Quinn, P.J. 1990, ApJ, 361, 381
Bender, R., 1988, A&A, 202, L5
Bender, R., Surma, P. 1992, A&A, 258, 250
Bertin, G., Bertola, F., Buson, L.M., Danziger, I.J., Dejonghe, H., Sadler, E.M., Saglia, R.P., de Zeeuw, P.T., Zeilinger, W.W. 1994, A&A, 292, 381
Bertola, F., Bettoni, D. 1988, ApJ, 329, 102
Bertola, F., Bettoni, D., Buson, L.M., Zeilinger, W.W. 1990, In: R. Wielen (ed.) "Dynamics and Interactions of Galaxies". Springer-Verlag, Berlin, p. 249
Bertola, F., Buson, L.M., Zeilinger, W.W. 1988, Nature, 335, 705
Bertola, F., Buson, L.M., Zeilinger, W.W. 1992, ApJ, 401, L79
Bertola, F., Cinzano, P., Corsini, E.M., Pizzella, A., Persic, M., Salucci, P. 1996, ApJ, 458, L67
Bettoni, D. 1984, The Messenger, 37, 17

Bettoni, D. 1989, AJ, 97, 79
Bettoni, D., Galletta, G. 1997, A&AS, 124, 61
Bettoni, D., Buson, L.M., Maira, L., Bertola, F. 1995, In: Richter O.G., Borne K. (eds.) ASP Conf. Ser. 70, "Groups of Galaxies". ASP, San Francisco, p. 95
Bettoni, D., Fasano, G., Galletta, G. 1990, AJ, 99, 1789
Bettoni, D., Galletta, G., Sage, L.J. 1993, A&A, 280, 121
Binney, J.J., May, A. 1986, MNRAS, 218, 743
Braun, R., Walterbos, R.A.M., Kennicutt Jr., R.C. 1992, Nature, 360, 442
Braun, R., Walterbos, R.A.M., Kennicutt Jr., R.C., Tacconi, L.J. 1994, ApJ, 420, 558
Bregman, J.N., Hogg, D.E., Roberts, M.S. 1992, ApJ, 387, 484
Caldwell, N., Kirshner, R.P., Richstone, D.O. 1986, ApJ, 305, 136
Carollo, C.M., Franx, M., Illingworth, G.D., Forbes, D.A. 1997, ApJ, 481, 710
Ciri, R., Bettoni, D., Galletta, G. 1995, Nature, 375, 661
Comins, N., Lovelace, R.V.E., Zeltwanger, T., Shorey, P. 1997, ApJ, 484, L33
Davies, R.L., Birkinshaw, M. 1988, ApJS, 68, 409
Emsellem, E., Arsenault, R. 1997, A&A, 318, L39
Evans, N.W., Collett, J.L. 1994, ApJ, 420, L67
Fisher, D., Illinghworth, G.D., Franx, M. 1994, AJ, 107 160
Franx, M., Illinghworth, G.D. 1988, ApJ, 327, L55
Franx, M., Illinghworth, G.D., Heckmann, T. 1989, ApJ, 344, 613
Friedli, D. 1996, A&A, 312, 761
Galletta, G. 1996, In: R. Buta, D.A. Crocker, B.G. Elmegreen (eds.) ASP Conf. Ser. 91, "Barred Galaxies". ASP, San Francisco, p. 429
Garcia-Burillo, S., Sempere, M.J., Bettoni, D. 1997, ApJ, submitted
Hau, G.K.T., Thomson, R.C. 1994, MNRAS, 270, L23
Held, E.V., Capaccioli, M., Cappellaro, E. 1992, In: S.S. Holt, S.G. Neff, C.M. Urry (eds.) AIP Conf Proc. 254 "Testing the AGN Paradigm". AIP, New York, p. 613
Hernquist, L., Barnes, J.E. 1991, Nature, 354, 210
Howard, S., Carini, M.T., Byrd, G.G., Lester, S. 1997, AJ, submitted
Jedrzejewsky, R., Schecther, P.L. 1988, ApJ, 330, L87
Jore, K.P., Broelis, A.H., Haynes, M. 1996, AJ, 112, 438
Kormendy, J. 1984, ApJ, 287, 577
Kuijken, K., Fisher, D., Merrifield, M.R. 1996, MNRAS, 283, 543
Lovelace. R.V.E., Chou, T. 1996, ApJ, 468, L25
Lovelace, R.V.E., Jore, K.P., Haynes, M.P. 1997, ApJ, 475, 83
Macchetto, F., Pastoriza, M., Caon, N., Sparks, W.B., Giavalisco, M., Bender, R., Capaccioli, M. 1996, A&AS, 120, 463
Mehlert, D., Saglia, R.P., Bender, R., Wegner, G. 1997, A&A, submitted
Merrifield, M.R., Kuijken, K. 1994, ApJ, 432, 575
Plana, H., Boulesteix, J. 1996, A&A, 307, 391
Prada, F., Gutierrez, C.M., Peletier, R.F., McKeith, C.D. 1996, ApJ, 463, L9
Rix, H.-W., Franx, M., Fisher, D., Illingworth, G. 1992, ApJ, 400, L5
Rix, H.-W., Kennicutt Jr., R.C., Braun, R., Walterbos, R.A.M. 1995, ApJ, 438, 155
Rubin, V.C. 1994a, AJ, 107, 173
Rubin, V.C. 1994b, AJ, 108, 456
Rubin, V.C., Graham, J.A., Kenney, J.D.P. 1992, ApJ, 394, L9
Schweizer, F. 1990, In: R. Wielen (ed.) "Dynamics and Interactions of Galaxies". Springer-Verlag, Berlin, p. 60
Schweizer, F., van Gorkom, J.H., Seitzer, P. 1989, ApJ, 338, 770
Sellwood, J.A., Merritt, D. 1994, ApJ, 425, 530
Thakar, A., Ryden, B. 1996, ApJ, 461, 55
Thakar, A., Ryden, B., Jore, K.P., Broeils, A.H. 1997, ApJ, 479, 702
Walterbos, R.A.M., Braun, R., Kennicutt Jr., R.C. 1994, AJ, 107, 184
Wozniak, H., Pfenninger, D. 1997, A&A, 317, 14

MAKING SPIRALS WITH COUNTER-ROTATING DISKS

The case of NGC 4550

D. PFENNIGER
Geneva Observatory, University of Geneva
CH-1290 Sauverny, Switzerland

Abstract. A single merger scenario for making galaxies such as NGC 4550 possessing equal coplanar counter-rotating stellar disks is investigated by collisionless N-body technique. The scenario is successful in producing an axisymmetric disk made of two almost equal counter-rotating populations. The final disk shows a clear bimodal line profile in the outer part, which demonstrates that disk-disk mergers do not always produce ellipticals.

1. The Puzzle of NGC 4550

The galaxy NGC 4550, an E7/S0 lenticular galaxy in the Virgo Cluster, was discovered by Rubin et al. (1992) to be a galaxy consisting of two coeval, coplanar, and counter-rotating stellar disks. Many other cases of counter-rotation are now known in ellipticals and also in spirals, but what makes the case of NGC 4550 particular (Kuijken, Fisher, & Merrifield 1996) is that the mass ratio of the counter-rotating disks is nearly 1/1. In the case of ellipticals counter-rotation is so frequent (e.g. Schweizer 1998) that the merger or accretion origins seem the most likely explanations for such kinematic misalignments.

The difficulty with a merger scenario for NGC 4550 is that strong disk mergers usually lead to ellipticals and the destruction of the disks, a "truth" often believed to be general since the seminal paper of Toomre & Toomre (1972). Thakar & Ryden (1996) have shown that over several Gyr a series of well correlated small *gaseous* merger events can lead to a massive counter-rotating gaseous disk. But then this scenario requires to preserve special correlations over several Gyr, contrary to a single event. Thus, it remains open whether a single merger of two equal mass spirals can result in some circumstances to a galaxy like NGC 4550.

In a forthcoming paper (Pfenniger & Puerari 1998) we will present in more details the general conditions leading to moderate heating. In addition to mass and energy, one must also consider the angular momentum in the final budget. An input of specific angular momentum tends to cool the system, while inputs of specific energy and mass tend to heat it.

2. Simulations

2.1. MERGER SCENARIO

Since NGC 4550 is a rare object, a generic process is not required; exceptional conditions can be acceptable and should be expected. After various considerations we have retained: **1.** A nearly circular orbit of the initial disks, supposing that the excess energy of galaxies coming from infinity has already been absorbed by some outer matter. **2.** Initial disks with opposite spins (see Fig. 1). Counter-rotating disks are ideal for minimizing shocks in outer gaseous disks. Also the retrograde disk is less affected by tidal interactions. **3.** Nearly or exactly coplanar disks in the orbital plane. This may look a rather improbable situation, but a favorable factor to align the disk spins is to have initially flattened dark matter (the torques on misaligned disks is then high). Several arguments, such as the high frequency of warped outer HI disks with straight line of nodes, do suggest flattened dark mass distributions in spirals (see Pfenniger & Combes 1994).

2.2. INITIAL CONDITIONS AND RUNS

The simulations presented here were run in Geneva. Independently, Puerari run similar simulations in Mexico. We intend to publish jointly both sets of simulations since they lead to the same conclusions despite different choices of initial parameters. In particular Puerari's simulations include round massive halos, showing that the shape of the halos does not change the qualitative results, and his disks start on more elongated orbits.

In view of the poor understanding of gaseous processes in galaxies leading to exaggerate viscosity with the SPH technique (see Thakar & Ryden 1996), we simulate first only the collisionless gravity part. Obviously the resulting heating is enhanced with respect to simulations including gas.

Our two initial disks are identical, except for the initial coordinates. They consist of a bulge, with a scale-length of 1 kpc, an exponential disk with horizontal and vertical scale-lengths of 3 and 0.5 kpc, and a flaring ($h_z = 0.03R$) massive "dark" disk with a constant surface density up to 10 kpc, decreasing as R^{-1} between 10 and 30 kpc, and as R^{-2} between 30 and 60 kpc. The mass ratios of the three components are respectively (0.25:1:5), and the particle masses are all equal. The particle velocities

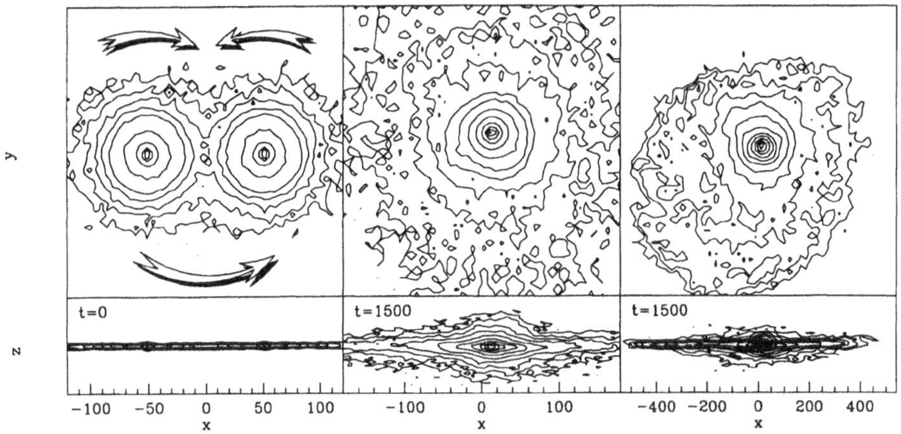

Figure 1. Merging of coplanar disks. The particle density is shown with iso-contours separated by 2 mag. The arrows indicate the disk spins and the disk sense of rotation.
Left: Initial conditions at t=0 Myr.
Middle: After the disk merging (t=1500 Myr), the remnant disk with counter-rotation.
Right: Large-scale view of the particle distribution with marked asymmetries.

are found by solving the Jeans equation for each component in the total potential, assuming velocity dispersions ellipsoids parallel to cylindrical coordinates. The potential of the mass distribution is calculated with a set of $1.25 \cdot 10^6$ particles on the Geneva PM polar grid code.

A rapid run of coplanar counter-rotating disks merger is performed with the PM code with $2.5 \cdot 10^6$ particles. The result is positive but not fully convincing since during the merging process substantial changes of positions of the mass distribution occur within the grid, which has a position dependent resolution. Therefore, we select out a subset of $1.25 \cdot 10^5$ particles, the mass of which is multiplied by a factor 20 to keep the same total mass. The particle subset is then run with the Barnes-Hut (1989) TREECODE (with an opening angle of θ=0.5), which does not imply any geometrical assumptions. In several experiments we verify that slight initial disk inclinations with the orbital plane (5–10°) still lead to counter-rotating coplanar disks.

2.3. RESULTS

Here we just describe the strictly coplanar disk merger. Fig. 1, left, shows the initial particle distribution with the senses of rotation. The disks are on a prograde near circular orbit. The tidal perturbation creates immediately (at ~200 Myr) a bar in each disk, persisting until the disks merge. The merging process conserves the disks fairly well, the less damaged one being the retrograde one. Fig. 1, middle, shows the inner remnant disk at t=1500 Myr, still containing 80% of the initial total mass, almost circular and barless. Fig. 1, right, shows the large-scale particle distribution, at

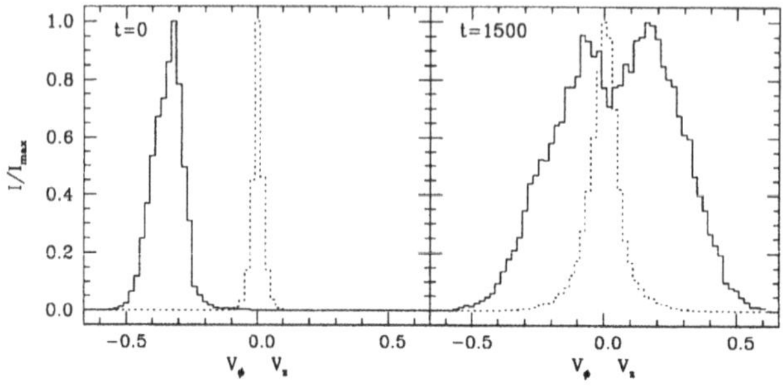

Figure 2. Tangential (solid) and vertical (dotted) line profiles, averaged in the interval 30<R<100.
Left: In the initial retrograde disk, **Right:** In the final remnant.
The final bimodal tangential velocity dispersion is clearly visible.

t=1500 Myr. The excess angular momentum is transported by 10-20% of the mass, mostly of the prograde disk, to large distances, 100–500 kpc.

Fig. 2 shows how the average line profiles as seen in the edge-on disks change from unimodal to bimodal. The bimodal distribution is crucial in order to distinguish counter-rotating disks from a hot population with a zero net rotation. The ratios of final to initial velocity dispersion amount to ~2.4 in the radial direction, and ~2 in the vertical direction. The final mass distribution follows well a $R^{1/4}$ profile. With a relatively limited heating the final disk resembles both in shape and kinematics a typical S0/E7 galaxy, with, in addition, marked counter-rotating populations.

At some moments during the merging process the system as seen edge-on looks like a single galaxy but with two bulges. Such a case of "two-bulge" looking galaxy is PGC 57064 in the Hercules cluster.

3. Conclusions

We have shown that counter-rotating co-spatial disks can be made by a single spiral-spiral merger. The required initial conditions are somewhat peculiar, but favored if the dark matter distribution is flat; however, this is not a necessary condition. The strong reaction of the disks, expelling 10–20% of the mass to 100–500 kpc, means that dark matter in such systems must be distributed in a pronounced asymmetric way for several Gyr. Thus, in many spirals the outer (dark) mass distribution should still be chaotic.

References

Barnes J.E., Hut P. 1989, ApJS 70, 389
Kuijken K., Fisher D., Merrifield M.R. 1996, MNRAS 283, 543
Pfenniger D., Combes F. 1994, A&A 285, 94
Pfenniger D., Puerari I. 1998, in preparation
Rubin V.C., Graham J.A., Kenney J.D., 1992 ApJ 394, L9-L12
Schweizer F. 1998, in Galaxies: Interactions and Induced Star Formation, Saas-Fee Advanced Course 26, Friedli D., Martinet L., Pfenniger D. (eds.), Springer, p. 105
Thakar A.R., Ryden B.S. 1996, ApJ 461, 55
Toomre A., Toomre J. 1972, ApJ 178, 623

STELLAR COUNTER-ROTATION ALONG THE HUBBLE SEQUENCE: A PROBE FOR GALAXY FORMATION SCENARIOS

F. PRADA
Instituto de Astrofísica de Canarias
Present address: Instituto de Astronomía, UNAM
Apdo. Postal 877, Ensenada, Baja California, C.P. 22830, Mexico

AND

C.M. GUTIÉRREZ
Instituto de Astrofìsica de Canarias
38200, La Laguna, Tenerife, Spain

Abstract. We present some preliminary results of an on-going project to investigate the phenomenon of counter-rotation in galaxies. The analysis of two Sb and two barred galaxies shows the presence of stellar counter-rotating components in the central 1 to 3 kpc of each galaxy. This, along with similar structures found in ellipticals and early-type spirals, demonstrate that the counter-rotation extends along the Hubble sequence. The physical origin of these features is briefly discussed stressing the possible relevance of this phenomenon in understanding the general problem of galaxy formation.

1. Introduction

It is now well established that some elliptical galaxies contain counter-rotating cores (Franx & Illingworth 1988; Rix & White 1992; Bender et al. 1994), and S0/Sa's can display extended retrograde motions in their discs (Rubin et al. 1992; Rix et al. 1992; Bertola et al. 1996). However very little is known about late type spirals and barred galaxies (see Bertola for a review). The study by Prada et al. (1996) reveals the presence of a retrograde bulge in the nearby Sb galaxy NGC 7331 and no signs of extended counter-rotating components have been found in barred systems. This presentation is focussed on some observational results on the stellar counter-rotation in

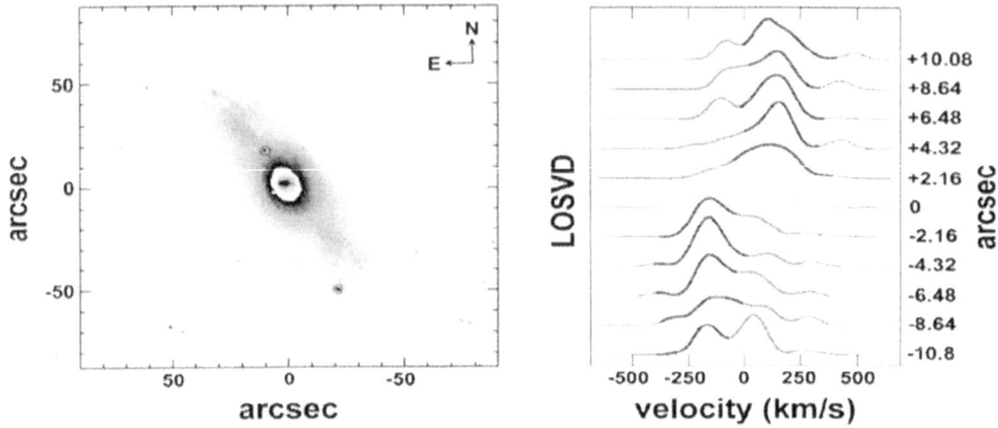

Figure 1. Left: a greyscale I-band image of the barred galaxy NGC5728; Right: the LOSVD along the main bar major axis of NGC 5728

galaxies. The main aims of this project are to study the dynamical properties of retrograde components, to understand the formation mechanism of stellar counter-rotating components along the Hubble sequence.

2. The observations, the algorithm and the parametrization of LOSVD

For a sample of two Sb galaxies (NGC 7331, NGC 2841) and two barred galaxies (NGC 5728 SBa; NGC 5005 SBbc) long-slit observations were performed with the ISIS spectrograph on the 4.2m WHT telescope at La Palma. The resolution was 25 and 50 km s^{-1} (depending of the object observed) in the spectral region around the CaII IR triplet (~8550 Å). The slit was placed along the major axis of the spirals and along the main bar major axis in the barred galaxies. More details of the observations and data reduction will be presented in future publications. The stellar line-of-sight velocity distributions (LOSVD) were determined from the long-slit CaII triplet absorption spectra by means of a two-dimensional unresolved Gaussian decomposition algorithm. This is the method used to determine the stellar kinematical structure of NGC 7331 (see Prada et al. 1996). Details about the algorithm can be found in that paper and in Prada et al. 1998.

The LOSVDs of these galaxies clearly shows two distinct components that can be well-parametrized by two Gaussians (see Fig. 1). The parameters of each Gaussian gives for each position in the galaxy recession velocities, dispersions and relative flux of the two components. The comparison of the relative flux with the photometry of the galaxy allows us to identify kinematical with photometric structures (see Figure 2).

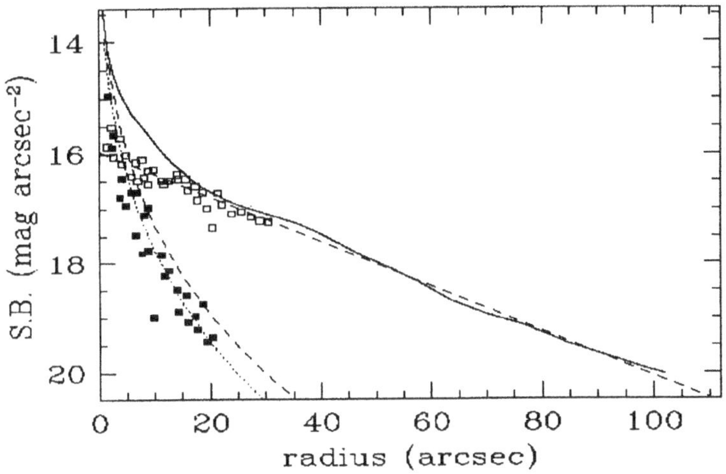

Figure 2. The major axis surface brightness profile of NGC 7331. The profile has been decomposed into an exponential disk and an $r^{1/4}$ component. We have also plotted the relative fluxes of the direct component (open symbols) and the counter-rotating component (dark symbols). Notice that the points of the retrograde component follows well the profile of the bulge and the direct component data agree with the disk profile.

3. Discussion and conclusions

The four galaxies presented here show extended counter-rotation in the central 1 to 3 kpc. The two Sb galaxies show a boxiness and a position angle twist of $\sim$ 10-15° which makes very likely that we are seeing a triaxial body in projection (e.g. Statler 1991). The retrograde components can be fairly warm ($v/\sigma \sim 0.5 - 1$) and boxy. They can be associated with the bulge (e.g. NGC 7331) and/or with the centre of a non-axisymmetric structure in the galaxy (e.g. NGC 5728). These results along with the previous work on the ellipticals and early-type spirals demonstrate that the counter-rotation is a phenomenon common along the Hubble sequence. Having established this, next question is to find out the origin of this. Two basic mechanisms have been proposed; the first involves accretion or mergers. For instance the counter-rotating cores in elliptical galaxies have been explained as a merger with a small dwarf galaxy (e.g. Balcells & Quinn 1990), while in this scenario the counter-rotating disks could be produced by accretion of a satellite with a retrograde orbit, however numerical simulations show that in most of the cases this will heat up the disk (see Thakar & Ryden 1996). The second kind of mechanisms involves only the internal evolution of the galaxy. This includes the process of galaxy formation itself from primordial clouds. The problem here is that it requires an efficient mechanism to change the angular momentum drastically as a function of radius. Dynamical instabilities like the disruption of a bar or the distribution of retrograde orbits in the bar (e.g. Wozniak & Pfenniger 1997), infall of gas (Thakar & Ryden 1996) and structures formed later from processed gas (Bender & Surma 1992) have been proposed as well. These scenarios have to explain why the relative fraction of rotating/counter-rotating stars in galaxies like NGC 4550 and NGC 7217 is so constant.

Finally, we would like to finish enphasizing that:

1/ The phenomenon of stellar counter-rotation seems to be present throughout the Hubble sequence.

2/ These retrograde components display a wide variety of dynamical properties and morphology.

3/ The physical origin of counter-rotation remains unclear, being needed better observations (2D spectroscopy), statistics, and more plausible models which can form such retrograde features along the Hubble sequence.

We are grateful to Tim de Zeeuw for discussions and useful comments on the text. This research is based on observations obtained at the WHT, operated by the Isaac Newton Group of telescopes at the Observatorio del Roque de los Muchachos at La Palma of the Instituto de Astrofísica de Canarias.

References

Balcells M., & Quinn, P. J., (1990), *ApJ*, **361**, 381
Bender, R., & Surma, P., (1992), *A&A*, **258**, 250
Bender, R., Saglia, R. P., & Gerhard, O. E. (1994), *MNRAS*, **269**, 785
Bertola, F., (1997), this meeting
Bertola, F., Cinzano, P., Corsini, E. M., Pizzella, A., Persic, M., & Salucci, P. (1996), *ApJ*, **458**, L67
Franx, M., & Illingworth, G., (1988), *ApJ*, **327**, L55
Prada, F., Gutiérrez, C. M., Peletier, R. F., & McKeith, C.D., (1996), *ApJ*, **463**, L9
Prada, F., Gutiérrez, C. M., & McKeith, C. D. 1998 ApJ (in press)
Rix, H.-W., & White, S. D. M., (1992), *MNRAS*, **254**, 389
Rix, H.-W., Franx, M., Fisher, D., & Illingworth, G., (1992), *ApJ*, **400**, L5
Rubin, V., Graham, J., & Kenney, J., (1992) , *ApJ*, **394**, L9
Statler, T. S. (1991), *ApJ*, **102**, 882
Thakar. A, & Ryden. B., (1996), *ApJ*, **461**, 55
Wozniak, H. & Pfenniger, D., (1997) *A&A*, **317**, 14

SHELLS, RIPPLES AND TAILS

D. CARTER
Liverpool John Moores University
Astrophysics Research Institute, Byrom Street, Liverpool, L3 3AF, UK.

1. Introduction

It has been known since the early simulations of Wright (1972), and Toomre & Toomre (1972), that interactions between galaxies can give rise to quite spectacular morphological features, including spiral structure in disk galaxies, and extensive tails. It appears that long tidal tails only arise from interactions involving disk galaxies (Toomre & Toomre 1972), and thus the presence of two opposed long tidal tails in a number of disturbed galaxies such as the Antennae (Whitmore & Schweizer 1995), NGC 3921 (Schweizer 1996), and NGC 7252 (Schweizer 1982), has led to the interpretation of these galaxies as disk-disk mergers in progress. These systems can be modeled rather successfully, as the work of Hibbard & Mihos (1995) has shown.

2. Shells and Ripples

The features that we know as shells were remarked upon by Arp (1966) in some peculiar elliptical galaxies such as NGC 474. They were found in large numbers by D.F. Malin (Malin 1979; Malin & Carter 1980, 1983) who had developed the techniques of photographic amplifaction and unsharp masking of photographic plates, which he applied to the plate collection of the UK Schmidt telescope at Siding Springs. Similar features were noted by Schweizer (1980) in the envelope of the peculiar elliptical NGC 1316, Schweizer used the term "ripples" to describe them. Malin & Carter (1983) found that some 10% of isolated elliptical galaxies have shells; Schweizer & Seitzer (1988) find rather higher percentages, using 4 meter rather than Schmidt telescopes, and they find shells around some S0 galaxies as well.

A striking feature of the most spectacular shell ellipticals, such as NGC 3923 and NGC 1344, later classified as Type I shell ellipticals by Prieur

(1987, 1988) is the alternating appearance of the outer shells. These shells are aligned along the major axis, the outermost is on one side of the nucleus, the next one in on the opposite side, the third on the same side as the first and so on. This feature provided a most important clue to their origin. This alternating appearance breaks down at small radii for systems of many shells, as pointed out by Thomson (1991).

2.1. MERGER MODELS

Quinn (1982, 1984) proposed that shells result from mergers of a low mass galaxy with a giant elliptical. The shells result from phase wrapping in the potential of the elliptical. Quinn's original simulations were of radial mergers of a cold disk component with an elliptical potential, which gave rise to features rather like Type I shells.

These models were developed further by Dupraz & Combes (1986) and Hernquist & Quinn (1987, 1988, 1989) who show that shells can be formed in encounters with a variety of impact parameters, impact energies and progenitors. Indeed they can be formed by mass transfer in encounters which do not involve mergers. If the impact parameter of the encounter is large, then the shell system is formed by spatial wrapping around the nucleus, rather than phase wrapping, and has an "all around" appearance, represented in nature by NGC 474 and NGC 2865 which are classified as Type II shell systems by Prieur (1988). Dupraz & Combes (1986) investigated mergers in non-spherical potentials with a variety of shapes, and find that prolate potentials can focus the shells along the major axis, so that a prolate potential will form a Type I shell system with a greater range of impact angles than an oblate potential. The discovery of minor axis rotation in NGC 3923, the prototype Type I shell elliptical (Carter, Thomson & Hau 1997) supports a prolate potential for this galaxy and thus to some extent the merger model for shell formation in this galaxy.

The simulations of Dupraz & Combes (1986) suggest that there should be a few oblate galaxies with shells similar to Type I shells aligned along the minor axis, although this geometry does require a very particular impact angle. NGC 3051 (Malin & Carter 1983) may be an example of such a galaxy, although the geometry of this system changes with radius.

The merger model for shell formation has problems explaining some features, including the high surface brightness of the inner shells, the existence of shells deep in the potential of some galaxies, and the unexpected bi-symmetry of the innermost shells in NGC 3923 (Prieur 1990). Dupraz & Combes (1987), using the Chandrasekhar approximation, estimate the effect of dynamical friction and claim that dynamical friction can account for the small radii of the innermost shells. The applicability of the Chandrasekhar approximation is in some doubt, and a more detailed calculation of the effect of dynamical friction in shell formation is required urgently.

2.2. MAJOR MERGERS

Schweizer (1980), following a suggestion of A. Toomre, proposed that shells could be formed in near equal mass disk-disk mergers, giving rise to elliptical remnants. Indeed galaxies such as NGC 7252 (Hibbard & Mihos 1995) show shells which can be modeled very successfully by disk-disk mergers. NGC 3921 (Schweizer 1996) is another merger remnant which appears to show shells. Moreover Balcells (1997) finds two faint opposed tails, usually considered as a signature of a disk-disk merger, in the shell elliptical NGC 3656. This type of merger is probably only capable of making Type II shell systems, the issue of whether or not it can make an elliptical galaxy is addressed by Bender (1995) and many others.

2.3. THE WEAK INTERACTION MODEL

An alternative model of shell formation was proposed by Thomson and Wright (1990) and further developed by Thomson (1991), motivated by the problems with the merger model discussed above. According to their model the shells are formed in a thick disc component of an oblate galaxy during a mildly hyperbolic interaction with another galaxy. The passage of this galaxy excites a one-armed spiral density wave in the thick disc component. When viewed edge-on, this density wave naturally reproduces the interleaved pattern seen along the major axis in Type I shell galaxies. When viewed face-on or obliquely the wave forms a Type II shell galaxy. In this model the difference between Type I and Type II shell systems is one of viewing angle, whereas in the merger models it is driven by the shape of the primary potential and the impact parameter. One advantage of this model is that it has no trouble in producing the inner shells. A weakness of the weak interaction model is that there must be a thick disk component in the first place. This model does not require a rotating galaxy however, only a rotating component in which the density wave is excited.

The merger and weak interaction models make different predictions about the colours, distribution and velocities of the shells. The weak interaction model predicts that the shells will have the same colour as the body of the galaxy at that radius, whereas in the merger model the colour will depend upon the colours of the secondary progenitor. Fort et al. (1986) find shell colours somewhat bluer than the main body of the galaxy in two out of three galaxies studied. Turnbull (1997) presents the first results from a programme of much more accurate photometry of shells and the underlying galaxies.

The most clear-cut discriminant between these two models would be a measurement of the velocity of the material in the shells in a Type I shell system. In the interaction model these are composed of stars in a rotating thick disk component, even in an underlying galaxy which is not rotating. In the merger model they are an expanding density wave, and in a Type I system their velocity will be in the plane of the sky. This measurement has proved too difficult to make to date, owing to the low surface brightness of the shells and their low contrast against the underlying galaxy.

2.4. OTHER MODELS OF SHELL FORMATION

A rather different model for the formation of shells was proposed by Fabian, Nulsen & Stewart (1980) and expanded by Bertschinger (1985) and Williams & Christiansen (1986). In this picture shells are formed as a result of star formation in shocked gas in the interstellar medium of the galaxy, the shocks being caused by outflow from the nucleus. With the measurement of shell colours, which are not much bluer than the main body of the galaxy, the failure to detect either ionized or neutral gas associated with the shells except in a very few cases, and the discovery of the interleaving of shells on opposite sides of the nucleus in many shell galaxies, this picture has fallen into disfavor, although Lowenstein et al. (1987) reconcile this model with the last of these observations by proposing that the shells are composed of stars formed off-centre in an elliptical potential as a result of a blast wave, and that these stars then phase-wrap forming the same structures as in the merger models. The lack of very blue colours in the shells still argues against this model.

3. The Fate of Gas in Interactions

3.1. NUCLEAR GAS AND STARBURSTS

Gas of course behaves differently from stars in interactions, and when gas collects regions of star formation occur. Hernquist and Weil (1992) use a hybrid N-body and hydrodynamic code to model the behavior of a gas disk in a disk galaxy which merges with a much more massive elliptical. The gas collects quickly in a disk or ring in the centre of the potential well, whilst the stars form shells over a longer period. This explains the finding of Carter et al. (1988), that some 10% of shell ellipticals show strong Balmer absorption lines in their nuclei. Galaxies such as 0140−658 and 1241−339 have spectra similar to the most extreme of the "E+A" galaxies in distant clusters (Dressler & Gunn 1982; Zabludoff 1997). Lavery and Henry (1988) argued that interactions between galaxies were a major cause of this phenomenon in distant clusters.

Hau, Carter & Balcells (1998) examine the structure and nature of a starburst induced by a merger in NGC 2865 (Fig. 1), a Type II shell elliptical and a mild example of an "E+A" galaxy (Carter et al. 1988, Bica & Alloin (1987). This galaxy has a rapidly rotating central disk containing both young, metal-rich stars and gas. The starburst in this galaxy appears to be 0.4–1.7 Gyr old, rather older than the dynamical age of the shells. NGC 2865 appears to be the result of a merger between a gas-rich spiral galaxy and another galaxy, probably an elliptical, of 2 or 3 times the mass of the spiral. The metal abundance inferred for the starburst is higher, by a factor of between 2.5 and 5, than that of the underlying elliptical. This kind of abundance is not characteristic of a gas rich spiral, but it is possible to get around this problem if the star formation episode takes place over a period of time, so that some enrichment can occur.

3.2. GAS AT LARGE RADII

Schiminovich et al. (1994,1995), and Schiminovich (1997) find HI at large radii associated with NGC 5128, NGC 2865 and NGC 474. Curiously, in NGC 5128 and NGC 2865 the HI forms a ring or shell like structure displaced *outwards* from the optical shells. This is hard to explain in the merger and phase wrapping picture of shell formation, but may result from gas already at large radii in the secondary galaxy, which is stripped very early in the formation of a type II shell system (which all of these are) by spatial wrapping.

4. Kinematic Shells and Kinematically Decoupled Cores

In a few galaxies, such as NGC 7626 (Balcells & Carter 1993) kinematic shells can be seen at small radii. These are dynamically cold velocity anomalies, as if a sheet of cold material is wrapped around the nucleus. These would seem to be late stages in the formation of Kinematically Decoupled Cores, another signature of a recent merger in an elliptical (Bender 1990). There is a strong connection between shells and KDCs, Forbes (1992) finds that all of the 9 well established KDCs and a further 4 out of the 6 "possible KDCs" in his sample possess shells.

NGC 474 (Balcells, Hau & Carter 1998) is particularly interesting in that the velocity profiles are double-peaked near the nucleus, suggesting that the velocity structure of the components has survived. In this case both velocity components are comparatively narrow ($\sigma < 150\,\mathrm{km/s}$), possibly reflecting the nature of the progenitors. NGC 474 shows photometric shells, kinematic shells, double velocity peaks and a rapidly rotating kinematically decoupled core, and is a particularly interesting and complicated case.

Figure 1. NGC 2865 - A type II shell elliptical with a mild E+A spectrum in its nucleus. This is a deep R band CCD image from the prime focus of the Anglo-Australian Telescope

5. Future Developments

Observationally the measurement of accurate colours and velocities of the material in the shells is of vital importance, the former is feasible with modern detectors, the latter will probably be a project for the next generation of ground-based telescopes, and even then only with extreme care in the background subtraction. Explanation of the inner shells and kinematic shells in galaxies such as NGC 474 will require a more detailed treatment of the effects of dynamical friction during shell formation, or if this doesn't work then we may be required to examine alternatives to the merger model, such as Thomson's (1991) weak interaction model, in more detail. Whilst there remains some doubt about whether mergers are required to explain shells, long tails are clear signatures of mergers involving disk galaxies. The relationship between these phenomena is vital to our understanding of the origin of shells.

References

Arp, H.C. (1966) *Ap.J. Suppl.* **14**, 1.
Balcells, M. (1997) *Preprint.*
Balcells, M., & Carter, D., (1993) *Astron. Astrophys.* **279**, 376.
Balcells, M., Hau, G.K.T., & Carter, D. (1998) in preparation.
Bertschinger, E. (1985) *Ap.J suppl.* **58**, 39.
Bender, R. (1990) *"Dynamics and Interactions of Galaxies"* ed. R. Wielen, Springer-Verlag, Berlin, p232.
Bender, R. (1995) *"New Light on Galaxy Evolution"* eds. R. Bender & R.L. Davies, Kluwer, Dordrecht, p181.
Bica, E., & Alloin, D. (1987) *Astron. Astrophys. suppl.* **70**, 281.
Carter, D., Prieur, J.-L., Wilkinson, A., Sparks, W.B., & Malin, D.F. (1988) *M.N.R.A.S.* **235**, 813.
Carter, D., Thomson, R.C., & Hau, G.K.T. (1997) *M.N.R.A.S.* in press.
Dressler, A., & Gunn, J.E. (1982) *Ap.J.* **229**, 42.
Dupraz, Ch., & Combes, F. (1986) *Astron. Astrophys.* **166**, 53.
Dupraz, Ch., & Combes, F. (1987) *Astron. Astrophys.* **185**, L1.
Fabian, A.C., Nulsen, P.E.J., & Stewart, G.C. (1980) *Nature* **287**, 613.
Forbes, D., (1992) *Ph.D. Thesis, Cambridge University.*
Fort, B.P., Prieur, J.-L., Carter, D., Meatheringham, S.J., & Vigroux, L. (1986) *Ap.J.* **306**, 110.
Hau, G.K.T., Carter, D., & Balcells, M. (1997) *M.N.R.A.S* submitted.
Hernquist, L., & Quinn, P.J. (1987) *Ap.J.* **312**, 1.
Hernquist, L., & Quinn, P.J. (1988) *Ap.J.* **331**, 682.
Hernquist, L., & Quinn, P.J. (1989) *Ap.J.* **342**, 1.
Hernquist, L., & Weil, M. (1992) *Nature* **358**, 734.
Hibbard, J.E., & Mihos, J.C. (1995) *Astron. J.* **110**, 140.
Lavery, R., & Henry, J.P. (1988) *Ap.J.* **330**, 596.
Löwenstein, M., Fabian, A.C., & Nulsen, P.E.J. (1987) *M.N.R.A.S.* **229**, 129.
Malin, D.F. (1979) *Nature* **277**, 279.
Malin, D.F., & Carter, D. (1980) *Nature* **285**, 643.
Malin, D.F., & Carter, D. (1983) *Ap.J.* **274**, 534.
Prieur, J.-L. (1987) *These, L'Universite Paul Sabatier de Toulouse.*
Prieur, J.-L. (1988) *Ap.J.* **326**, 596.
Prieur, J.-L. (1990) *"Dynamics and Interactions of Galaxies"* ed. R. Wielen, Springer-Verlag, Berlin, p72.
Quinn, P.J. (1982) *Ph. D. Thesis, Australian National University*
Quinn, P.J. (1984) *Ap.J* **279**. 596.
Schiminovich, D., van Gorkom, J.H., van der Hulst, J.M., & Kasow, S. (1994) *Ap.J. Lett.* **423**, L101.
Schiminovich, D., van Gorkom, J.H., van der Hulst, J.M., & Malin, D.F. (1995) *Ap.J. Lett.* **444**, L77.
Schiminovich, D., (1997) *this volume.*
Schweizer, F. (1980) *Ap.J.* **237**, 303.
Schweizer, F. (1982) *Ap.J.* **252**, 455.
Schweizer, F. (1996) *Astron.J.* **111**, 109.
Schweizer, F., & Seitzer, P., (1988) *Ap.J.* **328**, 88.
Thomson, R.C. (1991) *M.N.R.A.S.* **253**, 256.

Thomson, R.C., & Wright, A.E. (1990) *M.N.R.A.S* **247**, 122.
Toomre, A., & Toomre, J. (1972) *Ap.J.*, **178**, 623.
Turnbull, A.J. (1997) *this volume.*
Whitmore, B.C., & Schweizer, F. (1995) *Astron.J* **109**, 960.
Williams, R.E., & Christiansen, W.A. (1986) *Ap.J* **291**, 80.
Wright, A.E. (1972) *M.N.R.A.S.*, **157**, 309.
Zabludoff, A. (1997) *this volume.*

GLOBULAR CLUSTER SYSTEMS OF ELLIPTICAL GALAXIES

STEPHEN E. ZEPF
Dept. of Astronomy, Yale University, New Haven, CT 06520

AND

KEITH M. ASHMAN
Depts. of Physics, Univ. of Kansas, Lawrence, KS, 66045 and Baker University, Baldwin, KS 66006

Abstract. We review the observed properties of globular cluster systems and their implications for models of galaxy formation. Observations show that globular clusters form in gas-rich mergers, and that bimodal metallicity distributions are common in the globular cluster systems of ellipticals, with the metal-poor population more extended than the metal-rich one. These are three of the four predictions of the simple merger model of Ashman & Zepf (1992). The fourth prediction concerns the properties of the globular cluster systems of spirals, and is still to be tested by observation. Adopting Occam's razor, the confirmation of the fundamental predictions of the merger model from both young and old globular cluster systems is strong evidence that typical elliptical galaxies formed from the mergers of spiral galaxies. However, the simplifying assumptions of the Ashman-Zepf merger model limit its applicability to certain complex situations such as the formation of cD galaxies. We conclude this review by introducing new observational and theoretical programs that will further the understanding of the physical mechanisms of globular cluster and galaxy formation.

1. Introduction

The dramatic revolution in the understanding of globular cluster systems and their implications for galaxy formation can be traced by following the role the subject has played at major international meetings on mergers of galaxies. At the Heidelberg meeting in 1989, globular cluster systems (GCSs) were the focus of the "appointed skeptic" (van den Bergh 1990), whereas in Kyoto in 1997, globular clusters appeared prominently in the introductory talk as some of the strongest evidence that now quiescent elliptical galaxies formed from the past mergers of spiral galaxies (Schweizer 1998). This review will discuss the theoretical predictions and observational evidence that led to this revolution. We will then point to new directions that promise further advances in our understanding of globular cluster systems and their implications for the formation history of their host galaxy.

2. What the Merger Model Predicted

Ashman and Zepf (1992; hereafter AZ) carried out a detailed study of the relationship between galaxy mergers and globular cluster systems. If elliptical galaxies are the products of spiral galaxy mergers, globular clusters must form in such mergers (Schweizer 1987). This is because elliptical galaxies have a higher specific frequency (number per unit luminosity) of globular clusters than do spirals (van den Bergh 1990). By considering globular cluster formation efficiency and the gas content of the progenitor spiral galaxies, AZ showed that a sufficient number of globular clusters could form in major mergers to explain the specific frequency of globular clusters systems around ellipticals. This led to the conceptually simple prediction that, if elliptical galaxies are formed in galaxy mergers, ongoing galaxy mergers contain young globular clusters. In addition, AZ described the expected properties of young globular clusters, such as their blue colors, bright luminosities, and compact sizes. They also pointed out that *HST* observations of ongoing mergers like NGC 7252 and NGC 1275 should clearly reveal such objects. Thus the presence of young globular clusters in ongoing mergers constitutes a testable prediction that can be used to *refute* the merger model.

Perhaps the most unique prediction of the Ashman-Zepf model was that the globular cluster systems of elliptical galaxies formed in mergers have bimodal metallicity distributions. The progenitor spirals are expected to have a halo of metal-poor clusters, like those observed in the Milky Way and M31. More generally, in the context of the simple merger picture one expects such globular clusters to have formed out of gas that has experienced little metal enrichment. In contrast, globular clusters that form in the merger itself are expected to form out of the relatively enriched gas in the spiral disks. Thus these "merger-produced" globular clusters will have significantly higher metallicities than their counterparts in the halos of the progenitor spirals. For elliptical galaxies with "normal" specific frequencies (roughly double those of spirals), AZ showed that the number of metal-rich globular clusters formed in the merger must be comparable to the number of metal-poor globular clusters contributed by the progenitor spirals. Thus the overall metallicity distribution of the globular cluster systems of ellipticals is predicted to be bimodal, with roughly equal numbers of clusters in each "peak" of the distribution.

The third prediction of the Ashman-Zepf model is that, in an elliptical galaxy, the metal-rich cluster system will be more centrally concentrated than the metal-poor cluster system. This difference mirrors the different histories of the gas out of which the two globular cluster subsystems form. Compared to the low-metallicity halo gas that produces the metal-poor globular clusters, the enriched gas responsible for the metal-rich clusters is likely to have undergone more dissipation. This may arise as gas collapses from the halo to the disk in the progenitor spirals, or during the merger process itself. Consequently, the metal-rich cluster system is predicted to be more spatially concentrated than the metal-poor system. The effect may be increased by the tendency of mergers to "puff up" the stellar components of the progenitor galaxies. This would affect the metal-poor globular clusters, but not the dissipative gas out of which the metal-rich clusters form.

3. What the Observations Showed

Observational studies of globular cluster systems have advanced rapidly, so the fundamental predictions of the Ashman-Zepf model can now be tested. This section is devoted to a review of the relevant observations and the comparison of these to the model predictions.

3.1. YOUNG GLOBULAR CLUSTERS

The prediction that globular clusters form in gas-rich mergers has now been repeatedly confirmed, as reviewed in many places (e.g. Whitmore, these proceedings, Schweizer 1997, Ashman & Zepf 1998). Specifically, *HST* observations of gas-rich mergers have uncovered a wealth of objects with bright luminosities, blue colors, and compact sizes, as predicted by AZ. These properties are all consistent with those expected of young objects with masses and sizes of the globular clusters in the Galaxy. Moderate resolution spectroscopy has further confirmed that these objects are composed of a young stellar population consistent with those given by standard stellar population models. This agreement includes the strength of the Balmer lines as compared to LMC clusters (Bica & Alloin 1986), or to updated stellar population models and with improved stellar libraries, which were not available when the first spectra were published (e.g. Schweizer & Seitzer 1993, Zepf et al. 1995a). Finally, high resolution spectroscopy of a few of the most nearby examples provides velocity dispersions indicative of masses that are typical of Galactic globular clusters, and agree well with those calculated from the observed colors and luminosities, combined with stellar population models (Ho & Fillipenko 1997).

The prediction that globular clusters are formed in merger-induced starbursts is therefore confirmed. However, it is also important to recognize that well established physical processes will act to destroy some fraction of the initial young globular clusters (Fall & Rees 1977 and subsequent papers). Such destruction is also suggested by the higher ratio of light in clusters to total light seen in galaxy mergers compared to any old system, even the halo of the Galaxy or cD galaxies (Zepf et al. 1998). Moreover, the net effect of these processes is to preferentially destroy low mass clusters, so it is possible to begin with a power-law mass function, and end-up with a lognormal mass function, like that observed in the galaxy (e.g. Gnedin & Ostriker 1997).

3.2. BIMODAL COLOR DISTRIBUTIONS

The second prediction of the AZ merger model is that the globular cluster systems (GCSs) of elliptical galaxies will be composed of metal-poor populations from the progenitor spirals and metal-rich populations formed during the merger(s) that formed the elliptical. As described by Zepf & Ashman (1993) and Ashman & Zepf (1998), the metal-rich population will typically be significantly redder than the metal-poor population because most mergers occur at moderate or high redshift, and therefore metallicity differences dominate the broad-band colors.

Bimodality in the color distribution of the GGCs of elliptical galaxies was discovered by Zepf & Ashman (1993). Using the best available data and mixture-modeling algorithms, they showed that the color distributions of NGC 4472 and NGC 5128 were likely to be bimodal. Data of much higher quality for a number of elliptical galaxies have since become available. This large body of evidence indicates that bimodality is the norm for bright ellipticals (e.g. Ashman & Zepf 1998). Two typical examples are shown in Figure 1. In these and many other cases, mixture-modeling algorithms confirm objectively the significance of the bimodality apparent to the eye. The reader is referred to the Ashman & Zepf (1998) book for plots of many more examples of color distributions of GCSs.

3.3. COLOR GRADIENTS

The third prediction of the AZ model is that the the metal-rich population is more spatially concentrated than the metal-poor population. Color gradients, which were suspected at the time of AZ and have now been confirmed (Ashman & Zepf 1998 and references therein), are a natural result of the AZ model, although not unique to it. The key to testing the merger prediction is to compare directly the distributions of the blue and red populations around elliptical galaxies. This was first achieved by the Geisler et al. (1996) study of the NGC 4472 system. They showed that the metal-rich population is more spatially concentrated than the metal-poor population, thereby confirming the third prediction of the AZ merger model.

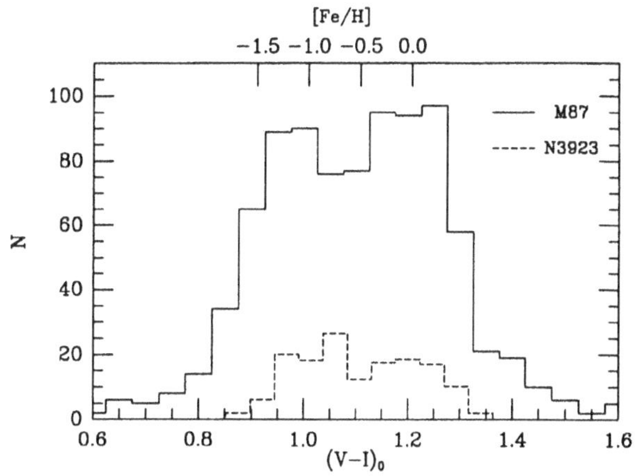

Figure 1. Color distributions for the GCSs of M87 from Whitmore et al. (1995) and NGC 3923 from Zepf et al. (1995b). The color distributions of both GCSs are bimodal, although the number of clusters observed in M87 is much greater

3.4. OTHER CONSIDERATIONS

While the fundamental predictions of the Ashman-Zepf model have been confirmed, there are specific cases where the quantitative agreement between prediction and observation breaks down. This is currently most clear for the number and precise color of the metal-poor population in elliptical galaxies (e.g. Zepf et al. 1995b, Forbes et al. 1997). In the simplest merger picture, the color and specific frequency (S_N) of the blue population is constant because all progenitor spirals are assumed to have the same halo population of metal-poor clusters. Higher specific frequencies of clusters in some ellipticals are therefore attributed to increased formation (or survival) efficiency of metal-rich clusters created in the mergers that made these galaxies. However, the metal-poor globular cluster populations of ellipticals do not appear to have identical metallicities, and some high S_N systems have an enhanced frequency of metal-poor clusters. Perhaps the clearest case is M87, for which the specific frequency of metal-poor clusters is about seven: much higher than the observed specific frequency of metal-poor globulars in spirals. However, this may not be universal, as the high S_N of NGC 3311 appears to be due solely to metal-rich clusters (Secker et al. 1995).

It is clear that these observations indicate the breakdown of the simplifying assumptions of the model. Brighter ellipticals are unlikely to be the result of the merger of only two spirals; they are too massive. Further, progenitor spirals will not have identical cluster populations before the merger. Finally, accretion of smaller satellites is likely to play a role as well, as seems to be the case for the halo population of the Galaxy. The question is whether a more sophisticated and realistic treatment of the merger process is likely to preserve these two fundamental predictions of the Ashman-Zepf model. While a complete answer to this question requires detailed modeling (see Section 4), it seems inevitable that metallicity bimodality and the spatial concentration dichotomy *will* be preserved, at least in some ellipticals. This is because the merger model requires that major spiral-plus-spiral mergers are involved at some point in the formation of history of an elliptical. The number of such mergers and the likelihood that later mergers may be predominantly stellar are both unimportant, since *typically* globular clusters in the progenitor spiral halos will be metal-poor, and those formed in any gas-rich merger will be relatively metal-rich and spatially concentrated.

4. New Observational and Theoretical Paths

4.1. THEORY

One way to advance the model beyond the formation of an elliptical from the merger of two spirals is to place the model in a more specific cosmological context. In this way, the full merging history of elliptical galaxies can be followed statistically for a variety of possible models. Semi-analytic models of galaxy formation and evolution (e.g. Kauffmann et al. 1993, Cole et al. 1994) are well-suited for this program, and one of us (SEZ) is working with G. Kauffmann to implement this approach. A second area ripe for advancement is the theoretical understanding of the formation of globular clusters, as the wealth of new observational evidence in nearby merging galaxies has not yet been turned in to advances in the understanding of the formation of dense, bound stellar systems like globular clusters.

4.2. OBSERVATION

The kinematics of globular cluster systems provides valuable information about the formation history and mass distribution of the host galaxy. Until recently, it had proven to be difficult to obtain more than a few tens of clusters around a given galaxy. However, the increase in efficiency and areal coverage of multi-object spectrographs, as well as higher quality imaging for object selection, has opened up this field, even with 4-m class telescopes.

One of the best studied cases is M87 (Cohen & Ryzhov 1997 and references therein). Here, we will focus on our work with R. Sharples and others on spectroscopy of globular clusters around NGC 4472 (Sharples et al. 1997). The most exciting result to come from our study is the tentative detection of kinematical differences between the metal-rich and metal-poor cluster populations in this galaxy. Specifically, the velocity dispersion of the metal-rich clusters appears to be higher than that of the metal-poor clus-

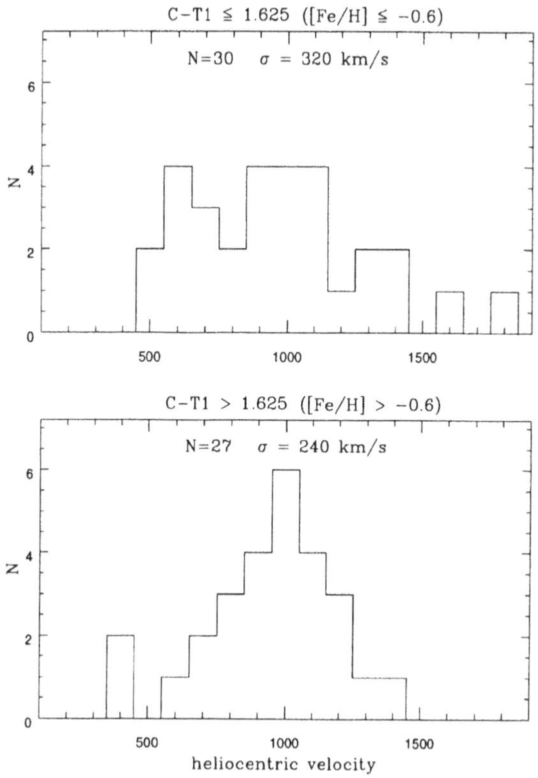

Figure 2. A comparison of the velocity dispersions of the metal-poor and metal-rich cluster populations in NGC 4472. An F-test rejects the hypothesis that these two have the same dispersion at the 86% confidence level.

ters (Figure 2). Moreover, the metal-poor clusters appear to rotate along the major axis, whereas the metal-rich clusters do not. At a basic level, the kinematics provide further physical evidence for the division of the cluster system into metal-poor and metal-rich subsystems that was originally based on an analysis of the colors alone. Furthermore, since the metal-rich cluster population is more spatially concentrated but has less rotation than the metal-poor cluster system, significant angular momentum transport must have occurred. This is seen in merger simulations, but is contrary to alternative models of episodic formation histories without mergers.

5. Conclusions

The merger model of Ashman & Zepf (1992) made four predictions for properties of globular cluster systems that had not yet been tested by observation, and which were generally contrary to the standard picture at that time. Three of these four have now been confirmed by observation, while the fourth has yet to be tested. Although limitations of the original model have also been revealed by other observations, the striking agreement with many of the predictions suggests that the model is mostly correct. For it not to be correct, the young clusters observed in mergers would have to be irrelevant for globular cluster populations as a whole, even though their properties are exactly those expected for young globular clusters. Moreover, the bimodal metallicity distributions observed in elliptical galaxy GCSs would have to have formed in an episodic process other than mergers, that is also efficient at transporting angular momentum. Given these constraints, any model that successfully accounts for the observations is likely to be similar physically to the merger model. Globular clusters have greatly improved our understanding of the formation history of galaxies, and with further photometric observations and new spectroscopic data, they promise to continue to do so.

Acknowledgements

We thank our many collaborators on the various projects described above. Some of the research described above was supported by grants GO-06092.01 -94A, AR-06405.01-95A, and an award to SEZ from the Dudley Foundation. SEZ also acknowledges the support of a Hubble Fellowship during much of this work.

References

Ashman, K.M., & Zepf, S.E. 1992, ApJ, 384, 50
Ashman, K.M., & Zepf, S.E. 1998, *Globular Cluster Systems* (Cambridge: Cambridge University Press), in press
Bica, E., & Alloin, D. 1986, A&A, 162, 21
Cohen, J.G., & Ryzhov, A. 1997, ApJ, 486, 230
Cole, S., Aragon-Salamanca, A., Frenk, C.S., Navarro, J., & Zepf, S.E. 1994, MNRAS, 271, 781
Forbes, D.A., Brodie, J.P., & Grillmair, C.J. 1997, AJ, 113, 1652
Geisler, D., Lee., M.G., & Kim, E. 1996, AJ, 111, 1529
Gnedin, O.Y., & Ostriker, J.P. 1997, ApJ, 474, 223
Kauffmann, G., White, S.D.M., & Guiderdoni, B. 1993, MNRAS, 264, 201
Ho, L.C. & Filippenko, A.V. 1997, ApJ, 466, L83
Schweizer, F. 1987, in Nearly Normal Galaxies, ed. S.M. Faber (New York: Springer), 18
Schweizer, F. 1997, in The Nature of Elliptical Galaxies, eds. M. Arnaboldi, G.S. Da Costa, & P. Saha (San Francisco: ASP), 447
Schweizer, F. & Seitzer, P., 1993, ApJ, 417, L29
Sharples, R.M., Zepf, S.E., Bridges, T.J, et al. 1997, MNRAS, submitted
Secker, J., Geisler, D., McLaughlin, D., & Harris, W.E. 1995, AJ, 109, 1019
van den Bergh, S. 1990, in Dynamics and Interactions of Galaxies, ed. R. Wielen (Berlin: Springer), 492
Whitmore, B.C. 1998, these proceedings
Whitmore, B.C., et al. 1995, ApJ, 454, L73
Zepf, S.E., & Ashman, K.M. 1993 MNRAS, 264, 611
Zepf, S.E., Carter, D., Sharples, R.M. & Ashman, K.M. 1995a, ApJ, 445, L19
Zepf, S.E., Ashman, K.M., & Geisler, D. 1995b, ApJ, 443, 570

GLOBULAR CLUSTERS IN ELLIPTICAL GALAXIES: CONSTRAINTS ON MERGERS

DUNCAN A. FORBES
School of Physics and Astronomy,
University of Birmingham, Birmingham, B15 2TT, UK

1. Introduction

There exists a relationship between globular cluster mean metallicity and parent galaxy luminosity (e.g. Brodie & Huchra 1991; Forbes et al. 1996), which appears to be similar to that between stellar metallicity and galaxy luminosity. The globular cluster relation has a similar slope but is offset by about 0.5 dex to lower metallicity. The similarity of these relations suggests that both the globular cluster system and their parent galaxy have shared a common chemical enrichment history. If we can understand the formation and evolution of the globulars, we will also learn something about galaxy formation. With this aim in mind we have created the SAGES (Study of the Astrophysics of Globular clusters in Extragalactic Systems) project. Project members include Brodie, Elson, Forbes, Freeman, Grillmair, Huchra, Kissler–Patig and Schroder. We are using *HST* Imaging and Keck spectroscopy to study extragalactic globular cluster systems. Further details are given at http://www.ucolick.org/~mkissler/Sages/sages.html.

2. Results and Discussion

van den Bergh (1975) has argued that ellipticals have too many globular clusters per unit starlight (called specific frequency, S_N) to be due to the simple merger of spirals. Spiral galaxies have $S_N \sim 0.5$. This is increased to $S_N \sim 2$ if we take into account the different mass-to-light ratios of spirals relative to ellipticals. A typical elliptical galaxy, with $M_V \sim -21$, has $S_N \sim 4$. Hence there is a factor of two difference. This discrepancy gets larger for more luminous ellipticals. However, it has been suggested that new globular clusters can form in the gas associated with the merger event (Schweizer 1987; Ashman & Zepf 1992). Recently proto–globular cluster candidates have been found in a number of merging systems, largely due to the resolving power of *HST*. A literature summary is given in Table 1.

TABLE 1. Proto-Globular Clusters

Galaxy	Merger Type	% Increase
NGC 4038/9	S + S	~100
NGC 7252	Sc + Sc	~70
NGC 3610	S + S	~70
NGC 3256	S + S	~100
NGC 3921	Sc + S0	~40
NGC 5128	E + S	~20
NGC 5018	E + S	~10
NGC 1316	E + S	0–30
NGC 1275	E + S	$\leq$10

From Table 1 it appears that the number of newly created proto-globulars varies with the progenitor types (i.e. the gas content) and that the percentage increase in the total cluster population is 100% or less. This is in contrast to the discussion above which requires 100% or *more* new globular clusters to counter van den Bergh's objection. A further issue to consider, is whether these cluster candidates will ever resemble globular clusters as we known them – there is some circumstantial evidence that they may not (Brodie et al. 1997).

Although the initial evidence was weak, there are now a handful of convincing cases for bimodal globular cluster color (metallicity) distributions in ellipticals (see Forbes, Brodie & Grillmair 1997). This indicates that some ellipticals have more that one globular cluster population. The observed metallicity distributions rule out a simple monolithic collapse and provide a strong constraint for any globular cluster formation model.

Bimodality has been taken as support for the merger picture of Ashman & Zepf (1992). Their model would predict the ratio of metal-rich to metal-poor (N_R/N_P) globulars to be about 1 for an elliptical with $S_N \sim 4$ and about 4 for $S_N \sim 16$. Furthermore the metal-poor peak should have a metallicity around [Fe/H] ~ -1.5, i.e. to match that of a typical spiral. Some examples of galaxies that do not fit this picture include: NGC 5846, although the ratio $N_R/N_P \sim 3$, S_N is not high but rather low at 2.8 (Forbes, Brodie & Huchra 1997). In the case of NGC 4472, the metal-poor peak contains a total of about 4000 globulars (Geisler et al. 1996). This would require about 10–20 L^* spirals. At the other extreme is NGC 3311 (Secker

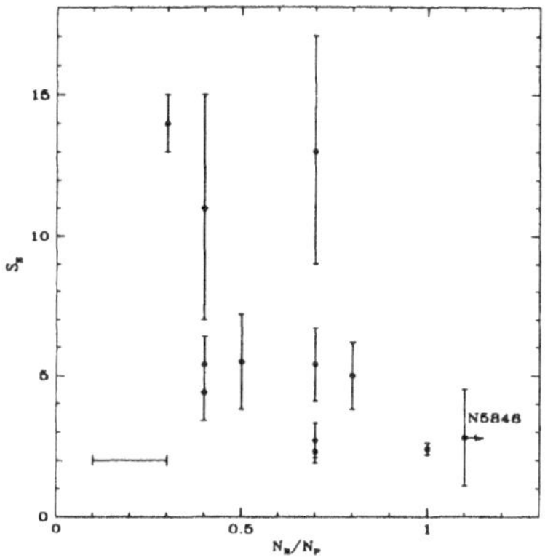

Figure 1. The specific frequency (S_N) versus the ratio of metal–rich to metal–poor globular clusters. The merger picture predicts a trend of increasing S_N with increasing N_R/N_P ratio, i.e. opposite to that seen.

et al. 1995) which has virtually no globulars with [Fe/H] = –1.5, i.e. none from L^* type spirals.

In Fig. 1 we show S_N against N_R/N_P for 12 large ellipticals (4 of which are cD galaxies). This figure includes two galaxies from the poster of Geisler & Lee (this conference). If the overabundant globular cluster systems (i.e. those with high S_N values) are due to the creation of more (metal–rich) globular clusters in a merger event, then we would expect a trend of increasing S_N with increasing N_R/N_P ratio. Fig. 1 shows the opposite trend, so that high S_N galaxies actually have a larger fraction of metal–poor globulars.

So although large ellipticals have bimodal metallicity distributions, when examined in detail they do not fit the expectations of the merger picture. What is the situation for low luminosity ellipticals? These are much harder to study given that such galaxies have fewer globular clusters and so any bimodality will be more difficult to detect. Kissler–Patig, Forbes & Minitti (1997) have combined *HST* and ground–based imaging of NGC 1427 to obtain what is probably the best photometrically–studied globular cluster system in a low luminosity elliptical. Its color distribution is shown in Fig. 2, indicating that it is unimodal. If NGC 1427 formed by the gaseous merger of two disk galaxies we would expect a bimodal distribution and yet it is clearly unimodal. Unless there is an age–metallicity conspiracy that hides the two globular cluster populations we are forced to conclude that NGC 1427 has only one population and does not fit the merger picture.

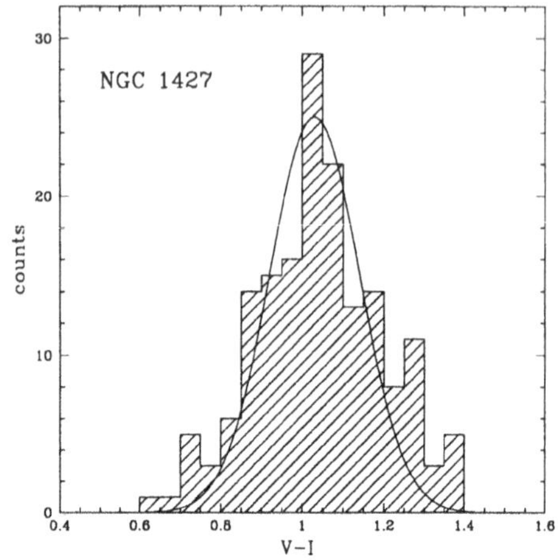

Figure 2. Globular cluster color distribution for NGC 1427 (M_V = -20.5, S_N = 5). The Gaussian shows the broadening due to photometric errors. The color distribution is consistent with a single, unimodal metallicity.

3. Concluding Remarks

It is clear that spiral+spiral mergers are occurring today at redshift $z = 0$, and that such gaseous mergers may be forming proto–globular clusters; however they may not be forming in sufficient quantities to resemble a 'normal' large elliptical. Gaseous mergers at $z \geq 1$ predict bimodal metallicity distributions but these are not seen in small ellipticals, and the details do not match the predictions for the large ellipticals. The presence of bimodal distributions rules out a monolithic collapse, so perhaps we should re–consider a two phase collapse (see Forbes, Brodie & Grillmair 1997).

References

Ashman K.M., & Zepf S.E. 1992, ApJ 384, 50
Brodie, J. P., & Huchra, J. 1991, ApJ, 379, 157
Brodie, J. P., et al. 1997, AJ, submitted
Forbes, D. A., et al. 1996, ApJ, 467, 126
Forbes D.A., Brodie J.P., & Grillmair, C.J. 1997, AJ 113, 1652
Geisler, D., Lee, M. G., & Kim, E. 1996, AJ, 111, 1529
Kissler–Patig, M., Forbes, D. A., & Minitti, D. 1997, MNRAS, submitted
Schweizer, F. 1987, Nearly Normal Galaxies, ed. S. Faber (New York: Springer–Verlag), p 18
Secker, J., et al. 1995, AJ, 109, 1019
van den Bergh, S. 1975, ARAA, 13, 217

FUNDAMENTAL PLANE AND MERGER SCENARIO

K. BEKKI
Astronomical Institute, Tohoku University, Sendai, 980-77, Japan

1. Introduction

The Fundamental Plane (FP) is one of the most important universal relations in early type galaxies because it contains valuable information about the formative and evolutionary process of galaxies (Djorgovski & Davis 1987, Dressler et al. 1987). The commonly used form of the scaling relation in the FP is described as $R_e = \sigma^A I^B$, where R_e, σ, and I are effective radius, central velocity dispersion, and mean surface brightness of elliptical galaxies, respectively. The exponents A, B are considered to be 1.56 ± 0.07 and -0.94 ± 0.09 in the FP derived by K band photometry, respectively, and these values deviate significantly from the values $A = 2.0$ and $B = -1.0$ expected from virial theorem (Pahre et al. 1995; Djorgovski, Pahre, & de Carvalho 1996). This apparent deviation requires that the ratio of dynamical mass (M) to luminosity of elliptical galaxies (L) depends on M as $M/L \propto M^{\alpha}$ ($\alpha = 0.12 \pm 0.03$ for K band). Possible interpretations for the required dependence of M/L on M are generally considered to be divided into the following two. One is that the required dependence of M/L on M results from the fact that the mean stellar age and metalicity of elliptical galaxies depend systematically on M. The other is that the required dependence reflects the M dependence of structural and kinematical properties of elliptical galaxies ("nonhomology"). Although we should not neglect the importance of stellar populations in generating the M dependence of the M/L (Renzini & Ciotti 1993), we here consider that the origin of the required M dependence of M/L is closely associated with the structural and kinematical properties dependent on M or L in elliptical galaxies.

The purpose of this contribution paper is to elucidate the origin of the luminosity-dependent structural and kinematical properties ("nonhomology") in elliptical galaxies and thereby to explore the physical meaning of the FP. We here focus particularly on the star formation history of elliptical galaxies and accordingly investigate the important roles of the star formation history in generating the structural and kinematical nonhomology of elliptical galaxies. In investigating the non-homologous nature in elliptical galaxies, we adopt the "merger hypothesis" in which elliptical galaxies are formed by galaxy mergers between two late-type spirals (Toomre & Toomre 1977). In particular, we consider galaxy mergers between two disk galaxies with a gas mass fraction larger than 0.2 observed in the present typical late-type spirals in order to mimic elliptical galaxy formation by galaxy mergers at higher redshift.

2. Model

We construct models of galaxy mergers between star-forming gas-rich disk galaxies with equal mass by using Fall-Efstathiou model (1980). The details of the model are given by Bekki & Shioya (1997). We adopt the Schmidt law (Schmidt 1959) with exponent γ as the controlling parameter of the rate of star formation. The γ is set to be 2.0 for all the simulations in the present study. The amount of gas consumed by star formation for each gas particle in each time step, $\dot{M}_{\rm g}$, is given as:

$$\dot{M}_{\rm g} \propto C_{\rm SF} \times (\rho_{\rm g}/\rho_0)^{\gamma-1.0} \quad (1)$$

where $\rho_{\rm g}$ and ρ_0 are the gas density around each gas particle and the mean gas density at 0.48 radius of an initial disk, respectively. The $C_{\rm SF}$ in the equation (1) is the parameter that controls the rapidity of gas consumption by star formation: The larger the $C_{\rm SF}$ is, the more rapidly the gas particles are converted to new stellar particles.

We here investigate fundamental roles of the rapidity of star formation ($C_{\rm SF}$) in determining structural and kinematical properties of merger remnants. To be specific, we investigate the $C_{\rm SF}$ dependence of the following three nonhomology parameters: $k_{\rm S} \propto R_{\rm e}/R_{\rm g}$, $k_{\rm K} \propto (1.0 + c_{\rm v} \times (v_{\rm m}/\sigma_0)^2)$, $k_{\rm M} \propto M_{\rm t}/L \propto M_{\rm t}/M_{\rm b}$. In the above equations, the $R_{\rm e}$, $R_{\rm g}$, $v_{\rm m}$, σ_0, $M_{\rm t}$, $M_{\rm b}$, and L represent effective radius, gravitational radius, maximum rotational velocity, central velocity dispersion, total mass, total baryonic mass, and luminosity of a merger remnant, respectively. The value of the parameter $c_{\rm v}$ depends on the details of radial distribution of kinematical properties of galaxies. In the present study, we adopt 0.81 for the $c_{\rm v}$, which is the same as that of Prugniel & Simien (1996). These three nonhomology parameters must cooperate to satisfy the following relation inferred from the FP.

$$K_{\rm FP} \propto k_{\rm S} {k_{\rm K}}^{-1} k_{\rm M} \propto {\sigma_0}^2/R_{\rm e}/I_{\rm e} \propto M/L \propto L^{\alpha} \quad (2)$$

where $I_{\rm e}$ is mean surface luminosity of elliptical galaxies. The $K_{\rm FP}$ represents "total" nonhomology of galaxies, which includes the above three different types of nonhomology. If elliptical galaxies do not have non-homologous nature, the $K_{\rm FP}$, which corresponds to the conventionally used M/L, is constant for all elliptical galaxies.

3. Results

Fig. 1 describes the $C_{\rm SF}$ (= 7.0, 3.5, 1,75, 0.7, and 0.35) dependence of the nonhomology parameters, $k_{\rm S}$, $k_{\rm K}$, and $k_{\rm M}$ in the merger remnants. In order to show more clearly characteristics of the $C_{\rm SF}$ dependence of $k_{\rm S}$, $k_{\rm K}$, and $k_{\rm M}$, we preset the best fitted line derived by least square fitting of each set of experimental data to assumed relations like as $k_{\rm S}$ ($k_{\rm K}$ and $k_{\rm M}$) $\propto {C_{\rm SF}}^{x}$, where the exponent x is described below. We can observe clear trends in the $C_{\rm SF}$ dependence of $k_{\rm S}$, $k_{\rm K}$, and $k_{\rm M}$ as follows. First, as is shown in the left panel of Fig. 1, $k_{\rm S}$ is appreciably larger for models with larger $C_{\rm SF}$. The reason for this dependence is that the rapidity of star formation of mergers basically determines how much amount of stellar component is transferred to the central region of the remnants during mergers, which is a key factor for final mass distribution of stellar component in merger remnants. The dependence of $k_{\rm S}$ on $C_{\rm SF}$ is described as $k_{\rm S} \propto {C_{\rm SF}}^{0.25}$. Second, we can observe in the middle panel of Fig. 1 that the larger the $C_{\rm SF}$ is, the smaller the $k_{\rm K}$ is. This is because total amount of gaseous dissipation during merging, which is smaller for models with larger $C_{\rm SF}$, principally determines how efficiently initial total potential energy of a galaxy merger can be converted into rotational energy rather than random kinetic energy during galaxy merging and thus how strongly the merger remnant is dynamically supported by global rotation. The dependence of $k_{\rm K}$ on $C_{\rm SF}$ is described as $k_{\rm K} \propto {C_{\rm SF}}^{-0.08}$. Third, $k_{\rm M}$ is larger for models with larger $C_{\rm SF}$. This is because in the model with larger $C_{\rm SF}$, less amount of stellar component is transferred to the central region owing to less amount of kinetic energy dissipated away by gaseous dissipation during merging. The dependence of $k_{\rm M}$ on $C_{\rm SF}$ is described as $k_{\rm M} \propto {C_{\rm SF}}^{0.11}$. Thus, the dependence of $K_{\rm FP}$ on $C_{\rm SF}$ is described as $K_{\rm FP} \propto {\sigma_0}^2/R_{\rm e}/I_{\rm e} \propto {C_{\rm SF}}^{0.34}$.

Dependence of mass-to-light ratio on the galactic luminosity implied by the scaling relation of the FP requires that $K_{\rm FP}$ ($\propto M/L$) should depend on L as $K_{\rm FP} \propto L^{0.14}$ (e.g., Pahre et al. 1995; Djorgovski et al. 1996). In the present study, the $K_{\rm FP}$ is found to depend on the $C_{\rm SF}$ as $K_{\rm FP} \propto {C_{\rm SF}}^{0.34}$. By using the above two dependences of the $K_{\rm FP}$, we can obtain the result that the $C_{\rm SF}$ should depend on L as $C_{\rm SF} \propto L^{0.41}$ for explaining the origin of the FP slope. This expected dependence of $C_{\rm SF} \propto L^{0.41}$ implies that if more luminous elliptical galaxies are formed by galaxy mergers with more rapid star formation, the slope of the FP can be at least qualitatively explained in the present merger model. This result furthermore demonstrates that although a specific relation between star formation history of galaxy mergers

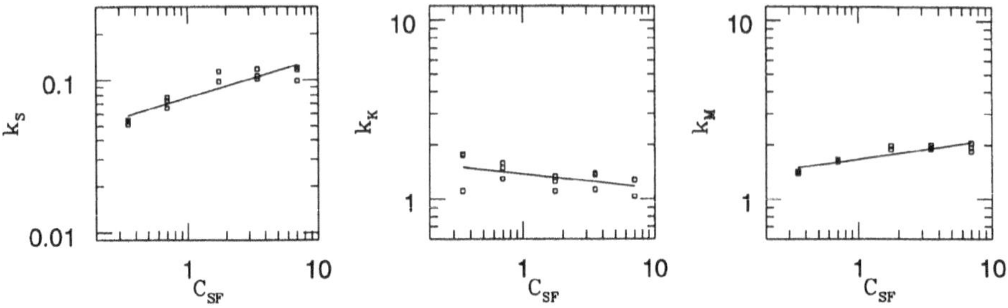

Figure 1. Dependence of nonhomology parameters, k_S, k_K, and k_M on C_{SF}. Physical meanings of the three nonhomology parameters are described in the manuscript. The parameter values projected onto xy, xz, and yz plane are plotted by open squares for five models with different C_{SF} in the Model 1. A solid line in each panel indicates the best fitted line derived by least square fitting procedure for each set of results.

and galactic luminosity is required, the origin of the FP slope can be closely associated with the star formation history of elliptical galaxies formed by dissipative galaxy merging. Although the present merger model is indeed rather idealized and less realistic for *real* dissipative galaxy mergers with star formation at higher redshift universe, it appears to have succeeded in demonstrating that the origin of the FP can reflect a close physical relation between galactic luminosity and star formation history of merger progenitor galaxies, in particular, the rapidity of star formation of galaxies.

References

Bekki, K., & Shioya, Y. 1997, ApJ, 478, L17
Djorgovski, S., & Davis, M. 1987, ApJ, 313, 59
Djorgovski S., de Carvalho R., & Han M-S., 1988, in Extragalactic distance scale ed. S. Van den Bergh, and J. P. Christopher, ASP Conf. Ser. Vol. 4, p. 329
Djorgovski, S., Pahre, M. A., & de Carvalho R. R. 1996, in Fresh Views of Elliptical Galaxies, ed. A. Buzzoni, R. Renzini, and A. Serrano , ASP Conf. Ser. Vol 86, p129
Dressler, A., Lynden-Bell, D., Burstein, D., Davies, R. L., Faber, S. M., Terlevich, R. J., & Wegner, G. 1987, ApJ, 313, 42
Fall, S. M., & Efstathiou, G. 1980, MNRAS, 193, 189
Pahre, M. A., Djorgovski, S., & de Carvalho R. R. 1995, ApJ, 453, L17
Prugniel Ph., & Simien F. 1996, in Fresh Views of Elliptical Galaxies, ed. A. Buzzoni, A. Renzini, and A. Serrano, ASP Conf. Ser. Vol. 86, p. 151
Renzini, A., Ciotti, L., D'Ercole, A, & Pellegrini, S. 1993, ApJ, 419, 52
Schmidt, M. 1959, ApJ, 344, 685
Toomre, A., & Toomre, J. 1972, ApJ, 178,623

MASS DISTRIBUTION OF THE E0 GALAXY NGC 6703 FROM ABSORPTION LINE PROFILE KINEMATICS

O.E. GERHARD
Astronomisches Institut, Venusstr. 7, CH-4102 Binningen
G. JESKE
Landessternwarte, Königstuhl, D-69117 Heidelberg
AND
R.P. SAGLIA AND RALF BENDER
Inst. für Astronomie, Scheinerstr. 1, D-81679 München

Absorption line velocity profiles (VPS) contain important information on the anisotropy and mass distribution of elliptical galaxies (e.g., Gerhard 1993, Merritt 1993). Here we briefly present results of an extensive analysis of the E0 galaxy NGC 6703 (Gerhard et al. 1997). This work is part of an observational and theoretical program aimed at understanding the orbit structure and dark matter content of ellipticals at intermediate radii (a preliminary account is given in Saglia et al. 1997).

1. Data and Analysis

NGC 6703 is an E0 galaxy at a distance $D = 36h_{50}^{-1}$ Mpc ($h_{50} \equiv H_0/50$km/s / Mpc). From a Jaffe profile fit, its absorption-corrected magnitude is $M_{\rm B} = -21.07$, or luminosity $L_{\rm B} = 4.16 \times 10^{10} h_{50}^{-2} L_{\odot,\rm B}$. Its effective radius is $R_{\rm e} = 30'' = 5.2\, h_{50}^{-1}$ kpc.

Our kinematic data for NGC 6703 extend to $2.6R_{\rm e}$. The galaxy shows little rotation (≈ 0 km/s for $R < R_{\rm e}$, $\approx 20 - 30$ km/s for $R > R_{\rm e}$). The velocity dispersion drops from the central ≈ 190 km/s to ≈ 140 km/s at $R_e/2$, slowly declining to about 110 km/s in the outer parts. The h_3 and h_4 values are everywhere close to zero.

With a new non-parametric technique we determine the DF $f(E, L^2)$ directly from the VP data. Monte Carlo tests using simulated data with the spatial extent, sampling, and error bars of the NGC 6703 data show that smooth DFS can be recovered to an RMS accuracy of $\sim 10\%$, and the anisotropy parameter $\beta(r)$ to ~ 0.1, in a *known* potential. Similar tests analyzing models in *different* potentials show that, from data like ours for

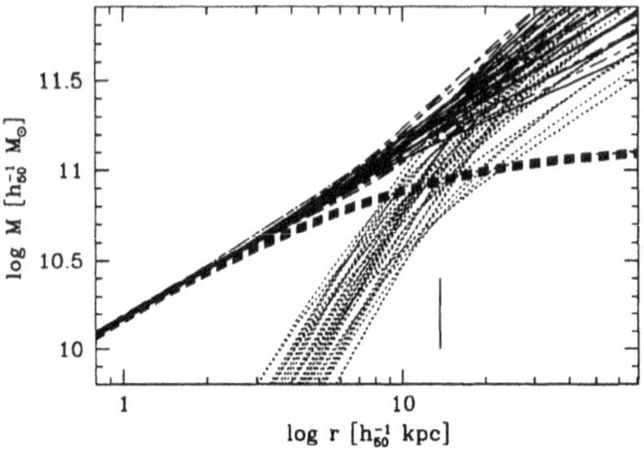

Figure 1. Luminous, dark, and total mass as a function of radius for acceptable models of NGC 6703 (short dashed, dotted, and dash–dashed or full lines, respectively). Mass distributions consistent with the data at 87% (1.5σ) and at 95% c.l. (2.0σ) are coded by full and dash–dashed lines, respectively. The vertical line denotes the position of the last kinematic data point. At this radius, the M/L ratio has increased by at least a factor of 1.6 from the center.

NGC 6703, an asymptotically constant halo circular velocity v_0 can be determined with an accuracy of $\pm \lesssim 50\,\mathrm{km\,s^{-1}}$, at 95% c.l.

2. Results

For NGC 6703 we thus determine the true circular velocity at $2.6R_e$ to be $250 \pm 40\,\mathrm{km\,s^{-1}}$ (95% c.l.), corresponding to a total mass inside $78'' = 13.5\,h_{50}^{-1}$ kpc of $1.6 - 2.6 \times 10^{11} h_{50}^{-1}\,M_\odot$. No model without dark matter fits the data; however, a *maximum stellar mass* model in which the luminous component provides nearly all the mass in the centre does. In such a model, the total luminous mass inside $78''$ is $9 \times 10^{10}\,M_\odot$ and the integrated B-band M/L out to this radius is $M/L_\mathrm{B} = 5.3 - 10$ (95% c.l.), rising from the central $M/L_\mathrm{B} = 3.3$ by at least a factor of 1.6.

The anisotropy of the stellar orbits in NGC 6703 changes from near-isotropic at the centre to radially anisotropic ($\beta = 0.3 - 0.4$ at $30''$, $\beta = 0.2 - 0.4$ at $60''$) and is not well-constrained at the outer edge of the data, where $\beta = -0.5 - +0.5$, depending on variations of the potential in the allowed range.

References

Gerhard O.E. 1993, *MNRAS*, **265**, 213
Merritt D. 1993, *ApJ*, **413**, 79
Gerhard O.E., Jeske G., Saglia R.P., Bender R. 1998, *MNRAS*, in press
Saglia, R.P., Bender, R., Gerhard, O.E., Jeske, G. 1997, in *Dark and Visible Matter in Galaxies and Cosmological Implications*, Eds. M. Persic & P. Salucci, *ASP*, **117**, 113

SHELL FORMATION IN NGC 474

A.J. TURNBULL
University of Hertfordshire, Hatfield, Herts, UK
D. CARTER
Liverpool John Moores, Byrom Street, Liverpool, UK
T.J. BRIDGES
Royal Greenwich Observatory, Cambridge, UK
AND
R. C. THOMSON
University of Hertfordshire, Hatfield, Herts, UK

We present broad band photometry (B and R) of the classic shell galaxy NGC 474. Preliminary results indicate that the shells have a similar colour to and follow the same trend of colour with radius as the underlying galaxy.

1. Introduction

NGC 474 is a classic shell galaxy originally catalogued as peculiar by Arp (1966), appearing in his atlas as Arp 227, together with its nearby spiral companion NGC 470. Both galaxies have the same redshift, as measured in HI of 2372 and 2374 km s^{-1} respectively (de Vancouleurs et al. 1991) and, assuming both lie at a distance of 31 Mpc, have an apparent separation of some 45 kpc. The shells around NGC 474 are unusually bright, making this a prime target for detailed photometric studies.

Several models have been put forward to explain the origin of shells, of which the most successful are the merger model (Quinn 1984; Dupraz & Combes 1986; Hernquist & Quinn 1988, 1989) and the interaction model (Thomson & Wright 1990; Thomson 1991). According to the merger model, shells are formed by the phase wrapping of stars accreted from the disrupted companion galaxy. In this case, the shell colours may be different to that of the underlying host galaxy. According to the interaction model, shells are density waves induced in a thick disk population during an interaction (not a merger) with another galaxy. Consequently, the shell colours are expected to be the same as that of the underlying host galaxy.

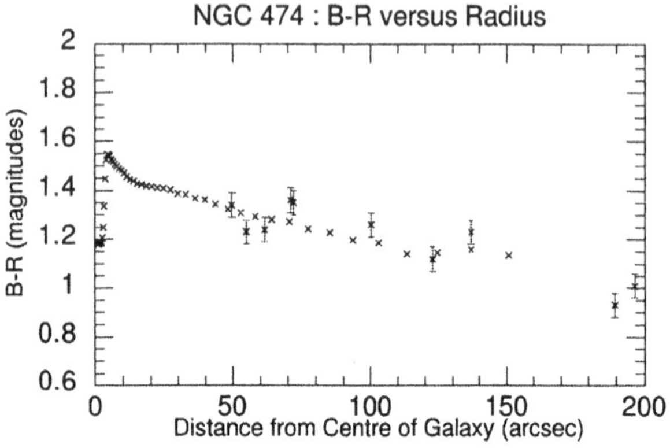

Figure 1. Derived B-R Colours for NGC 474 and Shells.

2. Observations and Data Reduction

The images of NGC 474 were obtained at prime focus on the WHT(B) and INT(R) on La Palma. Standard IRAF data reduction methods were used. The underlying galaxy profiles were obtained using the ISOPHOTE package in IRAF. Using a fixed size aperture (6″x6″) and residual sky subtraction the shell magnitudes in B and R were determined from the galaxy subtracted image. The derived galaxy colours (crosses) and shell colours (crosses with estimated errors) are plotted in Fig. 1.

3. Discussion

There appears to be a significant colour gradient with the host galaxy becoming bluer at larger radii. Each shell's B-R colour is similar to that of the underlying galaxy at the same radius. This is naturally explained by the interaction model whereas the merger model requires the special condition that the merged galaxy is of the same colour as NGC 474. A parallel N-Body study is underway.

References

Arp, H. 1966, *ApJS*, **14**, 1
de Vancouleurs et al. 1991, *Third Reference Catalog of Bright Galaxies* (RC3)
Dupraz, C., & Combes, F. 1986, *A&A*, **166**, 53
Hernquist, L., & Quinn, P. 1988, *ApJ*, **331**, 682
Hernquist, L. & Quinn, P. 1989, *ApJ*, **342**, 1
Thomson, R.C. 1991, *MNRAS*, **253**, 256
Thomson, R.C. & Wright, A.E., 1990, *MNRAS*, **247**, 122
Quinn, P. 1984, *ApJ*, **279**, 596

FORMATION OF BOXY/PEANUT-SHAPED BULGES IN SPIRAL GALAXIES: ACCRETION OR BAR INSTABILITY?

M. BUREAU AND K.C. FREEMAN
Mount Stromlo and Siding Spring Observatories
Private Bag, Weston Creek P.O., ACT 2611, Australia

1. Introduction

Boxy/peanut-shaped bulge galaxies constitute at least 20-30% of all spirals. Distribution functions and numerical simulations studies have shown that the slow decay of the orbit of a companion into a larger spiral can lead to the formation of a boxy/peanut bulge. However, the bar-buckling instability now appears a more likely process. Thin bars either do not form or, as shown by N-body simulations, buckle and settle with an increased thickness, appearing boxy or peanut-shaped when seen edge-on. This project aims to determine the dynamical state of a sample of 30 edge-on spirals, 3/4 of which have boxy/peanut-shaped bulges, half having close-by companions.

2. Observations and Discussion

We have obtained high S/N long-slit spectroscopy along the major axis of all our sample galaxies. Recent or ongoing accretion events are detectable as irregularities in the position-velocity diagrams (PVDs) obtained, while a double-peaked PVD denotes the presence of a bar. Of the 16 boxy/peanut galaxies with extended emission lines, 11 display a clear bar signature. Those galaxies also possess bulge morphologies resembling those of N-body simulations. Only 2 galaxies (with extreme bulge morphologies) show signs of accretion, and 3 have regular PVDs but are also very dusty. Furthermore, none of the galaxies with a normal spheroidal bulge show signs of a bar.

Our results point to a picture where most boxy/peanut-shaped bulges are due to the presence of a thick bar seen edge-on, and only a few extreme cases are formed through accretion. The possibility of hybrid scenarios where a bar is excited by an encounter and then buckles remains.

THE CHARACTER OF EMBEDDED RINGS

TAPAN K. CHATTERJEE
Facultad de Ciencias, Fisico-Matematicas,
Universidad A. Puebla, A.P. 1316, Puebla, Mexico

It is well known that, under favorable conditions, tidal interactions between a spiral galaxy and a more compact elliptical leads to the formation of embedded rings in the disk (e.g. Chatterjee 1979, *Bull. Astron. Soc. India*, **7**, 32); in addition there are also nuclear rings whcih do not seem to have a tidal origin (e.g. Buta 1986, *ApJ*, **306**, 768). The two types of rings occur in different regions of the spirals, which can be explained on the basis of the tidal hypothesis by extending the previous research work of Chatterjee (1984, *Ap&SS*, **106**, 309). In this context, using the same theory, we study many normal on-axis collisions (as rings are best formed for this orientation) between a disk galaxy, modeled as an exponential disk with a polytropic $n = 0, 3, 4$, bulge (mass being equally distributed amongst the polytropic indices), and a compact spherical intruder, modeled with a polytropic $n = 2, 3, 4$, distribution (mass being equally distributed amongst the polytropic indices). The internal energy changes suffered by the disk, have a direct bearing on the sharpness of the rings, and are directly obtained from the relationship, for the fractional change in internal energy of the disk, as $f_E = \gamma\beta$, where β is a function of the galactic models and the collision parameter $\gamma = G(\sigma_s/V^2) \propto \sigma_s/V^2$ (where σ_s is the projected velocity of collision). We find that sharp rings form for $f_E \geq 0.5$; beyond this value of f_E the intensity contrast of the rings diminish. However, fairly sharp rings (from an observational point of view) form for $f_E \geq 1/3$, which corresponds to $\gamma \geq 0.01$; which corresponds to a density maximum near a region $\sim (1/3)R$ (R being the radius of the disk galaxy), so that rings of tidal origin are not expected to be prominent at a distance interior to $\sim 1/3$ of the radius of the disk galaxy. They are not expected to form interior to $\sim 1/10$ of the radius of the disk.

TRENDS IN GALAXY FORMATION AND EVOLUTION IN THE CONTEXT OF THE VIRIAL AND FUNDAMENTAL PLANES

TAPAN K. CHATTERJEE AND V.B. MAGALINSKY
Facultad de Ciencias, Fisico-Matematicas,
Universidad A. Puebla, A.P. 1316, Puebla, Mexico

The kinetic description of gravitating systems has acquired vital importance in the context of trends in galaxy formation and evolution as evidenced by the existence of the virial and fundamental planes. The fundamental plane deviates for brighter and fainter ellipticals; until the brightest cluster members (BCMs), whose structures have been most modified by interactions, seem to occupy a fundamental plane with a different slope as compared to normal ellipticals. Extending the work of Magalinsky (1972, *AZh*, **49**, 1017; *Sov.Astron.-AJ*, **16**, 830), the Vlasov equation is applied to study small perturbations (considered as protogalaxies) of the exact solution corresponding to a spatially homogeneous medium in expansion. It is found that a perturbation attains a saturated size whose scale length, as a function of a reduced parameter of evolution (in terms of the characteristic frequency of dispersion of momenta, τ), $R(\tau) \propto K.E./P.E. \propto (K.E.)^2/\sigma \propto (\Delta V)^2/\text{Proj.density} \propto \sigma^2/I$, which has the parametric form of the virial plane. The subsequent evolution is characterized principally by the variation of the energy due to the gravitational interactions between stars (considered as mass points), given by the potential energy such that the harmonic mean separation scale (between stars) characterizes this evolution. In this stage of the evolution the harmonic scale separation has the parametric form, $\langle r^{-1}\rangle \propto (K.E.)^{1/2}$, and $\langle r^{-1}\rangle \propto (P.E.)$ such that $\langle r^{-1}\rangle \propto (K.E.)^{1/2}/(P.E.) \propto \sigma/I$. Notice that this is the parametric form of the fundamental plane of evolved ellipticals since the harmonic scale separation determines a physically significant scale.

MASS STRUCTURE OF Sa SPIRALS: NGC 2179 & NGC 2775

E.M. CORSINI, M. SARZI, P. CINZANO AND F. BERTOLA
Dipartimento di Astronomia, Università di Padova, Italy
A. PIZZELLA
European Southern Observatory, Santiago, Chile
AND
M. PERSIC[1], P. SALUCCI[2]
[1]*Osservatorio Astronomico,*[2]*SISSA, Trieste, Italy*

Recent analysis of extended rotation curves of late-type spirals have confirmed that these objects have detectable amounts of dark matter (DM) already in the optical region, with dark-to-luminous mass ratio scaling inversely with luminosity. Since currently available rotation curves of early-type spirals are often fragmentary and not extended enough, the question remains open as to whether dark halos are unequivocally present in Sa galaxies. To address this point we have observed a sample of 7 Sa's measuring major axis velocities and velocity dispersion for both stars and ionized gas. Here we present results for two objects, NGC 2179 and NGC 2775. We have constructed detailed dynamical models which, for each galaxy, explain the observed kinematics for both stellar and gaseous component and which are consistent with observed photometry. The best fit model for NGC 2179 involves an oblate isotropic rotator bulge, a thin exponential disk with the same mass-to-light ratio, and a pseudo-isothermal DM halo.The mass-to-light ratio in the inner galaxy was found to be $M/L_{\rm R} = 4.5\ (M/L_{\rm R})_\odot$ reaching $M/L_{\rm R} = 7.5\ (M/L_{\rm R})_\odot$ at the optical radius ($R_{\rm opt}$). For NGC 2775 the best fit model involves an oblate isotropic rotator bulge and a thin exponential disk, with $M/L_{\rm r} = 3.5\ (M/L_{\rm r})_\odot$, and $M/L_{\rm r} = 4.7\ (M/L_{\rm r})_\odot$ respectively. No DM halo was needed to explain the data extending to 0.7 $R_{\rm opt}$. In the inner regions the gas rotates as fast as stars but with negligible velocity dispersion. This rules out the case where the gas kinematics is dominated by random motion, and leads us to speculate we are seeing gas rotating on a non-equatorial plane, resulting from a past external acquisition, possibly from the companion galaxy NGC 2777.

MULTIMODAL COLOR DISTRIBUTIONS IN THE GLOBULAR CLUSTER SYSTEMS OF GIANT ELLIPTICAL GALAXIES

D. GEISLER
Kitt Peak National Observatory, National Optical Astronomy Observatories
950 N. Cherry Av., Tucson, AZ 85719 USA

AND

M.G. LEE
Seoul National University
Seoul 151-742, KOREA

We report on new observations of the globular cluster systems (GCSs) of two galaxies: M86 (NGC 4406) in Virgo and NGC 4696, the central giant elliptical (gE) in the Centaurus cluster. Previous observations in M86 showed no evidence for bimodality, but using only $(V-I)$ for small cluster samples. The NGC 4696 GCS is unstudied. We used the integrated Washington $(C-T_1)$ color. This metallicity index is more than twice as sensitive to [Fe/H] as $(V-I)$. In M86 we have about 1100 good GC candidates, and about 650 in NGC 4696, with mean internal metallicity errors ~ 0.15 dex. Both of these GCSs are found to have *bimodal metallicity distributions* (MDs). Our data strengthen previous results that MDs for the GCSs of gEs are widespread. The evidence for 2 separate populations in these galaxies is corroborated by examining the surface density distributions: the metal-rich clusters are more centrally concentrated than their metal-poor counterparts. The overall radial metallicity gradient present in the M86 GCS is due to the varying radial mix of the 2 populations. The existence of 2 GC populations signifies that there were 2 distinct epochs or events of cluster formation in a gE. The simple collapse model of gE formation is ruled out.

THE GLOBULAR CLUSTER SYSTEM OF NGC 1399

Optical HST imaging and ESO (ground-based) near-IR imaging

PAUL GOUDFROOIJ
Space Telescope Science Institute, Baltimore, USA†
M. VICTORIA ALONSO
Observatorio Astronomico Cordoba, Cordoba, Argentina
AND
DANTE MINNITI
Lawrence Livermore National Laboratory, Livermore, USA

We report on an ongoing study of the optical–near-IR colors of globular clusters (GCs) in E galaxies. The motivation is that *(i)* *HST* images give the necessary resolution to discriminate against foreground stars and background galaxies, while the photometry goes very deep; *(ii)* Near-IR observations reach only the brightest clusters, but provide a *much* larger color baseline which is very useful to *e.g.*, identify intermediate-age clusters such as those found in the LMC and in NGC 5128 (cf. Minniti et al. 1996, *ApJ*, **467**, 221), and to measure more accurate metallicities, particularly at the metal-rich extreme of the metallicity distribution.

J, H, K′ imaging was obtained at the ESO 2.2 m telescope, covering 2.′3 × 2.′5, overlapping with the WF4 chip of the *HST* images. The images reach $J = 21.5$, $H = 20.5$, and $K' = 19.5$. At the distance of NGC 1399, the GCs are not resolved. A DAOPHOT sharpness vs. roundness diagram is used to discriminate GCs from foreground stars and background galaxies in the *HST* images. The optically selected list of GCs is then correlated with the near-IR photometry. Only the brightest GCs in the fields are detected in the near-IR mosaics: 20 GCs in J and H, and 10 in K′. In a later stage we plan to include JHK′ colors for the GCs for which optical photometry was obtained by Kissler-Patig et al. (1997, *A&A*, **319**, 470).

The detected (bright) NGC 1399 GCs are found to span a wide color range, $1.7 < B-H < 4.2$, implying a wide range in metallicity. Using the integrated color vs. [Fe/H] relation of Galactic GCs to estimate metallicities, the range implied is $-2.5 < \mathrm{[Fe/H]} < 0.5$. Indeed, some of these bright GCs in NGC 1399 are redder than the reddest Milky Way GCs.

†P. Goudfrooij is affiliated to the Space Science Department, European Space Agency

THE NATURE OF THE DUSTY IONIZED GAS IN NGC 5846

(and other elliptical galaxies ?)

PAUL GOUDFROOIJ
Space Telescope Science Institute, Baltimore, Maryland, USA[†]

AND

GINEVRA TRINCHIERI
Osservatorio Astronomico di Brera, Milano, Italy

We present new optical imagery and *ROSAT* HRI X-ray imagery of the elliptical galaxy NGC 5846. A filamentary dust lane is detected in its central region, with a morphology strikingly similar to that observed for the optical nebulosity *and* the X-ray emission (cf. Fig. 1). A physical connection between the different phases of the interstellar medium therefore seems likely. The energy deposited from the hot gas into heating of the dust grains is consistent with the temperature distribution of the X-ray-emitting gas, which is found to be lowest in the dusty regions. The optical extinction of the dust is consistent with the Galactic extinction curve. We argue that the dust as well as the optical nebulosity are products of an interaction with a small, gas-rich galaxy, *not* remnants of a cooling flow. A full account of this work is currently in press in *Astronomy and Astrophysics*, and a preprint is available through http://www.stsci.edu/science/preprints/prep1191/prep1191.html.

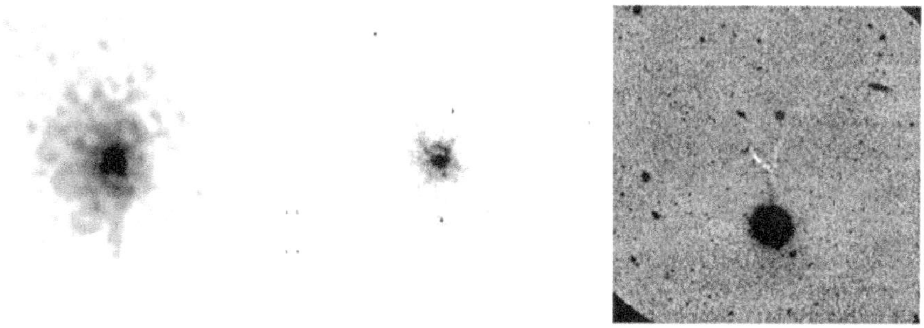

Figure 1. Grey-scale reproductions of the distributions of X-ray emission [using *ROSAT* HRI data] (left), Hα+[N II] emission (middle), and A_V of dust extinction in the central 200 × 200 arcsec of NGC 5846. North is up and east is to the left.

[†]P. Goudfrooij is affiliated to the Space Science Department, European Space Agency

THE ORIGIN OF HIGH SPECIFIC FREQUENCY GLOBULAR CLUSTER SYSTEMS

M.G. LEE
Department of Astronomy, Seoul National University
Seoul 151-742, KOREA

AND

D. GEISLER
KPNO/NOAO
950 N. Cherry Av., Tucson, AZ 85719 USA

There are known to be several giant elliptical galaxies with high globular cluster specific frequencies, which possess about three or more times the normal number of globular clusters for their luminosity. The origin of high specific frequency globular cluster systems is not yet known.

We have performed a definitive test of the idea of an intracluster origin for the high specific frequency by investigating the globular cluster system in NGC 4696. NGC 4696 is a giant elliptical, but not a cD galaxy, located at the dynamical center of the rich Centaurus cluster. The intracluster origin scenario predicts about 6 times more globular clusters than normal for its luminosity.

The luminosity function of the globular clusters shows a peak at $T_1 = 24.5 \pm 0.1$. Comparing this value with that of the Galactic globular clusters, we estimate the distance to NGC 4696, obtaining $(m - M)_0 = 32.1 \pm 0.2$ for the reddening of $E(B - V) = 0.12$ ($d = 26$ Mpc). Incompleteness is not significant up to $T_1 \sim 24.8$ mag. Using the luminosity function of the globular clusters, we estimate the total number of globular clusters to be $N_{\rm T} = 4100 \pm 200$. From this value and the absolute magnitude of NGC 4696 ($M_{\rm V}^{\rm T} = -22.1$), we derive a value for the globular cluster specific frequency: $S_{\rm N} = 6$. Then the excess number of globular clusters in NGC 4696 compared with $S_{\rm N} = 4$ case is ≈ 1400. This value is much lower than that expected by the intracluster origin theory, $\approx 20,000$. This result rules out the hypothesis of an intracluster origin for the high specific frequency globular cluster systems.

ANGULAR MOMENTUM TRANSFER DUE TO GALACTIC WINDS AND COOLING FLOWS

V. MISSOULIS
Astronomical Institute, National Observatory of Athens
P.O.Box 20048 - GR 11810 Athens - Greece

We examine a model of galaxy formation where the bulge is formed at very early stages and this burst of star formation leads to a galactic wind which interacts with a huge surrounding gaseous envelope.

The numerical solutions of the spherically symmetric situation (Missoulis 1994 and references therein) indicate that if the surrounding envelope is constituted mainly of cold clouds, the clouds will evaporate and their matter will be mixed with the hot and metal-enriched gas of the wind. If the surrounding is constituted mainly of diffuse gas, a cold shell will be formed and cloud and star formation are possible.

In a rotating system these phase transitions will lead to angular momentum transfer inwards: In the first case the earlier layers of the wind tend to evaporate clouds from the inner regions. The latter layers tend to evaporate clouds from the outer regions. When the cooling flow starts, the earlier layers with small specific angular momentum are located outside the latter layers. In the second case the angular momentum is transferred when the newly formed clouds from the outer regions start falling towards the center of the system and evaporate at the inner regions. The angular momentum transfer is significant only if the eccentricities of the trajectories of the clouds are large (for the present situation $e \simeq 1$ is expected).

It is clear that in both cases the central regions acquire large specific angular momentum and this is a possible explanation of the origin of Freeman's type II spiral galaxies.

Acknowledgements: The author is grateful to the LOC for financial support.

References

Missoulis, V. 1994, *Astron. Reports*, **38**, 12

STELLAR POPULATIONS IN HIGH-z GALAXY MERGERS

YASUHIRO SHIOYA AND KENJI BEKKI
Astronomical Institute, Tohoku University
Aoba, Sendai 980-77, JAPAN

We investigate the nature of stellar populations of major galaxy mergers between late-type spirals considerably abundant in interstellar medium by performing numerical simulations designed to solve both the dynamical and chemical evolution in a self-consistent manner. We particularly consider that the star formation history of galaxy mergers is a crucial determinant for the nature of stellar populations of merger remnants, and therefore investigate how the difference in star formation history between galaxy mergers affects the chemical evolution of galaxy mergers.

We found that the rapidity of star formation, $C_{\rm SF}$, which is defined as the ratio of the dynamical time-scale to the time-scale of gas consumption by star formation, is the most important determinant for a number of fundamental characteristics of stellar populations of merger remnants. The mean stellar metallicity ($\langle Z_* \rangle$) of the model with larger $C_{\rm SF}$ is larger than that of the model with smaller $C_{\rm SF}$. If galaxy mergers with larger $C_{\rm SF}$ become more luminous elliptical galaxies, the relation between $C_{\rm SF}$ and $\langle Z_* \rangle$ corresponds to *the mass - metallicity relation* which is the conventional interpretation of *the color - magnitude relation.* A negative metallicity gradient fitted by a power-law is reproduced by a dissipative galaxy merger with star formation. In our models, the heavy elements ejected from stars are more efficiently trapped by the stellar component than by the gas component ('chemical segregation'). This result may provide a solution for the *"iron abundance discrepancy problem"*. More detailed discussions are published in Bekki & Shioya (1997a,b).

References

Bekki, K., & Shioya, Y. 1997a, *ApJ*, **486**, 197
Bekki, K., & Shioya, Y. 1997b, *ApJ*, in press

ORTHOGONAL GASEOUS DISKS IN THE E5 GALAXY IC 4889

J.C. VEGA BELTRÁN
TNG, Osservatorio Astronomico di Padova, Italy
E.M. CORSINI AND F. BERTOLA
Dipartimento di Astronomia, Università di Padova, Italy
AND
A. PIZZELLA
European Southern Observatory, Santiago, Chile

IC 4889 is an apparently normal E5 elliptical (Jedrzejewsky 1987, *MNRAS*, **226**, 747) with an extended ionized gas emission (Macchetto et al. 1996, *A&AS*, **120**, 463). We performed long-slit spectroscopic observations along different position angles: 0°, 10°, 90° and 135°.

Here is a brief description of the observed gaseous and stellar velocity field: (i) an overall counterrotation of the ionized gas with respect to the stars is present along the major axis ($PA = 0°$). Moreover in the inner regions ($\pm 10''$) both stars and gas reverse their sense of rotation; (ii) both inner *gaseous* and *star counterrotation* appear at $PA = 0°$ and $PA = 10°$; (iii) the minimum and the maximum velocity gradients for the outer gaseous component are measured at $PA = 135°$ and at $PA = 90°$ respectively. The opposite happens for the inner gaseous component; (iv) both inner and outer star components do not show velocity gradient along the minor axis ($PA = 90°$).

The observed gaseous and stellar kinematics can be understood by considering IC 4889 as a triaxial elliptical with a counterrotating stellar core and two orthogonal gaseous structures that have settled onto the two allowed equilibrium planes. The inner gaseous disk rotates in the same plane as that of the stars namely the plane perpendicular to the minor axis, while the outer disk is settled onto the plane perpendicular the major axis. A multiple acquisition event of external material is claimed to explain the presence of such kinematically decoupled structures.

GASDYNAMICS AND STARBURSTS IN INTERACTING GALAXIES

J. CHRISTOPHER MIHOS
Department of Astronomy
Case Western Reserve University

Abstract. The onset of gaseous inflows and central activity in interacting galaxies is driven largely by induced bars in the host galaxies. The stability of galaxies against growing bar modes is a direct function of their structural properties — galaxies with central bulges or low disk surface densities are more stable against central starbursts than are bulgeless or disk-dominated systems. Low surface brightness galaxies prove less prone to bar formation and central starbursts than do normal high surface brightness galaxies. This stability of LSB disks also resolves many of the dynamical pitfalls encountered when attempting to link poststarburst "E+A" galaxies to interactions involving normal high surface brightness galaxy progenitors.

1. Introduction

Overwhelming evidence indicates that galactic collisions can lead to a large scale redistribution of gas in galaxies, driving strong nuclear inflows and fueling central activity (starburst and/or AGN) in many interacting systems. However, a one-to-one correlation between interactions and central starbursts is not evident — many interacting systems show only modest star forming activity, distributed throughout the body of the galaxy. What, then, determines the gasdynamical and star forming response of a galaxy to a gravitational encounter? Detailed N-body simulations of interacting systems have shown that the onset of gaseous inflows is intimately tied to the formation of global bars, which act to drive gas inwards to the central regions. As such, the question of induced *star* formation becomes one of induced *bar* formation — that is, the onset of inflow and activity is determined by a galaxy's stability against growing bar modes.

I describe how the structural properties of galaxies can influence the gasdynamical and star-forming response of galaxies to an interaction. I focus first on major mergers and the effects of central bulges, then turn to more subtle "flyby" encounters and the role of disk surface density in driving starburst activity. We find that differences in galaxy structure lead to significantly different responses; much of the variance in the properties of interacting systems can be traced to differences in the progenitor galaxies.

2. Gasdynamics in Major Mergers

Major mergers of equal mass disk galaxies are thought to result in the most dramatic starburst events. The "ultraluminous" infrared galaxies (ULIRGs) are prime examples of this process, where $\sim 10^{12}\ M_{\odot}$ of gas has been driven into their central regions, fueling intense ($L_{\rm IR} > 10^{12}\ L_{\odot}$) activity (see, e.g., Sanders & Mirabel 1996). To study the evolution of such mergers, we employ N-body models to follow the combined gravitational, hydrodynamic, and star-forming evolution of galaxies experiencing a merging event (Mihos & Hernquist 1994ab; 1996).

We contrast models in which the merging galaxies have different structural properties — in particular, galaxy models with and without central bulges. We employ a system of units wherein the disk mass $M_{\rm d} = 1$, the scale length of the disk $h = 1$, and the gravitational constant $G = 1$. In both models, the galaxies consist of an exponential disk of stars and gas ($M_{\rm gas} = 0.1$) embedded in a spherical dark matter halo with mass $M_{\rm h} = 5.8$ and core radius $\gamma = 1$, truncated at $r = 10h$. In the model which includes a central bulge, the bulge possesses a Hernquist (1990) profile, with mass $M_{\rm b} = 1/3$ and scale length $a = 0.2$. Rotation curves for the different models are shown in Figure 1ab. Star formation is included via a simple Schmidt law: SFR$\sim \rho_{\rm gas}^{1.5}$ (see Mihos & Hernquist 1994b). The galaxies are placed on (initially) parabolic orbits, with a Keplerian pericenter of $R_{\rm p} = 2.5$. One disk is exactly prograde, the other is inclined by 71° to the orbital plane. Figure 2 shows the inflow and star forming properties of each model; images of the models can be found in Mihos & Hernquist (1994a, 1996).

Even though the interaction parameters are identical, the star forming response of the two models is dramatically different. The galaxies without bulges rapidly develop strong bars — the $m = 2$ mode in the stellar disk dominates the mass distribution shortly after the galaxies first collide. Gas is compressed along this bar, forming a gaseous bar which slightly leads the stellar bar. This offset between the stellar and gaseous bars results in a net torque on the gas, driving the strong inflow of gas into the nuclear regions. At this time, starburst activity is triggered in each nucleus while the galaxies are still widely separated. These starbursts deplete the gas, so

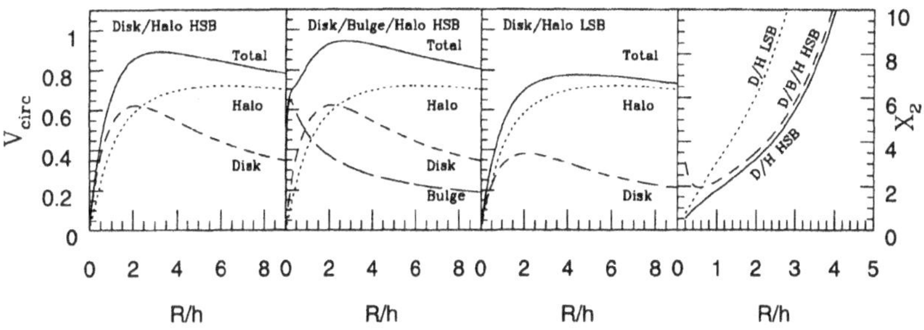

Figure 1. Rotation curves of galaxy models. The first three panels show the contribution of different components to the total rotation curve of each galaxy, while the final panel shows the Toomre X_2 bar stability parameter for each model.

that when the galaxies ultimately merge, they are gas poor and lack the fuel to power any strong starburst associated with the final merging. As such, these models are poor representations of ULIRGs, which are gas-rich, late stage mergers with strong central activity. Evidently a major merger alone is not a sufficient condition to trigger ultraluminous activity; some other criteria is necessary.

In contrast, the merger involving galaxies with bulges has a very different history of inflow and starburst activity. The presence of a central bulge acts to stabilize the galactic disks against the growth of bar instabilities; instead, the galaxies form tightly wound spiral arms which provide a weaker torque on the disk gas. As a result, the gas inflow occurs in two stages. Initially, the gas moves inwards, but "hangs up" at a radius of a few kpc, where the bulge dominates the mass distribution and the disk torques are weaker. This weak inflow results in only a modest enhancement of the star formation rate, and the gas is not strongly depleted. When the galaxies do finally merge, the accompanying strong torques result in a second phase of inflow — the gas in both galaxies is very quickly driven into the center of the merger, and an extreme starburst event is triggered. Unlike the bulgeless merger, this merger with bulges has properties (morphology, gas content, starburst strength) which compare favorably with observed ultraluminous infrared galaxies. It is the internal dynamics of the merging galaxies which is the necessary criterion for the formation of ULIRGs.

As these models demonstrate, the detailed response of galaxies to a merger depends critically on their stability against the onset of global bar modes. This stability has been characterized by the Toomre X_2 parameter (Toomre 1981): $X_2 = \kappa^2 R / 4\pi G \Sigma_{\rm disk}$. If $X_2 < 1$ (for a linearly rising rotation curve) or $X_2 < 3$ (for a flat rotation curve), disks are susceptible to growing $m = 2$ modes. Fig 1d shows X_2 for the different models — by

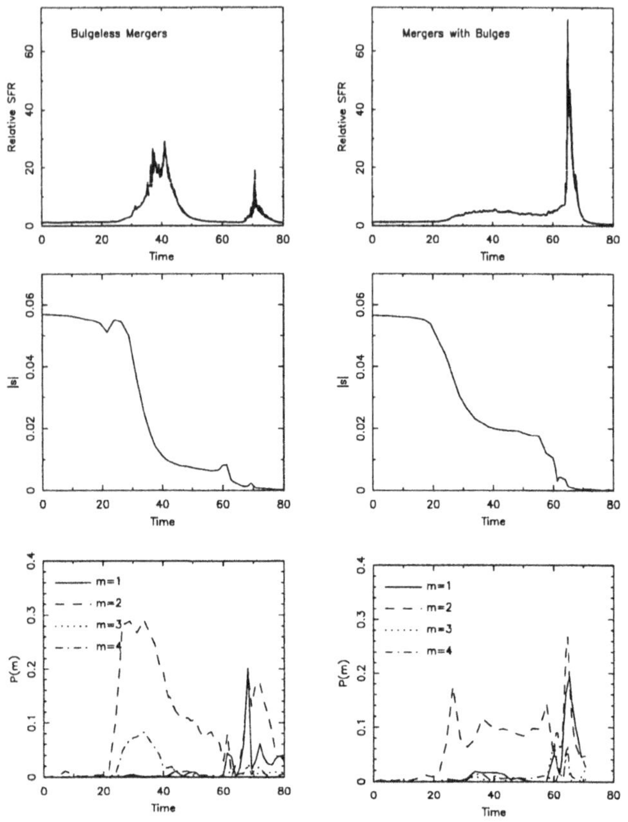

Figure 2. Gas inflow and starburst activity in major mergers. Left panels show the evolution of the bulgeless galaxy models, while the right panels show the evolution of the models with central bulges. Top panels show the star formation history in the models (assuming a Schmidt law for star formation); middle panels show the spin angular momentum of the inflowing gas in the prograde galaxy; and bottom panels show a Fourier decomposition of the stellar mass distribution in the prograde disk. The rotation period at the half mass radius is $T_{\rm rot} \sim 15$. Initial collision occurs at $T = 24$; the final merging occurs at $T = 65 - 70$.

changing the shape of the rotation curve, the bulge acts to stabilize the inner disk against bar modes. Without bulges, colliding disk galaxies become bar unstable on a dynamical timescale, and experience early inflows and central activity. Adding a bulge inhibits bar formation and the associated early inflow, resulting in more dramatic activity when the galaxies ultimately merge. Clearly these two model represent "endpoints" of a distribution of bulge:disk ratios in galaxies; the response of individual systems will depend in detail on the their structural properties and progenitor type.

3. Flyby Encounters and LSBs

The previous merger models show the connection between inflow, starbursts, and disk stability. However, there is another path to disk stability besides central bulges, and that is through lowered disk surface density. If the density of the disk is sufficiently low (at fixed rotation velocity), perturbations cannot be amplified into strong bar modes, and bar-induced inflows are suppressed. Such may be the case in low surface brightness (LSB) disk galaxies, which have low disk surface densities and large dark matter contents (de Blok & McGaugh 1996, 1997).

To examine how disk surface density influences inflow and star forming activity during galaxy interactions, we look at the evolution of galaxies experiencing an equal-mass, non-merging "flyby" encounters. Again, two models are contrasted. The first, representing a high surface brightness (HSB) disk galaxy, is the bulgeless disk/halo galaxy model employed in the previous merger simulations. The second model, representing a LSB disk galaxy, is simply the HSB model with the disk surface density reduced by a factor of 2.5 — i.e., $\Delta\mu_0 \sim 1$ mag/arcsec^2 for a similar $(M/L)_*$. The rotation curve for this model is shown in Figure 1; the low disk surface density results in a greater stability against growing disk modes than in HSB disk galaxies (as shown by the higher value of X_2), save for the very central regions where the disk still contributes an appreciable amount of the total mass density. Finally, compared to HSBs, LSBs generally have a higher gas-to-baryonic mass ratio and flatter gas mass profiles (de Blok et al. 1996; McGaugh & de Blok 1997); this is reflected in the LSB model which possesses a flatter gas mass profile with $M_{\rm gas}/M_{\rm disk} = 1/3$ (see Figure 3).

The flyby interactions involve parabolic orbits with pericenter separation of $R_{\rm p} = 10h$. Figure 3 shows the evolution of the ISM component in the different models. In the prograde HSB encounter, the galaxy quickly develops a strong bar (see Mihos et al. 1997); gas is compressed along this bar and is rapidly driven inwards. By $T = 36$, only one half-mass rotation period for the disk, already $\sim$ 30 % of the gas has been driven into the inner kpc (assuming a Milky Way scaling for the model); the calculation was stopped here, but inflow continues along the strong bar in the model.

By contrast, the prograde LSB disk lacks sufficient self-gravity in the disk to amplify the perturbation of the interaction into a strong bar. Instead, the galaxy develops a milder oval distortion with strong spiral arms. Gas is compressed along these arms, and fragments into small clumps throughout the disk; presumably these would be sites of enhanced star formation. There is some mild inwards migration of gas in the system, but the mass distribution in the inner scale length is largely unchanged — without

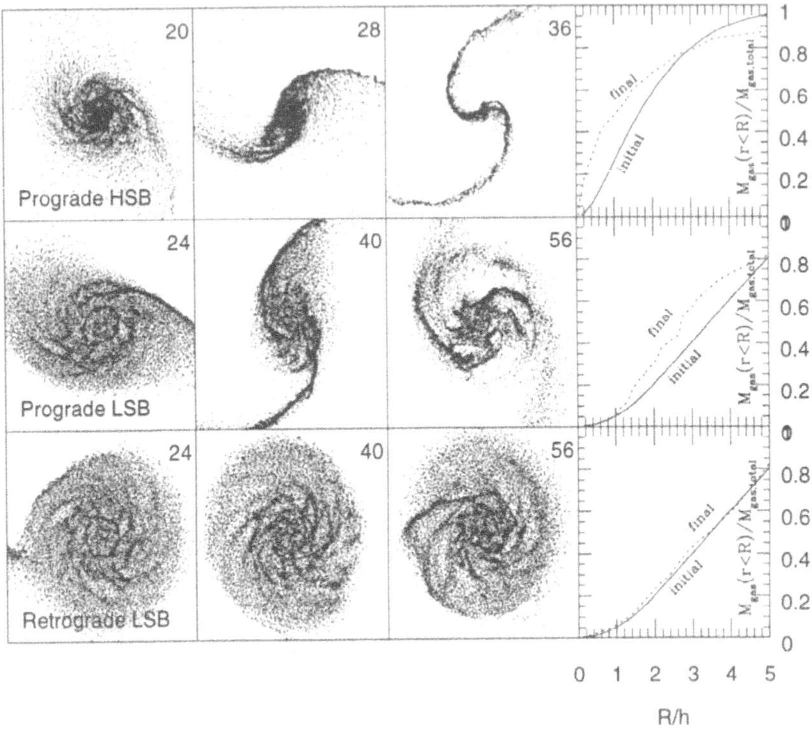

Figure 3. The evolution of the ISM in flyby interactions. In each strip, the first three panels show the morphology of the disk gas after the galaxies pass closest approach (at $T = 24$); the final panel compares the initial and final ($T = 36$ for the HSB disk, $T = 56$ for the LSB disks) cumulative gas mass profiles in each model. The rotation period at the half-mass radius for the disk is $T_{\rm rot} \sim 15$. Note the different timescales between the HSB and LSB models; the HSB model quickly develops a bar which drives rapid inflow. The LSB models evolve much more slowly.

a strong bar, little inflow into the central regions occurs. We emphasize that the prograde geometry is the "worst case" scenario for driving instabilities in disks, due to the resonance between orbital and rotational motion; other geometries will be much less damaging. For example, the last set of panels in Figure 3 shows a retrograde LSB encounter — very little evolution is observed, even during this relatively close encounter.

Clearly the low disk surface density and high dark matter content of LSB disks affords them a great deal of stability against bars and induced gas inflows. This stability actually strengthens with declining surface brightness (as surface density decreases and dark matter becomes ever more dominant; de Blok & McGaugh 1997), such that very low surface brightness galaxies may in fact be very stable systems, not easily destroyed or turned into starburst HII galaxies by casual interactions. Instead, mild evolution in surface brightness may result due to enhanced disk star formation.

Finally, we emphasize that the stability described here pertains to the *internal* amplification of perturbations into bar modes. This discussion is most apropos to mild interactions or secular evolution of LSBs. Mergers or continual interactions (such as in a cluster environment) may act to overwhelm the disk stability and drive additional activity even in these dark matter dominated systems.

4. The Progenitors of E+As

"E+A" galaxies are systems with spectral characteristics identifying them as having experienced a strong starburst in the past 10^9 years, after which star formation ceased entirely. The fact that such systems are observed in both the cluster and field environments (Zabludoff et al. 1996) argues in favor of formation mechanisms which are not cluster specific, such as galaxy interactions and mergers (e.g., Lavery et al. 1992). While this picture of mergers driving the formation of strong E+A systems works on a qualitative basis, upon closer scrutiny several inconsistencies appear when trying to invoke "normal" (i.e., HSB) spiral galaxies as the merging progenitors. Alternatively, I argue that LSB progenitors may solve a number of these inconsistencies. To wit:

1. *High starburst mass fractions:* Extreme E+A galaxies may have starburst mass fractions $M_{\rm burst}/M_* \gtrsim 0.2$ (Couch & Sharples 1987; Barger et al. 1996; Liu & Green 1996). Such high burst masses indicate the galaxies must have been *extremely* gas-rich; Milky Way-type spirals simply lack sufficient gas to fuel such a starburst. LSBs, on the other hand, are the most gas-rich objects in the local universe (McGaugh & de Blok 1997).
2. *Spatially extended burst populations:* In E+As the young stellar population is often spatially extended, and not confined to the nuclear region (e.g., Franx 1993; Caldwell et al. 1996). This is contrary to observations of local starburst galaxies, which are predominantly nuclear starbursts (e.g., M82, or the ULIRGs). However, the stability of LSB galaxies results in a *global* response to interactions, as gas fragments throughout the disk but is not driven into the galaxies' centers. Disk star formation, rather than nuclear starbursts, is the likely outcome.
3. *E+As in galaxy pairs:* The fact that some E+As are found in interacting pairs (Zabludoff 1996; Wirth 1996) raises a timescale problem — if the interaction caused the starburst, why is the system a *post*-starburst system, even though the interaction is still ongoing? Some mechanism

must act to shut off star formation, independent of the interaction phase. The low gas densities of LSBs, coupled with a threshold density for star formation (e.g., Kennicutt 1989, van der Hulst et al. 1993), may provide such a shutoff mechanism. If the initial collision drives disk gas above threshold, strong disk star formation ensues. As this star formation depletes the gas, the gas density drops back below threshold, and star formation is stopped. Because the cessation of star formation is linked to *local* dynamical conditions, the starburst can terminate irrespective of the dynamical phase of the interaction.

4. *Disky E+As:* Some E+A galaxies are disky systems (Caldwell et al. 1996; Wirth 1996; Franx, this volume). Such E+As can't form through major mergers, which destroy galactic disks. If interactions drive the formation of disky E+As, they must involve low mass accretions or flyby passages. With starburst efficiencies lower in these types of interactions, the gas reservoir must be extremely large, as found in LSBs.

Certainly not all E+As need arise from interacting LSB disks – many E+As show clear merger morphologies, or possess weaker burst strengths. However, the interdependence described here between galactic structure, disk stability, gas inflows, and starbursts suggests that the variety of progenitor galaxies available in the Universe necessarily dictates a variety of poststarburst E+A galaxies. The E+As which do not easily fit into the picture of interactions involving normal "Hubble-type" spirals may in fact follow from LSB progenitors.

References

Barger, A.J. et al. 1996, MNRAS, 279, 1
Caldwell, N. et al. 1996, AJ, 111, 78
Couch, W.J., & Sharples, R.M. 1987, MNRAS, 229, 423
de Blok, W.J.G., & McGaugh, S.S. 1996, ApJ, 469, L89
de Blok, W.J.G., & McGaugh, S.S. 1997, MNRAS, 290, 533
de Blok, W.J.G., McGaugh, S.S., & van der Hulst, J.M. 1996, MNRAS, 283, 18
Franx, M. 1993, ApJ, 407, L5
Lavery, R.J. et al. 1992, AJ, 104, 2067
Liu, C.T., & Green, R.F. 1996, ApJ, 458, L63
McGaugh, S.S., & de Blok, W.J.G. 1997, ApJ, 481, 689
Mihos, J.C., & Hernquist, L. 1994a, ApJ, 431, L9
Mihos, J.C., & Hernquist, L. 1994b, ApJ, 437, 611
Mihos, J.C., & Hernquist, L. 1996, ApJ, 464, 641
Mihos, J.C., McGaugh, S.S., & de Blok, W.J.G. 1997, ApJ, 477, L79
Kennicutt, R. 1989, ApJ, 344, 685
Sanders, D.B., & Mirabel, I.F. 1996, ARAA, 34, 749
Toomre, A. 1981, in The Structure and Evolution of Normal Galaxies, eds. S.M. Fall & D. Lynden-Bell (Cambridge: Cambridge University Press), 111
van der Hulst, J.M. et al. 1993, AJ, 106, 548
Wirth, G.D. 1996, Ph.D. thesis, UC Santa Cruz
Zabludoff, A. et al. 1996, ApJ, 466, 106

FUELING NUCLEAR STARBURSTS

JEROEN P.E. GERRITSEN
Kapteyn Instituut, Postbus 800
9700 AV Groningen, The Netherlands

AND

VINCENT ICKE
Sterrewacht Leiden, Postbus 9513
2300 RA Leiden, The Netherlands

Abstract. We present a numerical simulation of two merging equal-mass, gas-rich disk galaxies. Special emphasis is given to an accurate treatment of the interstellar medium physics and star formation with its feedback. We will explain how the negative feedback from young stars restricts the bulk of the star formation during the merger-induced starburst to the nucleus.

1. Star Formation in Numerical Simulations

The purpose of this study is to clarify the interaction between star formation and the interstellar medium (ISM) using numerical simulations. In particular we have adopted TREESPH, a hybrid N-body/Smoothed Particle Hydrodynamics code (Hernquist & Katz 1989) for our purposes. A full account of the modeling technique can be found in Gerritsen & Icke (1997, 1998).

The novelty of this work consists of the accurate treatment of the ISM physics. Briefly, the thermal balance in the ISM is regulated by a realistic treatment of stellar heating and radiative cooling, where the gas is allowed to cool down to $\sim 10\,$K. While giant molecular clouds (GMCs) can be identified in the simulations, star formation proceeds in unresolved subclumps of GMCs. Hence star formation is treated in a semi-empirical fashion: star clusters are allowed to form (with a fixed assumed initial mass function) from those gas clouds that remain Jeans unstable for longer than the local cloud collapse time. This approach has the advantage of providing a link between the large-scale (resolved) properties of the galaxy being modeled and the (unresolved) sub-process of star formation, removing some of the arbitrariness in the necessarily crude treatment of star formation. The evolution of the young star clusters is followed in detail, and synthetic spectra of evolving star clusters (Bruzual & Charlot 1993) are used to trace the evolution of both the resulting radiation field and supernova (SN) energy. Particular successes of the model are:

- a realistic multi-phase ISM (cold, warm and hot gas) develops naturally;
- the star formation rate (SFR) obtained follows a Schmidt law (SFR $\propto \rho^n$), with exponent $n \approx 1.3$ (ρ is the local gas density), in agreement with observations;
- the SN energy is redistributed over the ISM in such a way that realistic ISM structures (e.g., holes with realistic sizes) are formed.

1.1. GALAXY MODEL

Our galaxy model consists of an isothermal (particle) dark halo, an exponential stellar disk and a gas disk, and is based on observations of the Sc galaxy NGC 6503. Adopting a stellar mass-to-light ratio of $(M/L_{\rm B})^* = 1.75$ (Bottema 1989) yields a total stellar disk mass of $3.49 \times 10^9\,M_\odot$ (much smaller than a typical L^* galaxy). Radial scale length of the disk is 1.16 kpc and scale height is 0.19 kpc. The gas distribution is modeled to decline linearly with radius out to 8 kpc with a total gas mass of $1.26 \times 10^9\,M_\odot$. The total mass of the dark halo inside 12 kpc is $25.0 \times 10^9\,M_\odot$.

2. Star Formation during Equal-Mass Merger

As a merger scenario, we put two model galaxies on parabolic orbits, with a pericenter of 2.5 disk scale lengths (2.9 kpc); one of the disks moves on an exactly prograde orbit, the other disk is highly inclined. The evolution of the merger is detailed in Gerritsen (1997). Evolutions of similar mergers can be found in Barnes & Hernquist (1996) and Mihos & Hernquist (1996), although the evolution of the gas differs in these simulations, since gas and star formation are treated differently in those simulations.

After the start of the simulation, the galaxies move in space largely unperturbed until the first passage. Then large spiral arms develop due to swing amplification, and the galaxies no longer follow the Keplerian orbit. The main bodies of both galaxies are transformed into a bar. At this time the galaxies no longer move away from each other but start to move back. They have a second passage, after which their centers begin to merge.

Figure 1 shows the evolution of the global SFR and phase diagrams of the ISM during the merger. The SFR is enhanced after the first encounter and reaches then a maximum of eight times the pre-encounter rate. This

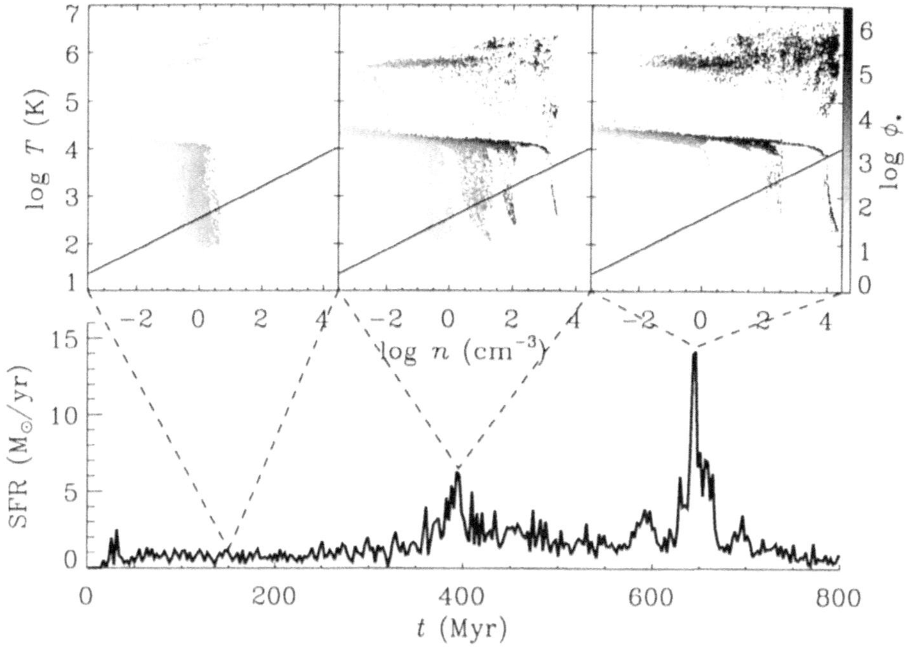

Figure 1. Evolution of the ISM and SFR during the merger. Each dot in the upper panels corresponds to a gas particle. Shown are the gas number density vs temperature, where the greyscale corresponds to the intensity of the stellar radiation field. Gas particles which fall below the solid line (Jeans criterion) may form stars on a dynamical timescale. The lower plot shows the evolution of the global SFR.

occurs when a bar causes the gas to flow into the nuclei of the separate galaxies as can be seen in the left panel of Fig. 2.

The SFR reaches a second peak just before the final merging of the nuclei of the individual galaxies (right panel of Fig. 2). During this burst the SFR is enhanced by a factor 20: most nuclear gas is converted into stars. Afterwards the SFR drops to very low levels.

3. Interstellar Medium

The top panels of Fig. 1 show the evolution of the ISM during the merger. In the left diagram the unperturbed ISM is visible. The cold gas is confined to a rather limited range in density. Gas below a critical density cannot cool and remains warm, which implies a thermal threshold to star formation.

The middle panel shows the ISM at first star forming burst. The density of the densest gas has increased by three orders of magnitude. Likewise the intensity of the stellar radiation field increased. Cold gas condensations appear at various distinct densities. In between gas occupies densities where

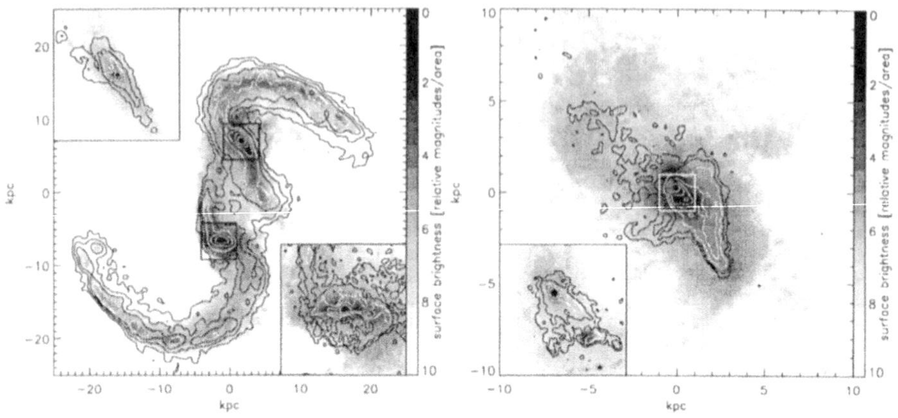

Figure 2. Distribution of gas and stars at first (left) and second (right) peak of the SFR. Greyscale represent the (absorption free) K-band stellar distribution with the intensity scale in magnitudes. The contours represent the gas surface density, with separations of one magnitude. Insets show close-ups as indicated by the boxes.

the stellar heating prevents its cooling and hence star formation.

This effect is at its extreme in the right panel, which shows the ISM at the peak SFR. The influence of the thermal threshold is especially strong now and limits ongoing star formation to either the merging nuclei or the tails. All gas not in those locations is kept warm and does not participate in star formation.

4. Conclusion

Dynamical and thermal processes in merging galaxies cooperate to produce nuclear starbursts. The dynamics drive gas into the nuclei. The energy produced when this high-density gas is converted into stars prohibits star formation outside the nuclear region. The stronger the starburst, the stronger this thermal threshold acts. Thus the strongest starbursts are nuclear.

References

Barnes J.E., Hernquist L., 1996, ApJ 471, 115
Bottema R., 1989, A&A 221, 236
Bruzual G.A., Charlot S., 1993, ApJ 405, 538
Gerritsen J.P.E., 1997, PhD Thesis, University of Groningen
Gerritsen J.P.E., Icke V., 1997, A&A 325, 972
Gerritsen J.P.E., Icke V., 1998, A&A, in press
Hernquist L., Katz N., 1989, ApJS 70, 419
Hernquist L., 1993, ApJS 86, 389
Mihos J.C., Hernquist L., 1996, ApJ 464, 641

MOLECULAR GAS AND STAR FORMATION IN INTERACTING AND ISOLATED GALAXIES

J.S. YOUNG
Department of Physics and Astronomy
Univ. of Massachusetts and FCRAO
Amherst, MA 01003 USA

Abstract. The results of the FCRAO Extragalactic CO Survey are used to examine the trends regarding the molecular gas distribution, the star formation efficiency, and the global gas surface densities (HI and H_2) in galaxies as a function of environment. Relative to a sample of isolated Sbc-Scd galaxies, the strongly interacting galaxies have more compact gas distributions, a higher mean value for the global star formation efficiency, and a larger fraction of gas in molecular form. Not only is the molecular gas redistributed during interactions, but evidence is presented for an enhanced conversion of atomic to molecular gas as well.

Among the merger remnants, the star formation efficiency is found to vary by almost two orders of magnitude. Part of this variation is shown to be a function of the merger age, in that the star formation efficiency increases with merger age. Additionally, a new result is presented showing that the strongly interacting galaxies exhibit a completely different trend in the star formation efficiency as a function of galaxy size, when compared with the isolated, paired, and cluster galaxies of similar dimensions.

1. Introduction

Just 20 years ago, galaxy-galaxy interactions were recognized as capable of significantly influencing the evolution of a galaxy through the triggering of bursts of star formation (Larson & Tinsley 1978). Since stars form in molecular clouds, a thorough understanding of galaxy-galaxy interactions and the subsequent bursts of star formation requires knowledge of the distribution and abundance of the molecular gas from which stars form in a galaxy. It is the inter-comparison of the gas and star-forming properties of interacting galaxies with those of isolated galaxies which can most clearly reveal the power of galaxy-galaxy interactions.

The molecular gas observations which will be used in this paper are taken from the FCRAO Extragalactic CO Survey (Young et al. 1995). These observations provide a unique data base among galaxies for the examination of the global gas content and star formation efficiency within and among galaxies as a function of morphology and environment. Using the 14 meter millimeter telescope of the Five College Radio Astronomy Observatory (HPBW=45″) we have observed 300 galaxies at 1,412 positions, with detections in 236 galaxies (79%), and major axis observations in 193 galaxies. The sample of 300 galaxies is not complete in any sense, since the galaxies were chosen to cover a wide range of morphological type, luminosity, and environment. The galaxies in the CO Survey were selected at either optical or IR wavelengths; most are bright spirals, with 260 galaxies ranging from type S0 through Sm. Among galaxies which are either brighter than 11th magnitude in the blue or brighter than 15 Jy at 100 μm, the CO Survey is 80% complete.

For over half of the galaxies observed, the 45″ HPBW of the 14 m telescope corresponds to a linear resolution of 5 kpc or less. Therefore, the molecular gas surface densities we are measuring correspond to regions within galaxies which are considerably larger than the sizes of individual giant molecular clouds, and therefore represent average surface densities over large regions. Distances to the galaxies in the CO Survey sample were derived assuming a Hubble constant of 50 km s^{-1} Mpc^{-1}. The H_2 masses used here were derived from major axis CO observations, assuming a constant CO→H_2 proportionality of 2.8×10^{20} cm^{-2} $[K(T_R)km\ s^{-1}]^{-1}$ (Bloemen et al. 1986). The global H_2 masses derived for luminous spirals have been shown to be statistically accurate to ~40% (Devereux & Young 1990), while the absolute value of the CO→H_2 proportionality for the Milky Way has been found to also hold for spirals as diverse as M31 (Sb) and M33 (Scd), by observations which resolve individual molecular clouds (*cf.* Young & Scoville 1991, Figure 1).

2. CO Distribution Results

Studies examining the molecular gas distributions in spiral galaxies were undertaken in the late 1970's and early 1980's, at a time when it was already known that the atomic gas distributions were similar among galaxies. This is perhaps the reason which led to the expectation that the molecular gas distributions would be similar among galaxies, and to the surprising result that few galaxies were like the Milky Way with its strong central peak, absence of gas between 1 and 4 kpc in radius, and molecular annulus between 5 and 10 kpc in radius (Burton et al. 1975; Scoville & Solomon 1975). Among the 193 galaxies whose CO radial distributions were observed as part of the FCRAO Extragalactic CO Survey (Young et al. 1995), most spiral galaxies exhibit a central CO peak and a smooth decrease as a function of radius. The implied surface density of molecular gas far exceeds that of the atomic gas in the centers of luminous spiral galaxies of all types. Overall, the mass of molecular gas in a galaxy is comparable to the mass of atomic gas, with as much as 10^{10} $M_{\odot}$ of molecular gas in the inner disk, and a similar amount of atomic gas distributed over a considerably larger surface area.

Only 10 galaxies in the FCRAO CO Survey were found to show central CO holes, or CO rings in the radial distribution of molecular gas. The galaxies with CO rings include types Sab-Sbc, with a peak at type Sb, while the type distribution for the entire CO Survey sample peaks at type Sc. The ring radii among these 10 galaxies range from less than 1 kpc to over 20 kpc. It is interesting that 2 of the galaxies with central CO holes were suggested in this meeting to have counter-rotating bulges.

In isolated and non-interacting galaxies, the molecular gas is found to be generally located within the inner half of the optical disk. There is also an indication that the CO extends over a larger fraction of the optical disk in Sc galaxies relative to Sa galaxies, or that the CO distributions in the Sa galaxies are more compact (Young et al. 1995). This result is interesting since the most compact CO distributions are found in interacting galaxies (Scoville et al. 1986), and given the possibility that some Sa galaxies may grow their bulges by the accretion of satellite galaxies.

3. Star Formation Efficiency Results

One of the most fundamental ways to describe the current epoch star formation in a galaxy is through the comparison of the current star formation rate with the mass of gas available to form stars. This yield of young stars per unit mass of molecular gas is what we call the star formation efficiency (SFE). Recent determinations of the SFE in hundreds of galaxies have shown that similar results are obtained on average, independent of whether one uses star formation rates traced by Hα emission [L(Hα)] or by the far-IR emission [L(IR)] observed by *IRAS* (Young et al. 1996). The results described below are based on both of these tracers of the high mass star formation rate.

Studies of the star formation efficiency in spiral galaxies of differing morphology indicate that there is no variation in the mean SFE [based on $L(\mathrm{IR})/M(\mathrm{H_2})$ or $L(\mathrm{H}\alpha)/M(\mathrm{H_2})$] as a function of Hubble type for morphological types Sa through Sc (Rengarajan & Verma 1986; Devereux & Young 1991; Young et al. 1996). For spiral galaxies of types Sa, Sb and Sc, the mean star formation efficiency within each type has a value of $L(\mathrm{IR})/M(\mathrm{H}_2)$ of 4 $L_\odot/M_\odot$. Thus, even though the galaxies have a diverse morphology and bulge size, the observed constancy of the mean SFE with spiral type indicates a remarkable similarity in the star formation process in the disks of these galaxies.

While similar mean values of the star formation efficiency are found for different morphological types among spiral galaxies, as described above, *different* mean values of the star formation efficiency are found for galaxies in different environments. Not only is the star formation rate elevated in interacting galaxies compared to isolated galaxies (Larson & Tinsley 1978; Lonsdale, Persson, and Matthews 1984; Joseph & Wright 1985), but the star formation efficiency is also elevated in merging systems relative to isolated galaxies (Young et al. 1986; Sanders et al. 1986; Solomon & Sage 1988; Tinney et al. 1990). For the galaxy samples investigated thus far, the star formation efficiency in strongly interacting galaxies has been found to be elevated by a factor of ~5-20 relative to isolated galaxies, and this enhanced SFE is independent of whether one uses star formation rates traced by Hα emission or by the far-IR emission observed by *IRAS* (Young et al. 1996). Interestingly, studies of the SFE in galaxy pairs show that the SFE is not enhanced in many systems (Solomon & Sage 1988), and that a very close encounter is needed to produce an observable enhancement in the SFE.

Since the earliest investigations of the star formation efficiency in interacting galaxies, it has been known that both the SFE and the dust temperature are elevated in these galaxies (Young et al. 1986; Solomon & Sage 1988). This trend is illustrated in Figure 1, where the ratio $L(\mathrm{IR})/M(\mathrm{H}_2)$ is plotted versus the ratio of 60 to 100 μm *IRAS* flux densities for a sample of 200 galaxies covering a wide range of environment. These results are consistent with a scenario in which the dust in molecular clouds is heated by young stars, and as the star formation efficiency increases, the resulting increase in the energy density of the radiation field heats the dust to higher temperatures. It is interesting that the ratio of far-infrared luminosity to *atomic* gas mass does not show a tight correlation like that in Figure 1, supporting the suggestion that it is primarily the dust in *molecular* clouds which is radiating at the 30-40 K dust temperatures indicated by the *IRAS* data.

Among the merging and closely interacting galaxies, a wide spread in the SFE is observed, a spread which is wider than the factor of 10 observed spread in SFE for the isolated galaxies. This scatter is also considerably larger than the measurement uncertainties, and is probably related to the age of a merger (Young et al. 1986). Joseph & Wright (1985) have suggested that the age of a merger can be estimated qualitatively based on the presence of two distinct disks in the younger systems (such as NGC

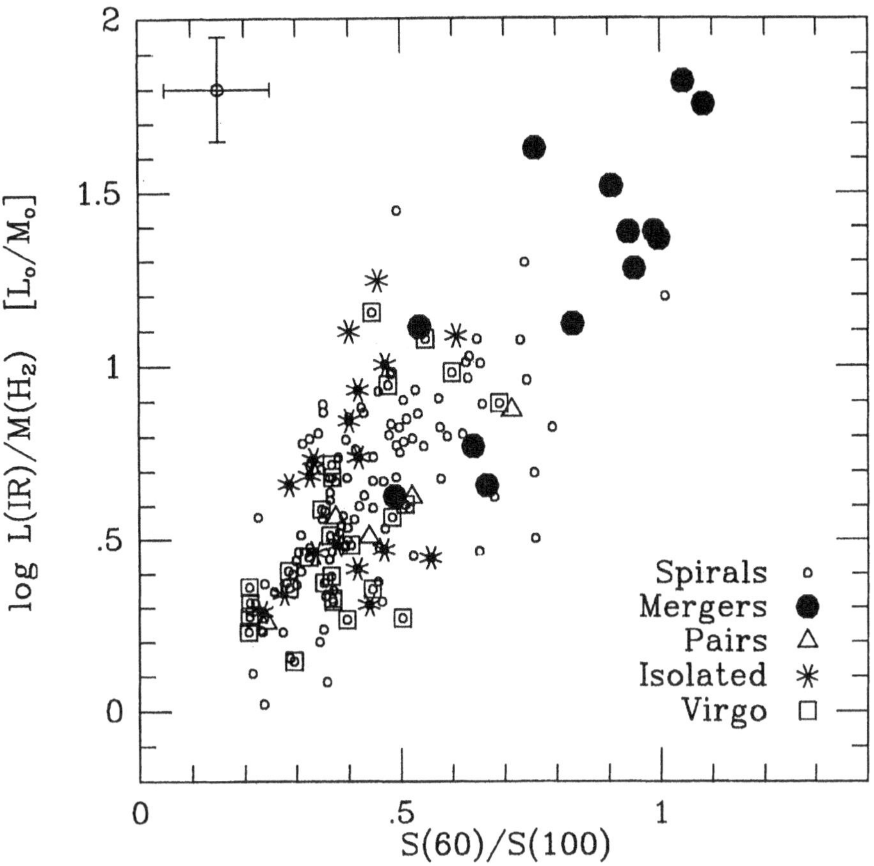

Figure 1. Comparison of the ratio $L(\mathrm{IR})/M(\mathrm{H_2})$ with the ratio of *IRAS* 60/100 μm flux densities for 200 spiral galaxies in the FCRAO Extragalactic CO Survey (Young et al. 1995). The galaxies are coded by environment as indicated. Note that the merging/interacting galaxies have the highest values of both the SFE and $S(60)/S(100)$ relative to the isolated galaxies, although the intermediate SFE values are shared by galaxies in all environments. Part of the scatter in the SFE for the merging and interacting galaxies results from an increase in the SFE with merger age (see text).

4038/39 and NGC 520), and that as a merger ages, the nuclei are found closer together, until the merger remnant resembles a single galaxy (such as NGC 3310). Since many of the galaxies in our sample are also in the sample of Joseph & Wright (1985), we have used their estimates of merger age in relation to the values of $L(\mathrm{IR})/M(\mathrm{H_2})$. We find that the lowest star formation efficiencies are found in the youngest systems, and the highest values in the oldest systems, indicating that the star formation efficiency increases with merger age (Young et al. 1986).

A new result regarding the star formation efficiency in interacting galaxies was uncovered in preparing the talk for IAU 186. It is clear from Figure 1 that there is a considerable spread in the star formation efficiency among galaxies found in similar environments. Some of this scatter in the SFE for closely interacting galaxies is related to the merger age, as described above. An additional source of scatter in the star formation efficiency among galaxies in similar environments is shown in Figure 2 to be galaxy *size*.

In Figure 2, the ratio $L(\mathrm{IR})/M(\mathrm{H}_2)$ is plotted versus the galaxy optical linear diameter [$D(25)$] for galaxies in different environments. Among the isolated galaxies, Virgo galaxies, pairs, and field spirals, there is a trend of decreasing mean star formation efficiency with increasing galaxy size. For 18 galaxies smaller than 16 kpc in diameter, the mean SFE is 9±1 $L_\odot/M_\odot$, while for 25 galaxies larger than 50 kpc in diameter, the mean SFE is 3±0.5 $L_\odot/M_\odot$. In sharp contrast to this trend in the SFE with galaxy size, the merging and closely interacting galaxies behave completely differently. At every galaxy size, the mergers have the highest star formation efficiencies for galaxies of that size, and there seems to be an upper limit to the SFE independent of galaxy size. The mergers with the lowest SFE values are found to be among galaxies of the largest size, and these are the young mergers referred to above. Even with these low SFE values, the star formation efficiencies in the large merging systems are as high as the highest values found in non-merging systems of the same size.

The reason for the observed trend of decreasing star formation efficiency with increasing galaxy size is not entirely clear. This effect could possibly be related to the triggering of star formation in normal galaxies. A triggering event such as an individual supernova explosion will impact on a much larger fraction of the area of a small galaxy than a similar event in a much larger galaxy. Additionally, there is a significant difference in the dynamics of a large galaxy versus a small galaxy. A smaller galaxy tends to have a rotation curve which barely turns over, leading to almost solid body rotation over most of the galaxy, while a larger galaxy with a flat rotation curve over almost the entire disk will experience shear throughout (Rubin et al. 1985). The effect of this shear could be to spread out the effects of a star formation trigger, and lessen its efficiency. Thus, given two galaxies with the same mass of gas, the above conditions could lead to higher star formation rates (and therefore higher efficiencies) in the smaller galaxies. Furthermore, the effect of a galaxy-galaxy encounter is to enhance the triggering of star formation on the galaxy wide scale, so that the large merging galaxies develop star formation efficiencies which exceed those of most non-interacting galaxies of the same size.

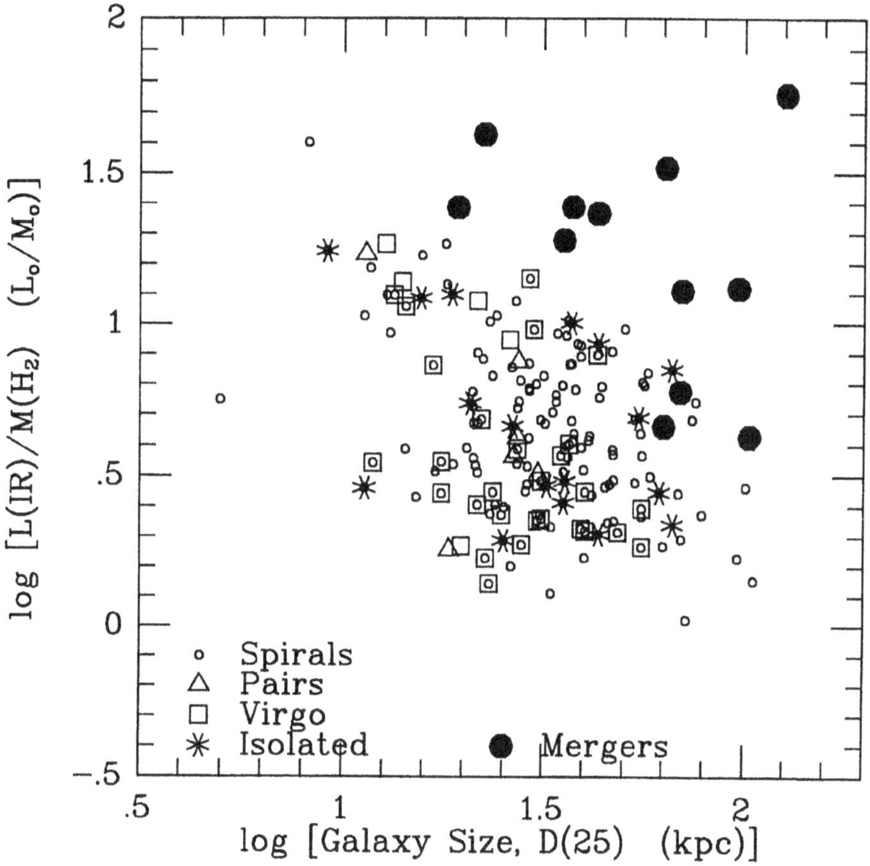

Figure 2. Comparison of the ratio $L(\mathrm{IR})/M(\mathrm{H_2})$ with the linear diameters $[D(25)]$ of galaxies for 200 spiral galaxies in the FCRAO Extragalactic CO Survey. The galaxies are coded by environment as indicated. Among isolated, paired and cluster galaxies, there is a trend of decreasing SFE with increasing galaxy size. The merging/interacting galaxies exhibit a completely different behavior in the SFE with galaxy size.

4. Molecular and Atomic Gas Compared

Since star formation occurs in molecular clouds, and molecular clouds form from atomic gas, the relative amounts of atomic and molecular gas will both have an influence on the future star formation in a galaxy. For this reason, it is important to examine the global H_2/HI ratio and the globally averaged molecular and atomic gas surface densities in order to constrain models which describe the gas and star formation in galaxies. Furthermore, it is necessary to examine galaxies in a range of physical environments in order to determine the prevailing conditions surrounding the most extreme variations in the atomic and molecular gas content.

Among non-interacting spiral galaxies, the global ratio of molecular to atomic gas mass shows a decreasing trend along the Hubble sequence (Young & Knezek 1989; Thronson et al. 1989), with more molecular than atomic gas in early type spirals (S0 and Sa) and more atomic than molecular gas in late type spirals (Sc). In determining the mean ratio of H_2/HI for galaxies of a specific morphological type, it is important to exclude galaxies which are in disturbed environments or in clusters, where the atomic gas may have been stripped from the disk. To this end, we have recently made a careful analysis of the H_2/HI ratio in galaxies for the entire FCRAO Extragalactic CO Survey. Within this sample, we have sufficient statistics to examine the ratio of molecular to atomic gas in isolated galaxies, in closely interacting and merging galaxies, in Virgo cluster spirals, and in non-Virgo spirals along the Hubble sequence.

Among isolated galaxies and non-interacting field galaxies, we confirm the trend of a decreasing mean value in the molecular to atomic gas ratio along the Hubble sequence, with $H_2/HI = 1.6$ in the Sa's and 0.4 in the Sc's. It is interesting that spiral galaxies of types Sa through Sc have similar mean molecular gas surface densities, so that the changing global ratio of molecular to atomic gas results primarily from variations in the HI content of spiral galaxies with type. Thus, the mean gas surface densities (HI+H_2) increase a factor of 2 from type Sa to Sc.

With regard to the effects of environment on the gas properties in galaxies, the molecular and atomic gas surface densities and H_2/HI ratio in the isolated galaxies (types Sbc-Scd) resemble those of the Sc galaxies. It is the mergers and close pairs which exhibit the most unusual gas properties among galaxies. The mean total gas surface density for the pairs and mergers is, not surprisingly, like that for 2 spirals (*i.e.* 2 Sc galaxies or an Sc and an Sa, but not 2 Sa galaxies). While the mean atomic gas surface density among the pairs and mergers is similar to that for an Sc or an isolated galaxy, the mean molecular gas surface density is a factor of 3 higher (*cf.* Braine & Combes 1993). Therefore, the mean H_2/HI ratio in pairs and mergers is 2, four times higher than that in the isolated (Sbc-Scd) galaxies and slightly higher than that in the Sa galaxies. This probably reflects the enhanced conversion of atomic to molecular gas in interactions as well as the removal of some of the more extended HI (Scoville et al. 1986; Mirabel & Sanders 1989). Thus, the dominant gas phase changes from HI in isolated Sbc-Scd galaxies to H_2 in interactions.

It has always seemed curious to me that only the galaxy mergers, and not the galaxy pairs, exhibit an enhanced star formation efficiency, since clearly many galaxy pairs (NGC 4038/39, for example) show enhanced rates of star formation. The above analysis indicates that the galaxy pairs do show elevated molecular gas surface densities. Therefore, the reason why no enhancement is found in the star formation efficiency is that the infrared surface brightnesses in the galaxy pairs are also elevated by the same amount.

5. Discussion and Conclusions

The results described here, with regard to the trends in the star formation efficiency and the molecular to atomic gas ratio as a function of morphology and environment, provide constraints for theories of the formation and evolution of galaxies, especially with regard to the Hubble sequence and to the effects of galaxy-galaxy interactions.

In spiral galaxies of types Sa-Sc, the mean star formation efficiency does not vary with Hubble type, while the ratio of molecular to atomic gas decreases by a factor of 4 among the same systems. These results can be understood in terms of the size scale on which gravity plays a role in cloud and star formation. To form the molecular clouds, atomic gas on a large scale must be assembled, so that perhaps it is not surprising that the H_2/HI ratio changes with galaxy morphology, or equivalently, with the mass distribution in a galaxy. Once molecular clouds are formed, however, the constant star formation efficiency with galaxy type indicates that star formation is primarily a local process, with gravity on the small scale playing the most important role. Only when a large scale perturbation is present, such as in a bar, a spiral arm, or a tidal encounter with another galaxy, does the star formation efficiency become elevated. This enhanced SFE generally reflects an enhanced concentration of molecular gas.

In merging and interacting galaxies, the gas and star formation are strongly influenced by the interaction, with enhancements in both the global ratio of molecular to atomic gas and the star formation efficiency in merging systems. Not only is the molecular gas redistributed and the atomic gas possibly stripped during an interaction, but enhanced conversion of atomic to molecular gas in interacting galaxies also seems to be required. The triggering of star formation on a galaxy-wide scale then elevates the star formation efficiency by a factor of ~5-20 over that observed in isolated galaxies. As a merger ages, we find that the SFE increases, probably reflecting the cumulative effects of both an increase in the luminosity from ongoing star formation, and the consumption of the molecular gas to form stars.

The future of star formation in the currently merging systems is not clear. This is in part because of the inability to measure the low mass end of the initial mass function. For the "extended" Miller-Scalo IMF (Kennicutt 1983), and given the present rates of high mass star formation and the global masses of interstellar gas in the galaxies we have studied, the gas will be cycled through stars in 10^8 to 10^{10} years. However, some galaxies (*i.e.* M82 and NGC 2146) will cycle the gas in the center in 10^9 years, while the gas in the outer parts can sustain the present rate of star formation for 10^{11} years. Thus, gas depletion may not occur everywhere, and not at all if only high mass stars form.

References

Burton, W.B., Gordon, M., Bania, T., & Lockman, F.J. (1975), *Ap.J.*, **202**, 30.
Devereux, N., & Young, J.S. (1990), *Ap.J.*, **359**, 42.
Devereux, N., & Young, J.S. (1991), *Ap.J.*, **371**, 515.
Joseph, R.D., & Wright, G.S. (1985), *M.N.R.A.S.*, **214**, 87.
Kennicutt, R.C. (1983), *Ap.J.*, **272**, 54.
Larson, R., & Tinsley, B. (1978), *Ap.J.*, **219**, 46.
Lonsdale, C., Persson, E., & Matthews, K. (1984), *Ap.J.*, **287**, 95.
Mirabel, F., & Sanders, D.B. (1989), *Ap.J.(Letters)*, **340**, L53.
Rengarajan, T.N., & Verma, R.P. (1986), *Astr.Ap.*, **165**, 300.
Rubin, V.C., Burstein, D., Ford, W.K., & Thonnard, N. (1985), *Ap.J.*, **289**, 81.
Sanders, D.B., et al. (1986), *Ap.J.(Letters)*, **305**, L45.
Scoville, N., & Solomon, P.M. (1975), *Ap.J.(Letters)*, **199**, L105.
Scoville, N., et al. (1986), *Ap.J.(Letters)*, **311**, L47.
Solomon, P.M., & Sage, L. (1988), *Ap.J.*, **334**, 613.
Thronson, H., et al. (1989), *Ap.J.*, **344**, 747.
Tinney, C., et al. (1990), *Ap.J.*, **362**, 473.
Young, J.S., Kenney, J., Tacconi, L., Claussen, M., et al. (1986), *Ap.J.*, **311**, L17.
Young, J.S., Xie, S., Tacconi, L., Knezek, P., et al. (1995), *Ap.J.Suppl.*, **98**, 219.
Young, J.S., Allen, L., Kenney, J., Lesser, A., & Rownd, B. (1996), *A.J.*, **112**, 1903.
Young, J.S., & Knezek, P. (1989), *Ap.J.(Letters)*, **347**, L55.
Young, J.S., & Scoville, N.Z. (1991), *Ann.Rev.Astr.Ap.*, **29**, 581.

LUMINOUS IR GALAXIES IN A MERGER SEQUENCE: BIMA CO IMAGING

Y. GAO AND R.A. GRUENDL
University of Illinois, Dept. of Astronomy
1002 W. Green St., Urbana, IL 61801, USA

AND

C.-Y. HWANG AND K.Y. LO
Academia Sinica, IAA
P.O. Box 1-87, Nankang, Taipei, Taiwan 11529, ROC

1. Introduction

The power output in luminous infrared galaxies (LIGs, $L_{\rm IR} \gtrsim 10^{11} L_\odot$, $H_0 = 75\ {\rm km\,s^{-1}\ Mpc^{-1}}$) can approach the bolometric luminosity of quasars and can be provided by either starbursts or dust-enshouded QSOs, or both. Most LIGs appear to comprise of mergers of gas-rich galaxies. So, intense bursts of star formation apparently result from interaction and merging of galaxies, but the exact physical processes involved in collecting the large amount of gas involved and in initiating the starbursts are not well understood.

In order to trace observationally the conditions in the interstellar medium (ISM) that lead to starbursts, we have used the newly expanded Berkeley-Illinois-Maryland Association (BIMA) millimeter-wave array to map the molecular ISM in a sample of LIGs chosen to represent different phases of the interacting/merging process. Most importantly, a few LIGs in the sample might be in a "pre-starburst" phase, so they provide an ideal laboratory for studying the conditions leading to starbursts (Lo, Gao & Gruendl 1997). Our emphasis is on the widely separated LIGs at early/intermediate stages of interaction for this reason, and also for complementary studies to previous CO imaging studies that have concentrated on relatively advanced merger systems (Scoville et al. 1991; Downes & Solomon 1997) in which the ISM has already been highly disrupted by the interaction and starbursts.

Our goal is to sample statistically the evolution of physical conditions of the molecular material in LIGs, and to delineate observationally the development of starburst activities along the merger sequence, using radio continuum, IR and other observations.

2. Interacting Luminous Infrared Galaxies

The observed decrease of $L_{\rm CO}$ with decreasing projected nuclear separation of a sample of interacting/merging LIGs can be interpreted as gas depletion resulting from the merger-enhanced starbursts (Gao & Solomon 1998).

Unlike the morphologically selected "Toomre sequence", the LIGs in our sample have a comparable gas content so that the galaxies in the early stage are likely to have the gas reservoir to reach the ultraluminous starburst phase. Therefore, they are likely to be a better sample for the statistical study of interaction/merger induced starbursts leading to ultraluminous IR galaxies.

3. CO Imaging of a Merger Sequence

We present our CO images in Fig. 1 and the gas properties along a merger sequence in Table 1.

In Table 1, we list the sample galaxies roughly in a merger sequence including a few late stage mergers from literature (marked with *), thus a study of the molecular gas properties at various phases of the merging process can be performed. These are the much needed observations to test those sophisticated simulations (e.g., Barnes, this volume).

Clearly, a sequence of merging is observed in the molecular gas traced by CO:
(1). The morphology of the molecular gas changes from weakly disturbed and separate gas disks to the disturbed or merged-common-envelope gas disks and finally to a single common gas disk for the double nuclei of the two galaxies.
(2). The spatial CO extent drops from $\sim$ 20 kpc for the early mergers to a few kpc for the intermediate mergers. Advanced mergers have typical nuclear CO concentration $\lesssim$ 1 kpc.
(3). The corrected face-on central gas surface density increases from a few times 10^2 $M_\odot$pc^{-2} to $> 10^3$ $M_\odot$pc^{-2} in our sample. However, advanced ultraluminous mergers typically have $> 10^4$ $M_\odot$pc^{-2}. A rapid increase of the nuclear gas surface density is evident along the sequence.
(4). The $L_{\rm IR}/M({\rm H_2})$ ratio (a measure of star formation efficiency, SFE) increases by roughly several times from the early mergers to the intermediate/advanced mergers. We can estimate the central SFE ratio, by scaling the far-IR luminosity and extent with those of the radio continuum emis-

TABLE 1. Luminous Infrared Galaxies in a Merger Sequence.

Source	$R^1_{\rm Sep}$ kpc	$L_{\rm IR}$ $10^{11}L_\odot$	$M(\rm H_2)^2$ $10^{10}M_\odot$	Beam $''$	kpc	CO[3]	$\Sigma_{\rm H_2}$ $10^3 M_\odot {\rm pc}^{-2}$	$L_{\rm IR}/M(\rm H_2)$ $L_\odot/M_\odot$
Arp302	25.8	4.1	8.0	6.0	3.7	u+e	0.4	5.0
N6670	14.6	3.8	5.5	4.8	2.7	u+e	0.3	6.9
U2369	13.1	3.9	3.4	6.1	3.6	u+e	1.2	11.5
Arp55	10.7	4.7	5.8	4.4	3.1	u+e	1.3	8.1
VV114*	6.0	4.2	5.1	3.7	1.4	e	3.3	8.2
N5256	4.8	3.1	2.7	4.6	2.3	u+e	>1.3	11.5
Mrk848	4.7	7.2	3.4	3.9	2.9	u+e	>1.0	21.2
Arp299*	4.5	6.4	1.4	2.3	0.4	e	11.3	45.0
N6090	3.5	3.0	2.4	1.9	1.0	e	3.0	12.5
N1614	2.0	4.1	1.6	5.8	1.7	u	>1.5	25.6
Mrk273*	1.0	13.2	2.7	2.3	1.5	u+e	5.5	47.6
Arp220*	0.4	15.0	3.0	1.0	0.4	e	80	50.0
Mrk231*	0.0	30.4	3.5	0.9	0.7	u+e	30	77.2

[1] Projected separation between the two galaxy nuclei. [2] 4.78 $L_{\rm CO}$ K km s^{-1} pc^2, the single-dish CO luminosity. [3] CO morphology, u≡ unresolved peak; e≡ extended structures resolved by the beam.

sion or the mid-IR emission (Hwang et al. 1998), which tends to increase more drastically than the global SFE along the sequence.

(5). We found that early stage mergers or pre-mergers like Arp 302 and NGC 6670 appear to have much smaller SFE throughout the entire interacting/merging disks, comparable to that of GMCs in the Milky Way disk. This strongly suggests that LIGs in the pre-merging stage are in a pre-starburst phase (Lo, Gao & Gruendl 1997).

(6). The starburst phase as indicated by large SFE's seems to start once the molecular gas disks begin merging and the ultraluminous starburst phase seems to be occurring only after the molecular gas disks merge into a common disk in late stage mergers.

(7). Gas seems to respond faster than the stellar component when merging advances to late stage (i.e., the gas disks merge faster than the stellar disks).

References

Downes, D., & Solomon, P.M. 1997, ApJ, submitted

Gao, Y., & Solomom, P.M. 1998, ApJ, submitted

Hwang, C.Y. et al. 1998, this volume

Lo, K.Y., Gao, Y., & Gruendl, R.A. 1997, ApJ, 475, L103

Scoville, N.Z. et al. 1991, ApJ, 370, 158

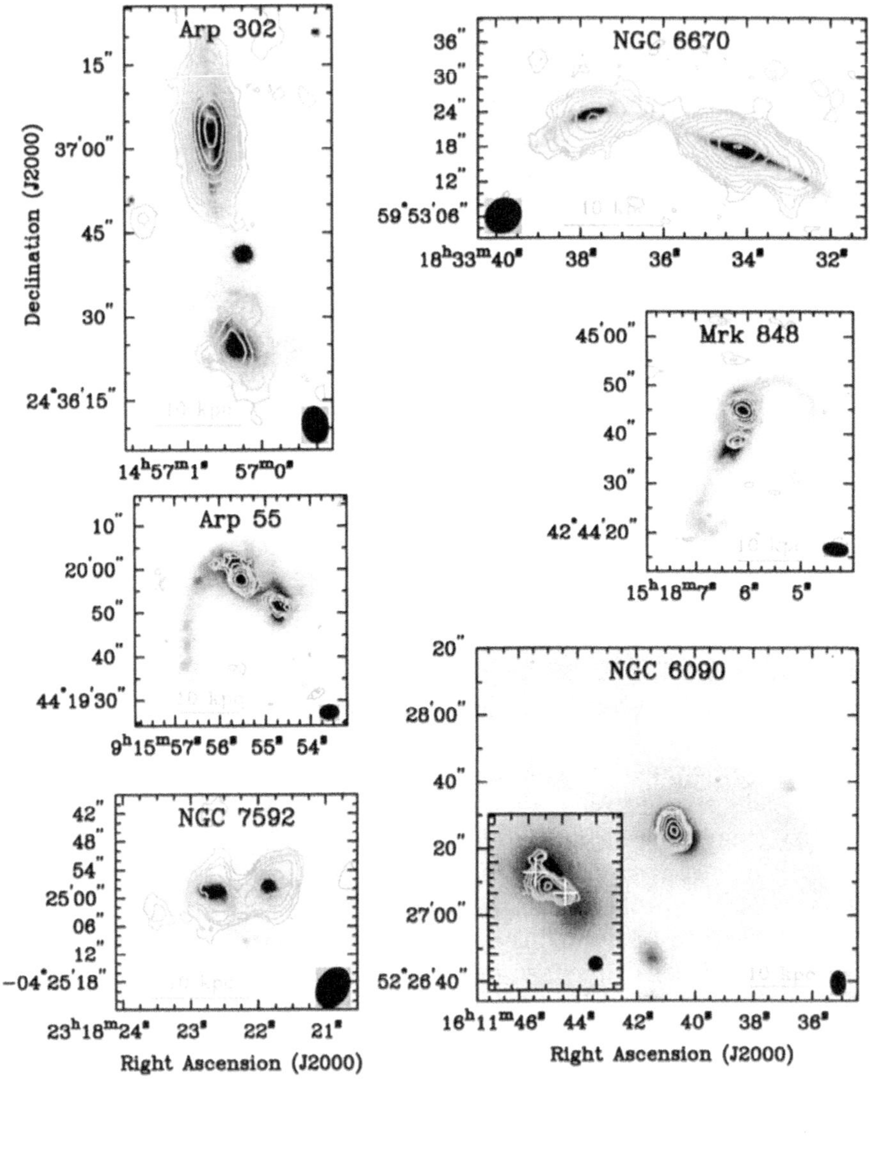

Figure 1. A merger sequence leading to ultraluminous starburst (CO contours overlaid on optical images). The insert in N6090 shows higher resolution ($\sim 2''$) CO contours with two radio continuum peaks marked.

DRAMATIC CHANGES IN MOLECULAR CLOUD PROPERTIES ACROSS THE ARP 299 MERGER

S. AALTO
Onsala Rymdobservatorium, Onsala, Sweden
Caltech, Pasadena, USA

S.J.E. RADFORD
NRAO, Tucson, USA

AND

N.Z. SCOVILLE AND A.I. SARGENT
Caltech, Pasadena, USA

Arp 299 is an IR-luminous ($L_{\rm IR} \approx 8 \times 10^{11}\ L_{\odot}$) merger system of two galaxies, IC 694 and NGC 3690, at a distance of 40 Mpc. Its proximity and richness in molecular gas make it suitable for studying the effects of gravitational interactions and starburst activity on molecular cloud distributions, dynamics, and physical conditions. Previous low resolution observations ($10 - 30''$) indicated that the CO/^{13}CO 1–0 intensity ratio is unusually high, > 20, suggesting an unusual population of molecular clouds (Aalto et al. 1991; Casoli et al. 1992).

Using the Caltech six-element OVRO array we have obtained high resolution maps of CO 1–0, ^{13}CO 1–0 and HCN 1–0 of the inner $1'$ of Arp 299 (Aalto et al. 1997). We have recently completed an OVRO CO 2–1 map of the same region and a ^{13}CO 2–1 map in a $33''$ field centered on IC 694. The synthesised beams are $1.''0 \times 0.''8$ for ^{12}CO 2–1 (uniform weighting); $3.''2 \times 2.''5$ for ^{13}CO 2–1 (natural weighting); $2.''5 \times 2.''2$ for ^{12}CO 1–0 (uniform weighting); $4.''3 \times 3.''6$ for ^{13}CO 1–0 (natural weighting); $5.''6 \times 5.''3$ for HCN 1–0 (natural weighting).

1. The CO, ^{13}CO, and HCN distributions

The CO distribution is dominated by compact structures at the centers of IC 694 and NGC 3690, and two bright CO peaks in the region where the disks of the two galaxies overlap. Extended CO emission connects the main high surface brightness regions, and emission also extends to the south east

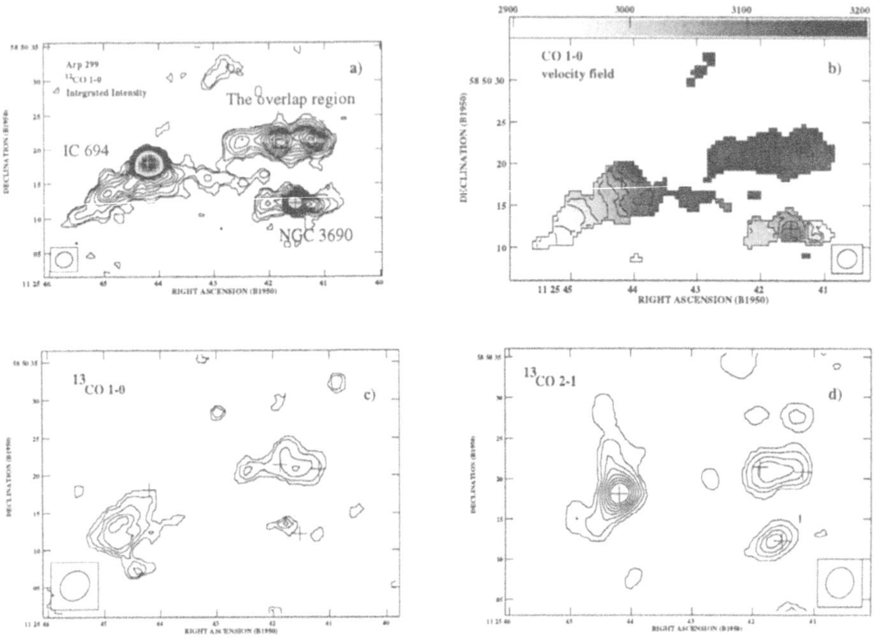

Figure 1. a) The CO 1–0 integrated intensity map (uniform weighting, contour levels are 0.9, 1.8, 3.6, 5.4, 7.2, 9, 10.8, 12.6, 14.4, 16.2, and 18 Jy beam^{-1} km s^{-1}, peak flux is 97 Jy beam^{-1} km s^{-1}. The crosses indicate 6 cm radio continuum positions (Gehrz et al. 1983). b) The CO 1-0 velocity field, the grayscale range from 2900 km s^{-1} to 3200 km s^{-1}. c) The ^{13}CO 1–0 integrated intensity map (natural weighting, contour levels are 0.4, 0.8, 1.2, 1.6, 2.0 Jy beam^{-1} km s^{-1}). d) The ^{13}CO 2–1 integrated intensity map smoothed to the ^{13}CO 1–0 resolution, (natural weighting, the lowest contour level is 1 Jy beam^{-1} km s^{-1} and the peak flux is 15 Jy beam^{-1} km s^{-1}. Note NGC 3690 and the overlap region appear at the edge of the OVRO 220 GHz primary beam and the intensities are therefore attenuated since the map has not been corrected.

of the IC 694 nucleus and to the east and west of the NGC 3690 nucleus. There is also emission north of the main features (Figure 1a).

The most striking feature of the ^{13}CO 1–0 map (Figure 1c) is the *absence of strong emission at the nucleus of IC 694* and at the other strong CO peaks. Emission is detected in the disk region of IC 694, in the overlap region, and in NGC 3690 (although not at the center). Unlike Casoli et al. (1992), we find significant variation in the ^{12}CO/^{13}CO 1–0 line ratio across Arp 299. From the very high value of 60$\pm$15 at the IC 694 nucleus, the ratio drops to 10 $\pm$ 3 in the disk region, a value quite typical for normal disks of galaxies. Within the overlap region, there is an east-to-west gradient in the CO/^{13}CO intensity ratio. The highest ratio is seen in the western, most active peak. Unlike ^{13}CO 1–0, HCN 1–0 is brightest in the two galaxy nuclei and not detected in regions of extended CO emission. The spatial correlation between the ^{13}CO emission and radio continuum emission peaks is also poor. There is no general lack of ^{13}CO in the Arp 299 system, which one would likely expect if faint ^{13}CO 1–0 emission was caused by abundance anomalies in low metallicity gas falling in from large radial distances. Instead, the extreme values of CO/^{13}CO 1–0 line ratios in high CO surface brightness regions are *mainly* due to excitation effects in the gas caused by unusually high gas kinetic temperatures in regions of active star formation. The most extreme gas properties are found in the nucleus of IC 694 which may be an AGN. In the *extended disk* of IC 694, the cloud properties may instead be similar to those of cold giant molecular clouds in the Galaxy.

2. Hot molecular gas in the IC 694 nucleus

The ^{13}CO 2–1 emission is dominated by the nucleus of IC 694, instead of its disk (Fig 1d) where ^{13}CO 1–0 is strongest. Hence, there are indeed ^{13}CO molecules in the inner region of IC 694, but they are highly excited. The observed ^{13}CO 2-1/1-0 integrated intensity ratio, 3.75 ± 0.5, is unusually high (values close to unity are often reported for centers of starburst galaxies). The observed ^{13}CO 2–1/1–0 ratio is close to the theoretical upper limit for optically thin, thermalized emission and sets lower limits of 100 K to the gas kinetic temperature and 5000 cm^{-3} to the gas density. It is unlikely that significant amounts of cold molecular gas prevails within the inner 3″, since the ^{13}CO 2–1/1–0 intensity ratio of such a gas component would be close to unity, lowering the average ratio. This means the molecular clouds of the IC 694 nucleus must be heated throughout, which may be an unusual situation. In centers of other galaxies, it is likely only the surfaces of the clouds that are heated, while their inner, denser regions remain cooler (see Aalto et al. 1994).

The ^{13}CO 1–0 and 2–1 lines are optically thin. The optical depth of the CO 1–0 line is likely to be moderate ($\tau \approx 1$) both because the CO 2–1/1–0 intensity ratio is greater than unity, ≈ 1.4 and because the CO/^{13}CO 1–0 intensity ratio is large. At a gas density of 10^4 cm^{-3} and a kinetic temperature of 100 K, the CO column density per velocity interval would be $\approx 10^{17}$ cm^{-2}(km s^{-1})$^{-1}$. The CO/^{13}CO abundance range is roughly 50-150 for the clouds in the inner region of IC 694. This value is rather typical of clouds in the disk of the Milky Way, but higher than in the Galactic Center.

Even though intensity is attenuated at the edge of the 220 GHz primary beam, the ^{13}CO 2–1 emission is still bright in NGC 3690 and the overlap region. This further emphasizes that ^{13}CO is abundant and widespread in the Arp 299 merger.

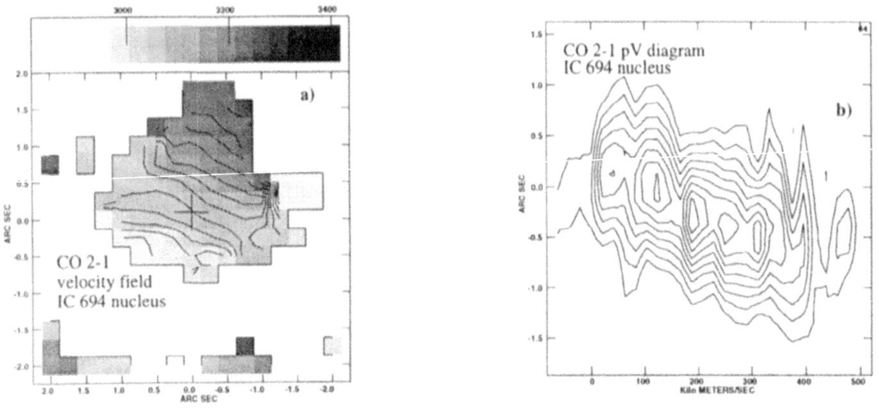

Figure 2. a) The CO 2–1 velocity field in the nucleus of IC 694 ($1.''0 \times 0.''8$ resolution). The grayscale ranges from 2950 to 3400 km s^{-1}. b) The position velocity (pV) diagram at PA 140° cut across the IC 694 nucleus.

3. A rotating molecular disk in the center of IC 694

The overall velocity field shows a monotonic shift from the blueshifted IC 694 disk to the redshifted overlap region and northern feature. Velocity gradients within the overlap region are small. The velocity field of NGC 3690 is complex. The structures both to the east and west of the nucleus appear blueshifted (Fig 1b).

The compact CO emission in the nucleus of IC 694 is now marginally resolved with a FWHM source size of $1.''3 \times 0.''8$, corresponding to a radius of 126 pc. The position angle (PA) of the central velocity field (Fig 2a) is close to 140°. If the central structure (Fig. 2b) rotates as a solid body, the projected velocity gradient is 200 km s^{-1} per 100 pc. In addition to the bright nuclear CO rotating at PA 140° there is fainter CO to the east and west that participates in a different velocity pattern with a PA of 90°.

References

Aalto, S., Black, J.H., Johansson, L.E.B., & Booth, R.S. 1991, A&A, 249, 323.
Aalto et al., 1994, A&A, **286**, 365
Aalto, S., Radford, S.J.E., Scoville, N.Z., Sargent, A.I. 1997, ApJ, **475**, L107
Casoli, F., Dupraz, C., Combes, F., 1992, A&A, **264**, 55
Gehrz, R.D., Sramek, R.A., & Weedman, D.W., 1983, ApJ, **267**, 551

STARBURSTS TRIGGERED BY CLOUD COMPRESSION IN INTERACTING GALAXIES

CHANDA J. JOG
Department of Physics, Indian Institute of Science
Bangalore 560012, India

Abstract. We propose a physical mechanism for the triggering of starbursts in interacting spiral galaxies by shock compression of the pre-existing disk giant molecular clouds (GMCs). We show that as a disk GMC tumbles into the central region of a galaxy following a galactic tidal encounter, it undergoes a radiative shock compression by the pre-existing high pressure of the central molecular intercloud medium. The shocked outer shell of a GMC becomes gravitationally unstable, which results in a burst of star formation in the initially stable GMC. In the case of colliding galaxies with physical overlap such as Arp 244, the cloud compression is shown to occur due to the hot, high-pressure remnant gas resulting from the collisions of atomic hydrogen gas clouds from the two galaxies. The resulting values of infrared luminosity agree with observations. The main mode of triggered star formation is via clusters of stars, thus we can naturally explain the formation of young, luminous star clusters observed in starburst galaxies.

1. Starburst Galaxies: Background

Starburst galaxies are characterized by a high infrared luminosity, $\sim 10^{10} - 10^{12} L_{\odot}$, seen typically over the central region of $\sim$ 1 kpc diameter; and they display a strong central concentration of molecular hydrogen gas. The infrared luminosity is attributed to a high rate of formation of massive stars. The stellar radiation is absorbed by dust in the gas clouds and re-radiated thermally in the infrared.

Starbursts are most often observed in interacting field galaxy pairs and barred galaxies, hence the interaction or the bar must be important in the triggering of the starbursts. For a brief review of the physical properties of starburst galaxies, as well as their implications for related fields, see for example Jog (1995).

1.1. PHYSICAL MECHANISM FOR TRIGGERING OF STARBURSTS

Ever since the discovery of starbursts, the triggering process for them has remained an important and a challenging problem. To understand the starburst phenomenon, it is useful to compare it with star formation in a normal, quiescent galaxy like the Milky Way galaxy. The molecular hydrogen gas in the galactic disk is distributed in giant molecular clouds (GMCs) with masses $\sim 10^5 - 10^6 M_\odot$ each. A typical GMC is self-gravitating and is in a near-virial equilibrium, and forms the site of massive-star formation but has a low star formation rate. Thus, the starburst galaxies pose two puzzles. First, what is the triggering mechanism for the observed burst of efficient star formation? We have addressed this question. Second, what is the origin of the central concentration of gas? Work on this question by a number of authors over the last ten years has clearly established that a galactic tidal interaction or a bar increases the dissipation in gas and this causes most of the gas from a galactic disk to fall into the central regions of the galaxy on timescales greater than the dynamical timescale (e.g., Norman 1992). The infall timescales are $\sim 3 \times 10^8$ yr for tidally interacting galaxies and $\sim 10^7$ yr for evolved mergers, and we use these values in our model. However, these gas infall models do not consider the physical details of triggering of star formation. Some of these models assume that cloud collisions enhance the star formation rate but do not give the details of triggering.

We have addressed the above first question of triggering of starbursts. We propose a novel yet simple *physical mechanism* for the triggering of starbursts by shock compression via high external pressure on a pre-existing disk GMC. The basic idea behind our mechanism is as follows. In an isolated galaxy, a disk molecular cloud is barely stable to begin with. If the ambient pressure around the GMC increases, this overpressure can cause a radiative shock compression of the outer layers of the GMC. The shocked outer shell of a GMC is shown to become gravitationally unstable, resulting in a burst of star formation. We treat the following two cases with different sources of high ambient pressure, these cover the different observed locations of starbursts in starburst galaxies:

Case 1: In tidally interacting galaxies and evolved mergers, the starbursts are observed to occur in the central regions of galaxies. Here the cloud compression is shown to occur due to the pre-existing, high-pressure central intercloud medium (Jog & Das 1992; Jog & Das 1996) - see Section 2.

Case 2: In the early stages of a physical collision between galaxies, the starbursts are observed to lie in the overlapping wedge regions. Here the cloud compression is shown to occur due to the hot, high-pressure remnant gas resulting from the collisions of atomic hydrogen gas clouds from the two galaxies (Jog & Solomon 1992) - see Section 3.

This is a simple physical mechanism, and inevitable, given the pre-existing high-pressure reservoir of the central intercloud medium in case 1, and the generation of the hot, high-pressure gas in case 2. In both the cases, we consider a galaxy with pre-encounter interstellar medium parameters as observed in the Milky Way galaxy, and then study the detailed evolution of this realistic interstellar medium following a galaxy encounter.

2. Triggering of Starbursts by Central Overpressure

The physical outline of the triggering mechanism for central starbursts in interacting galaxies (Jog & Das 1992) is as follows. The observations of the central disk region of ~ 1 kpc diameter of the Galaxy (Bally et al. 1988) have shown the existence of a molecular intercloud medium (ICM), with a high average molecular hydrogen number density of $50\,\mathrm{cm}^{-3}$. The ICM is distributed uniformly with a large volume filling factor ~ 1; and also has high internal, non-thermal random motions. The physical properties of the central molecular gas are very different from the molecular gas in the galactic disk (also see Section 2.2). We note that the average pressure in the central ICM in the Galaxy, P_{ICM}, is about ~ 10 times higher than P_{GMC}, the effective non-thermal gas pressure inside a disk GMC.

We show that as a typical disk GMC tumbles into the central region of the galaxy following a galactic tidal encounter or due to a bar, the pre-existing high pressure of the central ICM drives a strong isothermal/radiative spherical shock into the cloud. Hence, the post-shock gas accumulates into a thin, dense shell which moves inwards with a velocity equal to the shock velocity. The initial shock velocity is set by the ambient pressure of the central ICM, and at later stages the self-gravity of the shell accelerates the shell inwards. On solving the equation of motion of the shocked shell, we obtain $t_{\mathrm{c}}(r)$, the time taken for a shock to cross to a radius $r(< R_{\mathrm{c}})$ from R_{c}, where R_{c} is the radius of the cloud. The crossing time, t_{c}, increases as the shock is driven into the cloud.

The shocked shell of a GMC can develop gravitational instabilities. The stability of the thin, dense shell is analyzed using a plane-parallel geometry for the shell and a local, linear perturbation analysis - exactly as in the case of pressure-supported thin disks. This gives t_{g}, the growth time for the fastest growing mode, to be inversely proportional to the surface density of the shocked shell. Hence as the shock moves in and more mass is swept into the shell, t_{g} decreases.

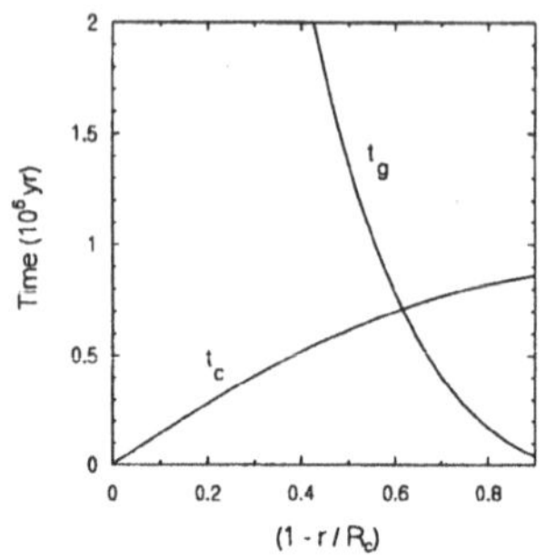

Figure 1. A plot of t_c, the shock crossing time, and t_g, the growth time for the fastest growing gravitational instability in the shocked shell, vs. the fractional cloud penetration depth. The instabilities set in when $t_g = t_c$; here $f_{\rm shell} = 0.85$ (from Jog & Das 1996).

The cloud compression continues until t_g becomes smaller t_c; at this point the shocked shell becomes gravitationally unstable and begins to fragment (Fig. 1). Further non-linear evolution of this fragmenting, high-density shell results in star formation. We define $f_{\rm shell}$ to be the fraction of the cloud mass that is compressed into a shocked shell at the onset of gravitational instabilities. We obtain $f_{\rm shell} = 0.85$ for the typical disk GMC and the central ICM parameters in the Galaxy.

The star formation in the shocked, high-density gas in the GMC shell is characterized by a high star formation or conversion efficiency, and a preferential formation of massive stars of a few $M_\odot$ each (Larson 1987; Jog & Solomon 1992). The net central infrared luminosity is shown to be linearly proportional to the fraction of cloud mass compressed ($f_{\rm shell}$), the efficiency of star formation in the shocked gas, and the rate of gas infall from the disk. The last factor depends only on the strength of the encounter and on the initial radial distribution of gas in the galaxy, whereas $f_{\rm shell}$ depends on the microscopic physics of cloud compression. It is interesting that the two effects are decoupled.

For a tidal/distant encounter or a barred galaxy, the resulting *lower limits* on the central infrared luminosity and the ratio of the infrared luminosity-to-gas mass are $\sim 6 \times 10^9 L_\odot$ and $\sim$ a few $L_\odot/M_\odot$ respectively. These results agree reasonably well with the observations of starburst galaxies such as M82 or NGC 253. The evolved mergers of galaxies, with their higher central gas concentrations, yield higher values of $\leq 10^{12} L_\odot$ and $\leq 100 L_\odot/M_\odot$, respectively. These agree with the observed values from ultraluminous galaxies such as Arp 220 or NGC 6240.

Our study has shown the viability of such a triggering mechanism. Future work should take account of the observed internal clumpy structure of clouds, and consider the shock formation in a clumpy cloud.

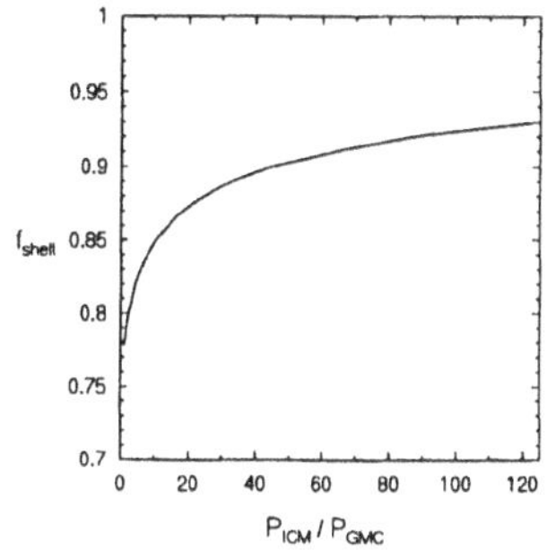

Figure 2. A plot of $f_{\rm shell}$, the fraction of cloud mass compressed at the onset of gravitational instabilities in the shell, vs. the ICM overpressure $P_{\rm ICM}/P_{\rm GMC}$ (from Jog & Das 1996).

2.1. A PARAMETER STUDY FOR CENTRAL STARBURSTS

Recent observations have shown the existence of a central molecular intercloud medium (ICM) with high pressure also in IC 342 (Downes et al. 1992), NGC 1808 (Aalto et al. 1994), Arp 220, and Mrk 231 (Downes & Solomon 1997). Thus the high-pressure central ICM seems to be a common gas component of spiral galaxies and especially of starburst galaxies.

Therefore, we have now done a parameter study of cloud compression which covers a wide range of realistic values for the disk cloud and the central ICM gas parameters. These cover a range for the ICM overpressure, $P_{\rm ICM}/P_{\rm GMC}$, from a few to 125. The aim was to see under which conditions shock compression of a GMC can occur (Jog & Das 1996). We show that $f_{\rm shell}$, the fraction of cloud mass shocked at the onset of gravitational instabilities, depends only on the external ICM pressure, the cloud density, and the cloud radius. We find that $f_{\rm shell}$ is high $\sim 0.75 - 0.90$, and is not sensitively dependent on the value of the ICM overpressure (Fig. 2). This somewhat surprising result comes about because the initial inward shock velocity is only proportional to the square root of the ICM pressure, hence even a small overpressure is sufficient to start the process of compression. At later stages of compression, the shock velocity is mainly set by the self-gravity of the shell, hence this process eventually results in the compression of a substantial fraction of the cloud mass. This is an important result because it shows that the starbursts triggered by central overpressure as proposed by Jog & Das (1992) is a general triggering mechanism.

Applying these results to IC 342 and NGC 1808, we show that in both cases $f_{\rm shell} > 0.9$. However, the net central luminosity from star formation depends on the gas infall rate. Thus, NGC 1808 with gas infall due to tidal interaction with NGC 1792 shows a strong central starburst, while IC 342 with a lower gas infall rate due to a weak bar shows only a moderate central starburst. Thus, all large, gas-rich spiral galaxies appear to be poised on the brink of starburst, and whether or not a strong central starburst occurs depends on the rate of gas infall from the galactic disk.

2.2. CENTRAL MOLECULAR INTERCLOUD MEDIUM – ICM

Our mechanism for triggering of starbursts via central overpressure (Jog & Das 1992; Jog & Das 1996) has highlighted the diffuse molecular ICM in the central regions of galaxies. The high-pressure ICM appears to be a common feature of spiral galaxies, and contains a substantial fraction of the central gas mass. Therefore it is important to understand its origin, energetics, and also its evolution following a starburst, as was stressed by Jog & Das (1992). The ICM is not self-gravitating but is instead probably confined by the galactic gravitational potential. The origin and the physical properties of the ICM seem to be determined by its *unique central* location in a galaxy. For example, the ICM can originate due to the strong central galactic tidal fields, especially in a barred potential (Das & Jog 1995); and also due to the high central, non-thermal random motions, which can result in disruptive cloud collisions.

To formulate these complex problems theoretically, we need to have more high-resolution, multi-transition m.m. wave data for different types of galaxies, and for starburst galaxies in different stages of evolution - as in Aalto et al (1994); and Downes & Solomon (1997).

3. Triggering Of Starbursts In Colliding Galaxies

The second case of triggering of starbursts involves the early stages of a physical collision/merger between two field spiral galaxies (Jog & Solomon 1992). We consider the detailed evolution of a realistic interstellar medium consisting of H_2 and H I in a galaxy during a galaxy collision. The H I clouds from the two galaxies collide; whereas the GMCs, with a mean free path typically larger than the size of the overlapping region, do not collide. Since the H I clouds collide at the typical relative velocity between field galaxies ~ 300 km s^{-1}, their kinetic energy of relative motion is thermalized and this results in a hot, ionized, high-pressure remnant gas. This gas has a pressure ~ 200 times higher than $P_{\rm GMC}$, the effective internal pressure in a disk GMC, hence it causes a radiative shock compression of the pre-existing GMCs in the overlapping wedge region. This results in a burst of massive-star formation in the shocked outer shells of GMCs, with the details of star formation the same as in Jog & Das (1992). In this case the hot remnant gas cools fast and hence the overpressure lasts for a short time, whereas in the earlier case the central ICM constitutes an "everpresent" high-pressure reservoir. The main results from this study are:

1. The starbursts occur in situ in regions of overlap/interpenetration between the two colliding galaxies. Thus we can naturally explain the starbursts observed in the regions of disk overlap as in Arp 244 (NGC 4038/39), and in Arp 299 (NGC 3690/IC 694).

2. The duration of starburst is equal to the disk crossing time $\sim$ a few $\times 10^7$ yr. Hence in a region of overlap, we would expect to see the progression of starburst on this time scale. It is interesting that exactly such age gradients have now been observed in Arp 244 (Vigroux 1997), and also for the high redshift galaxies (Abraham 1997).

3. Whether the central starburst or the extranuclear disk case is stronger depends on the geometry and the epoch of collision and also on the initial radial distribution of gas. In general, because of the higher geometric cross-section for a disk overlap, a starburst would occur there at first (as is observed in Arp 244), to be followed later on by a stronger central starburst.

4. We predict the formation of young globular clusters via shock compression of pre-existing GMCs in colliding galaxies (also see Section 4).

5. The absence of super-starbursts in the cluster galaxies at the present epoch is explained due to the high speed collisions that occur between cluster galaxies (Valluri & Jog 1990).

4. Formation of Young, Luminous Star Clusters

An important feature of our triggering mechanism is that the triggered star formation occurs via gravitational instabilities in the shocked outer shells of GMCs. Thus, the *main mode of star formation is via formation of clusters of stars*, which are gravitationally bound to begin with. Hence our model provides a physical mechanism for the formation of young globular or populous star clusters as we had proposed in Jog & Solomon (1992).

The study of the detailed shock propagation in a cloud and the complete coverage of the parameter space has now allowed us to obtain the detailed physical properties of such star clusters (Jog & Das 1996). The mass of an instability, $M_{\rm inst}$, formed in the shocked outer shell of a GMC gives an upper limit on the mass of the star cluster formed. The star clusters formed are predicted to be flattened. We have obtained $M_{\rm inst}$ as a function of the disk cloud mass and the central ICM pressure. We find that $M_{\rm inst}$ is a large fraction ($\sim 0.3 - 0.5$) of the GMC mass. This is because the fraction of a cloud mass shocked ($f_{\rm shell}$) is always high, and the number of fragments or instabilities per cloud is small. Thus, in our model of triggering by cloud compression, an efficient star formation is triggered simultaneously in a sizeable fraction of the cloud mass. The upper limit on the mass of the star clusters formed covers a range of values $\sim 10^4 - 10^6 M_\odot$.

There is a growing observational evidence especially from the *HST*, for the existence of young, luminous star clusters or super star clusters in starburst galaxies - as for example, in Arp 244 (Whitmore & Schweizer 1995), and in NGC 3921 (Schweizer et al. 1996). Our triggering mechanism provides a natural explanation for the origin of these young star clusters. Recently Ho & Filippenko (1996) have made direct, dynamical mass measurements for two young star clusters, one each in the dwarf galaxies NGC 1569 and NGC 1705, to be equal to $3.3 \times 10^5 M_\odot$ and $0.8 \times 10^5 M_\odot$, respectively. These lie in the range of masses of star clusters resulting from our mechanism. Also, the observed sizes of these ($\sim$ a few pc) agree with the sizes of instabilities obtained from our model. Thus, we can now attempt to compare the quantitative predictions from our model with the new high-resolution data on the young star clusters.

Acknowledgements: It is a pleasure to acknowledge the collaboration, and also more recent discussions, with Mousumi Das and Phil Solomon. I would like to thank the IAU, the Indian Institute of Science, and the Indian National Science Academy for financial support which enabled me to attend this exciting meeting.

References

Aalto, S., Booth, R.S., Black, J.H., Koribalski, B., & Wielebinski, R. 1994, *A & A*, **286**, 365
Abraham, R.G. 1997, these proceedings
Bally, J., Stark, A.A., Wilson, R.W., & Henkel, C. 1988, *ApJ*, **324**, 223
Das, M. & Jog, C.J. 1995, *ApJ*, **451**, 167
Downes, D., Radford, S.J.E., Guilloteau, S., Guelin, M., Greve, A., & Morris, D. 1992, *A & A*, **262**, 424
Downes, D., & Solomon, P.M. 1997, to appear in ApJ
Ho, L.C., & Filippenko, A.V. 1996, *ApJ*, **472**, 600
Jog, C.J. 1995, *BASI*, **23**, 135
Jog, C.J., & Das, M. 1992, *ApJ*, **400**, 476
Jog, C.J., & Das, M. 1996, *ApJ*, **473**, 797
Jog C.J., & Solomon, P.M. 1992, *ApJ*, **387**, 152
Larson, R.B. 1987, in Starbursts and Galaxy Evolution, eds. T.X. Thuan, T. Montmerle, & J. Trann Thanh Van, Edition Frontieres, Paris, 467
Norman, C.A. 1992, in Massive Stars in Starbursts, eds. C. Leitherer, T.M. Heckman, C.A. Norman, & N.R. Walborn, Cambridge Univ. Press, Cambridge, 271
Schweizer, F., Miller, B.W., Whitmore, B.C., & Fall, S.M. 1996, *AJ*, **112**, 1839
Valluri, M., & Jog, C.J. 1990, *ApJ*, **357**, 367
Vigroux, L. 1997, these proceedings.
Whitmore, B.C., & Schweizer, F. 1995, *AJ*, **109**, 960

THE STELLAR INITIAL MASS FUNCTION IN STARBURST GALAXIES

CLAUS LEITHERER
Space Telescope Science Institute
3700 San Martin Drive, Baltimore, MD 21218

Abstract. Starburst galaxies are currently forming massive stars at prodigious rates. I discuss the star-formation histories and the shape of the initial mass function, with particular emphasis on the high- and on the low-mass end. The classical Salpeter IMF is consistent with constraints from observations of the most massive stars, irrespective of environmental properties. The situation at the low-mass end is less clear: direct star counts in nearby giant H II regions show stars down to $\sim$1 $M_\odot$, whereas dynamical arguments in some starburst galaxies suggest a deficit of such stars.

1. Introduction

In this review I will define a starburst as a system in which high-mass stars *dominate* the energetics in terms of radiative and non-radiative heating. Stars falling under this definition have zero-age-main-sequence masses above 5 – 10 $M_\odot$. Such a galaxy would be recognized as having blue colors, strong nebular emission-line spectrum, high inferred supernova rate, and (if dust is present) high IR luminosity. If less massive stars form down to about 1 $M_\odot$ according to a Salpeter-like IMF, the gas consumption time-scale will be short in comparison with the Hubble time.

Three major starburst categories can be identified with this definition: *IR-luminous galaxies*, whose stellar population is enshrouded in dust (Sanders & Mirabel 1996); *nuclear starbursts*, whose optical morphology is similar to that shown by Seyfert galaxies but the primary energy source being a young stellar population; *starbursting dwarfs*, whose optical spectrum is H II region-like, often with an indication of an older, underlying population (Gallagher 1996; Telles & Terlevich 1997).

2. Techniques to derive the starburst IMF

Massive stars are often embedded in gas and dust, which leads to absorption of stellar radiation and re-emission at longer wavelengths. On the one hand, this allows an analysis via nebular recombination lines of the stellar far-UV radiation, which would otherwise be inaccessible; but on the other, it means a significant fraction of UV light is absorbed, and heavily dust-obscured starbursts may not be visible at all. Techniques have been developed to analyze the directly emitted or reprocessed UV or far-UV light. These techniques are often rather indirect and make use of complex theoretical stellar models. "Observationally" derived IMFs in starbursts can sometimes be heavily dependent on such models. The willingness to believe in a variation or the universality of the IMF depends on the trust one has in stellar atmospheric and evolutionary models. The main diagnostics are:

Far-IR fluxes — Some or most of the stellar UV flux is used to heat dust grains, which reach an equilibrium temperature of about 40 K. The associated far-IR emission peaks at about 60 μm (Soifer et al. 1987). The agreement between star-formation rates in normal spiral galaxies derived from Hα and far-IR luminosities suggests that the far-IR luminosity traces massive stars (Devereux & Young 1990).

Nebular emission lines — Nebular diagnostics are a major tool to study starburst populations. The subject has been reviewed by Stasińska (1996). The principle is to count photons, either from recombination lines or from excited transitions originating in the ISM, and then to infer the incident ionizing spectrum, which is then used to model the stellar population. The method is sensitive for stars with significant far-UV flux, i.e. temperatures above 30,000 K, or masses above about 20 $M_\odot$.

UV stellar-wind lines — The wavelength region between 912 Å and 3000 Å is dominated by stellar-wind lines. They result from absorption of photospheric far-UV photons, with an associated transfer of radiative into kinetic momentum. The profiles contain information on the stellar luminosity. Evolution models predict a stellar mass-luminosity relation so that the line profile is tied to the stellar mass (Mas-Hesse & Kunth 1991; Robert, Leitherer, & Heckman 1993; Leitherer, Robert, & Heckman 1995).

Near-IR red supergiant features — The most prominent stellar absorption bands of cool stars are in the 1 to 2.5 μm range (Lançon & Rocca-Volmerange 1992; Origlia, Moorwood, & Oliva 1993). Evolutionary synthesis models for this spectral range (e.g., Rieke et al. 1993; Lançon & Rocca-Volmerange 1996) provide constraints on the mass-to-light ratio (in combination with a dynamical mass) and on an additional diluting continuum due to non-thermal or dust emission.

3. IMF parameters

The concept of the IMF was introduced by Salpeter (1955) in a study of the stellar content in the solar vicinity. It is worth recalling that the mass ranges sampled observationally in starbursts and in the solar neighborhood by Salpeter (1955) are not the same: Salpeter's result applies to stars with masses below 10 $M_\odot$, whereas observations of starburst galaxies are usually restricted to higher stellar masses.

Excellent reviews have been written since Salpeter's pioneering work. I highlight Miller & Scalo (1979), Tinsley (1980), and Scalo (1986). The IMF is often parameterized as a power-law $\phi(m) \propto m^{-\alpha}$, where ϕ is the number of stars formed per unit space and time. ϕ is defined in the interval [$M_{\rm low}$,$M_{\rm up}$], where $M_{\rm low}$ and $M_{\rm up}$ are the lower and upper mass cut-offs, respectively. The slope α is 2.35 for a Salpeter IMF.

It is instructive to compare the starburst and the local field star IMF. The solar neighborhood IMF was derived by Miller & Scalo (1979) and subsequently revised by Scalo (1986). A power law with an exponent of 2.3 is an acceptable fit to the Miller-Scalo IMF for the mass range above 10 $M_\odot$. *For the mass range above 10 $M_\odot$, which is the one accessible for direct observations of starburst galaxies, the Scalo (1986) slope is consistent with a Salpeter slope of $\alpha = 2.35$.*

The mass range above 10 $M_\odot$ is the tip of the iceberg. Only a few percent of the total stellar mass in a starburst is in massive stars. A large, uncertain extrapolation from >10 $M_\odot$ (the range from which stellar light is observed) to 1 $M_\odot$ and below (where the mass is concentrated) is required to derive total star-formation rates and starburst masses.

Direct mass determinations can in principle be used to constrain the low-mass end of the IMF. Until very recently, obtaining stellar velocity dispersions in individual starburst clusters was technically infeasible. Massive stars which dominate optical and UV spectra have few suitable spectral lines. Furthermore, the velocities of their strong stellar winds by far exceed the expected velocity dispersions (about 10 km s^{-1} for a 10^5 $M_\odot$ starburst cluster). The most favorable conditions are in super star clusters where Ho & Filippenko (1996) measured velocity dispersions, demonstrating the existence of stars with masses down to the 2 – 3 $M_\odot$ range.

$M_{\rm up}$ becomes statistically meaningful only for masses below ~50 $M_\odot$. Otherwise stochastic effects prevail since for typical starburst masses there are insufficient numbers of stars above about 50 $M_\odot$ to contribute to the integrated light (Cerviño & Mas-Hesse 1994). If $M_{\rm up}$ is as low as 30 $M_\odot$, the ionizing photon output of the population is cut drastically, and a soft ionizing radiation field is often ascribed to a low value of $M_{\rm up}$.

To summarize the main points of this section, deriving the IMF in starburst galaxies is essentially focussed on three issues:

- determining the slope of the IMF above ~10 $M_\odot$;
- combining a mass and a light measurement to constrain $M_{\rm low}$;
- measuring $M_{\rm up}$ from the hardness of the radiation field.

4. Which stars form in different starburst types?

4.1. M 82 AND IR-LUMINOUS GALAXIES

In their pioneering study, Rieke et al. (1980) introduced the terminology of the "top-heavy IMF". Top-heavy means a higher proportion of RSGs over red giants and dwarfs than expected for a normal IMF. In terms of mass, a top-heavy IMF has an excess of stars in the mass range 10 – 20 $M_\odot$ over stars of 5 $M_\odot$ and less. Observational evidence for a top-heavy IMF in M 82 comes from the relatively high K-band luminosity of the nucleus, together with its relatively low dynamical mass. Satyapal et al. (1995; 1997) obtained spatially resolved near-IR spectroscopy of the central M 82 region. 12 unresolved starburst clusters contribute to the IR luminosity. They have an age dispersion of 6×10^6 yr and suggest a propagation velocity of the starburst of ~50 km s^{-1}. Satyapal et al. (1997) find that a starburst model with a Salpeter IMF from 0.1 to 100 $M_\odot$ fits the observational constraints of all clusters. It is interesting to note that starburst models can be found which do not require a deficit of low-mass stars.

M 82 is the low-luminosity extension ($L = 3 \times 10^{10}$ $L_\odot$) of the class of IR-luminous galaxies ($L > 10^{11}$ $L_\odot$). IR-luminous galaxies are the dominant extragalactic population in the local universe, and most (if not all) undergo an intense starburst (Sanders 1997). Wright et al. (1988) determined M/L ratios in a sample of interacting starburst galaxies. Unless most of the galaxy mass were concentrated in the starburst and the molecular gas were converted into stars at nearly 100% efficiency, the small M/L ratios suggest a deficit of stars with masses <5 $M_\odot$. Detailed studies of individual objects like NGC 3256 (Doyon, Joseph, & Wright 1994) point toward similar results.

A universal deficit of stars having masses <5 $M_\odot$ has important implications, such as a significant increase of the gas reservoir exhaustion timescale and the absence of a post-burst phase dominated by A stars.

The high-mass end of the IMF can be probed with nebular emission lines in the K band. IR-luminous galaxies are generally found to have a weak recombination spectrum, and therefore a soft stellar ionizing radiation field (Doyon, Puxley, & Joseph 1992; Goldader et al. 1997). The most provocative interpretation would be a deficit of very massive stars, i.e., a steep or even truncated IMF above 30 $M_\odot$. Alternative explanations are absorption of the ionizing photons by dust or mass-dependent dust extinction. The latter scenario would imply that even in the K band most of the massive stars are still buried in dust and not observable.

4.2. H II GALAXIES

The rich emission-line spectrum of H II galaxies allows the study of a key issue: the metallicity dependence of the IMF. Collisionally excited line ratios are sensitive to the electron temperature T_e. Since the equilibrium temperature of the H II region depends on metallicity (which affects the cooling) and on the IMF (which controls the heating), the challenge is to disentangle IMF- and metallicity-induced variations of T_e.

Stasińska & Leitherer (1996) found that the emission-line spectra of 100 giant H II regions and starburst galaxies in the metallicity range 0.25 $Z_\odot$ to 0.025 $Z_\odot$ are consistent with being powered by a stellar population following a Salpeter IMF up to ~100 $M_\odot$. Low values of $M_{\rm up}$ = 30 $M_\odot$ are not consistent with the data. At the metal-rich end, García-Vargas, Bressan, & Díaz (1995a,b) applied an independent set of stellar and nebular models to a sample of giant H II regions with above-solar metallicity. As in the case of metal-poor systems, a Salpeter IMF without truncation at the upper end is consistent with the observations.

Direct signatures of ionizing stars, via UV absorption lines, have been detected in numerous starburst galaxies with IUE (Kinney et al. 1993). *HST* resolution (both spectrally and spatially) is required for quantitative IMF studies. Detailed analysis of the UV line profiles have been done, e.g., for NGC 1741 (Conti, Leitherer, & Vacca 1996) and NGC 4214 (Leitherer et al. 1996). The strength and blueshift of C IV λ1550 and Si IV λ1400 require the presence of massive stars with masses of 50 $M_\odot$ and higher. This is consistent with the two galaxies being classified as Wolf-Rayet galaxies (Conti 1991).

4.3. ACTIVE GALACTIC NUCLEI

The importance of starbursts in the AGN phenomenon has been emphasized repeatedly (e.g., Perry & Dyson 1985; Cid-Fernandes & Terlevich 1995). Regardless of what powers the AGN, it is now fairly well established that luminous starbursts can be found within tens of pc of the AGN (e.g., Colina et al. 1997) and that the nuclear starburst can make a significant contribution to the overall energetics of the AGN (Fanelli et al. 1997; Heckman et al. 1997).

Do the same stars form in the vicinity of an AGN as in typical starbursts? González-Delgado et al. (in prep.) obtained *HST* GHRS spectra covering the central 500 pc of the Seyfert2 galaxy NGC 7130 and compared them to a starburst region in the Wolf-Rayet galaxy NGC 1741. NGC 1741 is a Magellanic-type irregular with quarter-solar metallicity. Its starburst is at the end of a central, kpc-size bar. In contrast, the (circum)-nuclear starburst in NGC 7130 has a chemical composition close to solar. Despite the different environment, the two spectra are virtually identical (except for the AGN emission-line contribution), suggesting rather similar stellar content. The strength of the absorption features in NGC 7130 excludes an additional source contributing to the starlight, such as a non-thermal continuum. Other observed galaxies in the sample (Mrk 477, NGC 5135, IC 3639) have very similar UV spectra, pointing toward similar massive-star content, and therefore similar IMF.

5. The universality of the starburst IMF

Is there a coherent picture emerging from the various pieces of evidence in different objects? In this section the main issues are summarized.

Temporal and spatial evolution — *Star formation in starburst clusters, if they are spatially isolated, occurs nearly instantaneously.* This is similar to the star-formation process in local high-mass star-formation regions where the observed age spreads are less than 2 – 3 Myr (Massey et al. 1995a,b). There is increasing evidence for starbursts propagating on scales of tens of parsecs at velocities of 10^1 to 10^2 km s^{-1}. Case studies are M 82 (Satyapal et al. 1995, 1997) or M 83 (Puxley, Doyon, & Ward 1997). Starbursts appear to be confined to localized regions which can then trigger other starbursts in the vicinity. *If observed at insufficient spatial resolution, propagating star formation will mimic a quasi-continuous starburst.* On larger scales, starbursts in IR-luminous galaxies extending over hundreds of pc have age spreads which may be comparable to the burst ages themselves (tens of Myr). Otherwise the synchronization mechanism would be difficult to understand.

The shape of the IMF — *Observations in different object classes generally lead to an IMF slope that is consistent with Salpeter's value ($\alpha = 2.35$).* The IMF slope in starburst galaxies agrees within the error bars with the slopes derived in local regions of high-mass star formation. Massey et al. (1995a,b) find a Salpeter IMF from number counts in the Galaxy, the LMC, and the SMC. The surveyed regions are Galactic open clusters and OB associations, and H II regions in the Magellanic Clouds, including 30 Doradus. Although not fulfilling the definition of a starburst, these regions are the closest local counterparts in existence. It is worth noting that most IMF determinations in starbursts refer to clusters. The field population of massive stars is yet unexplored and could well have a different IMF. Massey et al. derive a steeper IMF for massive field stars outside clusters in the Local Group of galaxies.

The high-mass end — The masses of the most massive stars determined in dust-poor starbursts agree with those found in Local Group galaxies from an analysis of the Hertzsprung-Russell diagram. *They are in the 50 to 100* $M_\odot$ *range.* The stellar ionizing radiation field in dusty IR-luminous galaxies is softer than expected for the IMF which is observed in dust-poor galaxies. It is unclear if this indicates a deficit of massive stars *formed* or *observed.* The first suggestion implies a truncation of the IMF, whereas the second an observational bias against detecting very massive stars, possibly due to obscuration effects. One might speculate that even at 2 μm the most massive stars are not observable in IR-luminous galaxies.

The low-mass end — The low-mass end of the IMF (below $\sim$5 $M_\odot$) in starburst galaxies is inferred from the observed M/L ratio. *Dynamical masses of starburst nuclei derived from rotation curves are relatively low, suggesting* $M_{\rm low} \approx 5\ M_\odot$. However, observations of M 82 at high spatial resolution have revealed the morphological complexity of the region, making its interpretation in terms of starburst modeling a challenge. Velocity dispersion measurements in individual starburst clusters are an alternative method for a mass estimate. Results obtained for super star clusters in NGC 1569 and NGC 1705 suggest $M_{\rm low}$ in the 1 – 3 $M_\odot$ range. A priori there is no reason for the low-mass IMF in starburst galaxies and in local sites of massive-star formation to be the same. Nevertheless, local regions may give some guidance. *HST* has pushed the detection limit of stars in 30 Doradus and other regions to about 2 $M_\odot$, with no indication for a truncation of the IMF (Hunter et al. 1995; 1997). Starburst galaxies and local regions show the same IMF at the high-mass end. Does this hold for the low-mass end as well?

The dependence on the environment — *There is little indication for a significant influence of the environment, including the metallicity, on the IMF.* Massive stars following a similar IMF form in the nuclei of starburst galaxies (e.g., NGC 7714), at the end of bars in irregular galaxies (e.g., NGC 1741), in dense super star clusters (e.g., NGC 1569), in candidate globular clusters in mergers (e.g., NGC 4038/39; Whitmore et al., in prep.), and around AGNs (e.g., NGC 7130). The derived IMF for the mass range 10 – 50 $M_\odot$ is similar to that in local H II regions.

The universality of the IMF in starburst galaxies in the local universe is encouraging for efforts to understand the spectro-photometric properties of star-forming galaxies at high redshift. It may not be unreasonable to assume a similar IMF for the first generation of stars as well.

Acknowledgements — Travel support to attend IAU 186 from the LOC and from the AAS is gratefully acknowledged.

References

Cerviño, M., & Mas-Hesse, J. M. 1994, A&A, 284, 749

Cid Fernandes, R., & Terlevich, R. 1995, MNRAS, 272, 423

Colina, L., García-Vargas, M. L., Mas-Hesse, J. M., Alberdi, A., & Krabbe, A. 1997, ApJ, 484, L41

Conti, P. S. 1991, ApJ, 377, 115

Conti, P. S., Leitherer, C., & Vacca, W. D. 1996, ApJ, 461, L87

Devereux, N. A., & Young, J. S. 1990, ApJ, 350, L25

Doyon, R., Joseph, R. D., & Wright, G. S. 1994, ApJ, 421, 101

Doyon, R., Puxley, P. J., & Joseph, R. D. 1992, ApJ, 397, 117

Fanelli, M. N., et al. 1997, AJ, 114, 575

Gallagher, J. S. 1996, in From Stars to Galaxies: The Impact of Stellar Physics on Galaxy Evolution, ed. C. Leitherer, et al. (San Francisco: ASP), 315

García-Vargas, M. L., Bressan, A., & Díaz, A. I. 1995a, A&AS, 112, 13

——. 1995b, A&AS, 112, 35

Goldader, J. D., Joseph, R. D., Doyon, R., & Sanders, D. B. 1997, ApJ, 474, 104

Heckman, T. M., et al. 1997, ApJ, 482, 114

Ho, L. C., & Filippenko, A. V. 1996, ApJ, 472, 600

Hunter, D. A., Light, R. M., Holtzman, J. A., Lynds, R., O'Neil, Jr., E. J., & Grillmair, C. J. 1997, ApJ, 478, 124

Hunter, D. A., Shaya, E. J., Holtzman, J. A., Light, R. M., O'Neil, Jr., E. J., & Lynds, R. 1995, ApJ, 448, 179

Kinney, A., Bohlin, R., Calzetti, D., Panagia, N., & Wyse, R. 1993, ApJS, 86, 5

Lançon, A., & Rocca-Volmerange, B. 1992, A&AS, 96, 593

——. 1996, New Astron., 1, 215

Leitherer, C., Robert, C., & Heckman, T. M. 1995, ApJS, 99, 173

Leitherer, C., Vacca, W. D., Conti, P. S., Filippenko, A. V., Robert, C., & Sargent, W. L. W. 1996, ApJ, 465, 717

Mas-Hesse, J. M., & Kunth, D. 1991, A&AS, 88, 399

Massey, P., Johnson, K. E., & DeGioia-Eastwood, K. 1995a, ApJ, 454, 151

Massey, P., Lang, C. C., DeGioia-Eastwood, K., & Garmany, C. D. 1995b, ApJ, 438, 188

Miller, G. E., & Scalo, J. M. 1979, ApJS, 41, 513

Origlia, L., Moorwood, A. F. M., & Oliva, E. 1993, A&A, 280, 536

Perry, J. J., & Dyson, J. E. 1985, MNRAS, 213, 665

Puxley, P. J., Doyon, R., & Ward, M. J. 1997, ApJ, 476, 120

Rieke, G. H., Lebofsky, M. J., Thompson, R. I., Low, F. J., & Tokunaga, A. T. 1980, ApJ, 238, 24

Rieke, G. H., Loken, K., Rieke, M. J., & Tamblyn, P. 1993, ApJ, 412, 99

Robert, C., Leitherer, C., & Heckman, T. M. 1993, ApJ, 418, 749

Salpeter, E. E. 1955, ApJ, 121, 161

Sanders, D. B. 1997, in Starburst Activity in Galaxies, ed. J. Franco, R. Terlevich, & A. Serrano, Rev. Mex. Astron. Astrofis. Conf. Ser., 6, 42

Sanders, D. B., & Mirabel, I. F. 1996, ARAA, 34, 749

Satyapal, S., et al. 1995, ApJ, 448, 611

Satyapal, S., et al. 1997, ApJ, 483, 148

Scalo, J. M. 1986, Fund. Cosm. Phys., 11, 1

Soifer, B. T., et al. 1987, ApJ, 320, 238

Stasińska, G. 1996, in From Stars to Galaxies: The Impact of Stellar Physics on Galaxy Evolution, ed. C. Leitherer, et al. (San Francisco: ASP), 232

Stasińska, G., & Leitherer, C. 1996, ApJS, 107,427

Telles, E., & Terlevich, R. 1997, MNRAS, 286, 183

Tinsley, B. M. 1980, Fund. Cosm. Phys., 5, 287

Wright, G. S., Joseph, R. D., Robertson, N. A., James, P. A., & Meikle, W. P. S. 1988, MNRAS, 233, 1

THE EVOLUTION OF YOUNG STAR CLUSTERS IN MERGING GALAXIES

BRADLEY C. WHITMORE
Space Telescope Science Institute
3700 San Martin Dr., Baltimore, MD 21218

1. Introduction

The formation of young star clusters in merging galaxies is, by now, well established. The new challenge is to use these young clusters as a tool to address some of the outstanding questions. For example, what fraction of these young clusters become globular clusters? Is this enough to explain the difference in the specific globular cluster frequencies for spirals and ellipticals? What is it about the collision between two gas-rich galaxies that triggers giant molecular clouds to form star clusters? Can the star clusters be used to age date merger remnants and establish a convincing evolutionary connection between merging spirals and elliptical galaxies? This review will focus on the last of these items.

2. A Brief History and Compilation of Recent Observations

A hint that young star clusters might be formed in mergers was provided by Schweizer's (1982) observations of six unresolved bluish knots in the merger remnant NGC 7252, which he suggested might be young star clusters formed during the merger. This idea was formally presented in Schweizer (1987) and Burstein (1987). Ashman and Zepf (1992) and Zepf and Ashman (1993) further developed these ideas, and made predictions about the bimodality of the metallicity histogram of the clusters. The most important pre- *HST* observation was made by Lutz (1991), who found a population of roughly a dozen blue, pointlike objects in the merger remnant NGC 3597, but was not able to resolve the objects.

The *HST* observations of NGC 1275 by Holtzman et al. (1992) was the primary catalyst in this active new field. They discovered a population of about 60 blue, pointlike objects and suggested that they were protoglobular clusters which formed $\leq$ 300 Myr years ago during a merger of NGC

1275 with another galaxy. Unfortunately, NGC 1275, the central cooling-flow galaxy in the Perseus cluster, is such a peculiar galaxy that it is not clear which of its peculiarities is responsible for the formation of the young clusters (e.g., see Richer et al. 1993).

Whitmore et al. (1993), using *HST* observations (before refurbishment) of the prototypical merger remnant NGC 7252, found a population of about 40 blue, pointlike objects with luminosities and colors nearly identical to those found in NGC 1275. Unlike NGC 1275, with all its peculiarities, NGC 7252 is a fairly isolated galaxy which therefore provided a much cleaner connection between the formation of young star clusters and merging galaxies. Whitmore & Schweizer (1995) followed this up with pre-refurbishment observations of another prototypical merger, NGC 4038/4039 (= the "Antennae Galaxy", see Color plate 5, p. xxiii). Over 700 young star clusters were found in this galaxy.

Since then, every major merger observed by *HST* involving at least one gas rich galaxy has revealed the presence of young star clusters. In addition, *HST* has shown that non-interacting starburst galaxies and barred galaxies can also produce young star clusters, but in much smaller numbers than the major mergers. The lesson appears to be that luminous young star clusters are found whenever there is vigorous star formation. Since the ultraluminous *IRAS* sources are essentially all mergers (Sanders et al. 1988), it is not surprising that mergers show the largest populations of young star clusters. Observations of cooling flow galaxies by Holtzman et al. (1996) show that young star clusters are generally not found in these galaxies unless there is also evidence for a recent accretion event.

Covering the full range of related topics is not possible in a review of this length. Instead, Table 1 provides a list of articles (with a preference for recent results), and the reader is referred to Whitmore (1995; cooling flows), Schweizer (1996; mergers), and Ho (1996; starbursts) for related reviews. In the following sections we will focus on recent work concerning the evolution of star clusters in merging galaxies.

3. The Evolution of Young Star Clusters in Merging Galaxies

A typical old globular cluster, as found in our Milky Way galaxy, has an absolute magnitude in the V band in the range $-10 < M_V < -5$ mag, and a color in the range $0.8 < (V - I) < 1.1$ mag. The ages are estimated at $\approx$15 Gyr. Simple spectral evolution codes can be used to extrapolate these numbers backward in time to estimate the luminosities and colors of globular clusters at any epoch. The result is that the clusters, when they are, say, 5 Myr old, should be roughly 6 magnitudes brighter and 1.0 mag bluer in $V - I$ (based on the Bruzual & Charlot 1996 models). This is exactly what we find for the star clusters in merging galaxies.

Fig. 1 shows a set of simulations from Whitmore et al. (1997) demonstrating how the luminosities and colors of young- and intermediate-age globular clusters can be used to age date merger remnants. While it is easy to distinguish the young star clusters (solid circles) from the old population (open circles) for recent mergers, it becomes progressively more difficult with increasing age.

3.1. ONGOING MERGERS (< 100 MYR)

The "Antennae" galaxy (NGC 4038/4039) is the youngest and closest of the prototypical merger remnants on Toomre's (1977) list. While the extensive dust in NGC 4038/4039 widens the color distribution in Fig. 2 dramatically, it is obvious that there are hundreds of clusters in NGC 4038/4039 that are much brighter and much bluer than normal globular clusters, as predicted by Fig. 1.

Whitmore and Schweizer (1995) found that the cluster luminosity function is approximately a power law ($\phi(L)dL \propto L^{-1.78\pm0.05}dL$) with no hint of a turnover down to the limiting magnitude at $M_V = -10$ mag, unlike the luminosity function for old globular clusters which is a Gaussian. One possible explanation is that clusters of all masses are originally formed, but the fainter and more diffuse clusters are destroyed by a variety of processes (e.g., evaporation, tidal disruption, etc) resulting in a more Gaussian shaped distribution after a few Gyr.

Preliminary results from much deeper *HST* images taken with a repaired *HST* (Whitmore et al. 1998; Figure 1) show well over 1000 young clusters. These images are deep enough to detect the brightest stars so it will be possible to study many of the clusters on a star-by-star basis.

Another ongoing merger with a large population of very young clusters is NGC 6052, where Holtzman et al. (1996) estimate ages of $\approx$ 5 Myr for some of the youngest clusters. Many of these clusters do not appear to be as compact as Galactic globular clusters and they suggest that they may be massive associations. Others have compact cores which Holtzman et al. estimate to have masses of $\approx 10^6\ M_\odot$, and hence are good candidates for young globular clusters.

3.2. INTERMEDIATE-AGE MERGERS (0.1 GYR - 1 GYR)

Estimates of the ages of the galaxies NGC 3921 (Schweizer et al. 1996) and NGC 7252 (Miller et al. 1997) fall in the range 0.3 - 0.7 Gyr, based on post-refurbishment *HST* images. NGC 3597 is a similar merger remnant which Holtzman et al. (1996) estimate has an age less than 0.5 Gyr (most likely $\approx$ 200 Myr). In these galaxies the clusters are typically only 2 magnitudes

TABLE 1. Observations of Galaxies with Young Star Clusters

Reference	Brief Description
	(Mergers)
Schweizer (1982)	NGC 7252 (ground-based)
Lutz (1991)	NGC 3597 (ground-based)
Holtzman et al. (1992)	NGC 1275
Whitmore et al. (1993)	NGC 7252
Crabtree & Smecker-Hane (1994)	NGC 7727 (ground-based)
Whitmore & Schweizer (1995)	NGC 4038/4039
Borne et al. (1996)	the "Cartwheel" galaxy
Holtzman et al. (1996)	NGC 3597, NGC 6052
Schweizer et al. (1996)	NGC 3921
Hilker & Kissler-Patig (1996)	NGC 5018
Miller et al. (1997)	NGC 7252
Whitmore et al. (1997)	NGC 1700, NGC 3610, + compilation
Stiavelli et al. (1997)	NGC 454
Surace et al. (1997)	9 ultraluminous infrared galaxies
Zepf et al. (1997)	NGC 3256
Lee et al. (1997)	UGC 7636 (companion to NGC 4472)
	(Starbursts)
Arp & Sandage (1985)	NGC 1569
Kennicutt & Chu (1988)	LMC (30 Dor)
Meurer et al. (1992)	NGC 1705
Conti & Vaca (1994)	He 2-10
Hunter et al. (1994)	NGC 1140
O'Connell et al. (1994)	NGC 1569
Meurer et al. (1995)	9 starbursts
O'Connell et al. (1995)	M82
Watson et al. (1996)	NGC 253
Ho & Fillppenko (1997)	NGC 1705, NGC 1569 (spectra)
Calzetti et al.(1997)	NGC 5253
	(Miscellaneous)
Barth et al. (1995)	NGC 1097, NGC 6951 (barred)
Holtzman (1996)	Abell 496, 1795, 2029, 2597 (cooling flows)

brighter and 0.3 mag bluer in $V - I$ than old metal-poor globular clusters (see Fig. 2). The luminosity functions are still power laws down to the limiting magnitude with values of $\phi(L)dL \propto L^{-2.1\pm0.3}dL$ for NGC 3921 (151 cluster and association candidates), and $\phi(L)dL \propto L^{-1.8\pm0.1}dL$ for

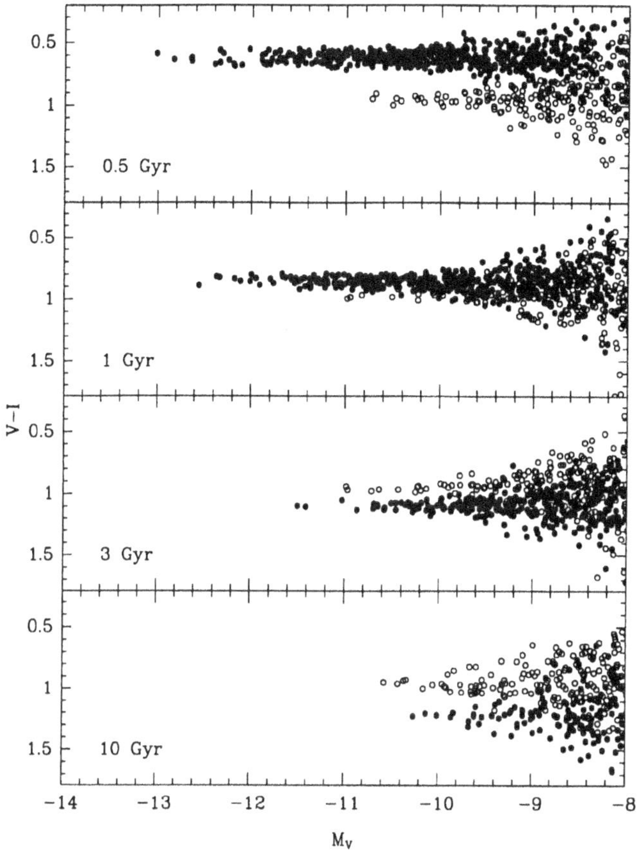

Figure 1. The evolution in the $V - I$ vs. V diagram of a young metal-rich population of globular clusters (filled circles) in the presence of a typical old metal-poor population of globular clusters (open circles). See Whitmore et al. (1997) for details.

NGC 7252 (499 cluster candidates).

van den Bergh (1995) has argued, based on the pre-refurbishment observations of the clusters in NGC 4038/4039 and NGC 7275, that the sizes of the young clusters are larger than typical values for old globular clusters (i.e., $R_{\rm eff} \approx 3$ pc). However, the more reliable measurements from post-refurbishment *HST* images yield upper limits of 5 pc for the median value in NGC 7252 (Miller et al. 1997), NGC 3921 (Schweizer et al. 1997), and NGC 3597 (Holtzman et al. 1996), hence this no longer appears to be a problem.

An important point is that by 0.5 Gyr the clusters are dozens of core crossing times old, hence they must be gravitationally bound and are not likely to disintegrate. Spectra of two of the brightest clusters in NGC 7252

(Schweizer & Seitzer 1993) and one cluster in NGC 1275 (Zepf et al. 1995) support this conclusion, showing strong Balmer absorption lines consistent with our age estimates of 0.5 Gyr. Finally, Ho and Filippenko (1996) have used high dispersion spectra from the Keck telescope to measure stellar velocity dispersions and determine masses of $3 \times 10^5\ M_\odot$ and $0.8 \times 10^5\ M_\odot$ for young clusters in NGC 1705 and NGC 1569. These masses are typical of normal globular clusters.

Similarities between the luminosity functions of young clusters, and the mass functions of giant molecular clouds (=GMCs), have lead several groups to suggest that the progenitors of the young star clusters found in mergers are GMCs, and to propose various models for their formation (e.g., Jog and Solomon 1992, Harris & Pudritz 1994, Schweizer et al. 1996, Elmergreen & Efremov 1997).

Estimates of the specific frequencies of globular clusters after 15 Gyr are 2.9 for NGC 3921 (Schweizer et al. 1996) and 2.5 for NGC 7252 (Miller et al. 1997). Hence it appears that these two galaxies will have specific frequencies typical of field ellipticals.

3.3. OLD MERGERS (1 - 5 GYR)

If young star clusters in merger remnants have roughly solar metallicity (see Whitmore et al. 1997) then Bruzual-Charlot (1996) models predict that 1 Gyr old clusters should have roughly the same colors as old globular clusters. However, they will still be 1 - 2 magnitudes brighter than old globulars. As the clusters continue to age the signatures will become even more subtle, with clusters predicted to be essentially identical in brightness and only $\Delta\ (V - I) \approx 0.2$ mag *redder* (due to higher metallicity) at an age of 5 Gyr.

Evidence for a population of star clusters with ages of 4 ± 2.5 Gyr has been found in NGC 3610, a galaxy with a high "fine structure" index (e.g., loops, shells, etc.) indicative of past merging activity (Schweizer & Seitzer 1992). A population of luminous red clusters which are ≈ 0.7 mag brighter and ≈ 0.2 mag redder in $V - I$ are reported in Whitmore et al. (1997). In addition, most of these red clusters are found near the center of NGC 3610 (see Figures 4 and 11 from Whitmore 1997), as predicted for clusters formed in mergers by Ashman & Zepf (1992) .

3.4. NORMAL ELLIPTICALS (> 5 GYR)

As shown in Fig. 1, and first discussed in Zepf & Ashman (1993), if elliptical galaxies are the products of mergers between spiral galaxies, and new metal-rich globular clusters are produced during these merger, we should expect bimodal color distributions for the globular clusters after a few Gyr.

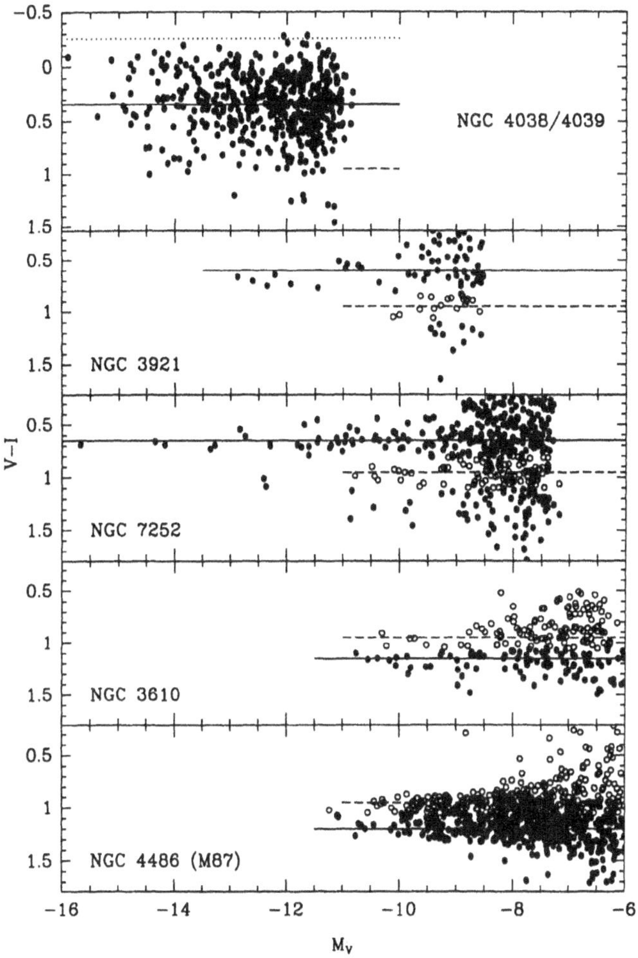

Figure 2. The $V - I$ vs. M_V diagram for five galaxies observed by *HST* suspected of being merger remnants. Compare these observations with the simulations shown in Fig. 1. See Whitmore et al. (1997) for details.

Evidence that the majority of old ellipticals do in fact have bimodal color distributions has accumulated over the past several years (see Zepf article in this volume), with perhaps the best example being M87 (Whitmore et al. 1995, Elson & Santiago 1996, Kundu et al. 1998), with a population of blue ($V - I$ = 0.95 mag) metal-poor clusters believed to be the original population of clusters, and a population of red ($V - I$ = 1.2 mag) clusters probably formed by gas rich mergers.

Fig. 3 shows in more detail than Fig. 1 how the luminosities and colors of the clusters can be used to follow the evolution of star clusters in merger

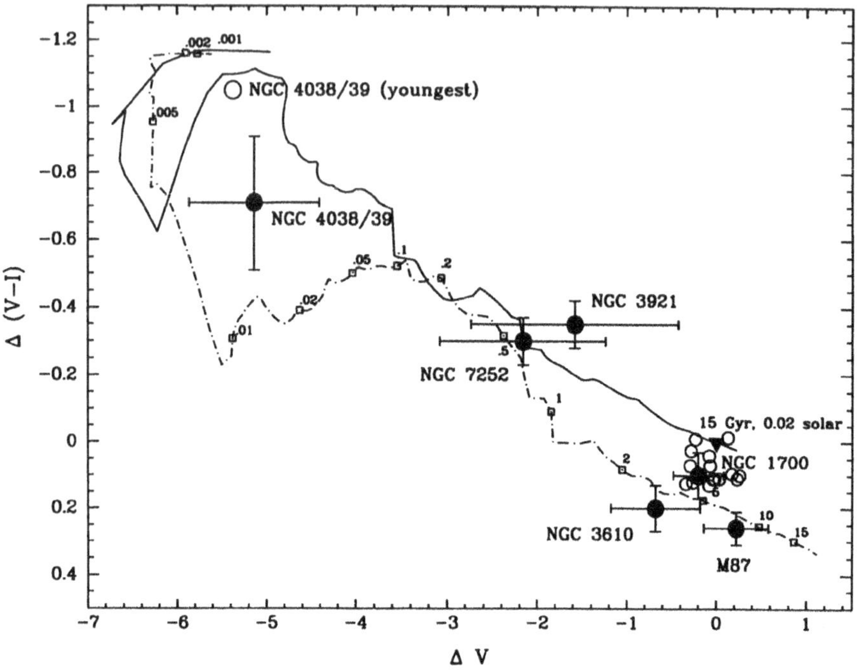

Figure 3. The $\Delta(V-I)$ vs. ΔV diagram for the six galaxies from Whitmore et al. (1997), superposed on Bruzual-Charlot (1996) tracks for a metal-poor population (solid line) and a solar metallicity population (dashed-dot line). The values are normalized to an old, metal-poor population (filled triangle). Typical elliptical galaxies are included as small circles. Ages in Gyr for the solar metallicity track are marked with squares.

remnants. The lines show the predictions from Bruzual-Charlot models for a metal-poor population (solid) and a solar metallicity population (dot-dashed). See Whitmore et al. (1997) for details.

Although the observed colors in M87 fit the model quite nicely, Forbes, Brodie, & Grillmair (1997) point out a problem with the relative numbers of red-to-blue clusters in M87. The specific frequency in M87 is extremely high, which would require several times more red clusters than blue clusters if the increase is due to the addition of metal-rich clusters formed during a gas-rich merger. This is an important result, and probably indicates that cD galaxies like M87 have had a more complicated formation history than the simple spiral + spiral = elliptical scenario. It is likely that M87 has accreted a large number of dwarf galaxies as well as the outer portions of galaxies throughout the galaxy cluster. Both would be rich in blue, metal-poor clusters. Furthermore, the initial formation of the seed galaxy that was to become M87 probably occurred very early, and may have involved galaxies with lower metallicity (i.e., high gas-fractions) hence forming metal-poor, blue globular clusters.

4. Summary

The fact that mergers between gas-rich galaxies can produce large numbers of young, very luminous, compact star clusters is now well established. The brightest of these star cluster have the right luminosities, colors, spectra, sizes, and distributions to be young globular clusters. Attempts to use the colors and luminosities to trace out an evolutionary sequence of merging galaxies are showing promising results (Fig. 3), although the numbers in the sample are still small.

References

Arp, H., & Sandage, A. 1985, AJ, 90, 1163
Ashman, K. M., & Zepf, S. E. 1992, ApJ, 384, 50
Barth, A. J., Ho, L. C., Filippenko, A. V., & Sargent, W.L.W. 1995, AJ, 110, 1009
Borne, K. D. et al. , L., in Science with the Hubble Space Telescope - II, eds. P. Benvenuti, F. D. Macchetto, & E. J. Schreier, 239.
Bruzual A. G., & Charlot, S. 1996, in preparation
Burstein, D. 1987, in Nearly Normal Galaxies, ed. S. Faber (Springer, New York) p. 47
Calzetti, D. et al. 1997, AJ, 114, 1834
Conti, P. S., & Vacca, W. D. 1994, ApJ, 423, L97
Crabtree, D. R.& Smecker-Hane, T.A. 1994, BAAS, 26, 1494
Elmergreen, B. G. & Efremov, Y. N. 1997, ApJ, 480, 235
Elson, R. A., & Santiago, B. X. 1996, MNRAS, 280, 971
Forbes, D., Brodie, J. P. & Grillmair, C. J. 1997, AJ, 113, 1652
Harris, W. E. & Pudritz, R. E. 1994, ApJ, 429, 177
Hilker, M., & M. Kissler-Patig 1996, A&A, 314, 357
Ho, L. C. & Filippenko, A. V. 1996, ApJ, 472, 600
Ho, L. C. 1996, in Starburst Activity in Galaxies, ed. J. Franco, R. Terlevich, & G. Tenorio-Tagle
Holtzman, J. A. et al. (the WFPC2 team) 1996, AJ, 112, 416
Holtzman, J. A. et al. (the WFPC team) 1992, AJ, 103, 691
Hunter, D. A., O'Connell, R.W., & Gallagher, J. S. 1994, AJ, 108, 84
Jog C. & Solomon 1992, ApJ, 387, 152
Kennicutt, R. C. & Chu. Y.-H. 1988, AJ, 95, 720
Kundu, A., Whitmore, B. C., Sparks, W. B., Macchetto, F. D., Zepf, S. E., & Ashman, K., 1998, AJ, submitted
Lee, M. G., Kim, E., & Geisler, D. 1997, AJ, 114, 1824
Lutz, D. 1991, A&A, 245, 31
Meurer, G.R., Freeman, K. C., Dopita, M. A. & Cacciari, C. 1992, AJ, 103, 60
Meurer, G.R., Heckman, T.M., Leitherer, C., Kinney, A., Robert, C., Garnett, D. R. 1995, AJ, 110, 2665
Miller, B. W., Whitmore, B. C., Schweizer, F., & Fall, S. M. 1997, AJ, in press
O'Connell, R. W., Gallagher, J. S., & Hunter, D. A. 1994, ApJ, 433, 65
O'Connell, R. W., Gallagher, J. S., Hunter, D. A., & Colley, W.N 1995, ApJ, 446, L1
Richer, H. B., Crabtree, D. R., Fabian, A. C., and Lin, D. N. C. 1993, AJ, 105, 877
Sanders, D. B. et al. 1988, ApJ, 325, 74
Schweizer, F. 1982, ApJ, 252, 455
Schweizer, F. 1996, in The Nature of Elliptical Galaxies, Proceedings of the Second

Stromlo Symposium, ed. M. Arnaboldi, G. S. Da Costa, & P. Saha
Schweizer, F., Miller, B., Whitmore, B. C., & Fall, S. M. 1996, AJ, 112, 1839
Schweizer, F., & Seitzer, P. 1992, AJ, 104, 1039
Schweizer, F., & Seitzer, P. 1993, ApJ, 417, L29
Schweizer, F. 1987, in Nearly Normal Galaxies, ed. S. Faber (Springer, New York) p. 18
Stiavelli, M., Panagia, N., Carollo, M., Romaniello, M., Heyer, I., Gonzaga, S. ApJL, in press
Surace, J. A. et al. 1997, ApJ, in press
Toomre, A. 1977, in The Evolution of Galaxies and Stellar Populations, ed. B. M. Tinsley & R. B. Larson (Yale, New Haven), p. 401
van den Bergh, S. 1995, Nature, 374, 215
Watson, A. et al. (WFPC2 team) 1996, AJ, 112, 534
Whitmore, B. C. 1995 in Clusters, Lensing, and the Future of the Universe. ed. V. Trimble (ASP, San Francisco), in press
Whitmore, B. C., & Schweizer, F. 1995, AJ, 109, 960
Whitmore, B. C., Schweizer, F., Fall, S. M., Miller, B. W., Leitherer, C., & Zhang, Q. 1998, in preparation
Whitmore, B. C., Miller, B. W., Schweizer, F., Fall, S. M. 1997, AJ, 114, 1797
Whitmore, B. C., Schweizer, F., Leitherer, C., Borne, K., & Robert, C. 1993, AJ, 106, 1354
Whitmore, B. C., Sparks, W. B., Lucas, R. A., Macchetto, F. D., & Biretta, J. A. 1995, ApJ, 454, L73
Zepf, S. E., & Ashman, K. M. 1993, MNRAS, 264, 611
Zepf, S.E., Ashman, K.M., English, J., Freeman, K.C., & Sharples, R.M. 1997, AJ, in prep.
Zepf, S. E., Carter, D., Sharples, R. M., & Ashman, K. M. 1995, ApJ, 445, L19

BRIGHT STAR CLUSTERS IN THE ANTENNAE ANALYZED WITH EVOLUTIONARY SYNTHESIS

UTA FRITZE – V. ALVENSLEBEN AND OLIVER KURTH
Universitäts–Sternwarte, Göttingen, Germany

1. Introduction

The Antennae galaxies (NGC 4038/39) are a pair of relatively nearby ($D \sim 19$ Mpc for $H_0 = 75$) gas-rich spirals of comparable mass in the process of merging with extensive dynamical and evolutionary synthesis modeling available. With *HST* WFPC1 Whitmore & Schweizer (1995) (WS95) detected 738 bright Young Star Clusters (YSCs) around the Antennae. YSCs have been detected in many interacting and starbursting galaxies, but the YSC system in the Antennae is the most populous one, well defining an exponential Luminosity Function (LF) over more than 5 mag above the completeness limit. Problems arising from crowding of the YSCs on a bright and variable galaxy background and from the PSF of WFPC1 may cause blending of YSCs and an overestimation of their effective radii.

A question with far-reaching consequences, e.g. for the formation of elliptical galaxies, is if the YSCs formed in this interaction-triggered starburst are open or (proto-) globular clusters (GCs).

Concentration parameters involving the tidal radius being inaccessible to observations as a distinction criterion, interest focuses on the LF. While in the Milky Way and nearby galaxies the LF is exponential for open clusters but Gaussian for *old* GC systems, *the LF of a young GC system is unknown.*

So, the question – if GCs are formed in mergers ? – gets related to the question as to the evolution of a GC system's LF. Vesperini (1997) shows that the strong dynamical effects on a GC system do *not* change the shape of its mass function, provided it starts from a Gaussian. The aim of this paper is to study the effects of the spectrophotometric evolution of the YSC system on its LF.

The formation of GCs is expected in gas-rich spiral-spiral mergers on the basis of the high star formation efficiencies in mergers and merger remnants (Fritze - v. Alvensleben & Gerhard 1994a,b, Kurth 1996) together with hydrodynamical modeling that requires high star formation efficiences for GC formation (Brown et al. 1995)

2. Evolution of Star Clusters and Metallicities

The evolution of star clusters for 5 different metallicities $10^{-4} \leq Z \leq 0.04$ is modeled as in Fritze - v. Alvensleben & Burkert (1995, FB95). Magnitudes, colours, stellar metallicity indicators, and synthetic spectra are calculated as a function of time and metallicity. Cluster metallicities can be predicted on the basis of the spiral progenitors' ISM abundances (Fritze - v. Alvensleben & Gerhard 1994a). Our prediction of $Z \sim 0.01$ for NGC 7252, an Sc-Sc merger like the Antennae, is confirmed by spectroscopy of the brightest cluster W3 in NGC 7252 by Schweizer & Seitzer (1993). For lack of spectroscopy and by analogy, we *assume* $Z \sim 0.01$ also for the YSCs in the Antennae. Once cluster metallicities are known precise age dating from $V-I$ or $U-V$ colours becomes possible.

Alone from their mean age of $\sim$ 1.3 *Gyr most of the YSCs in NGC 7252 are expected to be young GCs* (FB95).

3. Age Distribution of YSCs in the Antennae

The mean $\mathrm{V-I}$ colour together with an average internal reddening correction results in a mean age of the YSC population in the Antennae of 0.2 ± 0.2 Gyr, in agreement with global starburst age and dynamical time since pericenter (Kurth 1996, Barnes 1988).

Assuming a uniform age for the YSCs the time evolution of the LF simply is a shape-conserving shift towards fainter magnitudes. For a subsample of YSCs with effective radii $R_{\mathrm{eff}} \leq 10$ pc Fritze - v. Alvensleben (1996) shows that their LF after ~ 12 Gyr of evolution becomes compatible with a "normal" GC LF at least up to the turnover, though with some overpopulation of the faintest bins. Since these are most susceptible to depopulation by dynamical destruction and since the colour distribution also agrees with its Milky Way counterpart she argues that - at least for YSCs with $R_{\mathrm{eff}} \leq 10$ pc the LF does not preclude them from being GCs (at variance with van den Bergh's (1995) statement for the entire YSC sample).

In an ongoing starburst like in the Antennae, the age spread among star clusters may be comparable to their ages. Therefore we analyze the age distribution of the YSCs as derived from their individual observed colours in $V-I$ and, as far as available, $U-V$. Fig. 1 clearly reveals two distinct age populations: a large population of 399 (probably 481) YSCs

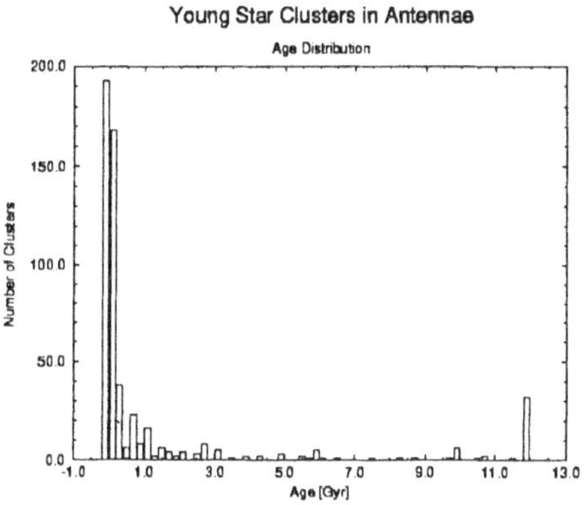

Figure 1. Age distribution of bright star clusters in the Antennae.

with ages from $0 - 2 \cdot 10^8$ yr from the present burst and a small population of 32 (probably 69) *old GCs* from the parent galaxies. The small number of interlopers is probably due to inhomogeneous internal reddening (see Fritze – v. Alvensleben 1998). Strikingly, the number of old GCs agrees with the number expected in case the Milky Way and M31 were merging instead of NGC 4038 and 4039.

The average as well as the distribution of effective radii is not significantly different for the old GC and the YSC subsamples. Statistical tests assure the fraction of blended pairs to be $< 10\%$ even among clusters with $R_{\rm eff} > 10$ pc (see Fritze – v. Alvensleben 1998 for details).

4. Evolution of the LF of the YSCs in the Antennae

Once we have individual ages of the YSCs our models also allow to calculate their individual fading until a common age of say 12 Gyr. Meurer (1995) already suspects that age spread effects might change the shape of the LF in the course of evolution. The brightest clusters tend to be the youngest and will therefore fade more, while part of the faintest clusters will fade less than average. We find that not only the LF of the subsample of YSCs with $R_{\rm eff} \leq 10$ pc evolves into a "perfectly normal" Gaussian, but also the LF of the *entire YSC sample* (see Fig. 2). The turnover occurs at $M_{\rm V_o} \sim -6.9$ mag, more than 1 mag brighter than the (evolved) completeness limit. The difference of ~ 0.2 mag to the typical elliptical galaxy GC systems' $M_{\rm V_o} \sim -7.1$ mag (Harris 1991, Ashman et al. 1995) is a consequence of the enhanced metallicity of this secondary cluster population.

We conclude: Properly accounting for age spread effects and the resulting differences in fading the observed exponential LF of the entire YSC system around the merging Antennae galaxies evolves into a "typical" Gaussian GC LF over 12 Gyr (cf. Fritze – v. Alvensleben 1998).

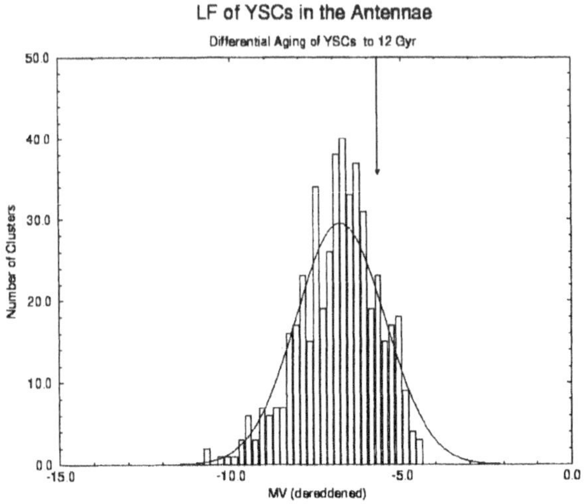

Figure 2. LF of YSCs in the Antennae as calculated from individual ages together with the resulting individual fading until 12 Gyr for every cluster. A Gaussian with $\langle M_{V_0} \rangle = -6.9$ mag and $\sigma(M_{V_0})) = 1.3$ mag is overplotted, normalized to the number of clusters in the histogram. The vertical arrow marks the evolved observational completeness limit.

Together with Vesperini's (1997) result that dynamical effects destroying as much as 60% of an original GC population do not change its mass function if this initially had a Gaussian shape our results indicate that – while there surely will be some open clusters among the YSC population – the bulk of the YSCs in the Antennae may well be young GCs. Their number is large enough for the Antennae to transform into an elliptical with "normal" specific GC frequency.

UFvA greatfully acknowledges a travel grant (Fr 916/4-1) from the DFG.

References

Ashman, K. M., Conti, A., Zepf, S. E., 1995, AJ **110**, 1164
Barnes, J. E., 1988, ApJ **331**, 699
Brown, J. H., Burkert, A., Truran, J. W., 1995, ApJ **440**, 666
Fritze – v. Alvensleben, U.,1998, A&A *submitted*
Fritze – v. Alvensleben, U., Burkert, A., 1995, A&A **300**, 58, (**FB95**)
Fritze – v. Alvensleben, U., Gerhard, O.E., 1994a, A&A **285**, 751
Fritze – v. Alvensleben, U., Gerhard, O.E., 1994b, A&A **285**, 775
Harris, W. E., 1991, ARAA **29**, 543
Kurth, O., 1996, Diploma Thesis, Univ. Göttingen
Meurer, G. R., 1995, Nature **375**, 742
Schweizer, F., Seitzer, P., 1993, ApJ **417**, L29
van den Bergh, S., 1995, Nature **374**, 215
Vesperini, E., 1997, MNRAS **287**, 915
Whitmore, B. C., Schweizer, F., 1995, ApJ **420**, 87
Zepf, S. E., Ashman, K. M., 1993, MNRAS **264**, 611

GAS AND DUST IN ULTRALUMINOUS GALACTIC NUCLEI

N.Z. SCOVILLE

California Institute of Technology
Astronomy 105-24, Pasadena, CA 91125

AND

M.S. YUN

National Radio Astronomy Observatory
P. O. Box 0, Socorro, NM 87801

Abstract. Millimeter-wave interferometry has clearly shown the existence of enormous masses ($10^9 - 10^{10} M_\odot$) of molecular gas concentrated in the nuclear regions ($R < 500$ pc) of many luminous and ultraluminous infrared galaxies. In these systems, molecular gas is an obvious source of fuel for nuclear starbursts and active galactic nuclei (AGN). In several ultraluminous systems (e.g., Arp 220 and Mrk 231), there now exists CO (2-1) interferometry at $\leq 1''$ resolution which reveals for the first time extremely dense, gaseous accretion disks on the scale 50-300 pc. Based on the low velocity dispersion of the molecular gas in the nuclear disks, we believe them to be extremely thin (10-50 pc). In addition, high brightness temperatures in the CO lines (10-20 K) imply that these disks are nearly uniformly filled with a continuous gas distribution, rather than being relatively isolated, self-gravitating GMCs. Although the gas is 'uniformly' distributed, the gas densities must be high, $> 10^4$ cm^{-3}. When viewed near the plane of the disk, the central AGNs, if they exist, will be totally obscured at optical and near infrared wavelengths. In Mrk 231, our line of sight is probably within 60° of the disk axis, but in Arp 220 the disk is closer to edge-on. In fact, recent near infrared imaging of Arp 220 with the NICMOS camera on the *HST* reveals totally opaque dust disks embedded within the central star clusters of both nuclei.

1. Introduction

The critical role played by dense molecular gas in the activity of galactic nuclei has been appreciated only in the last decade. This gas, which is virtually undetectable via 21 cm HI observations (except in rare cases of absorption), has now been studied at millimeter wavelengths by both single-dishes and aperture synthesis, the latter achieving $< 1''$ resolution in recent work. In the more luminous systems, the large masses of molecular gas are undoubtedly responsible for the prodigious starburst activity and very likely responsible for feeding accretion into central, pc-scale AGN accretion disks. The gas is important not only because it can form stars; it is also extremely dissipative and efficient in radiating bulk rotational energy and transferring angular momentum to larger radii. Despite its energetic environment, the molecular gas probably remains at temperatures of less than 100 K due to extremely effective cooling in molecular lines and the associated dust continuum. The bulk of the far-infrared luminosity can be characterized by color temperatures 40-80 K and the masses of dust derived from the far-infrared opacity are consistent with the molecular gas masses assuming reasonably standard gas-to-dust abundance ratios (100-500 by mass). Here we describe in detail the recent results on Arp 220 and Mrk 231 which, to some extent, represent earlier and later evolutionary phases of the ultraluminous galaxy phenomena.

2. Arp 220

Arp 220 is the prototypical, ultraluminous infrared galaxy with a luminosity at 8-1000 μm of 1.5 $\times 10^{12} L_{\odot}$, clearly placing it in the luminosity regime of quasars. In the visual, Arp 220 exhibits two faint tails which are probably the result of a past tidal interaction (cf Joseph & Wright 1985) and in the nucleus, high resolution near infrared and radio imaging reveals a double nucleus with spatial separation $0.95''$ (Graham et al 1990; Norris 1988). The projected separation of the nuclei corresponds to 330 pc; this double nucleus structure, together with the extended optical tails, suggests that the galaxy is in the final stage of galactic merging.

Single-dish observations of the CO in Arp 220 have revealed an extraordinarily high CO luminosity, corresponding to an H_2 mass of 2 $\times 10^{10} M_{\odot}$ assuming a Galactic conversion ratio (Sanders et al. 1991), and 3 mm aperture synthesis has revealed that the bulk of this CO luminosity originates from the central kpc (Scoville et al. 1991). Recently, this system has been mapped in CO (2-1) at $1''$ resolution (Scoville et al. 1997; Downes 1998). This new work reveals for the first time multiple components in the dense gas: peaks corresponding to each of the double nuclei (separated by $0.95''$ at $PA = 101°$), and a more extended disk-like structure at $PA = 53°$ sim-

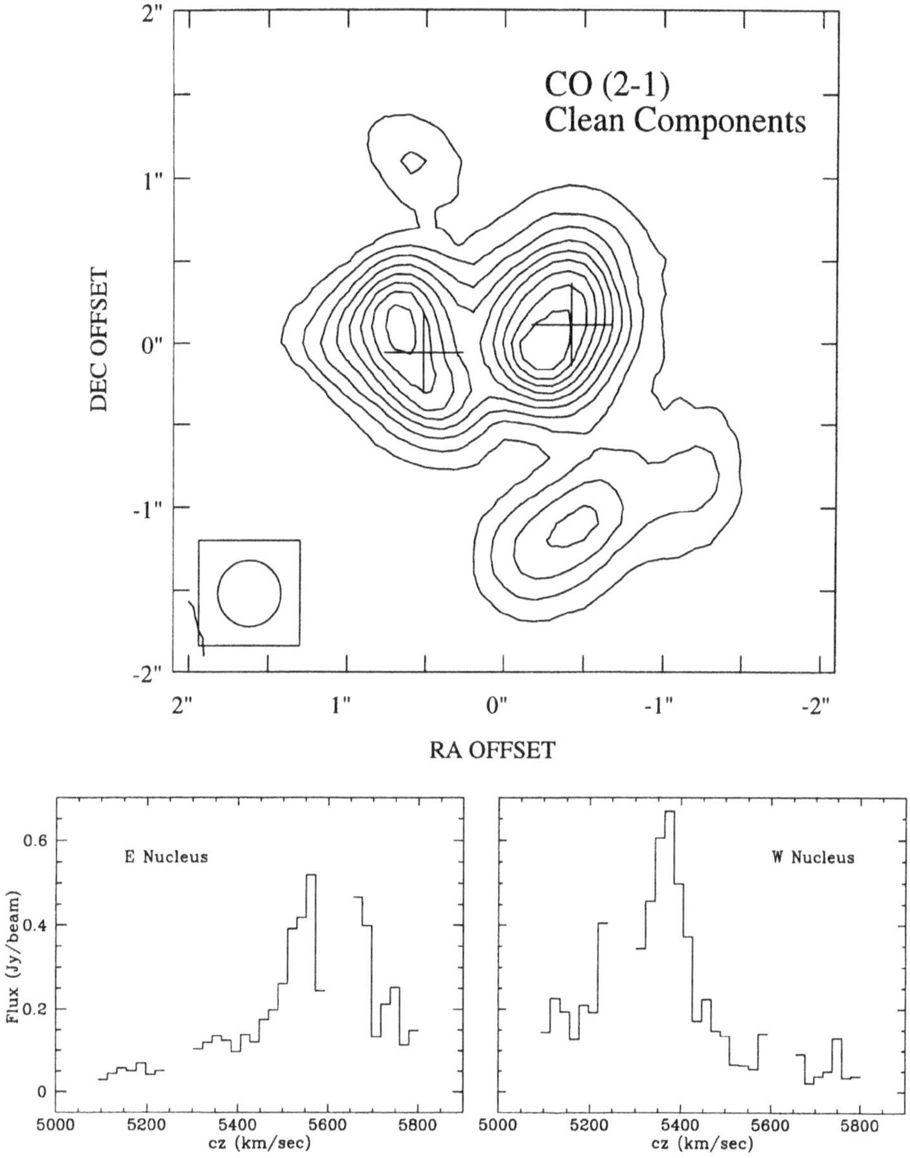

Figure 1. The integrated intensity of CO (2-1) emission from the brightest 'clean' components (Scoville et al. 1997b) is shown with the crosses indicating the positions of the two radio nuclei (Baan & Haschick 1992). Spectra of the clean components at the two nuclei are shown at the bottom. Their double horned shapes suggest gas disks within each of the nuclei.

ilar to the dust lane seen in optical images. Approximately two-thirds of the total CO emission (and presumably the H_2 mass) coincides with the compact double nucleus region.

Detailed modeling of the CO line profiles in Arp 220 using a Doppler image-deconvolution technique similar to that used previously for NGC 1068 (Scoville et al. 1983) yields a best-fit CO emissivity distribution and rotation curve which are mutually consistent – in the sense that, if the total mass distribution follows the CO emissivity, it produces the derived rotation curve. The implied CO-to-H_2 conversion ratio is 0.45 times the Galactic value if the bulk of the mass resides in the molecular gas rather than the stars. The total molecular gas mass for Arp 220 is then inferred to be approximately $\sim 9 \times 10^9 M_\odot$. An important result of the line profile modeling is that the intrinsic velocity dispersion in the extended disk is only 90 km s^{-1}. Assuming that the disk gas is entirely self-gravitating, its thickness (FWHM) is only 16 pc. The mean density in this disk is then $2 \times 10^4 cm^{-3}$, a value which is entirely consistent with the strong emission from high dipole moment molecules such as HCN and HCO^+ (cf Solomon et al. 1992). If the gaseous disk is only partially self-gravitating (i.e., the potential is dominated by a stellar disk), then the gas thickness can be a few times larger.

From the high brightness temperatures of the observed CO emission (17-21 K) and comparison with the infrared color temperature, it is clear that the area filling factor of the disk is very high (~0.25), and therefore that the gas must fairly uniformly fill the disks rather than exist in discrete, self-gravitating clouds. This represents a major change in our picture of the central gas distributions in merging galaxies: even in these highly disturbed systems, the gas has relaxed to a rotationally supported disk which, due to its extremely high surface density and relatively low velocity dispersion, must be very thin. Presumably, this rapid relaxation has occurred as a result of the very strong dissipation in the interstellar gas. Direct observational evidence for the existence of a thin disk in the center of Arp 220 has recently been provided by high resolution near-infrared imaging obtained with the NICMOS camera on the *HST* (Scoville et al. 1998). The near-infrared images clearly show the dominant western nucleus to be crescent-shaped, as though the central star cluster has been partially obscured by an embedded, opaque dust disk.

The fact that the locations and velocities of the nuclei are mutually consistent with their being situated in the molecular disk is suggestive that the nuclei are indeed orbiting with the disk plane. If they were above or below the disk and simply seen in projection against the disk, the positions and velocities would be unlikely to match those of the gas. The results are consistent with a geometry for the central region of Arp 220 in which

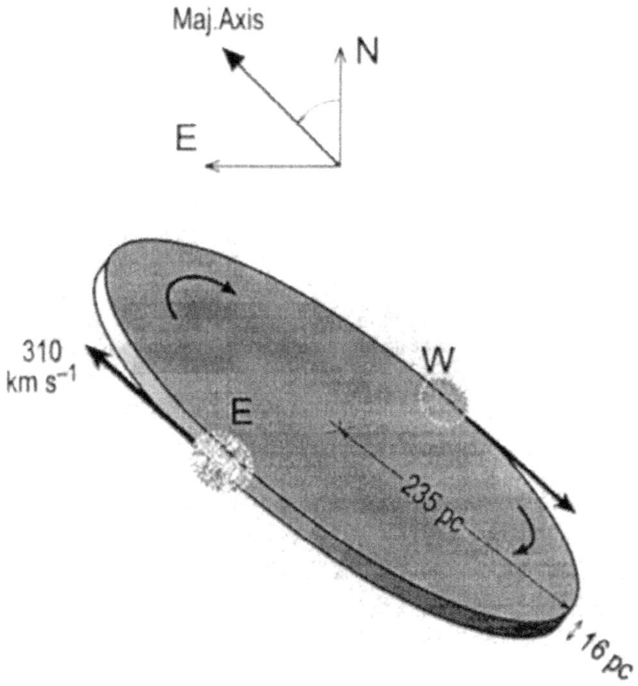

Figure 2. Schematic of the nucleus of Arp 220 showing the central molecular gas disk with the double infrared/radio nuclei orbiting at the outer edge of the nuclear gas disk. Based on the elongation of the molecular disk, the adopted major axis is at $PA = 45°$, and the disk is inclined at approximately 40° to the line of sight. The infrared nuclei then lie along a line between the major and minor axes of the gas disk (69° from the major axis in the disk plane). Approximately two-thirds of the total molecular emission arises from this disk, the remaining from a more extended disk with approximately 10 times lower surface density.

both the molecular gas and the double nuclei are situated in an inclined, rotationally supported disk. This geometry is shown schematically in Figure 2.

3. Mrk 231

At a distance of 174 Mpc, Mrk 231 is one of the most luminous objects in the local universe, with an infrared luminosity of $3.5 \times 10^{12} L_{\odot}$ (Soifer et al.

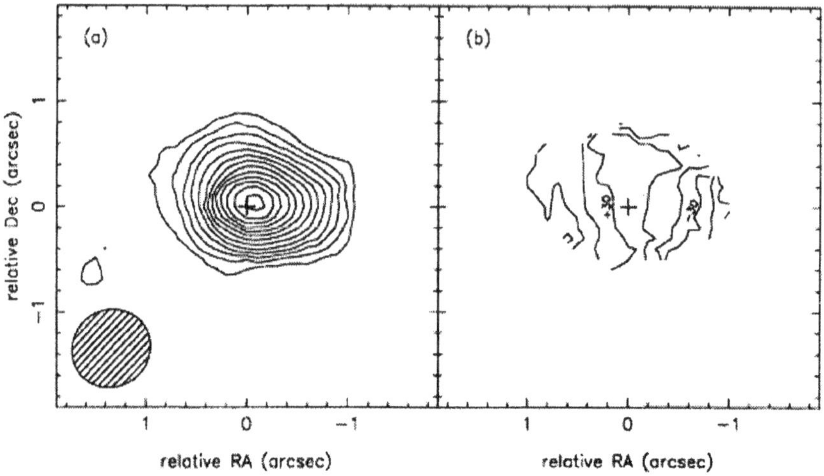

Figure 3. Velocity integrated flux density map (left) and mean velocity map (right) from the CO (2–1) observations of Mrk 231 at 1″ resolution (Bryant & Scoville 1996).

1989). The optical morphology indicates two linear, low surface brightness features that may be tidal tails (Hutchings & Neff 1987, Sanders et al. 1987). The optical spectrum of the nucleus reveals Seyfert I emission line features and a highly reddened continuum and several systems of low ionization absorption lines at velocities up to 8240 km s^{-1} relative to the systemic velocity (Boksenberg et al. 1977, Boroson et al. 1991).

Recently, Bryant & Scoville (1997) mapped the CO (2-1) emission at 0.8″ resolution. The source is well resolved with a deconvolved size of 0.3 ×1.0″ (Figure 3). The major axis radius corresponds to 420 pc. A velocity gradient is also seen in the CO (2-1) line and the kinematic major axis agrees in position angle with the CO isophotal major axis, suggesting as in Arp 220 that the molecular emission arises from an inclined disk. The peak beam-diluted brightness temperature of the CO (2-1) line is 14 K, indicating that the molecular gas also has a high area filling-factor.

Two factors suggest that the disk in the nucleus of Mrk 231 is more nearly pole on: the low visual extinction (A_V = 2 mag, Boksenberg et al. 1977) derived from the optical spectrum and the narrow CO line width (189 km s^{-1}, Bryant & Scoville 1997). The near pole-on geometry is further supported by the detection of the same molecular disk in the 21cm HI absorption but the absence of any absorption towards the central radio source in a VLBA experiment by Carilli et al. (1998). A useful lower limit to the molecular gas mass was derived by Bryant & Scoville (1997) based on the assumption that the CO line was optically thin and that the excitation temperature is such as to give the minimum gas mass consistent with the observed brightness temperature. This lower limit on the mass is 1.3 $\times 10^9$ $M_\odot$, assuming a standard Galactic abundance of CO relative to H_2. This lower limit to the mass is only a factor of 5-10 less than that derived under the standard assumptions of optically thick CO line emission.

4. Concluding Remarks

Several systems imaged at high spatial resolution in the molecular gas (e.g., VV114 and NGC 6240) bear striking similarities to Arp 220 and Mrk 231 (Yun et al. 1994, Bryant & Scoville 1997). Several exhibit CO peaks between double infrared/radio nuclei, clear rotational velocity gradients suggestive of a centrifugally supported disk, and relatively strong high-density molecular line tracers.

These dense molecular-gas accretion disks may play a critical role in the evolution of ultraluminous galaxy nuclei and in their energy release. The dense gas will undoubtedly promote extremely high rates of massive star formation and the disks are also efficient in removing angular momentum – thus feeding and building up a central active nucleus at rates up to 100 $M_\odot \mathrm{yr}^{-1}$. This latter process is probably most important in the most evolved and highest luminosity systems such as Mrk 231.

The characteristics of the molecular gas in the nuclear disk of Arp 220 are extraordinarily different from those in the centers of nearby lower luminosity galaxies. In the Galactic center, the molecular gas ($\leq 10^8 M_\odot$) is mostly contained in very massive, self-gravitating, molecular clouds (e.g., Sgr B2 or the Sgr A complexes) with relatively little intercloud molecular gas and a half-thickness for the cloud distribution of 60-120 pc. In Arp 220, the total gas mass is a factor $\geq$100 larger within a comparable radius, and it appears inescapable that this gas is much more uniformly distributed (i.e., less cloud-like) and confined to an extremely thin disk (16 pc half-thickness). Uniformity of the gas in this disk is suggested by the high area filling factor ($\simeq$ 0.25) of the molecular gas derived from the observed CO line brightness temperatures. And given the high area filling factor and small scale height, the mean free path of clouds within this layer should be only a few disk scale heights. If we conservatively estimate the cloud-cloud collisional mean free path as $\leq$50 pc, then the mean collision time is only 5×10^5 yr. With such short cloud-cloud collision times and high velocities, any preexisting clouds would be rapidly disrupted and their matter spread uniformly in the disk.

The two galaxies discussed here (Arp 220 and Mrk 231) may represent two critical phases in the merging galactic nuclei scenario which has been suggested to link the ultraluminous infrared galaxies with AGNs and quasars. Arp 220 is still very much dust-enshrouded but shows two distinct nuclei, whereas the latter has only one major nucleus and optical spectroscopy shows both the broad emission lines characteristic of a Seyfert 1 nucleus and broad absorption lines. In the context of the ultraluminous IR galaxy/quasar evolutionary models (e.g., Sanders et al. 1988), Arp 220 would represent an earlier phase just prior to the merging of the two galactic nuclei, and Mrk 231 the phase in which the quasar-like nucleus has been blown free of dust, at least above and below the central nuclear gas disk.

Acknowledgments
This research is supported in part by NSF Grant AST96-13717 and the Norris Planetary Origins Project. We thank Nanci Candelin for help in preparation of this manuscript.

References

Boksenberg, A., Carswell, R.F., Allen, D.A., Fosbury, R.A.E., Penston, M.V., & Sargent, W.L.W. 1977, *MNRAS*, **178**, 451
Boroson, T.A., Meyers, K.A., Morris, S.L., & Persson, S.E. 1991, *Ap.J.*, **370**, L19
Bryant, P.M. 1996, Ph.D. thesis, California Institute of Technology
Carilli, C.L., Wrobel, J.M., & Ulvestad, J.S. 1998, *Ap.J.*, in press
Downes, D. 1998, *IAU Symposium 184*, in press
Graham, J.R., Carico, D.P., Matthews, K., Neugebauer, G., Soifer, B.T., & Wilson, T.D. 1990, *Ap.J. (Letters)*, **354**, L5
Hutchings, J.B. & Neff, S.G. 1987, *Ap.J*, **92**, 14
Joseph, R.D. & Wright, G.S. 1985, *MNRAS*, **452**, 599
Larkin, J.E., Armus, L., Knop, R.1A., Matthews, K., & Soifer, B.T. 1995, *Ap.J.*, **452**, 599
Norris, R.P. 1988, *MNRAS*, **230**, 345
Sanders, D.B., Scoville, N.Z., & Soifer, B.T. 1991, *Ap.J.*, **370**, 158
Sanders, D.B., Soifer, B.T., Elias, J.H., Madore, B.F., Matthews, K., Neugebauer, G., & Scoville, N.Z. 1988, *Ap.J.*, **325**, 74
Scoville, N.Z., Evans, A.S., Dinshaw, N., Thompson, R., Rieke, M., Schneider, G., Low, F.J., Hines, D., Stobie, B., Backlin, E., & Epps, H. 1998, *Ap.J. (Letters)*, January 15, 1998 issue
Scoville, N.Z., Young, J.S., & Lucy, L.B., 1983, *Ap.J.*, **270**, 443
Solomon, P.M., Downes, D., & Radford, S.J.E. 1992, *Ap.J. (Letters)*, **387**, L55
Yun, M.S., Scoville, N.Z., & Knop, R.A. 1994, *Ap.J. (Letters)*, **430**, L109

PROTO–GLOBULAR CLUSTER CANDIDATES IN NGC 1275

JEAN P. BRODIE
Lick Observatory
University of California, Santa Cruz, CA 95060, U.S.A.

The discovery, with *HST* imaging, of proto–globular cluster candidates in NGC 1275 (Holtzman et al. 1992) was regarded by many as a major success of the merger model for globular cluster formation (e.g. Ashman & Zepf 1992) and has been cited in support of the idea that elliptical galaxies form from the merger of two or more spiral galaxies. A prediction of the Ashman & Zepf model was that newly–formed clusters should be observable in currently or recently merging systems. The NGC 1275 clusters constitute an important test of globular cluster formation models. NGC 1275 is the peculiar cD galaxy at the center of the Perseus cluster. It shows evidence for a merger history and may indeed be undergoing a merger at present. It also has one of the largest known cooling flows.

Spectroscopy with the Keck I telescope of 5 proto–globular cluster candidates in NGC 1275 (Brodie et al. 1997) revealed that the candidates are not HII regions, are clearly dominated by early A–type stars, and are not similar to young or intermediate age Magellanic Cloud or Milky Way open clusters.

The Balmer absorption lines were found to be too strong to be consistent with any of the standard IMF (Salpeter or Scalo), solar metallicity, Bruzual & Charlot stellar evolutionary models at any age. The preliminary Bruzual & Charlot (1997) models indicate that no appreciable increase in equivalent width can be achieved by changing the metallicity. However, a 2–3 $M_{\odot}$ IMF, adopted to simulate a flatter IMF, reproduces the observed equivalent widths and colors and indicates an age of $\sim$ 500 Myr for these objects. Preliminary Fritze v. Alvensleben & Kurth (1997) models are better able to reproduce the observed equivalent widths, with a best fit at 350 Myr, but the model continuum is redder than the observed spectra. The sense of the discrepancy is that the model predicts too much red light, consistent with the suggestion of a flatter cluster IMF.

Another problem with the assumption of a standard IMF for these objects is the fact that, based on their luminosities, the masses of these bright clusters are deduced to be $\sim 10^8 M_\odot$. Such high mass clusters would be difficult to form, requiring surprisingly massive progenitor gas clouds, and would be very unlikely to loose sufficient mass during their evolution to bring them into the mass range of normal old globular clusters. On the other hand, a pure A–star population, for example, would have a mass of $\sim 10^6 M_\odot$ at these luminosities, and a flatter than normal IMF would produce a mass somewhere in between.

Other key properties of the proto–globular cluster candidates are their spatial distribution and their velocity dispersion. These objects are extremely centrally concentrated. The entire sample (some 60 objects) is within 8 kpc of the nucleus and the brightest clusters are all within 2 kpc. We find a low (compared to the stellar value) velocity dispersion, $\sim$ 200 km/s, for our sample of 5 of the brighter candidates.

The spectroscopic information allows us to set some interesting constraints on the origin of these objects. We can clearly rule out formation in a continuous cooling flow. The spatial scale of the clusters is very much less than the cooling flow radius and their age is very much greater than the cooling time of $\sim$ 10 Myr. The star formation rate in excess of 400 $M_\odot$/yr, deduced from the cooling flow (Allen & Fabian 1997), and the absence of high mass stars (Smith et al. 1992) imply a steep IMF rather than the flat IMF deduced for the clusters. It is equally clear that these clusters did not form in widespread shocks from merging galaxies. They are centrally concentrated and, if they formed far from the center and later fell in, a high rather than a low velocity dispersion would be expected.

It appears, then, that the clusters formed in a discrete event some 500 Myr ago. This may have been induced by a merger which provided the fuel for a short–lived gas inflow episode. However, their properties are such that they may not represent formation processes that had any significant effect on the global properties of globular cluster systems. The resultant clusters are not distributed like old globular clusters in central cD galaxies, which are significantly more diffuse than the galaxy light. If they do indeed have an IMF which is biased against low mass stars, they may fade very rapidly. A pure A–star population would fade away in only $\sim 10^9$ yr.

References

Ashman, K.M., & Zepf, S.E. 1992, *ApJ*, **384**, 50
Allen, S.W., & Fabian, A.C. 1997, *MNRAS*, **286**, 583
Brodie, J.P., Schroder, L.L., Huchra, J.P., Phillips, A.C., Kissler-Patig, M., & Forbes, D.A. 1997, *AJ*, submitted
Holtzman, J.A., et al. 1992, *AJ*, **103**, 691
Richer, H., Crabtree, D., Fabian, A.., & Lin, D. 1993, *AJ*, **105**, 877
Smith, E.P. et al. 1992, *ApJ*, **395**, L49

CO OBSERVATIONS OF LUMINOUS IR GALAXY MERGERS

Y. GAO
Dept. of Astronomy, University of Illinois, Urbana, IL

AND

P.M. SOLOMON
Dept. of Phy. & Astronomy, SUNY at Stony Brook, NY

Luminous starbursts are observed to occur mostly as a result of a collision/merger in gas-rich galaxies, and most luminous infrared galaxies (LIGs) are indeed gas-rich mergers. In order to determine the relationship between the IR and molecular gas properties and the galaxy-galaxy interactions, we study LIG mergers in the intermediate merging process. We have observed nearly 20 LIG mergers and together with the CO data in the literature, we have found a correlation between the CO luminosity, $L_{\rm CO}$, and the projected separation of merger nuclei, $R_{\rm Sep}$, in > 50 LIG mergers. The correlation suggests the molecular content is decreasing as merging advances and is better established with $\sim$ 40 LIG mergers excluding ultraluminous ones, which resembles more a *volume-limited*, statistically complete sample of LIG mergers. In addition, an anti-correlation between $L_{\rm IR}/L_{\rm CO}$ (the measure of star formation efficiency, SFE) and $R_{\rm Sep}$ is evident. One interpretation is that the molecular gas content of LIG mergers is being rapidly depleted due to the merger-induced starbursts and the increase of SFE as merging progresses.

Both numerical simulations and observations have shown the vital role of the gas in triggering the starburst in galaxy mergers. Correlations we found here do suggest a sequence of merging can be traced by studying the molecular ISM in LIGs. This is because LIG mergers have roughly comparable initial gas reservoirs since they are all mergers of gas-rich spirals and the ultraluminous starburst phase can eventually be reached. In comparison, the "Toomre sequence" or optically selected samples of interacting pairs are mixtures of various gas content and many of them will never reach the ultraluminous phase and the merger sequence in terms of the gas content and the peak starburst (ultraluminous) phase is not well represented.

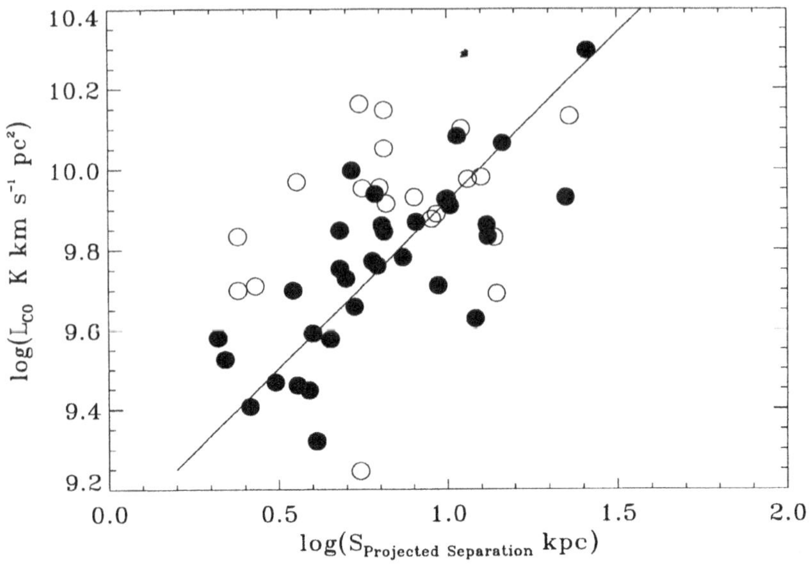

Figure 1. The correlation between L_{CO} and R_{Sep} for > 50 LIG mergers imlies molecular gas mass decreases as merging progresses. Open circles are for ultraluminous mergers

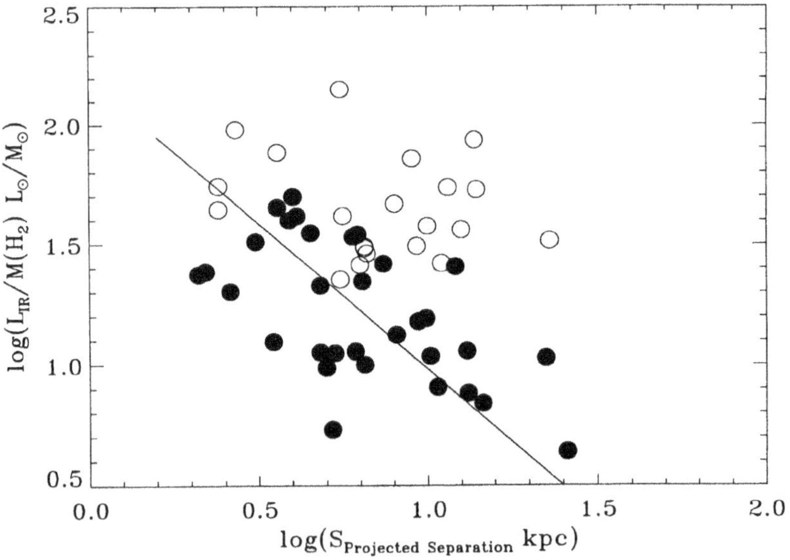

Figure 2. The anti-correlation between L_{IR}/L_{CO} (SFE) and R_{Sep} suggests enhanced SFE as merging advances.

STAR-FORMING ACTIVITY IN ARP-MADORE GALAXIES

A.M. HOPKINS AND L.E. CRAM
University of Sydney NSW 2006
Australia

1. Introduction & method

The rate of star formation is enhanced in many galaxies that show disturbed morphology and/or evidence of interaction. However, the physical explanation of this connection has proved elusive, since it appears that peculiar morphology of a specific type is neither necessary nor sufficient to promote star formation. To explore the relation between star formation rates and galaxy morphology we have selected a sample of galaxies from the *Arp-Madore Catalogue of Southern Peculiar Galaxies and Associations* (AMC) and the *Principle Galaxy Catalogue* (PGC), and estimated their star formation rate using the far-infra-red (FIR) power listed in the *QMW IRAS Galaxy Catalogue* (QMCIGC). There are 933 objects common to all three catalogues, potentially providing the necessary data for the peculiar galaxies.

We also identified a subset of 'normal' galaxies from the PGC and QMCIGC containing 523 objects to form a sample matched to the AM objects by absolute blue luminosity, absolute diameter, H I line width and redshift. With the exception of the redshift distribution, the two sample distributions were identical as characterized by the Kolmogoroff-Smirnov test. The redshift distribution of the 'normal' sample lies at smaller values than the AM galaxies, reflecting the relative rarity and possibly the intrinsic interest of AM-type objects.

From these samples we selected all of the objects having published blue (in PGC) and FIR (in QMCIGC) luminosities. We then determined a number of astrophysically significant quantities for the samples, including the ratio of FIR to blue luminosity as a measure of the *relative* rate of star formation. After inspecting the distribution of this ratio, we determined that if the ratio is greater than $10^{0.3}$, the galaxy may be very actively forming stars. Table 1 illustrates the propensity for galaxies from either sample to be an actively star forming object.

TABLE 1. Distribution of log(FIR/B-band)

log(FIR/B-band)	AM sample (%)	Control sample (%)
> 0.3	0.125	0.019
> 0.2	0.132	0.030
> 0.0	0.154	0.041

2. Results & Conclusions

There are 51/408 or 12.5% of the AM sample classified as active, and 7/370 or 1.9% of the control sample so classified. We conclude that AM galaxies are more likely to exhibit star-forming activity than galaxies without peculiarities. Of the Arp-Madore morphological categories, the ones with the highest proportion of activity belong to categories such as AM9 (M51-type with a companion at the end of a spiral arm) which are "recent" or "early" interactions. These are not the galaxies having the greatest tidal distortion. Our result suggests that increased star-forming activity sets in early in the interaction, and has faded by the time that tidal effects are fully manifest.

On the other hand, the great majority of AM objects show little or no sign of excess star forming activity. Also, some of the control sample galaxies have high rates of star formation but are not associated with the type of peculiarity leading to selection in the AM catalogue.

We considered differences between the AM sample and the control sample for subsamples selected (1) to be known Seyfert galaxies and (2) to be known members of ACO clusters. There is no significant increase in star-forming activity in the AM Seyferts (18 objects) relative to the control Seyferts (63 objects). There is a larger fraction of the AM galaxies (64/933=6.9% versus 9/523=1.7%) in clusters, and of the cluster members the activity is notably higher in the AM galaxies although only one has an index greater than 0.3.

The relatively large proportion of M51-type galaxies found to be among the most active AM objects is consistent with the occurrence of star formation at an early phase of an interaction. The proportion of AM galaxies showing a high rate of star formation is smaller for objects with well developed evidence of tidal interaction. This is consistent with the idea that star-forming time scales are of the order of 10^8 years, while tidal interaction time scales are of the order of 10^9 years.

GAS CONTENT OF MARKARIAN STARBURST GALAXIES

RAFIK KANDALYAN
Byurakan Astrophysical Obs., 378433 Byurakan, Armenia
E-mail: rkandali@bao.sci.am, rafik@ipia.sci.am

Abstract. The main results of this study can be summarized as follows: (a) The H I and CO linewidths are well correlated. Interaction between galaxies has little influence on the H I and CO line broadening. A rapidly rotating nuclear disk in the galaxy could lead to CO line broadening, while the H I line is less affected by the rotating disk. Molecular gas in Markarian galaxies is centrally concentrated. (b) For past and present star formation activity both H I and H_2 components of the gas are important. The atomic and molecular gas surface densities are well correlated with blue, FIR, and radio continuum surface brightnesses, but the H_2 surface density is better correlated than that of the H I. The two gas phases are also connected. (c) In general, galaxies with UV-excess (Markarian galaxies) are not distinguished by star formation properties from non-UV galaxies, however some second order differences may exist, like the relation between atomic surface density and radio continuum surface brightness.

1. Results

Our original sample of Mkn-*IRAS* galaxies contains 155 objects (Kandalian et al. 1995). In order to investigate the gas properties of Mkn-*IRAS* galaxies in our sample we have extracted from the literature all objects which were detected in the ^{12}CO(1-0) line (till July, 1997). The total number of CO detected Mkn galaxies is 61. Optical, FIR, H I and radio continuum data have been extracted from the literature (Martin et al. 1991,1997; Kandalian et al. 1997 and references therein). In order to study the gas kinematics of Mkn galaxies we have made a statistical investigation of the H I and CO linewidths, W. There is a good correlation between $W_{\rm H\,I}$ and $W_{\rm CO}$ (correlation coefficient $r = 0.72$ at significance $p < 0.0001$). Sofue et al. (1993) have suggested that tidal interaction could disturb the outermost but not innermost regions of a galaxy. As a conseauence $W_{\rm H\,I}$ for paired+interacting galaxies will be much broader than that for isolated ones and no difference will be observed in $W_{\rm CO}$ between these two types of galaxies. Our analysis shows that there are no significant differences between $W_{\rm H\,I}$ and $W_{\rm CO}$ for isolated and paired+interacting galaxies, although $W_{\rm H\,I}$ for the latter is slightly higher than that for the former. According to our data there are galaxies with $W_{\rm H\,I} - W_{\rm CO} < 0$. It is likely that when $W_{\rm CO} > W_{\rm H\,I}$ it indicates the existence of a rapidly rotating nuclear disk in the galaxy, and as a consequence, rotation curves of these galaxies could have a peak in their central region ($<$ 1kpc). It could also be due to an outflow from the nuclear region (expanding molecular gas). The clumpy structure of molecular gas could also lead to CO line broadening.

We have compared gas-to-luminosity relations for 5 samples, namely optically selected nearby galaxies (Nearby); optically selected starburst galaxies (Starburst); Markarian galaxies detected by *IRAS* (UV-IRAS); *IRAS* selected galaxies (IRAS) and galaxies which belonged in clusters (Cluster). Several important conclusions have been drawn: (a) For all samples, FIR surface brightness is more tightly correlated with the H_2 surface density than with H I surface density. (b) The tight correlation between H I surface density and FIR surface brightness for all samples indicates that the H I phase is also important for current or recent star formation. (c) Both phases of the gas are linked with indicators of past star formation, but the molecular phase does not now dominate in this relation for all samples as was the case for FIR emission. (d) There is tight correlation between $\sigma_{\rm R}$ and $\sigma_{\rm H_2}$ and for the "UV-IRAS" sample, the radio continuum surface brightness is also correlated with the H I surface density. (e) The galaxies with UV-excess are not distinguished in their star formation properties from other galaxies.

References

Kandalian R.A. et al. 1995, *Afz*, **38**, 639

Kandalian R.A., Martin J.-M., Bottinelli L., & Gouguenheim L. 1997, in preparation

Martin J.-M. et al. 1991, *A&A* **245**, 393

Martin J.-M., Kandalian R.A., Horellou C., Bottinelli L., & Gouguenheim L. 1997, in preparation

Sofue Y. et al. 1993, *PASJ*, **45**, 43

OCULAR GALAXIES: NGC 2535 AND ITS STARBURST COMPANION NGC 2536

E. BRINKS
Universidad de Guanajuato, Guanajuato, México
M. KAUFMAN
Ohio State University, Columbus, USA
D.M. ELMEGREEN
Vassar College, Poughkeepsie, USA
M. THOMASSON
Onsala Space Observatory, Onsala, Sweden
B.G. ELMEGREEN
IBM T. J. Watson Research Center, Yorktown Heights, USA
C. STRUCK
Iowa State University, Ames, USA
AND
M. KLARIĆ
Midlands Technical College, Columbia, USA

We obtained HI, radio continuum, and ^{12}CO $J = 1 \rightarrow 0$ observations at resolutions of $12''$ to $33''$ (= 2.9 – 8 kpc), and B, I, J, and K–band images, of the galaxy NGC 2535 and its small starburst companion NGC 2536. NGC 2535 has an ocular (eye–shaped) structure indicative of a recent, close, nonmerging encounter. Our observations reveal widespread high velocity dispersions (30 km s^{-1}) in the HI gas and five clouds with masses of $\sim 10^8$ $M_\odot$ in the tidal arms of NGC 2535. CO emission was detected at the center and on the tidal tail, but close to the center, of NGC 2535; no CO emission was detected from the companion. NGC 2535 has an intrinsically oval shape to the disk, an extended ($R = 48$ kpc) HI envelope and an outer elliptically–shaped HI arc that may be a gravitational wake produced by the passage of the companion within or close to the extended HI envelope. The starburst companion, NGC 2536, lies in a 2×10^9 $M_\odot$ clump of HI gas at the outer end of the tidal bridge from NGC 2535. A full account our results appears in Kaufman et al. (1997, *AJ*, **114**, 2323).

LUMINOUS INFRARED GALAXIES IN A MERGING SEQUENCE: ISO OBSERVATIONS

C.-Y. HWANG AND K.Y. LO
Institute of Astronomy and Astrophysics, Academia Sinica
P.O. Box 1-87, Nankang, Taipei, Taiwan 115, R.O.C.
Y. GAO AND R.A. GRUENDL
Department of Astronomy, University of Illinois, U.S.A.
AND
N.-Y. LU
IPAC, Caltech, U. S. A.

We report mid-infrared images of several luminous infrared galaxies (LIGs) taken with ISOCAM on the *Infrared Space Observatory (ISO)*. These LIGs were chosen to represent different phases of a merger sequence of galaxy-galaxy interaction with special emphasis on early/intermediate stages of merging. The molecular gas distribution of these LIGs has also been mapped at high spatial resolution (see contribution by Gao et al., this volume). The goal is to do a synoptic study of the evolution of physical conditions in these LIGs along the merger sequence.

The ISOCAM 15 μm (LW3) images of these LIGs show extended structures for the early and intermediate mergers. Individual galaxies are resolved, and the distribution of the mid-infrared emission has similar morphology as that of CO emission (see Gao et al., this volume); the results indicate that most of the IR luminosity of these LIGs is not from a central AGN. The ratios of the peak fluxes to the total fluxes of the 15 μm emission of these LIGs increase from around 0.05 for the early mergers to around 0.1 for the advanced mergers except for UGC 2369 and Mrk 848, which were found to have undergone some other merging processes. It is noted that there is a similar trend for the star formation efficiency ($L_{\rm IR}/M(\rm H_2)$; SFE), which also increases roughly by a factor of two from the early mergers to the intermediate/advanced mergers for these LIGs (see Gao et al., this volume).

FORMATION OF PLUMES IN THE HEAD-ON COLLISIONS OF GALAXIES

VLADIMIR KORCHAGIN
Institute of Physics, Stachki 194, Rostov-on-Don, Russia
TOSHIO TSUCHIYA
*Department of Astronomy, Faculty of Science,
Kyoto University, Kyoto 606-01, Japan*
AND
KEIICHI WADA
National Astronomical Observatory, Mitaka, Tokyo 181, Japan

We have studied the collisional interaction of disk galaxies using SPH and N-body numerical simulations. We follow the dynamics of a two-component star-gas disk centrifugally balanced by the gravity of a rigid halo. A companion galaxy is modeled self-consistently, and its two-component disk is balanced by the potential of the stellar halo. This model allows us to follow the dynamics of the gas component in the colliding galaxies simultaneously with the dynamics of the collisionless halo of the companion.

The specific goal of our paper was to study the formation of plumes in the head-on collisions of galaxies. Our results can be summarized as follows:
1. We found that the formation of a ring in the disk of a primary does not depend essentially on the admixture of gas. On the contrary, the presence of gas in the intruder is crucial for the formation of plumes connecting two interacting galaxies. A low-mass intruder with the gas mass about a few percent of the primary's gas content forms a well developed plume.
2. In agreement with previous studies we found that the amplitude of the outwardly propagating ring strongly depends on the relative mass of the intruder galaxy. The intruder galaxy with mass equal to 10 % of the mass of the primary does not form any noticeable ring structure in the disk of the primary, and the bridge becomes the main "fingerprint" of the interaction.
3. Finally we demonstrated that a direct collision forms a bridge of *stars* stripped from the stellar halo of the companion. The fate of the halo of an intruder depends on its mass. Most of the stars of a low-mass intruder are dispersed after the collision, but the central cores of the halos of higher mass intruders survive the collision.

SINGLE STELLAR POPULATIONS

Colors and Indices

O.M. KURTH, U. FRITZE-V.ALVENSLEBEN AND K.J. FRICKE
Universitätssternwarte Göttingen
Geismarlandstr.11, 37083 Göttingen, Germany

To determine metallicity and age distributions of globular clusters (GCs) in distant galaxies–now accessible, e.g. to Keck spectroscopy–it is important to have reliable calibrations of the color-metallicity, color-age, and index-metallicity relations.

We have calculated colors in UBVRIK and stellar atmospheric indices for single stellar population (SSP) models at various metallicities. We are using the evolutionary tracks from the Padova group (Bressan et al. 1993; Fagotto et al. 1994), theoretical color calibrations from Lejeune et al. (1997) and fit functions for atmospheric indices from Worthey et al. (1994). Our models give theoretical calibrations for GC colors and indices in terms of [Fe/H]. The theoretical colors and metallic indices at an age of about one Hubble time are in good agreement with the observations (Fig. 1).

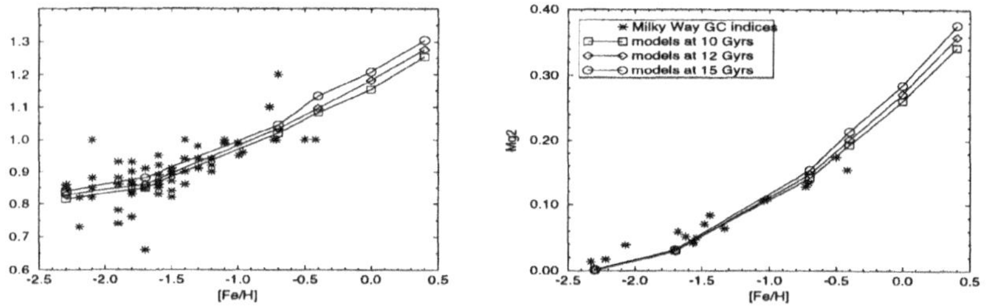

Figure 1. V-I (left) and Mg2 index (right) vs. metallicity for Milky Way globular clusters (stars) and for models at 10 (squares), 12 (diamonds) and 15 (circles) Gyrs. Metallicities and colors are from Harris 1996, indices from Burstein et al. (1984).

References

Burstein, D., Faber, S.M., Gaskell, C.M., & Krumm, N. 1984, *ApJ*, **287**, 586
Bressan, A., Fagotto, F., Bertelli, G., & Chiosi, C. 1993, *A&AS*, **100**, 647
Fagotto, F., Bressan, A., Bertelli, G., Chiosi, C. 1994, *A&AS*, **104**, 365, & **105**, 29+39
Harris, W.E. 1996, *AJ*, **112**, 1487
Lejeune, T., Cuisinier, F., & Buser, R. 1997, preprint
Worthey, G., Faber, S.M., González, J.J., & Burstein, D. 1994, *ApJS*, **94**, 687

NEW MODELS FOR MASSIVE STAR POPULATIONS IN YOUNG STARBURSTS

Summary of the models and applications

DANIEL SCHAERER
Space Telescope Science Institute, 3700 San Martin Drive
Baltimore, MD 21218, USA

AND

WILLIAM D. VACCA
Institute for Astronomy, Honolulu, HI 96822, USA

Using the latest stellar evolution models, theoretical stellar spectra, and a compilation of observed emission line strengths from Wolf-Rayet (WR) stars, we have constructed evolutionary synthesis models for young starbursts (Schaerer & Vacca 1997; see also Schaerer 1996). We provide detailed predictions of UV and optical emission line strengths for both the WR stellar lines and the major nebular hydrogen and helium emission lines, as a function of several input parameters related to the starburst episode.

Our models represent a significant improvement in the modeling of starburst regions; they are ideally suited to analysis of massive star populations in young starbursts, emission line galaxies, extragalactic H II regions, BCDG, etc. They can be used to determine the properties of the burst episode, such as the age, duration, star formation rate, IMF etc., and can be used to investigate the variation in these starburst properties as a function of environment and metallicity.

The results from our models are available in electronic form (see `http://www.stsci.edu/ftp/science/starburst`).

References

Schaerer D. 1996, *ApJ*, **467**, L17

Schaerer, D.,& Vacca, W.D. 1996, in *WR Stars in the Framework of Stellar Evolution*, 33rd Liège Int. Astroph. Coll., Eds. J.M. de Vreux et al. (Liège: Université de Liège), p. 641

Schaerer D., Vacca W.D.W. 1997, *ApJS*, in press

NIR LINE OBSERVATIONS OF STARBURST GALAXIES

H. SUGAI
Dept. of Astron., Kyoto Univ., Japan
M.A. MALKAN
Dept. of Astron., Univ. of California at Los Angeles, USA
M.J. WARD
Dept. of Physics and Astron., Univ. of Leicester, UK
R.I. DAVIES
Dept. of Physics, Nuclear Physics Laboratory, UK
AND
I.S. MCLEAN
Dept. of Astron., Univ. of California at Los Angeles, USA

We have obtained images of the H_2 and Brγ emission lines in the galaxy interacting system NGC 3690 + IC 694. We have also obtained simultaneous H- and K-band spectra for three of its 2μm continuum peaks. The most detectable line emission is concentrated at the continuum peaks. Therefore, the emission lines as well as stellar absorption lines can be used as tracers of the activity in the nuclei themselves. From the strong Brγ and marginal detection of Br10 at the nucleus of IC 694, we derive a large extinction for the fully ionized gas in this nucleus. If we adopt this extinction also for the [Fe II]1.64μm emission, the extinction-corrected [Fe II]1.64μm/Brγ ratio will lie at the higher end of starburst galaxies, and is typical for AGNs or AGN/starburst composites. This might imply that many SNRs are involved in the starburst at this nucleus, unless it includes an AGN. All of our results for Component C, including very little CO absorption in the K band, a large EW(Brγ), a small H_2/Brγ ratio, the effective temperature ($T_{\rm eff} \simeq$ 40,000K) derived from HeI 1.70μm/Br10 and HeI 2.06μm/Brγ, are consistent with a very young starburst.

We have also obtained an H_2 image of another galaxy interacting system Mrk 551, as well as its spectra. In contrast to NGC 3690 + IC 694, we have found that the H_2 emission peak is offset from the most luminous continuum peak by several hundred parsecs.

ASCA OBSERVATIONS OF LUMINOUS INFRARED STARBURST GALAXIES

H. WATARAI, K. MISAKI AND Y. TERASHIMA
Department of Astrophysics, Nagoya University, Japan

AND

T. NAKAGAWA
Institute of Space and Astronautical Science, Japan

We present recent results of X-ray observations of two luminous infrared galaxies, NGC3690+IC694 (Arp299) and NGC1614 obtained by the Japanese X-ray astronomical satellite *ASCA*. Both galaxies have quite high infrared luminosity ($> 10^{11} L_{\odot}$) and strong evidence of merger.

NGC3690+IC694(Arp299): The 0.5-10 keV spectrum of Arp299 is quite similar to that of the typical starburst galaxy, M82. Emission lines from heavy elements (e.g. Mg, Si, S etc.) are clearly seen in the spectrum, which means the existence of a thermal hot plasma in this galaxy. We tried spectral fitting with a three component model, which consists of the 2 kT Raymond-Smith (RS) component with variable abundance plus an hard component. Derived temperatures of the soft- and mid-components are similar to those of M82, and an apparently lower abundance of Fe than other elements in this galaxy is typical of what is seen in starburst galaxies. Although the X-ray luminosity of Arp299 is an order of magnitude higher than nearby starburst galaxies, the X-ray to far-infrared luminosity ratio ($L_X/L_{FIR} \sim 10^{-4}$) is similar to typical values of starbursts. The hard-component which shows very high temperature ($>$ 10 keV) is also seen in the spectra of both M82 and NGC253. These observational characteristics imply that Arp299 would be a typical starburst galaxy, and the huge far-infrared luminosity could be explained only by the starburst phenomenon.

NGC1614: Since photon statistics are limited, only a one-component model was used for spectral fitting, and both a power-law model and a 1kT RS model are acceptable for this galaxy. However, the X-ray luminosity and the derived L_X/L_{FIR} are almost comparable with those of Arp299, and thus, we conclude that NGC1614 would also be a pure starburst galaxy.

ULTRALUMINOUS INFRARED GALAXIES

D.B. SANDERS, J.A. SURACE, AND C.M. ISHIDA
Institute for Astronomy, University of Hawai'i
2680 Woodlawn Drive, Honolulu, HI 96822, USA

Abstract. At luminosities above $\sim 10^{11} L_\odot$, infrared galaxies become the dominant population of extragalactic objects in the local Universe ($z < 0.5$), being more numerous than optically selected starburst and Seyfert galaxies, and QSOs at comparable bolometric luminosity. At the highest luminosities, ultraluminous infrared galaxies (ULIGs: $L_{\rm ir} > 10^{12} L_\odot$), outnumber optically selected QSOs by a factor of $\sim$1.5–2. All of the nearest ULIGs ($z <0.1$) appear to be advanced mergers that are powered by both a circumnuclear starburst and AGN, both of which are fueled by an enormous concentration of molecular gas ($\sim 10^{10} M_\odot$) that has been funneled into the merger nucleus. ULIGs may represent a primary stage in the formation of massive black holes and elliptical galaxy cores. The intense circumnuclear starburst that accompanies the ULIG phase may also represent a primary stage in the formation of globular clusters, and the metal enrichment of the intergalactic medium by gas and dust expelled from the nucleus due to the combined forces of supernova explosions and powerful stellar winds.

1. Introduction

One of the major results of the Infrared Astronomical Satellite (*IRAS*) all-sky survey was the identification of a class of luminous infrared galaxies (LIGs: $L_{\rm ir} > 10^{11}\ L_\odot$; $H_{\rm o}$ = 75 km s^{-1}Mpc^{-1}, $q_{\rm o}$ = 0.5)[1], objects that emit more energy in the far-infrared/submillimeter than at all other wavelengths combined. Redshift surveys of complete samples of *IRAS* galaxies now agree that infrared selected galaxies become the dominant population of extragalactic objects at bolometric luminosities above $\sim 4\,L^*$ (i.e. $L_{\rm bol} > 10^{11}\ L_\odot$). Reasonable assumptions about the lifetime of the infrared

[1] $L_{\rm ir} \equiv L$(8-1000μm), computed from the observed infrared fluxes in all four *IRAS* bands according to the prescription in Perault (1987); see also Sanders & Mirabel (1996).

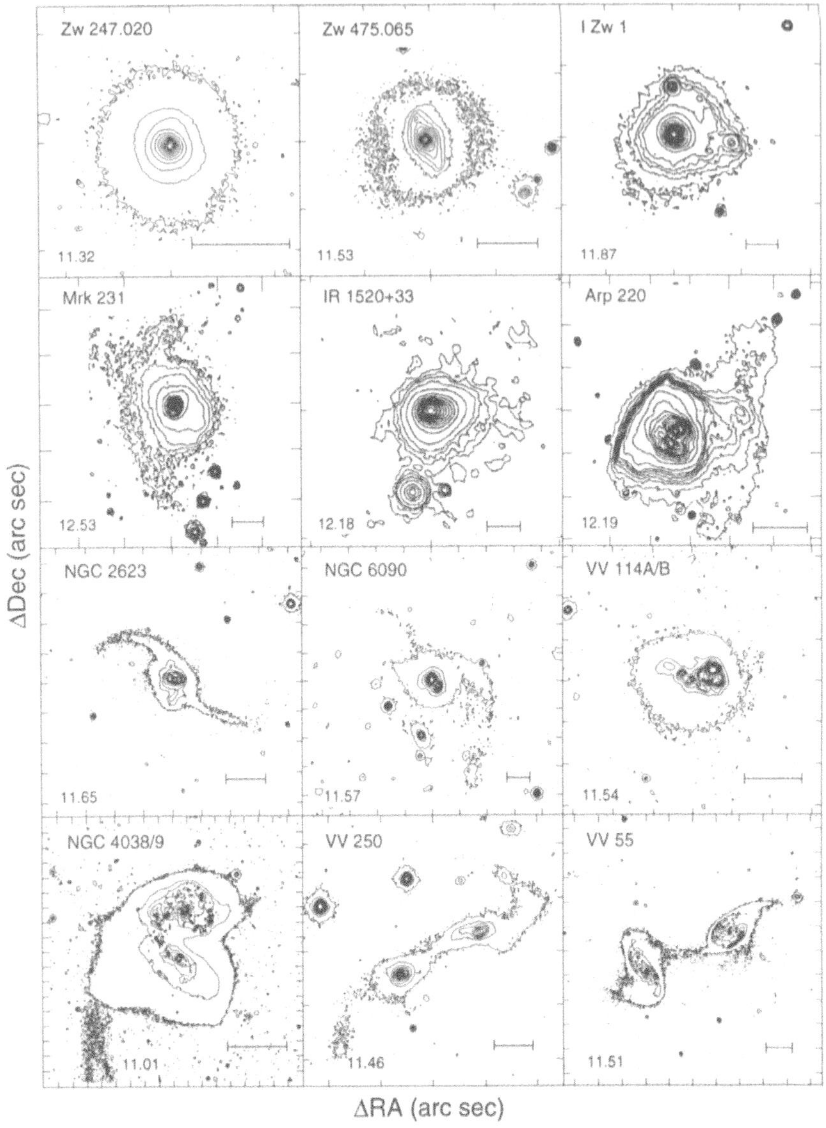

Figure 1. R-band images of a subset of 12 LIGs selected from the *IRAS* Revised Bright Galaxy Sample (RBGS: Sanders et al. 1998) and a complete sample of "warm" ULIGs (Sanders et al. 1988b). The scale bar represents 10 kpc, tick marks are at 20″ intervals, and the infrared luminosity ($\log L_{\rm ir}/L_{\odot}$) is indicated in the lower left corner of each panel. This subsample is chosen to illustrate the full range of morphologies and infrared luminosities found in the complete sample of LIGs and 'warm' ULIGs – from the most luminous ULIGs which appear to contain dominant single nuclei (e.g. Mrk 231, I Zw 1), to lower luminosity sources that are either pairs of distinct, tidally distorted disks in the early stage of merger (bottom row), or apparently single objects with elliptical-like radial light profiles that may be the most advanced and relaxed mergers (e.g. Zw 247.020, Zw 475.056). These ground-based data (typical seeing is 0.7″–1.2″) are currently being replaced with higher resolution *HST* and ground-based adaptive optics images at UV-to-nearIR wavelengths. As an example, see color plates 5-6 (pp. xxiii-xxiv) for new data on NGC 4038/39.

phase suggest that a *substantial fraction of all galaxies with* $L_B > 10^{10} L_\odot$ *may at some point in their lifetime pass through such a stage of intense infrared emission (Soifer et al 1987)*. This review focuses on providing an up-to-date summary of the observed morphological properties of ULIGs[2], those infrared-selected objects which represent an extreme phase of nuclear activity in galaxies, equivalent to the bolometric luminosity of optically selected QSOs.

2. LIGs and ULIGs: A Merger Sequence

Ground-based optical and near-infrared imaging of complete samples of the brightest infrared galaxies clearly show that a substantial fraction of LIGs are strongly interacting or merging spirals, and that the higher the luminosity the more advanced is the merger. Millimeterwave observations of have shown these spirals to be rich in molecular gas – $M(H_2) \sim 10^9 - 3\times10^{10} M_\odot$ (e.g. Sanders et al. 1991) – and that there is an increasing central concentration of this gas with increasing infrared luminosity. The representative subsample of LIGs shown in Figure 1 illustrates the signs of strong interactions/mergers (tidal tails, double nuclei, etc.) that are revealed in deep optical images of nearby LIGs. Comparison of the images with numerical simulations (e.g. Barnes & Hernquist 1992; Mihos & Hernquist 1994) allows these objects to be placed in a rough time sequence.

3. ULIG Properties

TABLE 1. Properties of ULIGs

Property	Median	Min	Max
redshift	0.05	0.018	0.136
$\log L_{\rm ir}$ [$L_\odot$]	12.2	12.0	12.65
$\log M(H_2)$ [$M_\odot$]	10.0	9.3	10.7
$M(H_2)$ at r <1kpc [%]	65	40	100
$\langle\sigma(H_2)\rangle$ at r <0.5kpc [$M_\odot pc^{-2}$]	4×10^4	1×10^4	1×10^5
$\langle\rho(H_2)\rangle$ at r <0.5kpc [$M_\odot pc^{-3}$]	3×10^2	1×10^2	1×10^3
$\langle A_V\rangle$ towards nucleus [mags]	800	400	2000
nuclear separation [kpc]	1.9	<0.02	9.3
tail length [kpc]	45	20	120
B-band luminosity [L^*_B]	2.5	1.1	4.4
K-band luminosity [L^*_K]	2.5	1.2	7.3

[2]Optical/near-IR spectroscopy of LIGs, and the nuclear gas and dust properties of ULIGs are discussed in conference papers by Veilleux and Scoville & Yun respectively.

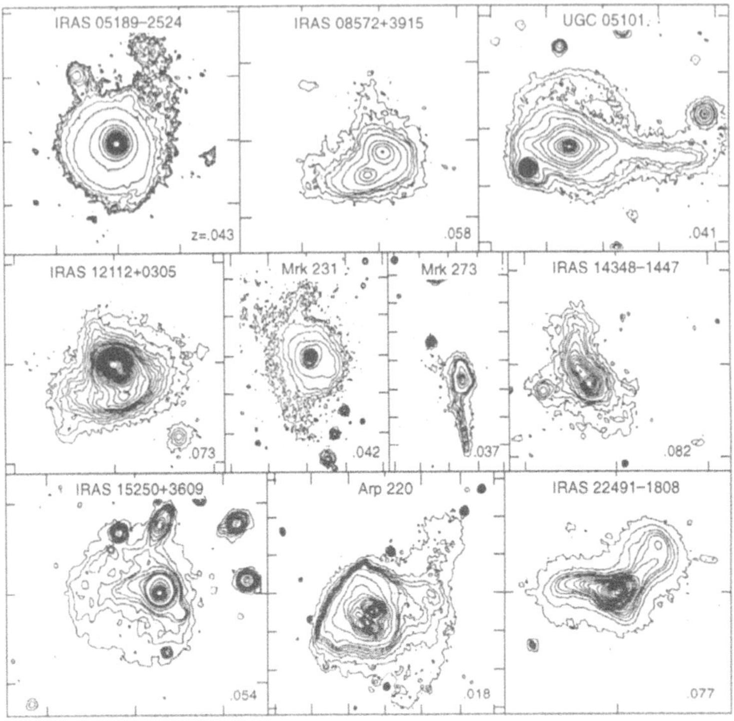

Figure 2. Optical (r-band) CCD images of the complete sample of 10 ULIGs from the original *IRAS* BGS (Sanders et al. 1988a). Tick marks are at 20″ intervals. Typical seeing for these ground-based images is ~0.8″–1.5″.

Nearly all ULIGs appear to be late-stage mergers (e.g. Sanders et al 1988a,b; Melnick & Mirabel 1990; Kim 1995; Murphy et al. 1996; Clements et al. 1996). Figure 2 illustrates the largely overlapping disks that are seen in a *complete* sample of the nearest and best-studied ULIGs. The true extent of faint tidal features plus greater detail in the inner disks of these ULIGs is better revealed in the higher resolution ground-based images and *HST* images of ULIGs shown in color plates 7-8 (pp. xxv-xxiv). Table 1 summarizes properties of the complete sample of 20 ULIGs from the *IRAS* Bright Galaxy Samples (Soifer et al 1987; Sanders et al 1995). The mean lifetime for the ULIG phase, estimated from the observed mean separation and relative velocity of the merger nuclei, is $\sim 2-4\times 10^8$ yrs.

4. The Nuclear Starburst-AGN Connection and Fate of ULIGs

The enormous central gas supplies present in ULIGs are clearly an ideal breeding ground for both powerful circumnuclear starbursts and AGN. In-

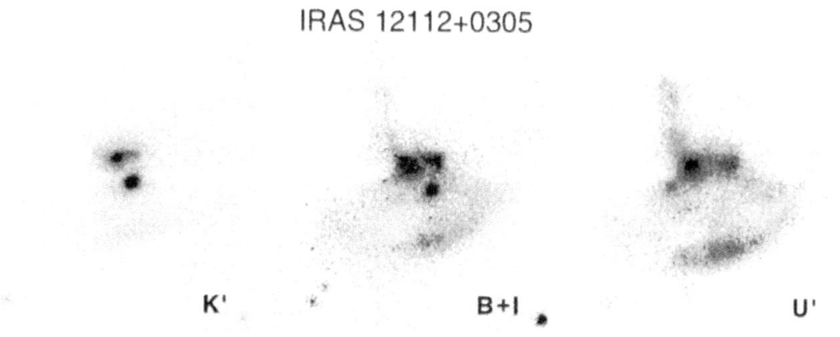

Figure 3. UV/Optical and near-infrared ground-based images of IRAS 12112+0305 (Surace 1998). The short wavelengths are dominated by knotty star formation, both in the central regions and along the extended tidal features. However most of the bolometric luminosity of this system appears to originate in the central knot (d <200 pc) that is completely obscured at U$'$, and that is either powered by an AGN, a superstarburst (which by itself would be much more powerful than the sum of the luminosity from all other starburst regions in this object), or a mixture of starburst and AGN.

deed those ULIGs that have been imaged with adaptive optics from the ground (see Fig. 3) or with *HST* (see Fig. 4) show evidence for a population of massive young ($\sim 10^7$ yrs) star clusters, although these clusters account for typically much less that half the ULIG bolometric luminosity (Surace & Sanders 1999). Most of the luminosity appears to be concentrated in one or two small (r <100 pc) regions centered on the putative nucleus (or nuclei) (e.g. Soifer, et al. 1998), and it is these compact regions which most likely harbor exotic superstarbursts, and/or a powerful AGN.

There is now substantial evidence that ULIGs are elliptical galaxies forming by merger-induced dissipative collapse (Kormendy & Sanders 1992), including $r^{1/4}$-law brightness profiles (e.g. Schweizer 1982; Wright et al. 1990; Kim 1995), newly-formed globular clusters (e.g. Fig. 4 and Surace et al. 1998), central gas densities ($\gtrsim 10^2\, M_\odot\, \mathrm{pc}^{-3}$ at $r \lesssim 0.5-1$ kpc: Scoville et al. 1991) that are as high as stellar mass densities in the cores of giant ellipticals, and powerful "superwinds" (Heckman et al. 1987; Armus et al. 1989) which will likely leave behind a largely dust free core. It seems reasonable to assume also, that this scenario might lead to a constant ratio of black hole mass to bulge mass in agreement with recent observational results (Kormendy & Richstone 1995; Magorrian et al. 1998).

Future infrared space missions and more sensitive submillimeter surveys should succeed in identifying more distant ULIGs, thus allowing a direct test of whether the infrared luminosity function evolves as steeply as that of QSOs, and whether ULIGs were more numerous at $z \sim 1-4$ when it is presumed that most of the ellipticals were formed from mergers of spirals.

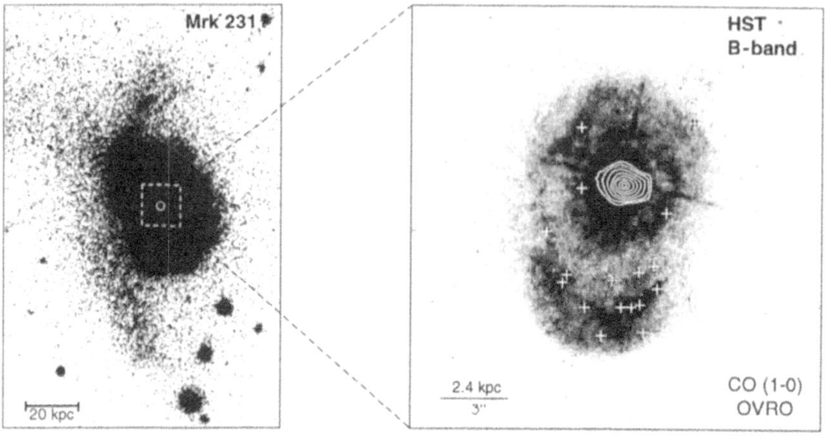

Figure 4. The advanced merger/ULIG/QSO Mrk 231 – Left panel: optical image (Sanders et al. 1987) and CO contour (Scoville et al. 1989). Right panel: *HST* B-band image and identified stellar clusters ('+') from Surace et al. (1998). The high resolution CO contours are from Bryant & Scoville (1996)

References

Armus, L., Heckman, T.M., & Miley, G.K. 1989, *ApJ*, **347**, 727
Barnes, J.E., & Hernquist, L. 1992, *ARAA*, **30**, 705
Bryant, P.M., & Scoville, N.Z. 1996, *ApJ*, 457, 678
Clements, D.L., et al. 1996, *MNRAS*, **279**, 459
Heckman, T.M., Armus, L., & Miley, G.K. 1987, *AJ*, **93**, 276
Kim, D.-C. 1995, Ph.D. Thesis, University of Hawaii
Kormendy, J., & Richstone, D. 1995, *ARAA*, **33**, 581
Kormendy, J., & Sanders, D.B. 1992, *ApJ*, **390**, L53
Magorrian, J., et al. 1998, *AJ*, **115**, 2285
Melnick, J., & Mirabel, I.F. 1990, *A&A*, **231**, L9
Mihos, J.C., & Hernquist, L. 1994, *ApJ*, **431**, L9
Murphy, T.W., et al. 1996, *A.J.*, **111**, 1025
Perault, M. 1987, Ph.D. Thesis, University of Paris
Sanders, D.B., & Mirabel, I.F. 1996, *ARAA*, **34**, 725
Sanders, D.B., Scoville, N.Z., & Soifer, B.T. 1991, *ApJ*, **370**, 158
Sanders, D.B., et al. 1987, *ApJ*, **312**, L5
Sanders, D.B., et al. 1988a, *ApJ*, **325**, 74
Sanders, D.B., et al. 1988b, *ApJ*, **328**, L35
Sanders, D.B., et al. 1995, *AJ*, **110**, 1993
Sanders, D.B., et al. 1998, *ApJS*, in preparation
Schweizer, F. 1982, *ApJ*, **252**, 455
Scoville, N.Z., et al. 1989, *ApJ*, **345**, L25
Scoville, N.Z., Sargent, A.I., Sanders, D.B., & Soifer, B.T. 1991, *ApJ*, **366**, L5
Soifer, B.T., et al. 1987, *ApJ*, **320**, 238
Soifer, B.T., et al. 1998, *ApJ*, submitted
Surace, J.A., & Sanders, D.B. 1999, *ApJ*, **512**, 162
Surace, J.A. 1998, PhD Thesis, University of Hawaii
Surace, J.A., et al. 1998, *ApJ*, **492**, 116
Wright, G.S., James, P.A., Joseph, R.D., & McLean, I.S. 1990, *Nature*, **344**, 417

SPECTROSCOPY OF LUMINOUS INFRARED GALAXIES

S. VEILLEUX
University of Maryland
College Park, MD 20742 USA

Abstract. A review of recent optical and infrared spectroscopic resu..s on luminous infrared galaxies is presented. The main emphasis is on the ultraluminous objects. Possible correlations with infrared luminosity are identified. These results are used to constrain the nature of the dominant energy source in luminous infrared galaxies, and to test whether these objects may represent an evolutionary phase between starburst galaxies and active galactic nuclei.

1. Introduction

Ultraluminous infrared galaxies (hereafter ULIGs: $L_{\rm ir}$[1] $\geq 10^{12}\ L_\odot$; $H_0 = 75$ km s^{-1} Mpc^{-1}, $q_0 = 0.5$) are an important class of extragalactic objects. Not only are they among the most luminous objects in the universe, with bolometric luminosities higher than many quasars, but they also dominate the top end of the galaxy luminosity function, with space densities that exceed those of optically selected quasars in the same luminosity range (Soifer et al. 1987). While exceptional in comparison with the majority of galaxies, ULIGs may have broader significance as representatives of a brief phase in the formation of massive galaxies and/or the genesis of active galactic nuclei (AGN; see review by Sanders & Mirabel 1996).

Assessing the relative importance of an AGN and massive stars for powering ULIGs is essential for understanding these objects as massive galaxies and/or quasars in formation, and for testing suggestions of an evolutionary connection between starburst and AGN phenomena. Considerable effort has therefore been invested in recent years in trying to determine the domi-

[1] $L_{\rm ir} \equiv L$(8-1000 μm), computed from the observed infrared fluxes in all four *IRAS* bands according to the prescription outlined in Perault (1987); see also Sanders & Mirabel (1996)

nant energy source in these objects. In §2 and §3, we summarize the results derived from ground-based optical and near-infrared spectroscopy. In §4, these results are compared with those obtained at longer wavelengths with *ISO*.

2. Optical Spectroscopy

All ULIGs present a large concentration of activity in their nuclei, including strong emission lines characteristic of a starbursting stellar population and in some cases, broad or high-ionization emission lines that suggest the presence of a powerful AGN (e.g., Sanders et al. 1988; Leech et al. 1989; Armus et al. 1989; Allen et al. 1991; Ashby et al. 1992, 1995; Vader et al. 1993; Veilleux et al. 1995; Clements et al. 1996). Here, we focus our discussion on the results from our own optical study of an unbiased subset of 45 ULIGs from the '1 Jy' sample (Kim 1995; Kim & Sanders 1998; Kim, Veilleux, & Sanders 1998).

Using several emission-line diagnostic diagrams (e.g., Veilleux & Osterbrock 1987; Osterbrock et al. 1992; Dopita & Sutherland 1995), these objects were classified as H II galaxies (star-forming galaxies with spectra resembling those of normal H II regions), LINERs (Low-Ionization Nuclear Emission-Line Regions; Heckman 1980), Seyfert 2s (AGN with strong low- and high-ionization lines), and Seyfert 1s (AGN with broad, quasar-like recombination lines; FWHM $\gtrsim$ 2,000 km s^{-1}). These data were then combined with the results from a previous study of primarily lower luminosity infrared galaxies (Veilleux et al. 1995) to examine the spectral properties of luminous infrared galaxies over the range $L_{\rm ir} \approx 10^{10.5} - 10^{12.8}\ L_\odot$. We find that the fraction of luminous infrared galaxies with Seyfert characteristics increases rapidly with increasing $L_{\rm ir}$. A rough inspection of the rest of the '1 Jy' sample seems to confirm this trend (Fig. 1). For $L_{\rm ir} > 10^{12.3}\ L_\odot$, we find that nearly 60% of the ULIGs are optically classified as Seyfert 1s or 2s.

However, even in cases where an AGN is present, starburst activity may contribute a large fraction of the total bolometric luminosity. This is evident from our spectra of the circumnuclear regions in these objects which indicate the presence of a young ($\sim 10^7$ year) stellar population at all radii. We discuss in the next section how to further constrain the nature of the dominant energy source in these objects.

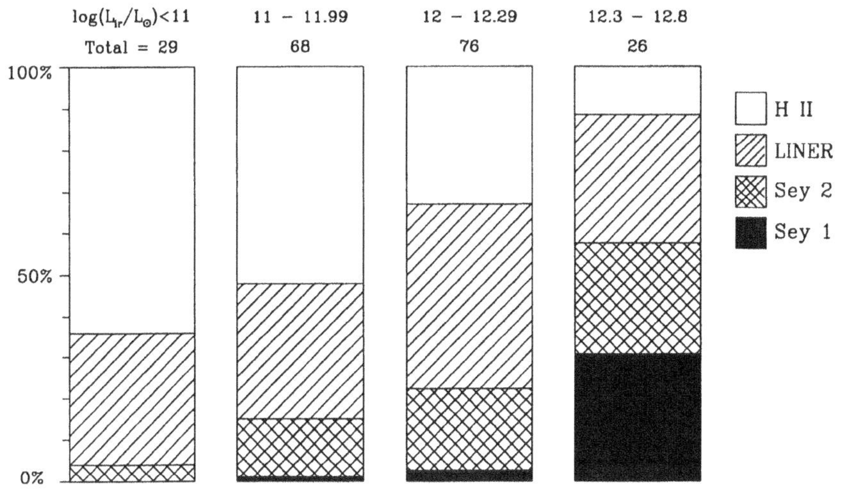

Figure 1. Optical spectral classification as a function of infrared luminosity for the objects in the '1 Jy' sample and the luminous infrared galaxies in the Bright Galaxy Survey (Veilleux et al. 1995, 1997b).

3. Near-Infrared Spectroscopy

Recent progress in infrared detector technology now provides a new approach to address the issue of the energy source in ULIGs. Infrared spectroscopy has the potential to more deeply probe the cores of ULIGs: for example, the extinction coefficient in the K band is nearly 10 times smaller than at optical wavelengths. This technique has proven very useful in the study of highly reddened broad-line regions (BLRs) in intermediate Seyferts (1.8's and 1.9's; Goodrich 1990; Rix et al. 1990) and has also had success finding obscured BLRs in some optically classified Seyfert 2 and radio galaxies (Blanco et al. 1990; Goodrich et al. 1994; Hill et al. 1996; Ruiz et al. 1994; Veilleux et al. 1997a). Low-resolution infrared spectra of a handful of ULIGs have shown tantalizing evidence for hidden BLRs (Armus et al. 1995; DePoy et al. 1987; Hines 1991; Nakajima et al. 1991a, b). Spectropolarimetry has lent support to some of these findings (Hines 1991; Hines et al. 1995; Hines & Wills 1993; Hough et al. 1991; Young et al. 1993). However, more recent spectroscopy of a large sample of galaxies with relatively low infrared luminosity did not reveal any new, previously unknown BLRs (Goldader et al. 1995, 1997a, b).

In an attempt to verify these results among ULIGs of higher luminosity, our group recently carried out a near-infrared spectroscopic study of the nuclear regions of 25 objects from the '1 Jy' sample (Veilleux et al. 1997b). These objects were selected for their lack of BLRs at optical wavelengths. The redshift range of our subsample ($z \sim 0.1 - 0.2$) allowed us to search for broad-line emission from the strong Paα $\lambda 1.8751$ μm feature (intrinsically 12 times stronger than the Brγ line used by Goldader et al. 1995) and for the high-ionization [Si VI] $\lambda 1.962$ μm emission line, two powerful AGN diagnostic lines which are generally inaccessible in lower redshift objects. The results of this study can be summarized as follows:

1. 70 – 90% (7 or possibly 9 out of 10) of the optically classified Seyfert 2s in our sample show signs of AGN activity at rest wavelengths $\sim$ 2 μm (i.e. BLR or [Si VI] emission; Fig. 2). The optical and near-infrared results taken together, therefore suggest that the total fraction of objects in the 1 Jy sample with signs of bonafide AGN is $\sim$ 25 - 30%, but reaches $\sim$ 50% for those objects with $\log[L_{\rm ir}/L_\odot] > 12.3$.
2. All 6 'warm' ($F[25\ \mu{\rm m}]/F[60\ \mu{\rm m}] > 0.2$) optically classified Seyfert 2 galaxies in our sample show either obscured BLRs or [Si VI] emission at near-infrared wavelengths, and present large Paα-to-infrared luminosity ratios. These results suggest that the screen of dust obscuring the cores of 'warm' Seyfert 2 ULIGs is often optically thin at 2 μm.
3. No obvious signs of an obscured BLR or strong [Si VI] emission are detected in any of the 15 optically classified LINERs and H II galaxies in our sample. The Paα-to-IR luminosity ratios and *IRAS* colors of the LINERs suggest that dust obscuration is significant in these objects and may be sufficient to hide a central AGN. The H II galaxies in our sample span a wide range of *IRAS* colors and are strong emitters of narrow Paα. Dust obscuration therefore appears to be relatively unimportant in many of these objects.
4. No correlation is found between *narrow-line* extinctions and the presence of an obscured BLR. The measurements derived from the narrow lines probably only reflect the properties of the circumnuclear gas rather than that of the dusty screen directly in front of the nucleus.
5. The obscured BLRs in our galaxies have dereddened broad-line luminosities which are similar to those of optically selected quasars of comparable bolometric luminosity. This is *not* what would be expected if these ULIGs were powered predominantly by a starburst. Most of the bolometric luminosity in these objects therefore appears to be powered by the same mechanism as that in optical quasars, namely mass accre-

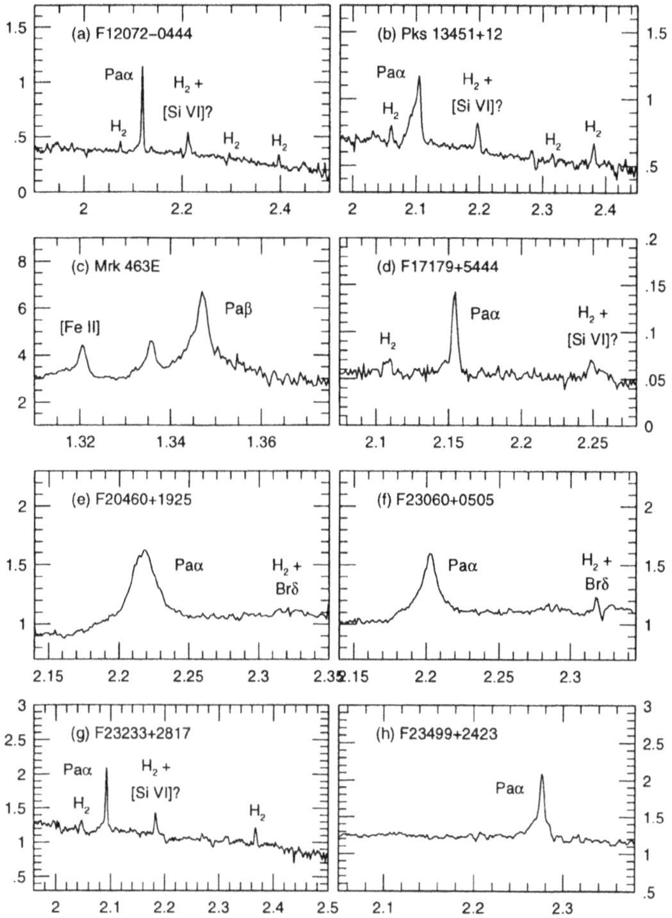

Figure 2. Reduced near-infrared spectra of 8 ultraluminous infrared galaxies optically classified as Seyfert 2s. f_λ is plotted versus $\lambda_{\rm observed}$. The units of the vertical axis are 10^{-11} ergs s^{-1} cm$^{-2}\mu$m^{-1}, while the wavelength scale is in μm. Broad Paα or Paβ emission with FWHM $\gtrsim$ 2,000 km s^{-1} is detected in Pks 1345+12, Mrk 463E, F20460+1925, F23060+0505, and F23499+2423. Possible [Si VI] detections are also indicated in these figures. (Veilleux et al. 1997b)

tion onto a supermassive black hole. The fraction of quasar-dominated ULIGs appears to increase with increasing infrared luminosity.

4. Comparisons with Recent *ISO* Results

The results from the *ISO* Central Program on ultraluminous infrared galaxies have recently been published (Genzel et al. 1998 and references therein). At first, the *ISO* results appear to contradict the results obtained from the ground: 70 – 80% of the ULIGs examined by *ISO* are predominantly powered by a starburst, while 20 – 30% are powered by a central AGN. However, a closer inspection of the *ISO* sample shows that nearly all of the objects in this sample have infrared luminosities less than $10^{12.3}$ $L_\odot$ and therefore fall within the *second* highest luminosity bin in Figure 1. At these lower luminosities, our ground-based survey indicates that the percentage of ULIGs with signs of AGN is only 20 – 30%, i.e. comparable to the fraction of ULIGs in the *ISO* sample which are powered by a central AGN. Unfortunately, objects with $\log[L_{\rm ir}/L_\odot] > 12.3$ are relatively faint and difficult to observe with *ISO*. Only four such objects were included in the *ISO* Central Program. Of these, two appear to be powered predominantly by an AGN (Mrk 231 and F23060+0505; Genzel et al. 1998). The number of objects is clearly insufficient to draw reliable conclusions. However, these preliminary results are consistent with the high percentage ($\sim$ 50%) of high-luminosity objects that show signs of an AGN at optical and near-infrared wavelengths.

If confirmed by more exhaustive studies, this trend with infrared luminosity might be a crucial element to our understanding of the origin and evolution of ULIGs. In one possible scenario, ULIGs represent a transition phase in the evolution of powerful nuclear starbursts into optical quasars (Sanders et al. 1988). The starburst is triggered by the tidal interaction between two gas-rich galaxies and, as the merger takes place, a large amount of gas is funneled toward the center, and the luminosity of the system increases dramatically (e.g., Mihos & Hernquist 1994). The physical conditions become favorable to the formation of a central massive black hole or to the fueling of a preexisting black hole; an active galactic nucleus results. According to this scenario, low-luminosity starburst-dominated ULIGs are the parent population of high-luminosity quasar-dominated ULIGs and optical quasars. Clearly, it will be important to quantify the possible spectroscopic trends with infrared luminosity among ULIGs to test this evolutionary scenario and shed new light on the starburst and AGN phenomena.

Acknowledgements

The ground-based study discussed in this paper was done in collaboration with Drs. D. B. Sanders and D.-C. Kim. The author gratefully acknowledges the financial support of NASA through grant number HF-1039.01-92A awarded by the Space Telescope Science Institute which is operated by the AURA, Inc. for NASA under contract No. NAS5-26555.

References

Allen, D. A., Norris, R. P., Meadows, V. S., & Roche, P. F. 1991, *MNRAS*, **248**, 528
Armus L., Heckman, T. M., & Miley, G. K. 1989, *Ap. J.*, **347**, 727
Armus, L., Neugebauer, G., Soifer, B. T., & Matthews, K. 1995, *A. J.*, **110**, 2610
Ashby, M., Houck, J. R., & Hacking, P. B. 1992, *Ap. J.*, **10**, 980
Ashby, M., Houck, J. R., & Matthews, K. 1995, *Ap. J.*, **447**, 545
Blanco, P. R., Ward, M. J., & Wright, G. S. 1990, *MNRAS*, **242**, 4P
Clements, D. L., et al. 1996, *M. N. R. A. S.*, **279**, 459
DePoy, D. L., Becklin, E. E., & Geballe, T. R. 1987, *Ap. J.*, **316**, L63
Dopita, M. A., & Sutherland, R. S. 1995, *Ap. J.*, **455**, 468
Genzel, R. et al. 1998, *Ap. J.*, **498**, 579
Goldader, J. D., Joseph, R. D., Doyon, R., & Sanders, D. B. 1995, *Ap. J.*, **444**, 97
——. 1997a, *Ap. J. Suppl.*, **108**, 449
——. 1997b, *Ap. J.*, **472**, 104
Goodrich, R. W. 1990, *Ap. J.*, **355**, 88
Goodrich, R. W., Veilleux, S., & Hill, G. J. 1994, *Ap. J.*, **422**, 521
Heckman, T. M. 1980, *A. & A.*, **87**, 142
Hill, G. J., Goodrich, R. W., & DePoy, D. L. 1996, *Ap. J.*, **462**, 163
Hines, D. C. 1991, *Ap. J.*, **374**, L9
Hines, D. C., et al. 1995, *Ap. J.*, **450**, L1
Hines, D. C., & Wills, B. J. 1993, *Ap. J.*, **415**, 82
Hough, J. H., et al. 1991, *Ap. J.*, **372**, 478
Kim, D.-C. 1995, Ph.D. Thesis, University of Hawaii
Kim, D.-C., & Sanders, D. B. 1998, *Ap. J. Suppl.*, **119**, in press
Kim, D.-C., Veilleux, S., & Sanders, D. B. 1998, *Ap. J.*, in press
Leech, K. J., et al. 1989, *M. N. R. A. S.*, **240**, 349
Mihos, J. C., & Hernquist, L. 1994, *Ap. J.*, **431**, L9
Nakajima, T., Carleton, N. P., & Nishida, M. 1991a, *Ap. J.*, **375**, L1
Nakajima, T., Kawara, K., Nishida, M., & Gregory, B. 1991b, *Ap. J.*, **373**, 452
Osterbrock, D. E., Tran, H. D., & Veilleux, S. 1992, *Ap. J.*, **389**, 196
Perault, M. 1987, Ph.D. Thesis, University of Paris
Rix, H.-W., Carleton, N. P., Rieke, G., & Rieke, M. 1990, *Ap. J.*, **363**, 480
Ruiz, M., Rieke, G. H., & Schmidt, G. D. 1994, *Ap. J.*, **423**, 608
Sanders, D. B., & Mirabel, I. F. 1996, AR&A, **34**, 725
Sanders, D. B., et al. 1988, *Ap. J.*, **325**, 74
Soifer, B. T., et al. 1987, *Ap. J.*, **320**, 238
Vader, J. P., Frogel, J. A., Terndrup, D. M., & Heisler, C. A. 1993, *A. J.*, **106**, 1743
Veilleux, S., Goodrich, R. W., & Hill, G. J. 1997a, *Ap. J.*, **477**, 631
Veilleux, S., et al. 1995, *Ap. J. Suppl.*, **98**, 171
Veilleux, S., & Osterbrock, D. E. 1987, *Ap. J. Suppl.*, **63**, 295
Veilleux, S., Sanders, D. B., & Kim, D.-C. 1997b, *Ap. J.*, **484**, 92
Young, S., et al. 1993, *M. N. R. A. S.*, **260**, L1

THE NUCLEAR INTERSTELLAR MEDIUM OF ULTRALUMINOUS INFRARED GALAXIES

P.P. VAN DER WERF
Leiden Observatory
P.O. Box 9513
NL - 2300 RA Leiden
The Netherlands
(pvdwerf@strw.leidenuniv.nl)

1. Introduction

Ultraluminous infrared galaxies (ULIRGs) form the dominant component of the local galaxy population at the highest luminosities (e.g., Sanders & Mirabel 1996 and references therein). They are major mergers of gas-rich galaxies and are characterized by a molecular interstellar medium (ISM) that is concentrated in the central kpc, where it forms a dynamically significant, or perhaps dominant, component (Scoville et al., 1991). Sanders et al. (1988) proposed that ULIRGs might harbor, or evolve into, dust-enshrouded quasars, formed in their obscured centers.

2. Near-infrared H_2 emission

Luminous vibrational H_2 emission at 2.12 μm, first detected in the nearby ULIRGs Arp 220 and NGC 6240 (Rieke et al., 1985), is a general feature of objects of this type (Goldader et al., 1995). A near-infrared spectrum of NGC 6240 (Fig. 1) illustrates the dominance of the H_2 lines, and the surprising weakness of Brγ, which is also a common feature among ULIRGs. As shown by Van der Werf et al. (1993), the H_2 emission in NGC 6240 peaks *between* the two remnant nuclei of the merging systems, and is excited by slow shocks in the dense central molecular medium. Recently, we imaged the H_2 $v = 1{\to}0$ S(1) emission in the nearby prototypical ULIRG Arp 220 (Van der Werf & Israel 1998, see Fig. 2) and here likewise the H_2 emission peaks between the two radio/near-infrared nuclei. Recent high resolution CO $J = 2{\to}1$ data confirm the location of the molecular gas between the two nuclei in these galaxies, as shown by Bryant (1996) and Scoville et al.

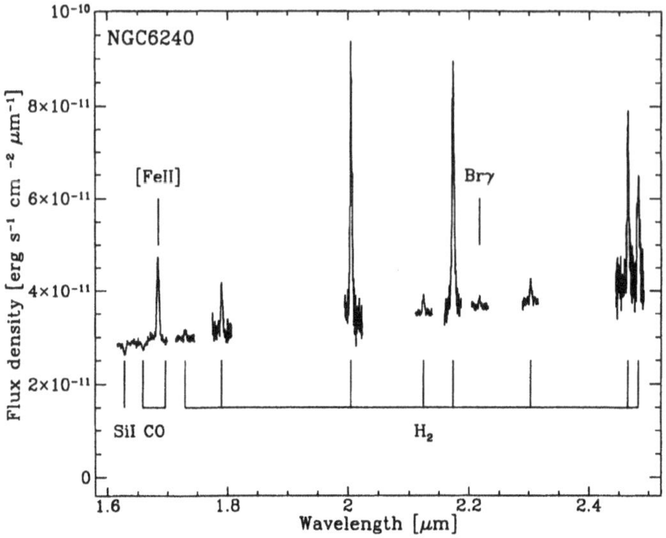

Figure 1. Spectra of the ULIRG NGC 6240 in the H- and K-bands, obtained with IRSPEC on the ESO New Technology Telescope, illustrating the presence of numerous luminous H_2 vibrational lines and the faintness of Brγ.

(1997, hereafter SYB97). The latter authors argue that the gas in Arp 220 is concentrated in a thin disk, where dissipation gives rise to an inward mass flux of $100 - 200\,M_\odot\,\mathrm{yr}^{-1}$.

This dissipation is traced directly by the near-infrared H_2 line emission, since the energy is dissipated by the shocks traced by this emission line. The total energy dissipated by the shocks must equal the total energy radiated away, which can be estimated from the observed H_2 $v = 1{\rightarrow}0$ S(1) emission line flux. This argument can be used to estimate an inward mass flux directly from the H_2 $v = 1{\rightarrow}0$ S(1) emission. The resulting value of $\sim 50\,M_\odot\,\mathrm{yr}^{-1}$ is in reasonable agreement with the value derived from CO, given the simplicity of the analysis.

It is tempting to speculate on a possible connection between this large inward mass flux to a position *between* the stellar nuclei, and the possible formation of a quasar nucleus at this location. On the other hand, the inflowing molecular gas may be consumed entirely by star formation, since the stellar nuclei, although not at the center of the molecular component, are still located *within* the molecular medium. The star formation rate implied by the far-infrared luminosity of Arp 220 (if entirely due to star formation) is $\sim 100\,M_\odot\,\mathrm{yr}^{-1}$, in remarkable agreement with the estimates of the inward gas mass flux. Higher spatial resolution observations will be able to solve this important issue.

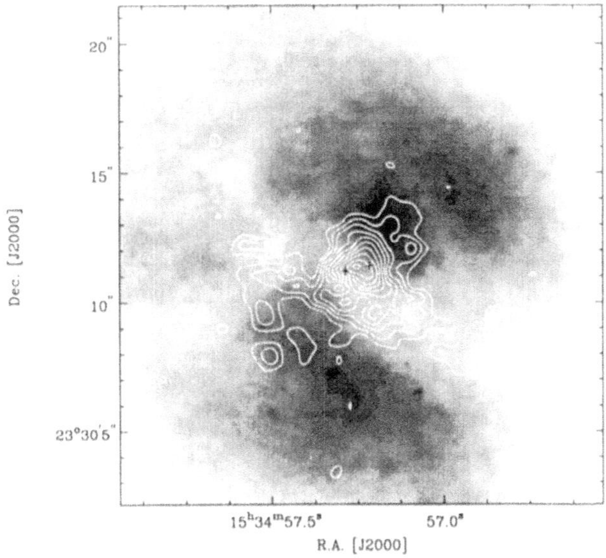

Figure 2. H_2 $v = 1 \rightarrow 0$ S(1) emission of Arp 220, shown in contours, overlaid on on F555W *HST* WFPC1 image. The two crosses denote the positions of the radio/near-infrared nuclei (Van der Werf & Israel, in preparation).

3. Nuclear star formation

The low hydrogen recombination line luminosities of ULIRGs has been used to argue against star formation as the main power source. The situation for Arp 220 has been analyzed in some detail by SYB97, who show that the observed $L_{\mathrm{Br}\gamma}/L_{\mathrm{FIR}}$ ratio implies that only stars with masses from 5 to 25 $M_{\odot}$ form, if star formation is to provide the total far-IR luminosity. In order to avoid this highly unusual situation, it is proposed that most of the bolometric luminosity of Arp 220 and other ULIRGs is produced by an obscured active nucleus. Alternatively, it has been proposed that Brγ suffers high extinction. Sturm et al. (1996) argue for an extinction $A_V \sim 50^m$ in front of the two nuclei, resulting in a suppression of Brγ by two orders of magnitude. However, this reasoning fails on several grounds.

1. The nuclei are visible in the near-infrared, at wavelengths as short as 1.3 μm (e.g., Mazzarella et al. 1992), so that the extinction at this wavelength is at most a few magnitudes; the near-IR colours indicate a visual extinction of at most $A_V \sim 6^m$.
2. Adopting the high extinction proposed by Sturm et al. (1996) leads to extinction-corrected fluxes for Brγ, [Ne II] 12.8 μm and [S III] 33 μm

fluxes which can all be used to estimate a star formation rate (SFR). However, the resulting values increase as the wavelength of the tracer used becomes shorter. This behavior clearly indicates that the extinction estimated by Sturm et al. (1996) is too high. On the other hand, an extinction correction of only a factor of 2 on Brγ would yield consistent SFRs from Brγ, [Ne II] 12.8 μm and [S III] 33 μm.

These arguments indicate that the extinction of Brγ is moderate, and that Brγ is *intrinsically* faint. However, this result *cannot* be used to argue against star formation as the main power source in Arp 220 (as was done by SYB97), because of the importance of Lyman continuum absorption by dust within the ionized regions. The *average* molecular gas density in the $\sim 10^{10}\,M_\odot$ nuclear molecular complex in Arp 220 is $n_{H_2} \sim 2 \cdot 10^4\,\mathrm{cm}^{-3}$ (SYB97). The strong emission from high dipole moment molecules such as CS, HCO^+ and HCN argues for even higher densities: $\sim 10^{10}\,M_\odot$ of molecular gas (i.e., *all* of the gas in the nuclear complex) has a density $n_{H_2} \sim 10^5\,\mathrm{cm}^{-3}$ (Solomon et al. 1990). At such densities the ionized nebulae created by hot stars are *compact* or *ultracompact* H II regions. However, in Galactic ultracompact H II regions 50 to 99% of the Lyman continuum is absorbed by dust (Wood & Churchwell 1989). As a result, all of the usual tracers of H II regions (recombination lines, fine-structure lines, free-free emission) will be *quenched*, not extincted. Therefore, given what is known about the ISM of Arp 220 and other ULIRGs, the intrinsic faintness of Brγ and other tracers of photoionized gas is entirely expected.

References

Bryant, P.M., 1996, Ph.D. Thesis, Caltech

Goldader, J.D., Joseph, R.D., Doyon, R., & Sanders, D.B., 1995, ApJ, 444, 97

Mazzarella, J.M., Soifer, B.T., Graham, J.R., Hafer, G., Neugebauer, G., & Matthews, K., 1992, AJ, 103, 413

Rieke, G.H., Cutri, R.M., Black, J.H., Kailey, W.F., McAlary, C.W., Lebofski, M.J., & Elston, R., 1985, ApJ, 290, 116

Sanders, D.B., Soifer, B.T., Elias, J.H., Madore, B.F., Matthews, K., Neugebauer, G., & Scoville, N.Z., 1988, ApJ, 325, 74

Sanders, D.B. & Mirabel, I.F., 1996, ARAA, 34, 749

Scoville, N.Z., Sargent, A.I., Sanders, D.B., & Soifer, B.T., 1991, ApJ, 366, L5

Scoville, N.Z., Yun, M.S., & Bryant, P.M., 1997, ApJ, 484, 702 (SYB97)

Sturm, E. et al., 1996, AA, 315, L133

Shaya, E.J., Dowling, D.M., Currie, D.G., Faber, S.M., & Groth, E.J., 1994, AJ, 107, 1675

Solomon, P.M., Radford, S.J.E., & Downes, D., 1990, ApJ, 348, L53

Van der Werf, P.P., Genzel, R., Krabbe, A., Blietz, M., Lutz, D., Drapatz, S., Ward, M.J., & Forbes, D.A., 1993, ApJ, 405, 522

Van der Werf, P.P., & Israel, F.P., 1998, in preparation

Wood, D.O.S., & Churchwell, E., 1989, ApJS, 69, 831

TRIGGERED STARBURSTS IN GALAXY MERGERS

Y. TANIGUCHI, Y. SHIOYA AND T. MURAYAMA
Astronomical Institute, Tohoku University
Aramaki, Aoba, Sendai 980-77, JAPAN

AND

K. WADA
National Astronomical Observatory of Japan
Osawa 2-21-1, Mitaka 181, JAPAN

Abstract. A unified formation mechanism of nuclear starbursts is presented; *all the nuclear starbursts are triggered by binary supermassive black holes made in the final phase of galaxy mergers.* Minor mergers cause both nuclear starbursts and hot-spot nuclei while major mergers cause (ultra)luminous infrared galaxies. We discuss the case of Arp 220 in detail.

1. Introduction

It has been considered that nuclear starbursts are driven by gas fueling toward the nuclear regions either by bar structure or by galaxy-galaxy interactions (Shlosman, Begelman, & Frank 1990; Barnes & Hernquist 1992). However, non-barred spirals account for $\simeq$ 50 % of the starburst galaxies and only 30 % of the starburst galaxies have companion galaxies (Balzano 1983; Keel & van Soest 1992). It is thus suggested that the presence of a bar or an interacting companion may not be a prime mover of the nuclear starbursts.

Recently, Taniguchi & Wada (1996, hereafter TW96) proposed that a supermassive black hole binary formed by a merger with a *nucleated* satellite galaxy triggers intense star formation in the central regions of spiral galaxies (see also Gaskell 1985; Hernquist 1989; Mihos & Hernquist 1994a; Hernquist & Mihos 1995). They stressed *the importance of definite triggering of the starbursts by the supermassive black hole binary relative to that of gas fueling caused by minor mergers.* It is known that (ultra)luminous starburst galaxies are associated with major mergers between two gas-rich (nucleated) spiral galaxies (Sanders et al. 1988). Taking both minor and major mergers into account, Taniguchi, Wada, & Murayama (1996) proposed a new unified formation scenario for nuclear starbursts. The minor merger scenario can also explain a starburst-AGN connection because a minor merger with a nucleated satellite drives circumnuclear starbursts and then leads to gas fueling onto the central engine as the merger proceeds (Taniguchi 1997). In this paper, we discuss the case of the archetypical, ultraluminous starburst galaxy Arp 220.

2. The Case of Arp 220

It is widely accepted that the dissipative collapse caused by a merger between two gas-rich galaxies is responsible for the intense nuclear starbursts in the ultraluminous infrared galaxies (ULIGs; Sanders et al. 1988; Kormendy & Sanders 1992). The most important observational property of the starbursts occurring in the ULIGs is that the diameters of the ionized nebulae are only $\sim$ 100 pc (Condon et al. 1991) and thus it is expected that a starburst core (i.e., a clusters of massive stars) is resided in the inner 10 pc regions in each nucleus. Although a major merger between two gas-rich galaxies can cause the efficient gas fueling into the central region of the merger system, such a gas system would fragment into several gas clumps because of the effect of self gravity (Shlosman & Noguchi 1993). Therefore the gravitational instability scenario may not explain the compact nature of the starbursts in ULIGs straightforwardly.

Here let us go back to the TW96 model. In order to generate one compact starburst region, the TW96 model needs a pair of supermassive black holes orbiting each other. Therefore, the presence of two compact starburst regions requires two merging nuclei, each of which contains a couple of nuclei (a supermassive black hole binary); i.e., *there are four nuclei in total in the double-nucleus ULIGs.* If this is the case, Arp 220 would come from a merger among several galaxies (a multiple merger) rather than from a merger between two galaxies. Since there are many compact groups of galaxies (e.g., Hickson 1982), it is not unlikely that some such groups have already merged into their remnants like ULIGs or elliptical galaxies because their merging timescale is generally shorter than the Hubble time (Barnes 1989; Weil & Hernquist 1996). Another merit of this scenario is that the TW96 model explains both the nuclear compact starbursts and the surrounding poststarburst, as observed in Arp 220 (Larkin et al. 1995), because the starburst regions move to the inner region as the supermassive black hole binary shrinks.

Finally, we mention that there is strong evidence for the multiple merger for Arp 220; the VLBI mapping of OH megamaser emission of Arp 220 shows that there are at least three bright OH megamaser spots in the nuclear region (Diamond et al. 1989). The eastern nucleus contains the two bright OH maser spots with a projected separation of 47.6 pc while the western one has only one component. Lonsdale et al. (1994) made a new VLBI measurement of OH emission of Arp 220 and found that the western (i.e., the brightest) component of the OH maser originates from a very compact region whose size is less than 1 pc. This measurement suggests strongly that the western component is attributed to pumping by far-infrared continuum emitted by a dusty torus around an active galactic nucleus rather than to that by the luminous starburst (see also, however, Skinner et al. 1997). If this is also the case for the two eastern OH maser components, Arp 220 contains three active galactic nuclei (i.e., three supermassive black holes) at least as suggested by Diamond et al. (1989). Though the western nucleus contains only one OH maser source, we may consider that the two nuclei aligns accidentally on our line of sight and thus they are not resolved spatially (see Fig. 1). In order to understand what happens in Arp 220 more precisely, we need a next generation millimeter/submillimeter arrays at Atacama.

References

Balzano, V.A. 1983, *ApJ*, **268**, 602

Baan, W.A., & Haschick, A.D. 1995, *ApJ*, **454**, 745

Barnes, J.E. 1989, *Nature*, **338**, 123

Barnes, J. E., & Hernquist, L. 1992, *ARA & A*, **30**, 705

Condon, J.J., Huang, Z.-P., Yin, Q.F., & Thuan, T.X. 1991, *ApJ*, **378**, 65

Diamond, P.J., Norris, R.P., Baan, W.A., & Booth, R.S. 1989, *ApJ*, **340**, L49

Gaskell, C.M. 1985, *Nature*, **315**, 386

Hernquist, L. 1989, *Nature*, **340**, 687

Hernquist, L., & Mihos, J.C. 1995, *ApJ*, **448**, 41

Hickson, P. 1982, *ApJ*, **255**, 382

Keel, W.C., & van Soest, E.T.M. 1992, *A & AS*, **94**, 553

Kormendy, J., & Sanders, D. B. 1992, *ApJ*, **390**, L53

Larkin, J.E., Armus, L., Knop, K., Matthews, K., & Soifer, B.T. 1995, *ApJ*, **452**, 599

Lonsdale, C.J., Diamond, P.J., Smith, H.E., & Lonsdale, C.J. 1994, *Nature*, **370**, 117

Mihos, J.C., & Hernquist, L. 1994, *ApJ*, **425**, L31

Sanders, D.B., et al. 1988, *ApJ*, **325**, 74

Shlosman, I., Begelman, M.C., & Frank, J. 1990, *Nature*, **345**, 679

Shlosman, I., & Noguchi, M. 1991, *ApJ*, **414**, 474

Skinner, C.J., Smith, H.A., Sturm, E., Barlow, M.J., Cohen, R.J., & Stacey, G.J. 1997, *Nature*, **386**, 472

Taniguchi, Y. 1997, *ApJ*, **487**, L17

Taniguchi, Y., & Wada, K. 1996, *ApJ*, **469**, 581

Taniguchi, Y., Wada, K., & Murayama, T. 1996, *Rev. Mex. de A & A*, **6**, 240

Weil, M.L., & Hernquist, L. 1996, *ApJ*, **460**, 101

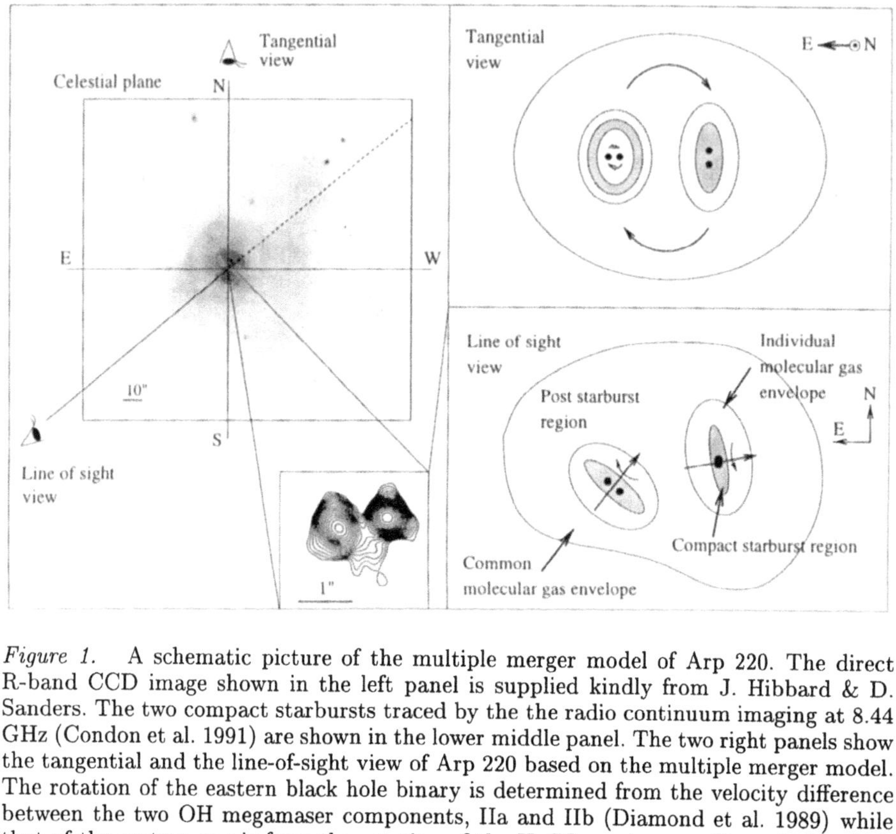

Figure 1. A schematic picture of the multiple merger model of Arp 220. The direct R-band CCD image shown in the left panel is supplied kindly from J. Hibbard & D. Sanders. The two compact starbursts traced by the the radio continuum imaging at 8.44 GHz (Condon et al. 1991) are shown in the lower middle panel. The two right panels show the tangential and the line-of-sight view of Arp 220 based on the multiple merger model. The rotation of the eastern black hole binary is determined from the velocity difference between the two OH megamaser components, IIa and IIb (Diamond et al. 1989) while that of the eastern one is from the rotation of the H_2CO masing gas (Baan & Haschick 1995). The global rotation of the E and W nuclei is from the NIR spectroscopy (Larkin et al. 1995).

INTERACTIONS, MERGERS, AND QSO ACTIVITY

ALAN STOCKTON
Institute for Astronomy, University of Hawaii
2680 Woodlawn Drive, Honolulu, Hawaii 96822

1. Introduction

Is QSO activity sometimes triggered by interactions or mergers of galaxies? Are all QSOs the result of interactions or mergers? Such questions, arising explicitly in the late 1970s, had their roots in the well-known *Stoking the Furnace?* section of Toomre & Toomre (1972), the *Feeding the Monster* paper of Gunn (1979), and the early observational evidence that weaker forms of nuclear activity, such as Seyferts, seemed often to be found in interacting systems (*e.g.*, Adams 1977). Reviews of the evidence regarding the relation of interactions and mergers to QSO activity have been given by Stockton (1990) and Heckman (1990). Some of the more important developments since these reviews are:

(1.) *Modeling the gaseous component in N-body simulations*—That strong interactions and mergers might "bring *deep* into a galaxy a fairly *sudden* supply of fresh fuel" (Toomre & Toomre 1972) has long seemed a plausible, but untested, speculation. The inclusion of a gaseous component in N-body simulations of mergers (Barnes & Hernquist 1991, 1992) has shown that much of the gas does, indeed, settle in towards the center of the merger remnant, although the models still fall short by about a factor of 10^7 of having the resolution range to follow gas from galactic scales to the scale of a supermassive black hole accretion disk.

(2.) *IR imaging of QSO hosts*—The advent of low-noise IR-array detectors of increasing size has led to a number of programs of IR imaging of QSO host galaxies (*e.g.*, Dunlop et al. 1993; McLeod & Rieke 1994). The scientific driver for these surveys is shown in Fig. 1 (see also McLeod & Rieke 1995): for QSOs with $z \sim 0.3$, observations in the J or H bands sample close to the peak in the spectral-energy distribution of stellar populations with ages $\gtrsim 1$ Gyr as well as close to where such populations show greatest contrast with respect to the QSO nuclear light. This ability to concentrate on the

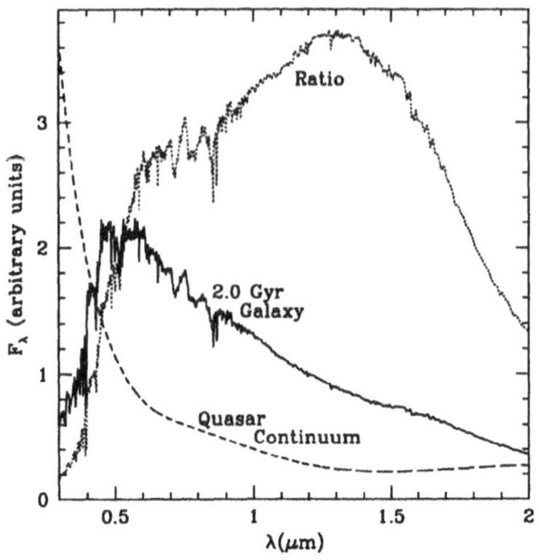

Figure 1. The advantage of near-IR observations for detecting older stellar populations in QSO host galaxies. The solid curve is a Bruzual-Charlot (1997) model for a 2-Gyr-old stellar population, the dashed curve is a median quasar continuum (Elvis et al. 1994), and the dotted curve shows the ratio of the two. At a modest redshift ($z \sim 0.3$), the best contrast with the QSO nucleus is achieved in the J and H bands.

older stellar populations, normally representing most of the luminous mass of the host galaxy, opens up many possibilities for carrying out statistical tests of hypotheses regarding QSO host-galaxy formation and evolution.

(3.) *HST imaging*—After a flurry of controversy, the major *HST* imaging programs on host galaxies of low-redshift QSOs (Bahcall et al. 1994, 1995*a*, *b*, 1996, 1997; Hutchings et al. 1994; Disney et al. 1995; McLeod & Rieke 1995) are in general agreement, both with each other and with most previous ground-based work. The *HST* images confirm that virtually all radio-loud QSOs are found in galaxies with (sometimes disturbed) elliptical-like light distributions, as expected. But they also show that radio-quiet QSOs often have similar light distributions, a result that went against the previous consensus, even though it had been foreshadowed in some ground-based work (*e.g.*, Hutchings & Neff 1992).

(4.) *Unified models*—To the extent that unification of quasars (*i.e.*, radio-loud QSOs) and FR II radio galaxies (Scheuer 1987; Barthel 1989; Antonucci 1993) holds at some non-trivial level, one can use radio galaxies as proxies for quasar hosts. If such a substitution is possible, one can study the inner regions of quasar hosts (at least from some lines of sight) without having to deal with the overwhelming glare of the quasar nuclear light, which in these cases is conveniently masked by the quasars' own internal occulting "disks."

While rigorous proof of the role of strong interactions in triggering QSO activity remains elusive, the circumstantial evidence in favor of this picture is now fairly compelling, and much of the focus of current research in this area has turned instead towards attempting to trace the evolutionary history of interactions/mergers that may result in QSOs. Here I shall describe some work my colleagues and I have been doing in two specific areas: (1) the nature of "transition" objects between ultraluminous IR galaxies and the classical QSO population, and (2) the morphologies of some specific cases of host galaxies of powerful radio sources at $z \sim 1$, which may offer some insight into merger and accretion processes at early epochs.

2. Far-Infrared Spectra and "Transition" Objects

2.1. INTRODUCTION

The one definite evolutionary scenario linking mergers to QSOs that has been proposed is that of Sanders et al. (1988a, b), which suggests ultraluminous IR galaxies as the parent population of at least a significant fraction of the classical radio-quiet QSOs. For $L_{\rm FIR} \geq 10^{12} L_{\odot}$, essentially all IR-luminous galaxies show clear morphological evidence for being merging systems (the fraction of apparent mergers drops quite steeply with luminosity: at $L_{\rm FIR} \sim 5 \times 10^{10} L_{\odot}$, only about 10% show evidence for strong interaction [Sanders & Mirabel 1996, and references therein]). Sanders et al. noted that both the bolometric luminosities and space densities of the ultraluminous objects were similar to those of radio-quiet QSOs.

Attempts to find true active nuclei in luminous IR galaxies have had varied results (Kim et al. 1995; Veilleux et al. 1995; Goldader et al. 1995; Veilleux et al. 1997; Surace et al. 1998). Most of the disagreement appears to be due to the different luminosity ranges covered by the different samples, coupled with a very steep dependence of strength of nuclear activity on IR luminosity (Kim et al. 1995; Veilleux et al. 1997).

If it were true that most or all radio-quiet QSOs follow this evolutionary path, and that the QSO luminosity is correlated with the luminosity of the IR-bright stage, then the observed correlation between IR luminosity and evidence for strong interaction would suggest that, while the more luminous QSOs may virtually all result from mergers, the fraction may be lower for lower luminosity objects. Indeed, if Seyfert galaxies are regarded as the low-luminosity continuation of the radio-quiet QSO sequence, the fact that the great majority of classical Seyferts are recognizably spiral galaxies indicates that most cannot have suffered roughly equal-mass mergers. On the other hand, one of the relatively few surprises emerging from *HST* imaging of low-redshift QSO hosts is that $\sim 70\%$ of radio-quiet QSO hosts have elliptical-like surface-brightness profiles (Bahcall et al. 1997), consistent with being merger remnants. Thus, while the distinction between Seyfert galaxies and radio-quiet QSOs is often thought to be an arbitrary division of objects with a continuum of properties, it may actually be closely correlated with the merger status of the object.

Figure 2 shows a FIR two-color diagram, based on *IRAS* data from Neugebauer et al. (1986) and Sanders et al. (1988*a*). The classical QSOs fall in loose clump centered around the power-law line in the diagram, while the ultraluminous IR galaxies populate the lower-right corner of the diagram. I shall refer to QSOs lying close to the region occupied by the ultraluminous IR galaxies as "transition objects." Several of these are labeled in Fig. 2. That choosing such objects by their FIR spectra has physical significance is suggested by two facts: (1) The five QSOs labeled in Fig. 2 not only all show tidal tails, but they are arguably the clearest and least ambiguous examples of tidal tails among low-redshift QSOs (Stockton & Ridgway 1991; Stockton et al. 1998*b*); and (2) at least three of the five are low-ionization broad absorption line (BAL) QSOs, which comprise only about 1% of optically-selected QSOs. Voit et al. (1993) argue that low-ionization BAL QSOs are heavily dust-enshrouded objects, and Egami et al. (1996) have already made the suggestion that similar objects found at high redshifts are related to ultraluminous IR galaxies.

2.2. WHAT IS THE NATURE OF THE "TRANSITION?"

As I have already mentioned, considerable effort has been put into determining whether ultraluminous IR galaxies are dominated by AGN or by starbursts. But demonstration of the presence of active nuclei alone would not confirm the hypothesis that ultraluminous IR galaxies are the progenitors of QSOs: one also has to deal with the relative *time scales* of the physical phenomena involved. To take the extreme cases, if the QSO activity lasts much longer than the ultraluminous IR phase, then objects will move from the ultraluminous IR galaxy region to the classical QSO region in a diagram like Fig. 2, but if the nuclear activity is short-lived compared with the time it takes to clear dust from the central region, they will live out their lives as QSOs in one region of the two-color plot. In the former case, the position of an object in Fig. 2 is dominated by its stage in an evolutionary sequence; in the latter, by its intrinsic characteristics. In a realistic scenario, it is quite possible that the time scales will be roughly comparable ($\sim 10^8$ years each), so the situation may well be more complicated than either of these extreme cases.

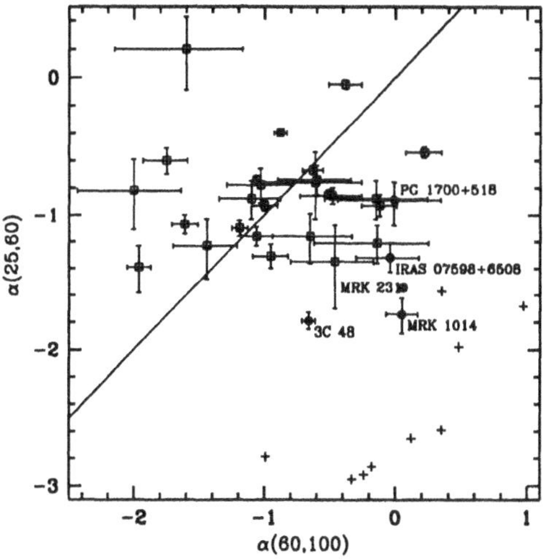

Figure 2. Far-IR two-color diagram. QSOs are generally shown as open squares with error bars, from *IRAS* data given by Neugebauer et al. (1986). Nine of the ten galaxies from the *IRAS* Bright Galaxy Survey with $L_{\rm bol} > 10^{12} L_{\odot}$ are shown as small crosses (Sanders et al. 1988*a*; the tenth is Mrk 231, included in the next sample). Five QSOs having colors intermediate between those of the bulk of the QSOs and the ultraluminous galaxies are indicated as closed circles and labeled. The solid diagonal line indicates the locus of power-law spectra.

2.3. EVIDENCE THAT TRANSITION OBJECTS ARE POST-STARBURSTS

One way to try to distinguish whether position in the FIR two-color diagram is dominated by evolution is to measure a parameter that correlates with time since the active nucleus turned on. Hutchings & Neff (1992) have attempted to classify QSO host galaxies roughly by age since a supposed merger event, based on their optical morphologies. However, given the large range in initial parameter space covered by possible mergers and relatively small number of total resolution elements available on the host galaxies, even Hutchings & Neff's coarse classification must suffer from some degeneracy. Moreover, the simulations of Mihos & Hernquist (1996) indicate that the peak of the starburst activity may occur anywhere from ~ 0 to $\sim 5 \times 10^8$ years prior to the final merger, depending mostly on the internal structures of the merging galaxies. One might expect that AGN triggering would correspond more closely in time with the starburst, since both are dependent on gas flows to the inner regions.

My graduate student, Gabriela Canalizo, has been concentrating on trying to date the starbursts themselves in transition QSOs and their close

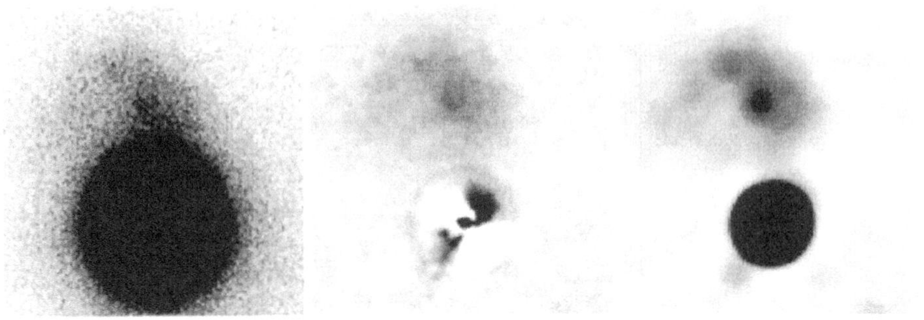

Figure 3. H-band image of PG 1700+518, obtained with the University of Hawaii Adaptive Optics System on the CFH Telescope. The left panel shows the recorded image, at a scale of $0\rlap{.}''035$ pixel^{-1}. The FWHM of the QSO is $0\rlap{.}''25$. The middle panel shows a subtraction of a model of the QSO, derived from an elliptical profile fit to the bottom half of the QSO profile, and the right panel shows a Gaussian restoration of a deconvolution of the image, using the same elliptical profile fit as the PSF kernel. Any symmetrical host galaxy is likely to be mostly removed in both the subtraction and the deconvolution. Each panel is $4\rlap{.}''2$ square.

companions. The idea is to estimate the time since the the most recent major starburst activity by obtaining spectra of the QSO host galaxy or companion and modeling the composite spectrum as a simple superposition of two populations: one due to the starburst and the other due to the older, pre-existing component. We have published one example, the asymmetric extension to the north of PG 1700+518 (Canalizo & Stockton 1997). Figure 3 shows a recent adaptive-optics image (Stockton et al. 1998*a*) of PG 1700+518, which shows that this feature is a close but discrete companion galaxy, with an obvious tidal tail, apparently in the process of merging with the host galaxy of PG 1700+518 itself. We currently have spectroscopic data for the four additional transition objects labeled in Fig. 2. All show the clear presence of components with early-type spectra, ranging in age from $\lesssim 2 \times 10^7$ years to $\gtrsim 10^8$ years, consistent with being post-starburst populations (Canalizo & Stockton 1998; see also Boroson & Oke 1984). The next, and more difficult, stage in this investigation will be to try to discern and date fading starburst components in QSOs lying closer to the power-law line in the FIR two-color diagram to see whether these host galaxies are consistent with the Sanders et al. (1988*a*) evolutionary scenario.

3. Quasars and Radio Galaxies at Higher Redshifts

3.1. INTRODUCTION

In observing interactions at low redshifts, we seem mostly to be seeing encounters between mature, well formed galaxies. As we go to higher redshifts, deep *HST* images show that galaxies tend to be more ragged and irregular. Even at a relatively modest $z \sim 1$, a large percentage of galaxies show asymmetric structure and apparently are still undergoing a major part of their disk formation. Accordingly, mergers increasingly take on the nature of successive accretion of various lumps of the still forming galaxies. The distinction between merger and accretion blurs into a continuum.

When the unusual elongated morphologies of several high-redshift radio galaxies were first discovered, the initial assumption was that these structures were the result of recent mergers (*e.g.*, Djorgovski et al. 1986). Once it was realized that the optical elongations were aligned with the radio structure (Chambers et al. 1987; McCarthy et al. 1987), this assumption no longer seemed tenable. While the final verdict on the physical mechanisms responsible for the alignment effect is not yet in, it seems clear that it is no longer safe to assume that any global distortion or large-scale transient structure *necessarily* means an interaction or merger. On the other hand, the ubiquity of aligned structure in high-redshift radio galaxies and quasars may tend to mask genuine tidal features or at least make their interpretation more difficult.

With these cautions in mind, I would like to discuss briefly two objects from an *HST* imaging program of a complete sample of $z \sim 1$ 3CR quasars and galaxies Susan Ridgway and I have been conducting, which we believe are likely examples of recognizable major interactions or mergers at high redshift.

3.2. INTERACTIONS AND MERGERS AT $z \sim 1$

3.2.1. *3C 280*

We have shown our deep *HST* WFPC2 and ground-based images of the radio galaxy 3C 280 in Ridgway & Stockton (1997; see also Best et al. 1997), and I refer the reader to that paper for an illustration of the morphology. Briefly, the radio core is centered on the E component of a very close ($\sim 0\farcs2$) double (a and b), well aligned with the radio axis. Also aligned with the radio axis, $\sim 1''$ W of the radio core, lies an object (c), elongated more-or-less transversely to the radio axis. From the W side of c, an almost semicircular arc extends to the N, proceeding E, and terminating close to the position of b. The arc is strong in both the continuum and in [O II] λ3727 emission. Its line-to-continuum ratio appears to be the same as that of c, for which we have found that $\sim 50\%$ of the continuum is due to nebular thermal emission (Ridgway & Stockton 1997, 1998). We have considered, and rejected, several possible origins for the arc, including gravitational lensing of background objects or of components of 3C 280 itself. The most likely remaining explanation is that it is a tidal tail, either from b or c. Mihos (1995) has discussed the visibility of merger signatures, including tidal tails, in *HST* WFPC2 images of high redshift objects. He concludes that typical tails would be marginally visible, if at all, in an exposure similar to ours (8800 s, F622W filter). However, we not only detect the apparent tail, but easily do so in a single 1100 s exposure. Mihos' simulations neglect two factors that may increase the surface brightness. Firstly, they do not include star formation during the interaction. Mihos argues that simulations show that starbursts "occur predominantly in the central regions," and that "star formation in the tidal debris is generally suppressed by the interaction." However, observations show that real tidal tails are often dominated by recent star formation (Stockton 1974; Schombert et al. 1990), so the problem may lie in the limitations of the simulations. Secondly, the surface brightness of a tail lying in the illumination cone of a quasar nucleus may be enhanced by scattering and/or photoionization. The strong line emission and evidence for a large contribution from nebular thermal emission suggests that this is likely the case for the arc of 3C280.

3.2.2. *3C 190*

Figure 4 shows our *HST* WFPC2 F675W and Keck Near-Infrared Camera images of the 3C 190 field. The most striking object in the *HST* image is the almost perfectly straight feature offset to the W of the quasar, reminiscent of a similar feature seen in the [O III] line image of the low-redshift quasar, 4C 37.43 (Stockton & MacKenty 1987). In 3C 190, however, the linear feature is strongly dominated by continuum radiation. In the IR image, we see two faint linear features: one is an apparent extension of the *HST* feature to the NW; the other, extending to the S, is faintly visible on the *HST* image as well. Together, these feature bear a remarkable similarity to the tails of the local merger NGC 7252 (Schweizer 1982).

In addition to this evidence for a major merger, there is a swarm of small objects to the N and NE of 3C 190, all within projected distances of $\lesssim 25$ kpc. These are likely to merge shortly with the 3C 190 host galaxy, and they thus may be examples of "building blocks" in a process of continuing galaxy formation (*e.g.,* Pascarelle et al. 1996). 3C 190 is a compact-steep-spectrum (CSS) radio source. The small sizes of CSS sources may mean that they are very young FR II sources or that they are confined by unusually high gas densities near the active nucleus (*e.g.,* Fanti et al. 1990). While total lifetimes of FR II radio sources are probably always small compared with a merger time scale, a high central gas density could well be related to merging events. We are currently obtaining deep Keck spectroscopy of components of this remarkable system.

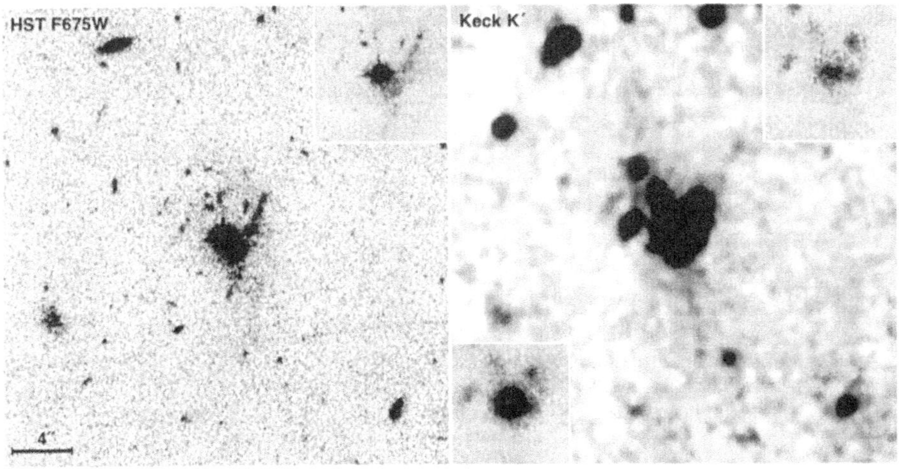

Figure 4. The field of 3C 190. The left-hand panel shows the *HST* image, and the right-hand panel shows the Keck K' image, after PSF subtraction. In both panels, the insets in the upper-right corners show the same image at lower contrast; the insert in the lower-left corner of the right-hand panel shows the K' image prior to PSF subtraction. The main panel K' image has been smoothed slightly to bring up low-surface-brightness detail.

Acknowledgements

I thank my colleagues, Gabriela Canalizo, Laird Close, and Susan Ridgway for allowing me to present some of our observations prior to publication. Some of these results are based on observations made with the NASA/ESA *Hubble Space Telescope*, obtained at the Space Telescope Science Institute, which is operated by AURA, Inc., under NASA contract NAS 5-26555, and on observations made at the W.M. Keck Observatory, jointly operated by the California Institute of Technology and the University of California. Support for this work was provided by NASA through Grant No. GO-05401.01-93A from the STScI, and from NSF Grant No. AST95-29078.

References

Adams, T.F. 1977, *ApJS,* **33,** 19
Antonucci, R. 1993 *ARA&A,* **31,** 473
Bahcall, J.N., Kirhakos, S., & Schneider, D.P. 1994, *ApJL,* **435,** L11
Bahcall, J.N., Kirhakos, S., & Schneider, D.P. 1995a, *ApJL,* **447,** L1
Bahcall, J.N., Kirhakos, S., & Schneider, D.P. 1995b, *ApJ,* **450,** 486
Bahcall, J.N., Kirhakos, S., & Schneider, D.P. 1996, *ApJ,* **457,** 557
Bahcall, J.N., Kirhakos, S., Saxe, D.H., & Schneider, D.P. 1997, *ApJ,* **479,** 642
Barnes, J.E., & Hernquist, L. 1991, *ApJ,* **370,** L65
Barnes, J.E., & Hernquist, L. 1992, *ARA&A,* **40,** 705
Barthel, P.D. 1989, *ApJ,* **336,** 606
Best, P.N., Longair, M.S., & Röttgering, H.J.A. 1997, *MNRAS,* in press

Bruzual A., G., & Charlot, S. 1997, in preparation
Boroson, T., & Oke, J.B. 1984, *ApJ,* **281,** 535
Canalizo, G., & Stockton, A. 1997, *ApJL,* **480,** L5
Canalizo, G., & Stockton, A. 1998, in preparation
Chambers, K.C., Miley, G., & van Breugel, W. 1987, *Nature,* **329,** 609
Disney, M.J., Boyce, P.J., Blades, J.C., Boksenberg, A., Crane, P., Deharveng, J.M., Macchetto, F., Mackay, C.D., Sparks, W.B., & Phillipps, S. 1995, *Nature,* **376,** 150
Djorgovski, S., Spinrad, H., Pedelty, J., Rudnick, L., & Stockton, A. 1987, *AJ,* **93,** 1307
Dunlop, J.S., Taylor, G.L., Hughes, D.H., & Robson, E.I. 1993, *MNRAS,* **264,** 455
Egami, E., Iwamuro, F., Maihara, T., Oya, S., & Cowie, L.L. 1996, *AJ,* **112,** 73
Fanti, R., Fanti, C, Schilizzi, R.T., Spencer, R.E., Nan, R.D., Parma, P., van Breugel, W.J.M., & Venturi, T. 1990, *A&Ap,* **231,** 333
Goldader, J., Joseph, R.D., Doyon, R., & Sanders, D.B. 1995, *ApJ,* **444,** 97
Gunn, J.E. 1979, in *Active Galactic Nuclei,* ed. C. Hazard & S. Mitton (Cambridge Univ. Press, Cambridge), p. 213
Heckman, T.M. 1990, in *IAU Colloquium 124, Paired and Interacting Galaxies,* ed. J.W. Sulentic, W.C. Keel, and C. Telesco (NASA CP-3098), p. 359
Hutchings, J.B., & Neff, S.G. 1992, *AJ,* **104,** 1
Kim, D.-C., Sanders, D.B., Veilleux, S., Mazzarella, J.M., & Soifer, B.T. 1995, *ApJS,* **98,** 129
McCarthy, P.J., van Breugel, W., Spinrad, H., & Djorgovski, S. 1987, *ApJL,* **321,** L29
McLeod, K.K., & Rieke, G.H. 1994, *ApJ,* **431,** 137
McLeod, K.K., & Rieke, G.H. 1995, *ApJL,* **454,** L77
Mihos, J.C. 1995, *ApJ,* **438,** L75
Mihos, J.C., & Hernquist, L. 1996, *ApJ,* **464,** 641
Neugebauer, G., Miley, G.K., Soifer, B.T., & Clegg, P.E. 1986, *ApJ,* **308,** 815
Pascarelle, S.M., Windhorst, R.A., Keel, W.C., & Odewhan, S.C. 1996, *Nature,* **383,** 45
Ridgway, S.E., & Stockton, A. 1997, *AJ,* **114,** 511
Ridgway, S.E., & Stockton, A. 1998, in preparation
Sanders, D.B., & Mirabel, I.F. 1996, *ARA&A,* **34,** 749
Sanders, D.B., Soifer, B.T., Elias, J.H., Madore, B.F., Matthews, K., Neugebauer, G., & Scoville, N.Z. 1988*a*, *ApJ,* **325,** 74
Sanders, D.B., Soifer, B.T., Elias, J.H., Neugebauer, G., & Matthews, K. 1988*b*, *ApJL,* **328,** L35
Scheuer, P. 1987, in *Superluminal Radio Sources,* ed. J. Zensus and T. Pearson (Cambridge Univ. Press, Cambridge), p. 104
Schombert, J.M., Wallin, J.F., & Struck-Marcell, C. 1990, *AJ,* **99,** 497
Stockton, A. 1974, *ApJ,* **187,** 219
Stockton, A. 1990. in *Dynamics and Interactions of Galaxies,* ed R. Wielen (Springer-Verlag, Heidelberg), p. 440
Stockton, A., Canalizo, G., & Close, L. 1998*a*, in preparation
Stockton, A., & Ridgway, S.E. 1991, *AJ,* **102,** 488
Stockton, A., Ridgway, S.E., & Kellogg, M. 1998*b*, in preparation
Surace, J.A., Sanders, D.B., Vacca, W.D., Veilleux, S., & Mazzarella, J.M. 1998, *ApJ,* **492,** 116
Toomre, A., & Toomre, J. 1972, *ApJ,* **178,** 623
Veilleux, S., Kim, D.-C., Sanders, D.B., Mazzarella, J.M., & Soifer, B.T. 1995, *ApJS,* **98,** 171
Veilleux, S., Sanders, D.B., & Kim, D.-C. 1997, *ApJ,* **484,** 92
Voit, G.M., Weymann, R.J., Korista, K.T. 1993, *Apj,* **413,** 95

RADIO SOURCE SURVEYS

Mergers at high redshifts?

PATRICK J. MCCARTHY
Carnegie Observatory
813 Santa Barbara St
Pasadena, CA 91101

1. Introduction

The early study of radio sources at visible wavelengths and the first empirical evidence that galaxies can have strong dynamical interactions are closely intertwined. Baade & Minkowski's (1954) model of Cygnus A as a pair of galaxies in collision, while now believed to be incorrect, presaged the merger-driven picture for the generation of radio sources (e.g. Heckman et al. 1986) by some 30 years. Morphological evidence for an association between mergers and radio loud AGN is seen in both the nearest radio galaxies (e.g. Cen A; Schweizer 1986) and in the most powerful sources at $z \sim 0.1 - 0.3$ (Stockton & Mackenty 1983; Hutchings et al. 1988; Heckman et al. 1986).

When the first radio galaxies with redshifts beyond ~ 2 were identified, their properties were viewed from a perspective that was heavily colored by the merger induced starburst/AGN paradigm. Thus Lilly & Longair (1984) and Spinrad & Djorgovski (1984) cited the complex morphologies, strong and spatially extended emission lines, UV excesses, and large amplitude velocity fields as signs of strong dynamical interactions. While this view may ultimately have considerable validity, we have learned that there a number of physical processes at play in producing the colors and morphologies of radio galaxies. From a variety of observations we now know with some confidence that there are both local and scattered sources of UV continuum (Dey et al. 1997; Cimatti et al. 1997), that recombination can produce much of the UV continuum in some objects (Dickson et al. 1995) and that there is a complex interplay between the extended radio source and its environment.

The discovery of the alignment of the UV continuum and emission-lines with the radio source axes lead people to reconsider strongly held views regarding these objects and ultimately, to question the stellar origin of the

UV continuum. Our present understanding of the rest-frame visible continuum is not much more sophisticated than our early 1980's understanding the UV continuum. Most who work in this field are comfortable with the idea that the bulk of the light at $\lambda > 4000\,\text{Å}$ arises from cool stars and that the radio galaxies form a fairly homogeneous population that is drawn from the larger population of massive ellipticals. The passively evolving old galaxy model for radio galaxy hosts is clearly laid out in Lilly (1988). There are now spectra of weak radio sources at $z > 1$ that show convincing evidence for stars with ages of more than 1 Gyr (Dunlop et al. 1996 ; Spinrad et al. 1997). There is at present no demonstration that the K-band light of radio galaxies with $z > 2$ is dominated by stars of any age. The latest generation of near-IR arrays on large apertures and the advent of NICMOS on *HST* are now providing us with data that has the potential to seriously challenge the status quo.

2. The MRC/1 Jy Radio Survey

Over the past several years Kapahi, van Breugel, Persson, Athreya and I have carried out a large survey of radio galaxies with flux densities near the peak of the 408 MHz source counts. The MRC/1Jy sample (McCarthy et al. 1996; Kapahi et al. 1997) is the largest sample of sources for which essentially complete optical and near-IR identifications are available. Its flux limit ($S_{408} > 0.95\,\text{Jy}$) is roughly a decade below and 3CRR (Laing et al. 1985) and while its redshift content ($\sim 75\%$) is less than optimal, it is nevertheless, well suited to a number of statistical investigations. In Figure 1 I show the run of $P(408)$ vs. z for the MRC/1Jy and 3CRR samples. The strong, but spurious, correlations within each sample is the Malmquist bias, the gap between the two tracks results from the different solid angle coverages. The combination of the 3CRR and MRC/1Jy samples allows us to span a decade in power at any z and a wide range of z for a fixed radio power. Athreya et al. (1998) and McCarthy (1998) have used the combined samples to examine the evolution and luminosity dependences of sources size and Athreya (1998) has determined the redshift dependence of the rest-frame radio spectral indices. The quasar sample has been compiled by Kapahi et al. (1997) and spectroscopy of nearly all of the QSR and BL Lac identifications were obtained by Hunstead et al. (1998). Baker & Hunstead (1996) have produced composite rest-frame UV spectra of the CSS, and core and lobe-dominated MRC/1Jy quasars.

3. Color and Luminosity Evolution of the Host Galaxies

The large formation redshift and passive evolution paradigm for radio galaxies is based largely on the uniformity of the $K-z$ relation and the evolution

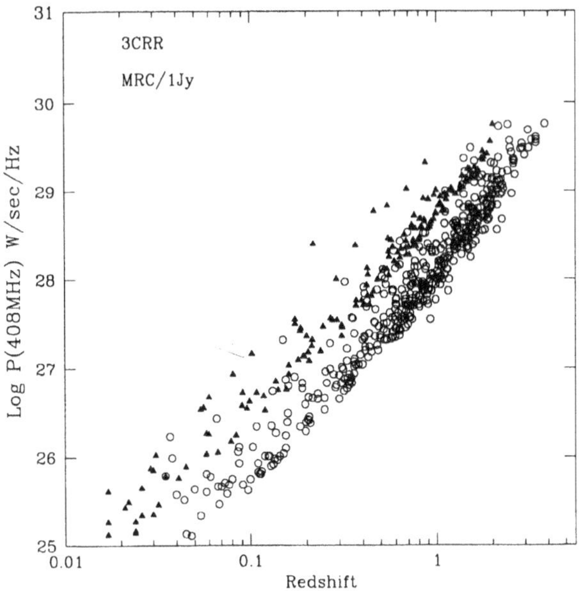

Figure 1. The run of 408MHz power with redshift for the 3CRR (filled symbols) and MRC/1Jy (open symbols) surveys.

of the red envelope in the visible and near-IR colors in the 3CRR sample (e.g. Lilly & Longair 1984; Eisenhardt & Lebofsky 1986). A similar behavior is seen in other samples (e.g. Dunlop et al. 1989). I illustrate this behavior for the MRC/1Jy sample in Figures 2 & 3 where I plot the K magnitudes and $r-K$ colors against z. The model curves show the apparent magnitudes and colors of passively evolving systems with formation redshifts from 10 to 2 in a low Ω, H_0 Universe. For any redshift below ~ 2 there are galaxies with $r-K$ colors are red as the passive evolution model and in $J-K$ the galaxies continue to get red to $z \sim 2.5$.

4. Velocity fields in the extended emission-line regions

Baum, Spinrad and I have measured the extranuclear velocity fields in the emission-line gas for a large sample of radio galaxies. At redshifts below ~ 0.8 the typical radio galaxy has a velocity field with an amplitude of $100 - 300$ km s^{-1}. At larger redshifts the amplitudes increase to typically 800 km s^{-1}, with a number of systems having gas moving at more than 1000 km s^{-1}s. Baum & McCarthy (1998) speculate that the apparently sudden change in velocity amplitudes reflects the change in galaxy environments.

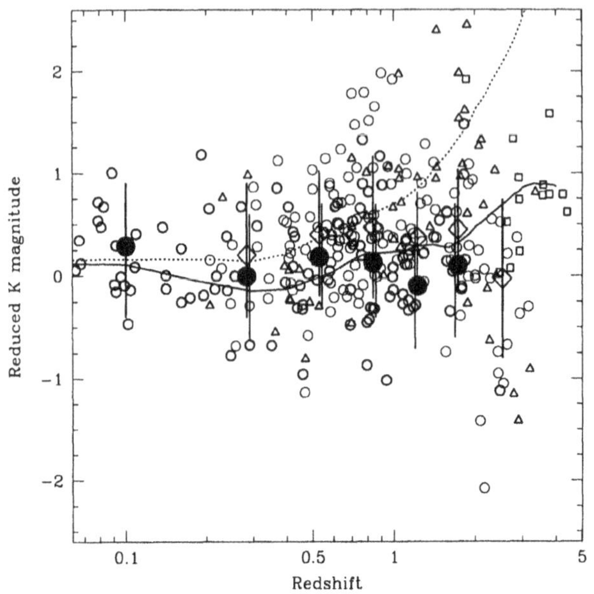

Figure 2. The K-band Hubble diagram for radio galaxies from the 3CRR, MRC/1-Jy surveys and a number of $z > 3$ radio galaxies taken from the literature. The dotted line is a no-evolution model with $H_0 = 50$ and $q_0 = 0.1$; the solid line is a BC96 passive evolution model with $z_f = 20$. A low order polynomial has be subtracted from both the data and the models to compress the scale.

At these redshifts radio galaxies are, on average, in richer environments than the low redshift FR II sources that comprise the bulk of the 3CRR and 1Jy sources with $0.2 < z < 0.5$. The merger picture for radio galaxies at these low redshift is quite consistent with the range of velocities seen in the ionized gas. The large amplitudes in the more distant systems are comparable to the velocity dispersions of rich clusters and suggest that galaxy collisions, rather than actual mergers, may be a more common form of dynamical interaction.

5. WFPC2 & NICMOS Imaging of Radio Galaxies with $z > 2$

Miley, Fosbury, van Breugel, Rottgering and I have imaged several $z > 2$ radio galaxies with NICMOS in the F160W bandpass. These images benefit both from the large reduction in background compared to the ground and from diffraction limited images. In several cases we find that the H-band light is dominated by a nuclear point source. In other cases we have detected resolved emission and have attempted to separate the contributions

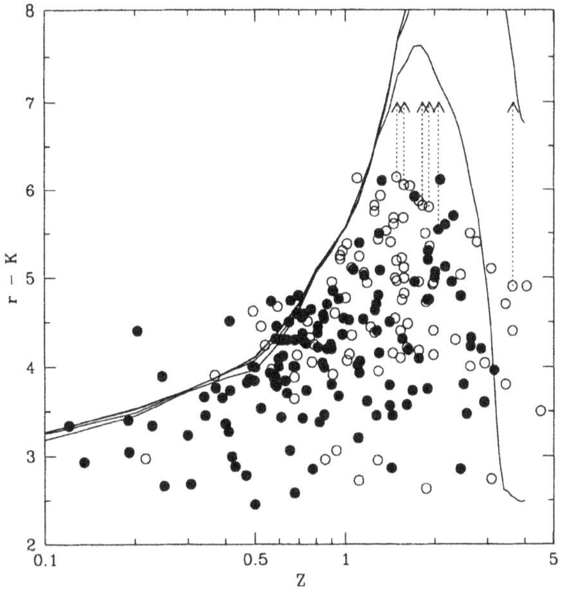

Figure 3. (a) The $r - K$ colors for galaxies in the MRC/1Jy sample. The filled symbols are objects with secure spectroscopic redshifts, the open symbols are objects for which the redshift has been estimated from the K magnitude. The solid lines are the same passive evolution model as above, but with formation redshifts ranging from 20 to 3.

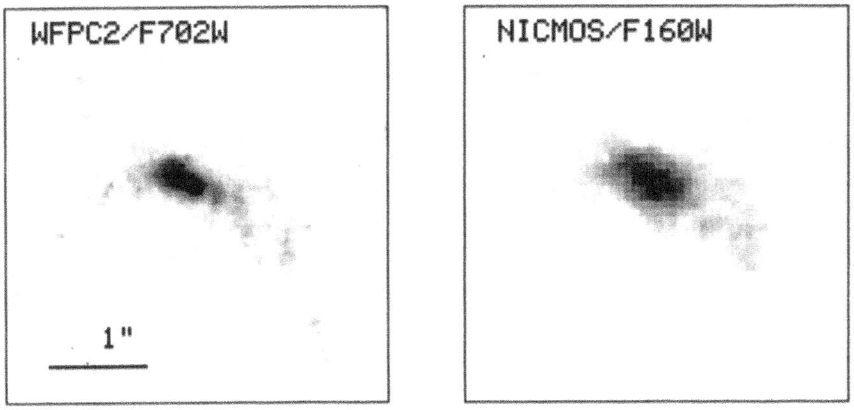

Figure 4. WFPC2/F702W and NICMOS/F160W images of the radio galaxy MRC 0943-242 ($z = 2.93$).

from a symmetric host galaxy and a component aligned with the radio axis. This separation is accomplished by using the WFPC2 and NICMOS images to iteratively solve for linear combinations of aligned and host galaxy contribution. In Figure 4 I show the WFPC2 and NICMOS images of MRC 0943-242 ($z = 2.93$). One can easily see that the F160W image contains an strong aligned component, the underlying symmetric component is less obvious. We find that the fractional contribution of the symmetric component, assumed to be stellar, is roughly 10% at F702W and 25% at F160W. In Figure 5 I show the spectral energy distribution of 0943-242 derived from ground-based imaging and from *HST*. The solid line is a Bruzual-Charlot model for the 1.5 Gyr population and it is fitted to the photometry of the symmetric component. The dotted line is the observed composite SED for a core dominated MRC/1Jy quasar, as derived by Baker & Hunstead (1996), the upper solid line is the sum of the two components. The good agreement between the $4''$ aperture photometry and the model suggests that 0943-242 contains a large contribution from quasar continuum that is scattered by a nearly grey scattering medium. Extrapolating the model fit from the H-band to K-band yields a 30% contribution from the symmetric component.

In the case of 0943-242 was are confident that much of the observed H and K light is not associated with stars in a normal elliptical galaxy. The objects found to be dominated by nuclear point sources must also contain substantial non-stellar contributions. Thus the meaning of the colors and K magnitudes of the $z > 2$ galaxies with high radio powers is no longer clear. Our thinking regarding the evolution of the stellar populations in terms of an early formation redshift and subsequent passive evolution needs revision.

In Figure 6 I show an F160W image of MRC 0406-244 recently obtained with NICMOS. At $z = 2.43$ this image contains a contribution from the nebular lines. The striking bipolar morphology revealed in this image is reminiscent of the wind-blown bubbles associated with Arp 220 and other ULIRGs. Deep WFPC2 imaging of this object by Rush et al. (1997) reveal an apparent double nucleus and continuum features suggestive of tidal tails. It is tempting to speculate that this object may by a high redshift radio-loud analogue of the powerful ULIRGs and SNe driven outflows.

6. Summary

Our understanding of powerful radio sources and their relationship to other classes of AGN and star forming galaxies is undergoing considerable evolution at the present. The paradigms that work well for the local population of radio sources, and merger-driven activity in particular, may not be valid at early epochs. Imaging and spectroscopic observations show that the

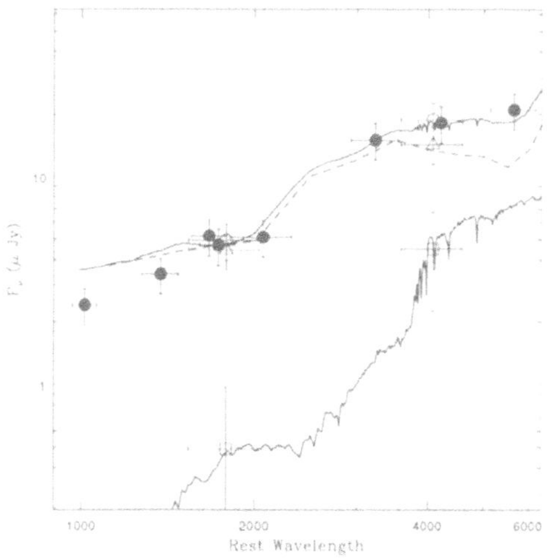

Figure 5. The derived SED for 0943-242. The open squares are the NICMOS and WFPC2 magnitudes for the host galaxy, the open triangles are the aligned component. The filled circles are the ground-based photometry. The model curves are described in the text.

Figure 6. A NICMOS/F160W image of MRC 0406-244 (z = 2.44). The large bubble-like structures are aligned with the radio source axis and are likely to be composed of [OIII]5007,4959 and Hβ emission lines. The total size of the bipolar structure is $3.5''$ or $\sim$ 30 kpc.

environments and dynamical states of radio galaxies evolve with look-back time. The application of NICMOS and ground-based near-IR imaging to the most distant radio galaxies has lead us to reexamine the basic assumptions that underly our interpretation of the apparent color and luminosity evolution of these objects and their relation to early type galaxies. I thank my collaborators, as listed in the text, for allowing the use of previously unpublished data.

References

Athreya, R. 1998, In *The Most Distant Radio Galaxies*, Kluwer Academic Press, H. Rottgering et al. eds., in press
Athreya, R., McCarthy, P., Kapahi, V., & van Breugel, W. 1998, in prep.
Baker, J., & Hunstead, R., 1996, 461, L59
Cimatti, A., Dey, A., van Breugel, W., Hurt, T., & Antonucci, R. 1997, 476, 677
Dey, A., van Breugel, W., Vacca, W., & Antonucci, R. 1997, ApJ, 490, 698
Dickson, R., Tadhunter, C., Shaw, M., Clark, N, Morganti, R. 1995, MNRAS, 273, 29
Dunlop, J., et al. 1989, MNRAS, 240, 257
Dunlop, J., et al. 1996, Nature, 381, 581
Eisenhardt, P. R. M. & Lebofsky, M. J. 1987, ApJ, 316, 70
Heckman, T., et al. 1986, ApJ, 311, 526
Hunstead, R., Baker, J., Kapahi, V., & Subramhanya, C. 1998, ApJS, in press
Hutchings, J. B., Johnson, I, & Pyke, R. 1988, ApJS, 66, 361
Kapahi, V. K., et al. 1997, ApJS, in press.
Laing, R., Longair, M., & Riley, J. 1983, MNRAS, 204, 151
Lilly, S. J., & Longair, M. S. 1984, MNRAS, 211, 833
Lilly, S. J. 1988 ApJ 340, 77
McCarthy, P., 1998, In *The Most Distant Radio Galaxies*, Kluwer Academic Press, H. Rottgering et al. eds., in press
McCarthy, P., Kapahi, V. K., van Breugel, W., Persson, S. E., Athreya, R. M., & Subrahmanya, C. R. 1996, ApJS, 107, 19
Rush, B., McCarthy, P., Athreya, R., & Persson, S. 1997, ApJ, 484, 163
Stockton, A. & Mackenty, J. W., 1983, Nature, 305, 678
Schweizer, F. 1986, Science, 231, 227
Spinrad, H., & Djorgovski, S. 1984, PASP, 96, 795
Spinrad, H., et al. 1997, 484, 581

UNIFIED SCHEME FOR SEYFERTS OR THE INFLUENCE OF INTERACTIONS?

D. DULTZIN-HACYAN
Instituto de Astronomía UNAM, México

IVETTE FUENTES GURIDI AND YAIR KRONGOLD
Facultad de Ciencias, UNAM, México

AND

PAOLO MARZIANI
Osservatorio Astronomico de Padova, Italia

1. Introduction

The first statistical study of the environment of Seyfert galaxies (of both types 1 and 2) was done by Petrosian (1982), who found that Sy2 galaxies had an excess of nearby companion galaxies with respect to field galaxies. Dahari (1985) made a similar study, and obtained that both types of Seyfert galaxies are found five times more often in interacting systems that "normal" galaxies. Fuentes-Williams & Stocke (1988) obtained the opposite result: that Seyfert galaxies have no significative excess of nearby companions, and MacKenty (1989) obtained the same result as Petrosian. The last papers appeared in 1995: Rafanelli, Violato & Baruffolo (1995) obtained a similar result as Dahari. Laurikainen & Salo (1995) made a careful attempt to understand the discrepancies; they concluded that the incompatibilities in the results were mainly due to control sample selection, and problems in background galaxy determination. Their work confirmed the result of the first study. However, in order to explain the discrepancies between previous results, they also used incorrectly defined control samples. Objections can be made against *all* of these works for this only reason. We have repeated the study using, for the first time, the digitized Palomar Sky Survey plates (POSS), an automated selection procedure, and correctly defined control samples (Osterbrock 1993).

2. Methodological improvements to study the environment of Seyfet galaxies and results

The sample of Seyfert galaxies was compiled from Véron-Cetty & Véron (1991). It consists of 72 Sy1 and 60 Sy2. Both samples are volume limited, and the $V/V_{\max}$ test assures completeness to a level of 92%. The redshifts are limited to $0.007 \leq z \leq 0.035$, and we selected galaxies with galactic latitudes $b \geq 45°$, in order to avoid extinction, and confusion due to galactic stars. Rich clusters were avoided.

For the control sample, the above criteria were also imposed. In this work we used complete samples for the first time, but an even more important methodological improvement is the definition of control samples of field galaxies which match the Seyfert galaxies in *all* respects except that they are not Seyfert galaxies. In oreder to achieve this, two control samples were defined, one for each type of Seyfert galaxies, because both the Hubble type and redshift distributions of the two types of Seyfert galaxies differ, and our aim was to match these distributions. The control samples were obtained by a process of multiple random extraction from more than 10,000 objects of the CfA catalog (Huchra et al. 1983). For each control sample we matched Hubble type, and redshift distributions. We did *not* match absolute magnitudes, since this would introduce instead of eliminate a bias (Seyfert galaxies are brighter due to the nucleus), instead we matched the diameters distribution. The control samples are complete (in volume) to a confidence level of up to 97%. Although the above mentioned similarities were long ago known to be required for a proper comparison, in previous works matching the distributions was impossible maintaining the same densities, due to the selection of small control samples from nearby galaxies. In this work the background densities between samples are statistically equal (we applied Mann-Whitney's U test). Other selection details, histograms, and statical tests will be given in forthcoming paper.

In order to look for companions, we analyzed more than 500 digitized POSS plates using the automated detection package FOCAS (Faint Object Classification and Analysis System; Jarvis & Tyson 1981). The correct use of FOCAS requires a fine tunning of several parameters which shall be discussed elsewhere. This procedure reduces to a minimum the *subjective* bias present in all previous works, which were done counting galaxies from the POSS prints "by eye". The search for companions was done in circular areas around each galaxy, with radii equal to three times the diameter of the galaxy. Obviously, some of the neighbors are background galaxies. The correct estimation of the background galaxy density is crucial. As in previous studies, we assumed a Poisson distribution of background galaxies, but in this work, the determination of the number density ρ, was made directly from the digitized POSS plates in regions of one sqaure degree surrounding each galaxy. The use of the Lick counts given by Shane & Wirtanen (1967) to estimate the projection effects can introduce an important bias (as in Rafanelli et al. 1995). The final result we obtained is a confirmation of the first result obtained by Petrosian (1982): *It is only Sy2 galaxies that have excess companions, but not Sy1 galaxies.* The excess factor for Sy2 with respect to their control sample is 1.8 (less than the value of 2.7 obtained by Laurikainen & Salo in 1995). Our result is relevant to a confidence level of 99.5% according to a chi square test. Other statistical test were also applied, but they will be discussed elsewhere.

3. Interpretation of the results

Two types of Seyfert galaxies were defined by Khachikian & Weedman (1971) on the basis of spectroscopic properties. Miller & Goodrich (1990) discovered that in plane-polarized light some Sy2 have the spectrum of a Sy1. The interpretation they proposed was that those galaxies have a Broad Line Region (BLR), whose radiation does not escape directly because of strong obscuration by a dusty torus around it, but which does escape along the axis of the torus, and is scattered towards us (and thus polarized) by free electrons above and below the obscuring torus. The extrapolation implying that *all* Sy2 galaxies are obscured Sy1 galaxies, oriented in such a way with respect to the observer, that the BRL is hidden is not so obvious, and yet this has become an almost unquestionable paradigm, known as the "Unified Scheeme for Seyferts". We want to stress that Miller & Goodrich (1990) clearly stated that many bright Sy2 observed by them do *not* show the broad emission lines (see also Villeux, Sanders & Kim 1997).There are many other indications of intrinsic differences between Sy1 and Sy2 galaxies. Just to mention a few: Heckman et al. (1989) discovered that Sy2 have larger $L_{\rm CO}/L_{\rm B}$ ratios than Sy1. Dultzin-Hacyan, Masegosa & Moles (1990) showed that while the mid-infrared ($\sim 25\mu$m) emission in Sy1 is synchrotron radiation (or dust re-emission of it), in Sy2 it is dust re-emission of starlight (see also Mouri & Taniguchi 1992; Mouri, Kawara & Taniguchi 1997). Norris & Roy (1997) showed that Sy2 (and not Sy1) follow the FIR *vs* radio correlation, as Starburst (SB) galaxies. Even in the X-rays, Sy2 have similar properties as SB galaxies (Muzhotsky 1997), except for the very few cases where the FeK$_\alpha$ line is found (these may be indeed obscured Sy1 nuclei).

An alternative to the unified scheeme was proposed to explain the different properties of the two types of Seyfert galaxies in Dultzin-Hacyan (1995). In this new scheeme, radiation due to accretion unto a black hole decreases, while the relative contribution of a circumnuclear starburst radiation increases, from Seyfert nuclei types 1 to 2. This scheeme is supported by the above mentioned observations and by statistical studies of the multifrequency emission of Seyferts (Mass-Hesse et al. 1995; Dultzin-Hacyan & Ruano 1996). It is also clearly supported by this work, because the difference in environment disagrees with the interpretation that the two types of Syferts represent similar objects seen from different viewing angles. Further discussion and interpretation of our result will be given in a forthcoming paper.

References

Dahari, O. 1985, ApJS, 57, 643
Dultzin-Hacyan, D., Masegosa, J. & Moles, M. 1990, A&A, 238, 28
Dultzin-Hacyan, D. 1995, RMA&A SC 3, 31
Dultzin-Hacyan,D. & Ruano, C.1996, A&A, 305, 719
Fuentes-Williams, T. & Stocke, J.T. 1988, AJ, 93, 1235
Huchra, J., Davis, M., Latham, D. & Tonri, J. 1983, ApJS 52, 89
Jarvis, B.J. & Tyson, H. 1981, AJ, 86, 476
Khachikian E.Ye. & Weedman D.W. 1971, Astrofizika, 7, 389
Laurikainen,E. & Salo, H. 1995, A&A, 293, 683
MacKenty, J.W. 1989, ApJ, 343, 125
Mass-Hesse, J.M., Rodríguez-Pascual, P.M., Sanz Fernández de Córdoba, L, Mirabel, I.F., Wamsteker, W., Makino, F. & Otani, C. 1995, A&A, 298, 22
Miller, J.S. & Goodrich, R.W. 1990, ApJ, 355, 456
Mouri, H. Kawara, K. & Taniguchi, Y. 1997, ApJ, 484, 222
Mouri, H. & Taniguchi, Y. 1992, ApJ 386, 68
Muzhotsky, R. 1997, RMA&A SC 6, 213
Norris, R.P. & Roy, A.L. 1997, RMA&A SC 6, 230
Osterbrock, D.E. 1993, ApJ, 404, 551
Petrosian, A.R. 1982, Astrofizika, 18, 548
Rafanelli, P., Violato, M. & Baruffolo, A. 1995, AJ, 109, 1546
Shane, C.D. & Wirtanen, C.A. 1967, Publ. Lick Obs., 22, 647
Véron-Cetty, M.P. & Véron, P. 1991, A Catalogue of Quasars and Active Nuclei, 5th ed. (ESO, Garching)
Villeux S., Sanders D. B. & Kim D.-C. 1997, ApJ 484, 92

RELICS OF NUCLEAR ACTIVITY: DO ALL GALAXIES HAVE MASSIVE BLACK HOLES?

ROELAND P. VAN DER MAREL

STScI, 3700 San Martin Drive, Baltimore, MD 21218, USA

Abstract. The distribution of black hole (BH) masses $M_\bullet$ in galaxies is constrained by photometric and kinematic studies of individual galaxies, and by the properties of the quasar population. I review our understanding of these topics, present new results of adiabatic BH growth models for *HST* photometry of elliptical galaxies with brightness profiles of the 'core' type, and discuss the implications of ground-based stellar kinematical data. It is not yet possible to uniquely determine the BH mass distribution, but the available evidence is not inconsistent with a picture in which: (i) a majority of galaxies has BHs; (ii) there is a correlation (with large scatter) between $M_\bullet$ and spheroid luminosity $L_{\rm sph}$ of the form $M_\bullet \approx 10^{-2} L_{\rm sph}$ (solar B-band units); and (iii) the BHs formed in a quasar phase through mass accretion with efficiency $\epsilon \approx 0.05$.

1. Introduction

Considerable evidence suggests that the energetic processes in active galaxies and quasars are due to the accretion of matter onto massive BHs. Lynden-Bell (1969) already suggested that BHs may also be present in quiescent galaxies, such as the Milky Way, M31 and M32. This spurred efforts to find BHs in nearby galaxies through kinematical studies, which have since increased steadily in sophistication, both observationally and theoretically. There are now convincing BH detections for at least a dozen galaxies, and new detections are reported at an ever increasing rate. The techniques for detecting BHs in individual galaxies have been reviewed by, e.g., Kormendy & Richstone (1995, hereafter KR95), Ford et al. (1998) and Richstone (1998). Here I address the more general question: *do all galaxies have BHs?*

2. Quasar counts and evolution

Integration of quasar number counts yields the comoving energy density in quasar light. Assuming that this energy is produced by accretion onto massive black holes, one obtains the total mass per cubic Mpc that is collected in black holes (Sołtan 1982). Division by the observed luminosity density of galaxies (Loveday et al. 1992) yields an estimate of the average black hole mass per unit luminosity: $\langle M_\bullet \rangle / \langle L \rangle = 2.0 \times 10^{-3} (0.1/\epsilon)$, where ϵ is the accretion efficiency (Chokshi & Turner 1992). [Throughout this paper, $H_0 = 80\,\mathrm{km\,s^{-1}\,Mpc^{-1}}$, mass-to-light ratios are in solar units, and luminosities are in the B-band.]

To address the BH mass *distribution*, one must model not only the total energy budget of the quasar population, but also its evolution. Tremaine (1996; also Faber et al. 1997, hereafter F97) presented a simple argument based on the typical quasar lifetime to show that a model in which every spheroid (bulge or elliptical) has a BH is consistent with the inferred $\langle M_\bullet \rangle / \langle L \rangle$. Haehnelt & Rees (1993, hereafter HR93) presented a more detailed model (in which BH formation is linked to hierarchical structure formation) to fit the distribution of quasars as function of magnitude and redshift. Their predicted BH mass distribution at the current time (their Fig. 8) is consistent with a fraction $f \approx 0.3$ of all galaxies having a BH.

The uncertainties in these estimates are considerable. The only conclusion that can be drawn with some confidence is that a fraction $f = 0.1$–1 of all galaxies is likely to contain BHs with $M_\bullet / L = 10^{-2}$–10^{-3}. The product of these quantities, $\langle M_\bullet \rangle / \langle L \rangle$, is better constrained than either quantity independently, but is still rather uncertain (if only due to the unknown ϵ).

3. BH detections

HR93 predict a direct connection between $M_\bullet$ and the galaxy formation redshift $z_{\rm form}$ (galaxies that form later have smaller BHs), but not between $M_\bullet$ and galaxy luminosity. Spirals form later than ellipticals, and are therefore predicted to have smaller BHs. Observations appear to confirm this; e.g., the (active) galaxies M87 and NGC 1068 have similar luminosities, but $M_\bullet$ is $10^{2.5}$ times larger in M87 (cf. Figure 1a below). Observations do not rule out a correlation between $M_\bullet$ and the *spheroid* luminosity of the host (KR95). This may indicate that $M_\bullet$ and $L_{\rm sph}$ depend similarly on a common underlying parameter (e.g., $z_{\rm form}$, as in the models of HR93), or alternatively, that there is a physical link between BHs and spheroids.

Figure 1a shows $M_\bullet$ versus $L_{\rm sph}$ for all currently available BH mass determinations inferred from: ($\times$) radio observations of water masers; ($\circ$) ionized gas kinematics of nuclear disks; ($*$) time variability of broad double-peaked Balmer lines; and ($\bullet$) stellar kinematical studies that included anisotropic modeling (studies with only isotropic models are discussed in §6).

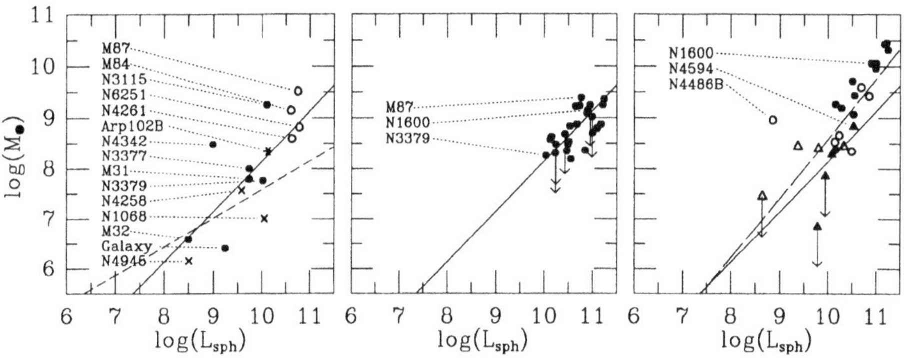

Figure 1. Measurements of black hole mass $M_\bullet$ versus spheroid luminosity $L_{\rm sph}$. Arrows indicate upper limits. **(a; left)** 'Secure' BH detections obtained with various techniques, as indicated by different symbols and discussed in §3. References are included in the bibliography; where necessary, spheroid/total luminosity ratios were estimated from the Hubble type and the relation in Simien & de Vaucouleurs (1986). **(b; middle)** $M_\bullet$ values for all core galaxies in F97 with $M_{\rm B} < -20$, inferred from models of adiabatic BH growth for *HST* photometry (§4; van der Marel 1998, in preparation). M87 and NGC 3379 are the two core galaxies for which secure kinematic determinations are also available (cf. panel a). NGC 1600 is discussed in §5. **(c; right)** $M_\bullet$ determinations from axisymmetric $f(E, L_z)$ models for ground-based stellar kinematical observations of elliptical galaxies (Mg98). Circles: 'core' galaxies; triangles: 'power-law' galaxies; open symbols: galaxies in Virgo. Galaxies already included in the left panel are omitted. For NGC 4486B and 4594 the presence of a BH has been suggested previously on the basis of isotropic models. The $M_\bullet$ values in panel a are typically believed to have $|\Delta \log M_\bullet| \leq 0.3$. The accuracy of the $M_\bullet$ measurements in panels b and c is dominated by systematic uncertainties in the model assumptions, as discussed in §4, 5 and 6. In particular, mild radial velocity anisotropy would lower the values in panel c. The solid line in each panel is the 'reference model' discussed in the §3. It has $M_\bullet = 1.4 \times 10^{-2} L_{\rm sph}$, and fits the constraints from quasar number counts for an assumed accretion efficiency $\epsilon = 0.05$. The long-dashed line in the rightmost panel is an alternative model. It has $M_\bullet = 6.0 \times 10^{-3} M_{\rm sph}$ (with $M_{\rm sph} \propto {L_{\rm sph}}^{1.18}$, cf. Mg98), and fits the constraints from quasar number counts for $\epsilon = 0.022$. The dashed line in the left panel indicates those $M_\bullet$ for which $r_\bullet = 0.1''$ at $D = 10\,{\rm Mpc}$.

There is indeed a correlation, but the scatter is large (~ 2 dex at fixed $L_{\rm sph}$) and selection bias may be important. The dashed line shows the $M_\bullet$ for which the BH sphere of influence, $r_\bullet \simeq GM_\bullet/\sigma^2$, extends $0.1''$ at a distance $D = 10\,{\rm Mpc}$ (σ is determined by $L_{\rm sph}$ through the Faber-Jackson relation). BHs below this line can be detected only in galaxies closer than 10 Mpc, and in galaxies in which kinematical tracers can be observed at resolutions $< 0.1''$ (e.g., water masers).

The solid line shows the predictions of one possible model that is consistent with the $\langle M_\bullet \rangle / \langle L \rangle$ from quasar number counts. This 'reference model' assumes that every spheroid has a BH with $M_\bullet \propto L_{\rm sph}$. Approximately 30% of the light from galaxies is due to spheroids (Schechter & Dressler 1987). So for an assumed accretion efficiency $\epsilon = 0.05$ this yields $M_\bullet = 1.4 \times 10^{-2} L_{\rm sph}$, which reproduces the trend in the data.

4. Surface-brightness profiles

HST observations of early-type galaxies reveal central surface brightness cusps that fall in two categories (F97), 'power-laws' (showing no clear break) and 'cores' (showing a clear break). Cusps can be explained as a consequence of the influence of a BH on surrounding stars (Young 1980, hereafter Y80). Properties of cusps around BHs depend on $M_\bullet$, initial conditions (Quinlan et al. 1995, hereafter Q95) and two-body relaxation (Bahcall & Wolf 1976). Cusps may also be due to processes unrelated to BHs (KR95), and they can also be destroyed (Quinlan & Hernquist 1997). Observed cusps therefore do not uniquely constrain the BH masses in galaxies.

Nonetheless, simple models of adiabatic BH growth for observed photometry of M87 (Young et al. 1978; Lauer et al. 1992; Crane et al. 1993) and several other galaxies imply BH masses that agree well with kinematic determinations. I have therefore started a study of adiabatic BH growth models for a large sample of galaxies with published *HST* photometry. These models may be particularly relevant for core galaxies, for which the observed break in the brightness profile may be associated with an originally homogeneous core. I have used the software of Q95 to fit the photometric models of Y80 to all core galaxies in the sample of F97 with $M_B < -20$ (van der Marel 1998, in preparation). The models fit well in the central few arcsec (RMS residual ~ 0.05 mag/arcsec^2), and the photometrically inferred $M_\bullet$ appear meaningful: the kinematically determined $M_\bullet$ for M87 and NGC 3379 (see Figure 1a) are reproduced to within 0.12 and 0.50 dex, respectively.

Figure 1b shows the results for the whole sample. The $M_\bullet$ are remarkably consistent with the kinematical detections in Figure 1a, and show a similar trend with L_{sph}. So despite their simplicity, it may well be that the Y80 models capture the essence of surface brightness cusps in core galaxies. The prevalence of these cusps would then imply that most or all core galaxies have BHs, with $M_\bullet \propto L_{sph}$ as suggested by Figure 1b.

5. Stellar kinematics and velocity dispersion anisotropy

Stellar motions often provide the only kinematical tool to study BH masses in quiescent galaxies, but the well-known degeneracy between $M_\bullet$ and velocity dispersion anisotropy (Binney & Mamon 1982) is still a major complication. This degeneracy can be resolved when high resolution *HST* data are available (e.g., van der Marel et al. 1997; Gebhardt et al. 1998), but such data are not yet available for many galaxies. Lower resolution ground-based data are plentiful, but more ambiguous to interpret. I use the case of NGC 1600, an E3 core galaxy with no significant rotation, to illustrate this.

Ground-based kinematical data with $\sim 2''$ resolution (Jedrzejewski & Schechter 1989) show a mildly peaked velocity dispersion profile, and *HST*

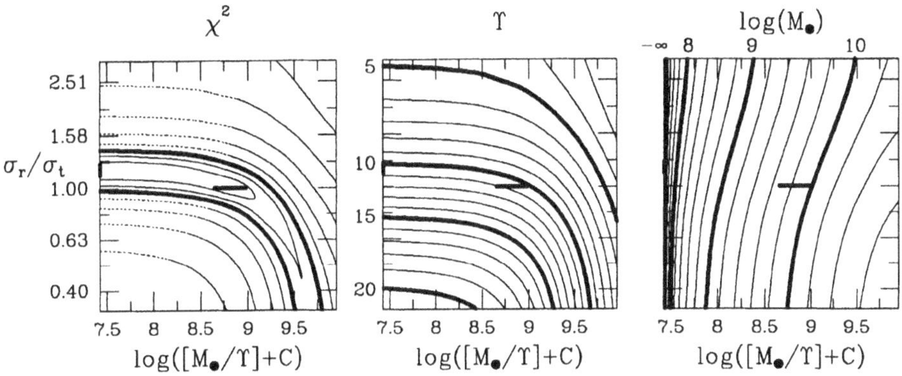

Figure 2. Predictions of spherical dynamical models for NGC 1600 that reproduce *HST* and ground-based photometry (F97; Peletier et al. 1990). The abscissa in each panel is $\log([M_\bullet/\Upsilon] + C)$, where $\log C \equiv 7.425$; i.e., approximately logarithmic in $(M_\bullet/\Upsilon)$ for $(M_\bullet/\Upsilon) \gg C$, and $M_\bullet = 0$ on the left boundary. The ordinate is σ_r/σ_t, increasing logarithmically from 1/3 to 3. For each $(M_\bullet/\Upsilon; \sigma_r/\sigma_t)$, the stellar mass-to-light ratio Υ was chosen to minimize the χ^2 of the fit to the kinematical data of Jedrzejewski & Schechter (1989). **(a; left)** Contours of χ^2. The first two contours correspond to the 1σ and 2σ level, as calculated from the statistic $\Delta\chi^2$. The heavy contours show the 3σ region. **(b; middle)** Contours of Υ, increasing linearly. **(c; right)** Contours of $M_\bullet$, increasing logarithmically. Heavy curves in panels b and c indicate contours for which the value of Υ or $\log M_\bullet$ is indicated. Contours of Υ and χ^2 are approximately parallel, so Υ is well determined independent of the anisotropy, $\log \Upsilon = 1.02$. By contrast, contours of $M_\bullet$ and χ^2 are approximately perpendicular, so there is a strong degeneracy between $M_\bullet$ and anisotropy. In all panels, a heavy horizontal bar indicates those isotropic models with a BH that are acceptable at the 1σ level, and a heavy vertical bar indicates those anisotropic models without a BH that are acceptable at the 1σ level.

photometry shows a shallow (F97), marginally significant (Byun et al. 1996; Gebhardt et al. 1996), surface brightness cusp. I construct spherical dynamical models (adequate for the present purpose) following the approach of van der Marel (1994, hereafter vdM94). I solve the Jeans equation for a given velocity anisotropy profile profile, $\sigma_r/\sigma_t(r)$ (where $2{\sigma_t}^2 \equiv \sigma_\theta^2 + \sigma_\phi^2$), project and convolve the results, and compare with the data in a χ^2 sense. The normalization of the dispersion profile is determined by the stellar mass-to-light ratio Υ, and its shape is determined by σ_r/σ_t and $M_\bullet/\Upsilon$.

Figure 2 shows the fitted χ^2, $M_\bullet$ and Υ in the $(M_\bullet/\Upsilon; \sigma_r/\sigma_t)$ parameter space, for models with constant σ_r/σ_t. The valley in the χ^2 contours shows that a one-parameter family of $(M_\bullet/\Upsilon; \sigma_r/\sigma_t)$ combinations fits the data. An isotropic model requires a very massive BH, $\log M_\bullet = 9.85$, but only modest anisotropy, $\sigma_r/\sigma_t = 1.18$, is required to fit the data without a BH. Models that are radially anisotropic at all radii may not correspond to a positive phase-space distribution function (DF) (vdM94), or may not be stable (Stiavelli et al. 1993), so I also constructed models in which the core is isotropic, and in which there is a smooth transition at the break radius of the surface brightness profile ($r_b = 3.12''$) to a value $(\sigma_r/\sigma_t)_{\rm main}$ characteristic of the main body. Models with $\log M_\bullet < 10.01$ are still all acceptable at the 1σ level. The best fit with no BH has $(\sigma_r/\sigma_t)_{main} = 1.45$, and the best fit with $\log M_\bullet = 9.15$, as suggested by adiabatic BH growth (Figure 1b), has $(\sigma_r/\sigma_t)_{main} = 1.30$. So in the absence of independent constraints on the velocity dispersion anisotropy, the data do not significantly constrain $M_\bullet$.

Our understanding of the velocity anisotropy in galaxies is only rudimentary. Ellipticals and bulges with power-law brightness profiles have low to intermediate luminosity (F97), and are flattened by rotation. The tensor virial theorem indicates that they may be isotropic. Any anisotropic model with the same average $(\sigma_R^2 + \sigma_\phi^2)/\sigma_z^2$ is also viable, but both M32 (van der Marel et al. 1998) and the Galactic Bulge (Evans & de Zeeuw 1994) have indeed been shown to be nearly isotropic. By contrast, core galaxies like NGC 1600 have intermediate to high luminosity, and little rotation (F97). Detailed studies of three individual galaxies (Merritt & Oh 1997; Rix et al. 1997; Gerhard et al. 1998) are consistent with mild radial anisotropy, $\sigma_r/\sigma_t = 1.2$–1.4, with a possible transition to isotropy at small radii. Such a velocity distribution can be produced by dissipationless collapse (van Albada 1982). Mild radial anisotropy is also consistent with studies of the ratio of major to minor axis kinematics (van der Marel 1991) and line-of-sight velocity profile shapes (Bender et al. 1994) in a larger sample of core galaxies, and is also seen in the Galactic halo in the solar neighborhood ($\sigma_r/\sigma_t = 1.5 \pm 0.2$; Beers & Sommer-Larsen 1995). So it appears that power-law galaxies may be approximately isotropic and that core galaxy may be mildly radially anisotropic, but neither result is firmly established.

6. Isotropic models for stellar kinematical data

Magorrian et al. (1998, hereafter Mg98) studied 36 (mostly) elliptical galaxies for which *HST* photometry and ground-based stellar kinematics have been published. Each galaxy was modeled with the Jeans equations, assuming an $f(E, L_z)$ DF (the axisymmetric generalization of a spherical isotropic model). Figure 1c shows the BH masses that best fit the observed kinematics. BHs are required in nearly all galaxies, with $M_\bullet \approx 6 \times 10^{-3} M_{sph}$ (long-dashed line), consistent with quasar number counts if the accretion efficiency $\epsilon = 0.022$. This is the first dynamical study that addresses a large sample in a homogeneous way while including *HST* photometry. It establishes the important fact that the presence of a BH in every spheroid is consistent with kinematical data, and that the required BH masses are consistent with quasar counts for a reasonable value of ϵ.

Nonetheless, the Mg98 results are not unique. Of the 29 galaxies that require a BH under the $f(E, L_z)$ hypothesis, 19 are core galaxies with similar data as for NGC 1600. Mg98 find $\log M_\bullet = 10.07$ for NGC 1600, but the results in §5 showed that all $M_\bullet$ smaller than this are equally acceptable.

So the Mg98 models may have overestimated the masses and/or prevalence of BHs. This would not violate the constraints from quasar counts: if one assumes a higher $\epsilon = 0.1$, one may decrease all $M_\bullet$ by a factor 4.5, or remove the BHs in 78% of the galaxies.

Two core galaxies in the sample have $M_\bullet$ determinations from independent sources. Neither is well fit by an $f(E, L_z)$ model. For M87, Mg98 infer the same $M_\bullet$ as inferred from *HST* gas kinematics, but only if the data outside $5''$ are ignored. For NGC 3379, Mg98 infer an $M_\bullet$ that exceeds the more accurate determination of Gebhardt et al. (1998) by a factor 7. Independent of whether one views these comparisons as reasonable or poor agreement, it leaves open the question whether $f(E, L_z)$ models return the correct result for galaxies that may not have a (significant) BH.

One may wonder whether the correlation between $M_\bullet$ and $L_{\rm sph}$ inferred by Mg98 can be explained if the $M_\bullet$ values were partly spurious. This is in fact the case. For galaxies that are radially anisotropic, isotropic models will fit the observed dispersion gradients by invoking BHs for which $r_\bullet \equiv GM_\bullet/\sigma^2$ is similar to the observational resolution. This predicts a correlation of $M_\bullet$ with distance of the form $r_\bullet \approx 2''$, which is not inconsistent with the Mg98 results. The more distant galaxies in the sample are the most luminous. So this predicts not only the correlation of $M_\bullet$ with $L_{\rm sph}$, but also that this correlation should be weaker for the galaxies in Virgo (which are all at the same distance), as seen in Figure 1c.

Actual measurements of the velocity anisotropy are required to establish whether or not the $M_\bullet$ inferred by Mg98 are correct. Either way, the $M_\bullet$ in Figure 1c are 4.5 times higher than those in Figure 1b, averaged over the 14 galaxies common to both samples. So either the photometric measurements are too low (not impossible, cf. the uncertainties discussed in §4), or the Mg98 results are too high (which would require mild radial anisotropy that is not inconsistent with our understanding of core galaxies, cf. §5).

7. Conclusions

Our understanding of the BH mass distribution is still incomplete, partly due to a lack of complete representative samples that cover quiescent and active galaxies of all Hubble types, and partly due to persistent uncertainties in the correct interpretation of photometric and kinematic data. However, it is clear that we are finding BHs in the correct mass range to explain quasar fueling and evolution, to within the uncertainties.

I thank Gerry Quinlan for kindly allowing me to use his adiabatic BH growth software. This work benefited from discussions with Eric Emsellem, Tod Lauer, John Magorrian, Scott Tremaine and Tim de Zeeuw. It was supported by STScI grant HF-1065.01-94A and an STScI Fellowship. STScI is operated by AURA Inc., under NASA contract NAS5-26555.

References

Bahcall J.N., Wolf R.A., 1976, ApJ, 209, 214
Beers T.C., Sommer-Larsen J., 1995, ApJS, 96, 175
Bender R., Saglia R.P., Gerhard O.E., 1994, MNRAS, 269, 785
Binney J., Mamon G.A., 1982, MNRAS, 200, 361
Bower G.A., et al., 1997, ApJL, in press; astro-ph/9710264 *[M84]*
Byun Y.-I., et al., 1996, AJ, 111, 1889
Chokshi A., Turner E.L., 1992, MNRAS, 259, 421
Crane P., et al., 1993, AJ, 106, 1371
Evans W., de Zeeuw P.T., 1994, MNRAS, 271, 202
Faber S.M., et al., 1997, AJ, in press; astro-ph/9610055 [**F97**]
Ferrarese L., Ford H.C., Jaffe W., 1996, ApJ, 470, 444 *[NGC 4261]*
Ferrarese L., Ford H.C., Jaffe W., 1998, ApJ, submitted *[NGC 6251]*
Ford H.C., et al., 1998, in Proc. IAU Symp. 184 (Kluwer), in press; astro-ph/9711299
Gebhardt K., et al., 1996, AJ, 112, 105
Gebhardt K., et al., 1998, AJ, submitted
Genzel R., Eckart A., Ott T., Eisenhauer F., 1997, MNRAS, 291, 219 *[Galaxy]*
Gerhard O.E., et al., 1998, MNRAS, submitted; astro-ph/9710129
Greenhill L.J., Gwinn C.R., Antonucci R., Barvainis R., 1996, ApJ, 472, L21 *[NGC 1068]*
Greenhill L.J., Moran J.M., Herrnstein J.R., 1997, ApJ, 481, L23 *[NGC 4945]*
Haehnelt M.G., Rees M.J., 1993, MNRAS, 263, 168 [**HR93**]
Jedrzejewski R., Schechter P.L., 1989, AJ, 98, 147
Kormendy J., Richstone D., 1995, ARA&A, 33, 581 [**KR95**]
Kormendy J., et al., 1996, ApJ, 459, L57 *[NGC 3115]*
Kormendy J., et al., 1996, ApJ, 473, L91 *[NGC 4594]*
Kormendy J., et al., 1997, ApJ, 482, L139 *[NGC 4486B]*
Lauer T.R., et al., 1992, AJ, 103, 703
Loveday J., Peterson B.A., Efstathiou G., Maddox S.J., 1992, ApJ, 390, 338
Lynden-Bell D., 1969, Nature, 223, 690
Macchetto F., Marconi A., Axon D.J., Capetti A., Sparks W., 1997, ApJ, 489, 579 *[M87]*
Magorrian J., et al., 1998, AJ, in press; astro-ph/9708072 [**Mg98**]
Merritt D., Oh S.P., 1997, AJ, 113, 1279
Miyoshi M., et al., 1995, Nature, 373, 127 *[NGC 4258]*
Newman J.A., et al., 1997, ApJ, 485, 570 *[Arp 102B]*
Pei Y.C., et al., 1998, ApJ, submitted *[M31]*
Peletier R.F., Davies R.L., Lindsey E, Illingworth G.D., Cawson M., 1990, AJ, 100, 1091
Quinlan G.D., Hernquist L., Sigurdsson S., 1995, ApJ, 440, 554 [**Q95**]
Quinlan G.D., Hernquist L., 1997, New Astronomy, submitted; astro-ph/9706298
Richstone D., 1996, BAAS, 189, 111.02 *[NGC 3377]*
Richstone D., 1998, in Proc. IAU Symp. 184 (Kluwer), in press
Rix H.-W., de Zeeuw P.T., Cretton N., van der Marel R.P., Carollo, 1997, ApJ, 488, 702
Schechter P.L., Dressler A., 1987, AJ, 94, 563
Simien F., de Vaucouleurs G., 1986, ApJ, 302, 564
Sołtan A., 1982, MNRAS, 200, 115
Stiavelli M., Møller P., Zeilinger W.W., 1993, A&A, 277, 421
Tremaine S., 1996, in 'Unsolved Problems in Astrophysics', (Princeton Univ. Press), 137
van Albada T.S., 1982, MNRAS, 201, 939
van den Bosch F., 1997, PhD thesis (Leiden University) *[NGC 4342]*
van der Marel R.P., 1991, MNRAS, 253, 710
van der Marel R.P., 1994, MNRAS, 270, 271 [**vdM94**]
van der Marel R.P., de Zeeuw P.T., Rix H.W., Quinlan G.D., 1997, Nature, 385, 610
van der Marel R.P., et al., 1998, ApJ, 493, in press; astro-ph/9705081 *[M32]*
Young P., et al., 1978, ApJ, 221, 721
Young P., 1980, ApJ, 242, 1232 [**Y80**]

ASCA OBSERVATIONS OF LUMINOUS INFRARED GALAXIES

Evolution from Starburst to AGN?

TAKAO NAKAGAWA, TSUNEO KII, RYUICH FUJIMOTO, TOSHIYUKI MIYAZAKI AND HAJIME INOUE
Institute of Space & Astronautical Science
3-1-1 Yoshinodai, Sagamihara, Kanagawa 229, Japan

YASUSHI OGASAKA AND KEITH ARNAUD
NASA/GSFC, Laboratory for High Energy Astrophysics
Greenbelt, MD 20771, USA

AND

RYOHEI KAWABE
Nobeyama Radio Observatory
Nobeyama, Minamimaki, Minamisaku, Nagano 384-13, Japan

1. Introduction

1.1. ORIGINS OF LUMINOSITY

One of the most important results of the *IRAS* survey is the discovery of a class of "Luminous Infrared Galaxies" (LIGs), which emit most of the energy in the infrared and are the dominant population in the local universe at luminosities above 10^{11} $L_\odot$ (e.g., Sanders & Mirabel 1996). All LIGs appear to be extremely rich in molecular gas, and many of them show evidence of recent interacting/merging activities. Hence it is now accepted that strong interactions of gas-rich galaxies triggers large central concentration of molecular gas, and makes optimal conditions for both enormous nuclear starbursts and building and/or fueling AGN. Actually, various observations show evidence of starburst activity as well as that of AGN in many LIGs (Sanders & Mirabel 1996).

Some of the "warm" (large $f_\nu(25\mu m)/f_\nu(60\mu m)$ ratio) LIGs (e.g., Mrk 231) show the closest appearance of classical quasars, such as broad emission lines and dominant optical nucleus. On the basis of these results, Sanders et al. (1988) proposed an evolutional connection between Ultra Luminous Infrared Galaxies (ULIGs, $L > 10^{12} L_\odot$) and optical quasars. They identified ULIGs with cold infrared colors ($f_\nu(25\mu m)/f_\nu(60\mu m) < 0.2$) as the initial, dust-enshrouded stage, and "warm" ($f_\nu(25\mu m)/f_\nu(60\mu m) > 0.2$) ULIGs as a transition stage to optical quasars.

1.2. WHY HARD X-RAY?

It has been controversial what is the dominant source of luminosity of LIGs. However, heavy dust extinction in LIGs makes it very difficult to determine how much of the total luminosity is due to starbursts and how much can be attributed to AGN.

Here we present a preliminary result of our on-going program to make hard X-ray observations of LIGs. The hard X-ray observations have the following merits.

1. We can observe heavily obscured objects with $N_{\rm H}$ as large as 10^{24} cm^{-2}
2. Hard X-ray observations are especially sensitive to AGN activities. The $L_{\rm X}/L_{\rm bol}$ ratio is ~ 0.1 for Seyfert 1 type objects but is 10^{-4} for starburst galaxies (Kii et al. 1997).
3. Seyfert 1 type objects have a power-law spectrum with a photon index ~ 1.7, and starburst galaxies show a soft spectrum characteristic of hot thin plasm. The difference of the spectrum enables us to make quantitative estimates of contributions of AGN and starburst activities.

Hence the hard X-ray observations are indispensable to reveal the origins of the luminosities in LIGs.

2. Results

We observed 10 LIGs (Table 1) by the Japanese X-ray satellite *ASCA*. The Hubble constant H_0 is assumed to be 75 km s^{-1} Mpc^{-1} to derive the luminosities. We saw clear evidence of AGN activities in 5 of the observed galaxies.

The most clear example is NGC 6240. Its soft (below 2 keV) X-ray spectrum is well represented by optically thin hot plasma emission, which is probably due to starburst activity. On the other hand, the spectrum above 3 keV is represented by a combination of flat power-law continuum and a complex Fe K line feature with very large equivalent width (total ~ 2 keV). This spectrum is very similar to that of the archetypical Seyfert 2 galaxy NGC 1068 (Ueno et al. 1994). Hence we conclude that, in NGC 6240, the direct emission from the AGN is completely blocked by a Compton thick ($N_{\rm H} > 10^{24}$ cm^{-2}) molecular torus and we see only a reflected component. The observed luminosity of the power-law component is $5 \times 10^8 L_\odot$. Assuming that $\Omega/4\pi \sim 0.25$ (Ω is the unobscured solid angle), we estimate that the intrinsic luminosity of the AGN component in NGC 6240 is $\sim 10^{11} L_\odot$,

TABLE 1. Observed Luminous Infrared Galaxies (LIGs)

Name	z	$\log(\frac{L_{\rm FIR}}{L_\odot})$	$\frac{f_\nu(25\mu m)}{f_\nu(60\mu m)}$	$\log(\frac{L_{2-10\rm keV}}{L_\odot})^{(1)}$	$\frac{N_{\rm H}}{10^{22}{\rm cm}^{-2}}$
Arp 220	0.018	11.9	0.08	$< 8.02^{(2)}$	...
NGC 6240	0.024	11.5	0.15	$10.89^{(3)}$	> 100
Mrk 273	0.037	11.8	0.11	$9.14^{(4)}$	20
UGC 05101	0.039	12.0	0.09	$< 8.32^{(5)}$	...
Mrk 231	0.042	12.1	0.27	$8.89^{(4)}$	6
IRAS 05189-2524	0.042	11.7	0.25	$9.88^{(4)}$	5
IRAS 08572+3915	0.058	12.1	0.23	$8.26^{(4)}$	...
IRAS 07598+6508	0.149	12.4	0.32	$< 9.02^{(5)}$	...
IRAS 20460+1925	0.181	11.8	0.52	$10.48^{(6)}$	3
IRAS 15307+3252	0.926	12.8	0.55	$< 10.61^{(6)}$	...

(1) Power law component with photon index of around 1.7
(2) Corrected for extinction with $N_{\rm H} = 10^{23}$ cm^{-2}. (3) Estimated intrinsic luminosity under the assumption that only the reflected component is observed.
(4) Corrected for extinction.
(5) Observed upper limit. Not corrected for extinction.
(6) Ogasaka et al. (1997)

which is luminous enough to power the whole far-infrared luminosity of NGC 6240. There is large uncertainty of this estimate of the intrinsic luminosity of NGC 6240, but we can conclude that a significant fraction of the total luminosity of NGC 6240 is due to the AGN activity.

We have detected clear sign of AGN activity in four more galaxies: IRAS 20460+1925 (Ogasaka et al. 1997), Mrk 273, IRAS 05189-2524, and Mrk 231. All of them show large extinction of $N_{\rm H} = 3-20\times10^{22}$ cm^{-2}. Even Mrk 231, which has been regarded as a Seyfert 1 galaxy, have large extinction which corresponds to $A_{\rm v} \sim 30$ mag. The AGN contribution to the total luminosity is significant in IRAS 20460+1925 and IRAS 05289-2524, but is small ($< 10\%$) in Mrk 273 and Mrk 231.

In the most archetypical LIG Arp 220, we detected no hard X-ray emission but only weak soft X-ray emission. Even if we assume that the AGN is heavily obscured by $N_{\rm H} = 10^{23}$ cm^{-2}, the upper limit for the power-law component is $L_{2-10\rm keV} < 1 \times 10^8 L_\odot$ or $L_{2-10\rm keV}/L_{\rm FIR} < 10^{-4}$. Hence we conclude that the contribution of AGN activity to the total luminosity in Arp 220 is negligible ($< 1\%$) (if any).

3. Discussion

We plot the relative AGN contribution ($L_{2-10\rm keV}/L_{\rm FIR}$) versus the infrared color ($f_\nu(25\mu$m$)/f_\nu(60\mu$m$)$) in Figure 1.

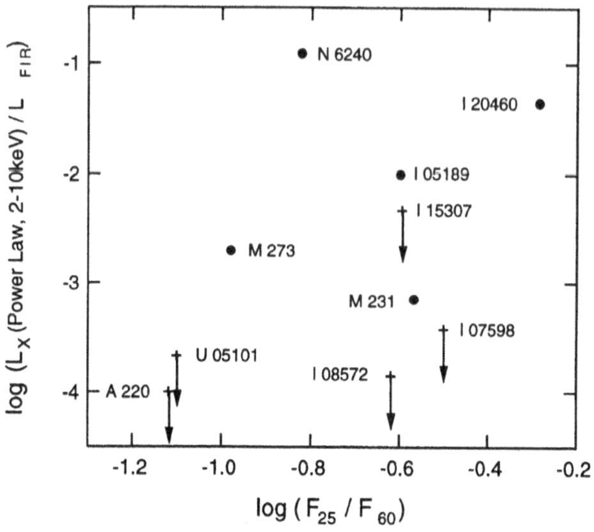

Figure 1. Infrared color $(f_\nu(25\mu m)/f_\nu(60\mu m))$ versus relative AGN activity $(L_{2-10\mathrm{keV}}/L_{\mathrm{FIR}})$

Following the evolutionary scenario by Sanders et al (1988), we can expect a positive correlation between the two parameters. However, Figure 1 shows no obvious correlation between the two parameters. Moreover, we see no obvious trend for the AGN contribution to increase in the more advanced mergers (such as Mrk 231). We hence conclude that the proposed evolutionary scenario cannot be applied to all of the LIGs, and it is not the evolutional stage as mergers but the local accretion rate onto the central engine which determines the AGN activity.

References

Kii, T., Nakagawa, T., Fujimoto, R., Ogasaka, Y, Miyazaki, T., Kawabe, R. & Terashima, Y. 1997, in X-Ray Imaging and Spectroscopy of Cosmic Hot Plasmas, ed. F. Makino & K. Mitsuda (Tokyo: Universal Academy Press), 161

Ogasaka, Y., Inoue, H., Brandt, W., Fabian, A., Kii, T., Nakagawa, T., Fujimoto, R., & Otani, C. 1997, PASJ, 49, 1790

Sanders, D. B., Soifer, B. T., Elias, J. H., Madore, B. F., Matthews, K., Neugebauer, G., & Scoville, N. Z. 1988, ApJ, 325, 74

Sanders, D. B. & Mirabel, I. F. 1996, ARAA, 34, 749

Ueno, S., Mushotzky, R. F., Koyama, K., Iwasawa, K., Awaki, H., & Hayashi, I. 1994, PASJ, 46, L71

CFHT ADAPTIVE OPTICS IMAGING OF ACTIVE GALAXIES

J.B. HUTCHINGS
National Research Council of Canada
Dominion Astrophysical Observatory
5071 W.Saanich Rd, Victoria, B.C., Canada

1. CFHT adaptive optics camera

The CFHT adaptive optics camera uses a visible light guide signal from a star to operate a bimorph mirror. The system is a unit that is operated by the observer and can be used with CCD or HgCdTe detectors. Pixel sizes are of order 0.04″. The amount of correction varies as the guide star brightness, the angular distance from it, and the natural seeing at the time. With good CFHT conditions, a guide star of 13 mag will give JHK images of FWHM near to the diffraction limit (0.1 to 0.15″) up to 20″ away. Correction is worse in the optical, but images of 0.2″ or better can be obtained in R and I-band. The camera performance is described by Rigaut et al (1998).

The camera is being used in programs on active galaxies ranging from the nuclear regions of Seyferts to high redshift QSOs and radio galaxies. Poor weather in 1997 plus delays in a 1K NIR detector have slowed progress. This poster summarizes some programs and results to date. The full details will be published in due course. Other participants in the programs are D. Crampton, S.L. Morris, E. Steinbring, and S. Chapman.

2. QSOs

This is a program planned to cover a range of QSO properties and redshifts. The AOB was used to obtain H-band and I-band images of the low luminosity, optically-selected, quasar 1055.3+019 at z=1.06. The FWHM of the stars were 0.11″ in H and 0.3″ in I during these observations, and the detectors used had 0.034″ pixels in H and 0.12″ pixels in I.

The appropriate off-axis point spread functions were measured and modeled. The QSO is clearly resolved in both wavebands, with significant ex-

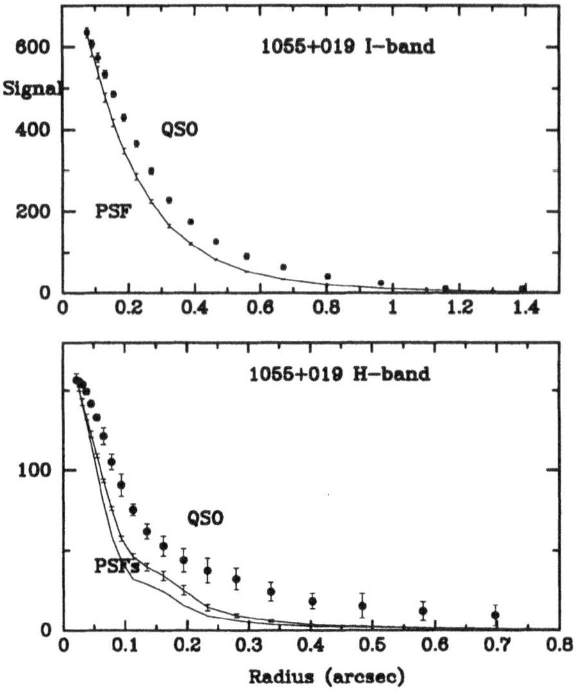

Figure 1. Azimuthally averaged profiles of $z = 1.06$ QSO and PSFs in NIR and visible light. H-band PSF shows off-axis correction to QSO.

tended flux within 0.1″ of the nucleus. Overall, it appears to be elongated in the direction of a knot and jet-like feature extending about 1 arcsec NE of the nucleus visible in the H-band image. There are no other evident close companions, but the host galaxy's complex luminosity profile and off-centered nucleus indicate recent tidal disturbance. The $I - H$ colours of the host galaxy correspond to an unreddened stellar population which is currently star-forming or is within a very short time of ending an initial starburst. The reddening must be small to match any models.

3. High z Radio sources.

The AOB is being used for NIR imaging of several radio sources with known optical structure from *HST* images. Early results with a 9″ field-of-view, low sensitivity camera were compared with *HST* images with similar resolution for two z∼1.1 3C sources. Luminosity profiles and colours were derived from the J, H and K′ images.

This program will be used to model the stellar populations and other continuum light sources in the objects, and compared with the results from the QSO imaging program.

4. Nuclei of nearby active galaxies.

Further programs are mapping the inner regions of Seyferts in the NIR to probe the dust that is seen in most objects, to look for structures such as nested bars or features associated with radio jets, to understand the fueling and nuclear processes, and associated circumnuclear star-formation. In these extended and bright structures, it is possible to perform useful deconvolution to improve the images.

Results on the Seyfert galaxy NGC 2992 have been combined with the *HST* optical band images with similar resolution. The nuclear region is on the edge of heavy dust obscuration and the NIR image reveals a curved "jet" behind the dust, as well as other knots in the region. The curve corresponds with a radio loop seen in the 6cm VLA map. The galaxy has indications of a strong outflow driven by star-formation activity in its inner regions.

References

Rigaut et al. 1998, *PASP*, in press

DYNAMICAL EXPLORATIONS OF NUCLEAR STRUCTURES IN BARRED GALAXIES

J. ANOSOVA
ASE, University of Texas at Austin, Austin TX 78712, USA
AND
G.F. BENEDICT
Astronomy Department, University of Texas at Austin, Austin TX 78712, USA

NGC 4314 (Benedict et al. 1996) has a complex nuclear morphology with recent star formation confined to a nuclear ring. *HST* observations resolve the nuclear ring into multiple sites of new star formation and resolve the associated dust lanes into discrete clouds. We construct dynamical models of this galaxy in order to provide plausible identification of the dynamical processes that led to the formation of the observed structure. We assume that the center of this galaxy contains a very massive double black-hole, surrounded by relatively low-mass particles - the star clusters as well as gas and dust complexes. Our previous work (Anosova & Anandarao 1994, Anosova & Tanikawa 1995) showed that the dynamical evolution of such a model produces many structures similar to those observed in galaxies of diverse types: spiral and elliptical galaxies, interacting galaxies, and various types of flows and jets. In the present work, we consider a number of such models with different initial parameters. We study their dynamical evolution of the gravitational N-body problem, taking into account strong interactions of bodies. Comparison of the evolution of our models with the observed structure, distributions and motions of stars, gas, and dust complexes in NGC 4314 shows good agreement. The model predicts the velocity fields observed in this galaxy.

References

Anosova J. & Anandarao B. 1994, *Astrophys. Space Sci.*, **220**, 83.
Anosova J. & Tanikawa K. 1995a, *Astrophys. Space Sci.*, **234**,191.
Benedict,G.F., Smith B.J., Kenney J.D.P. 1996, *AJ*, **111**, 1861.

THE FATE OF ULTRALUMINOUS MERGERS

A.C. BAKER AND D.L. CLEMENTS
Service d'Astrophysique, C.E.A. Saclay, Orme de Merisiers
Bât. 709, F91191 Gif-sur-Yvette CEDEX, France.

Galaxy formation is a diverse range of ongoing processes. Numerical simulations suggest that disk galaxies in collision pass through a massive burst of star formation, and produce 'elliptical–like' remnants similar to *bone fide* elliptical galaxies. The observed relative numbers of merging systems and elliptical galaxies are consistent with this picture (Toomre 1977). We here investigate further by studying the distribution of old stars in a sample of merging galaxies : the ultraluminous IR galaxies (ULIRGs). We selected ten ULIRGs from the literature (Clements & Baker 1996; Leech et al. 1994; Zhenglong et al. 1991; Melnick & Mirabel 1990) by two criteria: proximity (redshifts $z < 0.15$), for good spatial resolution; and confirmed signs of merging. We obtained deep K–band images ($t_{\text{int}} \sim 1000 - 2700$s) in good seeing using MAGIC on the Calar Alto 3.5m telescope. Our data have a field–of–view roughly equivalent to 10 – 20 effective radii for the galaxies. We have fitted analytic surface brightness profiles to the data for the exponential disk and the de Vaucouleurs $r^{1/4}$ elliptical descriptions. We strongly favour the elliptical–like description in 8 out of 10 cases, supporting the picture that collision and merger of classical spiral galaxies can produce classical elliptical galaxies, through tidal disruption, violent star formation accompanied by prodigious infrared emission, and gravitational relaxation. These data can also constrain the properties of the ULIRG (double) nuclei and hence the lifetime of the ULIRG phase (Baker & Clements 1998).

References

Baker, A.C., & Clements, D.L. 1998, in preparation
Clements, D.L. , & Baker, A.C. 1996, *A&A*, **314**, L5
Leech, K.J. et al. 1994, *MNRAS*, **267**, 253
Melnick, J. & Mirabel, I.F. 1990, *A&A*, **231**, L19
Toomre, A. 1977 in *The Evolution of Galaxies and Stellar Populations*, ed. B.M. Tinsley, R. B. Larson, (New Haven:Yale U.Obs.), **p.401**,
Zhenglong, Z., et al. 1991, *MNRAS*, **252**, 593

THE HOST GALAXIES OF IR LUMINOUS QUASARS

D.L. CLEMENTS
Institut d'Astrophysique Spatial
Bat 121, Universite Paris XI,
F-91405 ORSAY CEDEX, France

A.C. BAKER
Service d'Astrophysique
CEA Saclay, Orme de Merisiers Bat 709,
F91191 Gif-sur-Yvette CEDEX, France

AND

C.J. LIDMAN
European Southern Observatory
La Silla, Chile

A connection is suspected between quasars and ultraluminous IR galaxies (ULIRGs). Almost all ULIRG activity is triggered by mergers (e.g. Clements et al. 1996, *MNRAS*, **279**, 477). We thus investigate the relationship between quasars and ULIRGs by examining the relationship between FIR luminosity and host morphology in quasars. We use the $z < 0.3$ imaging survey of PG quasars by McLeod & Rieke (1994, *ApJ*, **420**, 58; **431**, 137), and ancillary data, to determine which quasars in the survey have disturbed hosts. This is combined with FIR data from Sanders et al. (1989, *ApJ*, **347**, 29). We apply Survival Analysis techniques to see if the FIR luminosity of the disturbed and undisturbed host quasars differ. We find, with 99.99% confidence, that the disturbed host quasars have higher FIR luminosity. We observe a small, but complete, sample of six quasars selected to have $L_{\rm FIR} > 10^{11.8}$ $L_{\odot}$, and a comparison sample of quasars with similar redshifts and B-band absolute magnitudes, but without detected FIR emission. At least 3 of the FIR luminous quasars are found in disturbed hosts while the non-FIR luminous quasars all lie in undisturbed hosts. The FIR luminosity in ULIRGs is triggered by interactions. If ULIRGs evolve into quasars we would expect FIR luminous quasars to be younger, and to share more characteristics with ULIRGs, especially disturbed host morphology. These studies are consistent with this model for quasar evolution.

THE NATURE OF ULTRALUMINOUS IRAS GALAXIES

D.L. CLEMENTS
Institut d'Astrophysique Spatial
Bat 121, Universite Paris XI,
F-91405 ORSAY CEDEX, France

W.J. SUTHERLAND
Oxford University Astrophysics
NAPL, Keble Road, Oxford, UK

AND

R.G. MCMAHON
Institute of Astronomy
Madingley Road, Cambridge, UK

We present the results of a continuing programme to investigate the properties of the Ultraluminous *IRAS* Galaxies (ULIRGs) in a large sample selected from the *IRAS* Faint Source Catalogue and identified with the APM galaxy catalogue (for details of the survey selection and initial results see Clements et al. 1996, *MNRAS*, **279**, 459). Observations now include optical imaging for a complete subsample to $B_J = 19.5$ (Clements et al. 1996, *MNRAS*, **279**, 477) and near-IR imaging for a complete subsample to $B_J = 19$ (Clements et al. in preparation). Moderate resolution optical spectroscopy, ISOPHOT photometry and ISOSWS spectroscopy have also been obtained for smaller subsamples of this survey. New redshift measurements have identified 11 new ULIRGs, extending the sample size to 102 ULIRGs. These observations have extended the maximum redshift to 0.44, and have also discovered a new broad-line ULIRG at $z = 0.351$. Initial analysis of our ISOPHOT data, while subject to substantial calibration uncertainties, shows a broad range of 160–120μm flux ratios which might indicate the presence of cold dust ($T \sim 20$ K) in some objects. Such dust would not have been detectable by *IRAS*. Our continuing multiwavelength studies of ULIRGs will be used to examine the nature of the power source in these objects, and their relationship to other extragalactic populations.

THE INTERACTING SEYFERT 2 GALAXY UGC 3995A

D. DULTZIN-HACYAN
Instituto de Astronomía, UNAM
P. MARZIANI
Osservatorio Astronomico di Padova
AND
M. D' ONOFRIO
Dipartimento di Astronomia, Università di Padova

Direct observational evidence of interaction driven inflow of gas toward a Seyfert nucleus remains scant. UGC 3995 is one of the rare systems of galaxies that, at a first glance, offers direct evidence of this phenomenon: a bright filament appears to connect the bar of the spiral galaxy UGC 3995A (whose nucleus shows a Seyfert 2 spectrum) to the nucleus of a smaller spiral companion (UGC 3995B) whose radial velocity exceeds by only 30 km s^{-1} that of UGC 3995A. Narrow (redshifted Hα) and broad band (R, B) images of UGC 3995, along with long slit spectra in the Hα region were collected at the 2.1 m telescope of the Observatorio Astronómico Nacional at San Pedro Mártir (SPM), México, in January-February 1995. From the B-R color map we infer that UGC 3995B is partially obscuring UGC 3995A. It is likely that a conspiracy of chance alignments and obscuration is giving the bar of UGC 3995A a disrupted appearance suggestive of stripping. The fairly regular profile of the HI 21 cm line suggests that the neutral hydrogen disks of both UGC 3995A and B have not yet undergone strong perturbations. However, while on the eastern side of UGC 3995A we observe HII region gas approximately following the rotation of the HI gas, on the western side we may be observing shock-heated "minority" gas whose motion is not exclusively rotational. We conclude that most of the morphological peculiarities observed in UGC 3995 cannot be easily related to infall of gas toward the Seyfert nucleus. Nevertheless, the kinematics of ionized gas in the western part of the disk of UGC 3995A has been affected by the intruder, probably because UGC 3995B is already physically in contact with UGC 3995A.

TESTING THE MERGER HYPOTHESIS OF POWERFUL RADIO GALAXIES

AARON S. EVANS
Caltech, California, USA

D.B. SANDERS
Institute for Astronomy, Hawaii, USA

AND

JOSEPH M. MAZZARELLA
IPAC-Caltech, California, USA

We present K′-band imaging and millimeter (CO) spectroscopy of a 60 and 100 μm flux-limited sample of 35 low redshift, powerful radio galaxies (LzPRGs: $P_{178\mathrm{MHz}} > 10^{23.5}$ W Hz^{-1} and $0.01 < z < 0.22$). These observations are being obtained to test the hypothesis that the radio activity in LzPRGs is triggered by the merger of gas-rich galaxies, as well as to look for evolutionary correlations between the degree of irregularity in the K′-band morphologies, the amount of star-forming molecular gas, and the radio morphologies.

Several intriguing results have resulted from this survey. Only 40% of the FR II (i.e, lobe-dominated, edge-brightened radio sources, typically with $P_{178\mathrm{MHz}} > 10^{25.3}$ W Hz^{-1}) galaxies show evidence of disturbed, high surface brightness features and multiple nuclei or close companions. Further, aside from the CO detection in the radio galaxy 3C 293 ($z \sim 0.04$), none of the FR II sources have been detected in CO. In contrast, 86% of the radio-compact and FR I (i.e, edge-darkened radio sources, typically with $P_{178\mathrm{MHz}} < 10^{25.3}$ W Hz^{-1}) galaxies show evidence of disturbed, high surface brightness features and multiple nuclei or close companions, and 50% have been detected in CO. Such a dramatic contrast suggests that either the more powerful FR II sources have different origins than the radio-compact and FR I sources (e.g., the progenitors of the FR II sources may be gas-poor relative to the progenitors of radio-compact and FR I sources), or that some compact and FR I sources evolve into FR II galaxies.

NEAR INFRARED SPECTROSCOPY AND THE SEARCH FOR CO EMISSION IN 3 EXTREMELY LUMINOUS IRAS SOURCES

AARON S. EVANS
Caltech, California, USA

D.B. SANDERS
Institute for Astronomy, Hawaii, USA

ROC M. CUTRI
IPAC-Caltech, California, USA

SIMON J.E. RADFORD
NRAO, Arizona, USA

PHIL M. SOLOMON
SUNY, New York, USA

DENNIS DOWNES
IRAM, Grenoble, FRANCE

AND

CARSTEN KRAMER
IRAM, Granada, SPAIN

Rest-frame 0.48–1.1 μm emission line strengths and molecular gas mass (H_2) upper limits for three luminous infrared sources – the hyperluminous infrared galaxies (HyLIGs: $L_{\rm ir} \geq 10^{13} L_\odot$ where $L_{\rm ir} \sim L(8-1000\mu{\rm m})$ IRAS F09105+4108 (z = 0.4417), IRAS F15307+3252 (z = 0.926), and the optically-selected QSO PG 1634+706 (z = 1.338) – are presented. Diagnostic emission-line ratios ([O III] λ5007/Hβ, [S II] λ6724/Hα, [N II] λ6583/Hα, and [S III] $\lambda\lambda$9069+9532/Hα) indicate a Seyfert 2-like spectrum for both infrared galaxies, consistent with previously published work. Upper limits on the molecular gas mass for all three sources are $M({\rm H_2}) < 1-3\times10^{10}h^{-2}\ M_\odot$ ($q_0 = 0.5$, $H_0 = 100h$ km s^{-1} Mpc^{-1}), less than the H_2 mass of the most gas-rich infrared galaxies in the local Universe. All three sources have $L_{\rm ir}/L'_{\rm CO} \sim 1300-2000$, the most extreme values for extragalactic sources measured to date. Given the relatively warm far-infrared colors for all three objects, much of their infrared luminosity may emanate from a relatively small quantity of hot dust near an AGN.

SEYFERT GALAXIES AND THEIR ENVIRONMENT

P. FOCARDI, B. KELM AND G.G.C. PALUMBO
Dipartimento di Astronomia, Università di Bologna, via Zamboni, 33, 40126 Bologna, Italy

The role played by environment on nuclear activity in galaxies is not clear and largely debated (see e.g. Barnes & Hernquist 1992, Kelm 1996). To overcome statistical uncertainties, environment properties of two large samples of Seyfert galaxies (Sy 1 and Sy 2 have been kept separated) have been computed and compared with equivalent size "normal galaxy" sample ones. Seyfert samples have been extracted from the Veron & Veron catalogue (Veron & Veron 1996), whilst "normal galaxies" have been randomly extracted from ZCAT (Huchra 1993). The samples are limited in cz ([1500-9500] km/sec) and contain 149 Sy 1, 173 Sy 2 and 160 "normal galaxies" (hundreds of random extractions from ZCAT). For each galaxy neighbors have been computed (from ZCAT) within two variables radii, R (isolation radius) and r (pair separation), which span [0.2 - 2] h_{100} Mpc. and [20-90]h_{100} kpc. respectively. Neighbors must lie also within 700 km/sec from the galaxy. In this way, for each value of R and r, environment of each galaxy has been "quantified".

The following results have been found for any value of R and r:

- Sy1 and Sy2 are significantly less isolated than "normal galaxies";
- Sy1 and Sy2 are marginally displaced from the mean locus of "normal galaxies" and tend to be more grouped;
- Sy1 and Sy2 are much more frequently members of galaxy pairs than "normal galaxies";
- Fractions of Sy1 and Sy2 in isolated galaxy pairs are 2 orders of magnitude larger than any analogous fraction of "normal galaxies".

References

Barnes, J.E. & Hernquist, L. 1992, *ARAA.*, **30**, 705
Kelm, B. 1996, *Ph.D. Thesis*, Università degli Studi di Bologna
Veron-Cetty, M.P. & Veron P. 1996, A Catalogue of Quasars and Active Galaxies, *ESO Scient. Rep.*, **17**
Huchra, J.P.et al. 1993, ZCAT Catalogue

RADIATIVE AVALANCHE DRIVEN BY SPHERICAL STARBURSTS

J. FUKUE
Astronomical Institute, Osaka Kyoiku University, Asahigaoka, Kashiwara, Osaka 582, Japan

M. UMEMURA
Center for Computational Physics, University of Tsukuba, Tsukuba, Ibaraki 305, Japan

AND

S. MINESHIGE
Department of Astronomy, Kyoto Univsersity, Sakyo-ku, Kyoto 606-01, Japan

As a solid physical mechanism to link between starburst activity and AGNs, we have proposed a *radiative avalanche*, where the mass accretion is driven by radiation drag exerted by stellar radiation from circumnuclear (ring-like) starburst regions (Umemura et al. 1997a,b). We here examine a radiative avalanche for a case of spherically distributed sources (Fukue et al. 1997). The results of the present spherical case are qualitatively similar to those of the ring-like case. For example, the drift timescale $t_{\rm drift}$ is found to be $t_{\rm drift} = 1.86\times10^9\ {\rm yr}\ (m+1)(b/100\ {\rm pc})^2(L_*/10^{12}L_\odot)^{-1}g(r)^{-1}$, where m is the power of rotational velocity ($v_\varphi \propto r^m$), L_* the starburst luminosity, b the radius of the starburst region, and $g(r)$ a function of the order unity. This timescale is sufficiently shorter than the viscous timescale of the relevant region. In contrast to ring-like starbursts, the advantage of spherical starbursts is the large solid angle subtended by the starburst region. Hence, the luminosity of the starburst region is fully available to avalanche on the surface of the disk. Furthermore, the radiation field of the starburst region has no angular momentum, which is in favor for the avalanche.

References

Fukue J., Umemura M., Mineshige S. 1997, *PASJ*, in press
Umemura M., Fukue J., Mineshige S. 1997a, *ApJL*, **479**, L97
Umemura M., Fukue J., Mineshige S. 1997b, *ApJ*, submitted

THE INTERPLAY BETWEEN THE NUCLEAR BARS, CENTRAL STARBURST, AND REMARKABLE OUTFLOW IN NGC 2782

S. JOGEE AND J.D.P. KENNEY
Yale University, New Haven, CT 06520-8101, U.S.A.

AND

B.J. SMITH
IPAC/Caltech, Pasedena, CA 91125

We show that the nearby peculiar interacting galaxy NGC 2782 (Arp 215) harbors a clumpy molecular CO bar. The gas bar has a radius of $\sim 7.5''$ (1.3 kpc), a mass of $\sim 2.4 \times 10^9$ $M_\odot$, and leads a nuclear stellar bar of similar extent (Jogee et al. 1997b). We estimate the gravitational torque exerted by the nuclear stellar bar and find large gas inflow rates ($\gg 1$ $M_\odot$ yr^{-1}) into the central 200 pc where the starburst activity peaks. We suggest that the nuclear gas bar will disappear on timescales ranging from few $\times$ 10^8 years to a Gyr, under the effect of star formation, dynamical friction, and gravitational torque. We also show that NGC 2782 harbors a starburst-driven outflow with a remarkable bubble morphology that has not been observed to date in any other starburst galaxy of comparable luminosity (Jogee et al. 1997a). The outflow is driving hot, warm, and possibly cold phases of the ISM out of the central kpc. The estimated outflow rate is less than the inflow rate, and this suggests the dynamical mass in the inner 100 pc might increase by several folds within a Gyr. These results might be relevant to theory (Shlosman et al. 1989) and simulations (Friedli & Martinet 1993) which propose nuclear bars and 'bars within bars' as efficient mechanisms for fueling central starbursts/AGNs.

References

Friedli, D., & Martinet, L. 1993, *A&A*, **277**, 27
Jogee, S., Kenney, J. D. P. & Smith B. J., 1997a, *ApJL*, in press.
Jogee, S., Kenney, J. D. P. & Smith B. J., 1997b, *ApJ*, submitted.
Shlosman, I., Frank, J., & Begelman, M. C. 1989, *Nature*, **338**, 45

MERGING GALAXIES WITH ACTIVE NUCLEI

W. KOLLATSCHNY
Universitäts-Sternwarte
Geismarlandstr. 11, 37083 Göttingen, Germany

We are investigating a sample of nearby interacting and merging galaxies showing starburst properties and/or nonthermal nuclear activity (e.g. Mrk 231, Mrk 739, Mrk 1027, NGC 2992, Arp 92).

We have obtained deep optical images (B,V,R) as well as deep X-ray images (*ROSAT* HRI/PSPC) for all these targets. Medium and high resolution optical spectra have also been taken at various slit positions to determine the excitation of the nuclear and extra-nuclear emission regions as well as the kinematics of the gas. We synthesize the observed absorption line spectra using population and evolutionary models to determine the nuclear and circumnuclear stellar components and to estimate the age of the starburst regions. There are indications for the nearly simultaneous triggering of starburst and nonthermal activity in Seyfert galaxies due to tidal effects. The goal of our present study is the search for interrelations between nuclear activity on the one hand and morphology, stellar composition and kinematics of the merging galaxies on the other hand.

One of our target galaxies is Mrk 266. The extreme properties of this double nucleus system have been demonstrated in many earlier publications (e.g. Petrosian, A.R. et al. 1980, *Afz,* **16**, 621; Osterbrock, D.E. & Dahari, O. 1983, *ApJ,* **273**, 478; Kollatschny, W. & Fricke, K.J. 1984, *A&A,* **135**, 171; Mazzarella, J.M. et al. 1988, *ApJ,* **333**, 168; Hutchings, J.B. et al. 1988, *AJ,* **96**, 1227; Wang, J. et al. 1997, *ApJ,* **474**, 659). We compare new optical images (including *HST* data) with very deep X-ray images (40 ksec). The similar morphologies of the $V - R$ colors and X-ray structures indicate their same physical origin. The north-eastern LINER nucleus is more compact in all frequences (optical, UV, X-ray, radio) in comparison to the south-western Seyfert 2 nucleus. In contrast to earlier observations three different components can bee seen in the X-ray data: the two merging nuclei and a strong extended northern region. A publication of our results is in preparation (Kollatschny et al.).

Acknowledgements: This work has been supported by the Deutsche Agentur für Raumfahrtangelegenheiten (DARA) grant 50 OR 94089.

THE DETECTION OF A LARGE, POWERFUL FR I RADIO GALAXY IN A SPIRAL HOST

MICHAEL J. LEDLOW
University of New Mexico, Institute for Astrophysics
Dept of Physics & Astronomy, Albuquerque, NM USA

FRAZER N. OWEN
National Radio Astronomy Observatory
Socorro, NM USA

AND

WILLIAM C. KEEL
University of Alabama, Dept of Physics & Astronomy
Tuscaloosa, AL USA

We report the detection of a FR I-like radio galaxy with a total extent of more than 200 kpc in a disk-dominated host. Traditional wisdom maintains that these types of radio sources are only found in elliptical hosts. We confirm the optical classification of this galaxy from deep, multicolor optical/NIR imaging and the detection of a spiral arm, an optical rotation curve, and line-ratios in the disk consistent with HII regions and star formation. At 20cm, we find a 36kpc knotty, jet extending into the southern lobe. At 3.6cm we detect a kpc-scale jet with the same position angle. With the exception of the radio source, this galaxy appears to be a fairly ordinary, dusty, star-forming spiral, with some evidence for a weak, obscured, AGN.

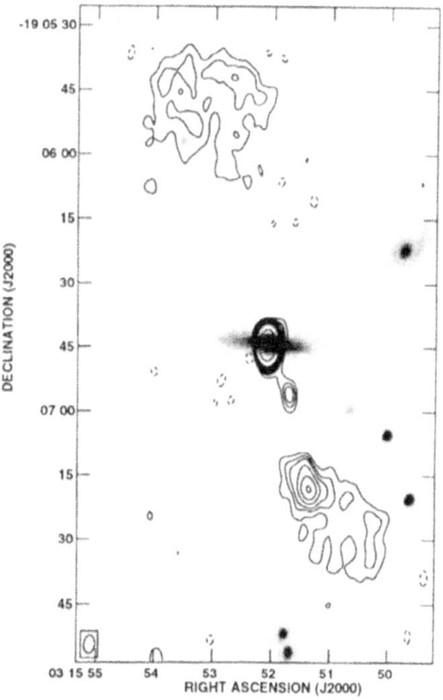

X-RAY STUDY OF ULTRALUMINOUS INFRARED GALAXIES

ASCA results of IRAS20551–4250 and IRAS23128–5919

K. MISAKI, Y. TERASHIMA, H. WATARAI AND H. KUNIEDA
Dept. of Physics, Nagoya University, Japan
K. IWASAWA
Institute for Astronomy, Cambridge, UK
AND
Y. TANIGUCHI
Astronomical Institute, Tohoku University, Japan

We observed two ULIRGs, IRAS 20551–4250 and IRAS 23128–5919 with the X-ray satellite *ASCA*. Both of them are merger, $100\mu m$ bright galaxies with $L_{\rm IR} \sim 10^{12} L_\odot$ and have a "warm" *IRAS* color (25μm/100μm ≥ 0.2), so the presence of an AGN would be expected.

The *ASCA* spectrum of IRAS 20551–4250 can be characterized by two components, one of which is a soft thermal component ($kT \sim 0.3$ keV) and the other is a hard power-law component ($\Gamma \sim 1.8$) suffering from absorption ($N_{\rm H} \sim 10^{22}{\rm cm}^{-2}$). A strong line feature seen around 1.3 keV (He-like Mg?) suggests that a higher temperature component exists. IRAS 20551–4250 may have multi-temperature thermal emission similar to the well-known nearby starburst galaxy M82. Adding another higher kT component, the power-law index would be smaller. The observed X-ray luminosity is $\sim 2.5 \times 10^{42}$ ergs/s in the rest frame 2–10 keV band (assuming $H_0 = 50$ km/s/Mpc), log $L_{\rm HardX}$/log $L_{\rm IR} \sim -3.4$ which is smaller than the typical value of Seyfert galaxies ($-1 \sim -2$). If the hard X-ray emission is scattered light (as a Compton-thick source), the intrinsic luminosity of the central engine is much higher than estimated. However, we may see the central engine directly, because a significant Iron K line is not clearly seen and time variability might be found especially in the hard-band. IRAS23128–5919 also shows a hard spectrum ($L_{\rm X} \sim 3 \times 10^{42}$ ergs/s), though the soft component could not be recognized in contrast to the former since photon statistics are limited.

Since these targets are similar in infrared luminosity as well as in hard X-ray but not in soft X-ray, we can suggest that $L_{\rm FIR}$ would be associated with the hard X-rays.

NEAR-INFRARED OBSERVATIONS OF A TYPE-2 QSO AT $z = 0.9$

K. NAKANISHI, M. AKIYAMA AND K. OHTA
Department of Astronomy, Kyoto University, Japan

AND

T. YAMADA
Astronomical Institute, Tohoku University, Japan

We report the results of near-infrared observations of a type-2 QSO, AX J08494+4454 at $z = 0.9$ which was identified in our optical follow-up observations of the *ASCA* Lynx deep survey. This object has a hard X-ray spectrum with an X-ray luminosity of about 1×10^{44} erg s^{-1} in 2-10 keV. The optical spectrum shows high-excitation and high-ionization lines but no significant broad $H\beta$ emission. These properties strongly suggest that this object is a "type-2" QSO (Ohta et al. 1996).

In order to examine the $H\alpha$ emission line profile, we made a near-infrared J-band spectroscopic observation of this object using UKIRT with CGS4 and found the FWHM of the $H\alpha$ emission line is $\sim$ 500 km s^{-1} confirming that this object is "type-2". Although the signal-to-noise ratio of the spectrum is low, we could not recognize a broad wing component of the $H\alpha$ emission line. A large [OI] 6300/$H\alpha$ line intensity ratio supports the idea that this object has an active nucleus.

A K-band image obtained with the QUIRC detector on the UH88 telescope shows that the object has a point-like structure, in contrast to the diffuse features seen in optical light (I-band). The optical-NIR color of this object is significantly redder than those of normal galaxies and QSOs. These properties may suggest the presence of a large amount of dust in this object. That is to say, we can see the nucleus directly at K-band, therefore this object appears like point source, on the other hand the nucleus is obscured in the I-band. Another possibility is that we see thermal emission from very hot dust ($\sim$ 1500 K°) associated with the nucleus at K-band.

CO OBSERVATIONS OF HIGH-z OBJECTS

K. OHTA, K. NAKANISHI, M. AKIYAMA AND T.T. TAKEUCHI
Dept. of Astronomy, Kyoto Univ., Kyoto 606-01, Japan
T. YAMADA AND Y. SHIOYA
Astronomical Institute, Tohoku Univ., Sendai 980-77, Japan
AND
K. KOHNO, R. KAWABE, N. KUNO AND N. NAKAI
Nobeyama Radio Observatory, Nagano 384-13, Japan

1. BR1202-0725 at z=4.7

We have made a CO(J=2-1) observations using the Nobeyama 45m telescope aimed at examining the physical properties of the molecular gas in this object. The upper limit obtained is 1.8 mK (3σ) at a velocity resolution of 100 km s^{-1}, which leads to an upper limit on the molecular gas mass of $5.3 \times 10^{11} M_{\odot}$, if we assume a line width of 250 km s^{-1} obtained in the CO($J = 5-4$) line (rest-frame) and the Galactic CO-to-H_2 conversion factor of 4.5 ($M_{\odot}$ K km s^{-1} pc^2). The line ratio between the 2-1 line and the 5-4 line as well as those from the 7-6 and the 4-3 lines (Omont et al. 1996, *Nature*, **382**, 428) imply that the mean gas density is as high as 10^{3-5} cm^{-3}, which is comparable to that in nearby star burst galaxies (e.g., Solomon et al. 1992, *ApJ*, **387**, L55).

2. Forming galaxy candidate cB58 at z=2.7

We have observed this object in the CO(J=3-2) line aimed at detecting a large amount of molecular gas, which is expected from the high star formation rate in this object. We have obtained an upper limit of 7.5 mK (3σ) at a velocity resolution of 25 km s^{-1}. The upper limit on the CO luminosity is 4.4×10^{10} K km s^{-1} pc^2 if we assume a velocity width of 300 km s^{-1}. A lower limit on the ratio of the Hα line luminosity to the CO luminosity is at the upper part of, but within the range of ratios for nearby galaxies. Results are presented in Nakanishi et al. (1997, *PASJ*, **49**, 535).

HIGH SPATIAL RESOLUTION NEAR-IR TIP/TILT IMAGING OF "WARM" ULTRALUMINOUS INFRARED GALAXIES

J.A. SURACE AND D.B. SANDERS
Institute for Astronomy, University of Hawaii
2680 Woodlawn Dr., Honolulu, HI 96822

We present results from high spatial resolution (FWHM $\approx$ 0.3-0.5 $''$) near-IR (1.6 and 2.1μm) imaging of a complete sample of ultraluminous infrared galaxies (ULIGs) chosen to have "warm" mid-IR colors ($f_{25}/f_{60} > 0.2$) characteristic of AGN. In conjunction with our WFPC2 imaging program (Surace et al. 1998), we have found that nearly all of these systems are advanced mergers with complex nuclear morphologies. The extended underlying galaxies are detected in each system at H and K$'$, and are found to have luminosities of a few L^*, similar to quasars (McLeod & Rieke 1994). Many of the circumnuclear star-forming knots seen at optical wavelengths have been detected. Based on model SEDs, their bolometric luminosities appear similar to those of the extended nuclear starbursts seen in other, less-luminous interacting systems (i.e. NGC 4038/9). Each ULIG is increasingly dominated at long wavelengths by a compact source which we identify as a putative active nucleus. The optical/near-IR colors of these putative nuclei are more extreme than the most infrared-active starburst galaxies, yet are identical to "far-IR loud" quasars which are in turn similar to optical quasars with significant hot (800 K) dust emission. Half of the ULIGs have dereddened nuclear near-IR luminosities comparable to those of QSOs, while the others resemble Seyferts; this may be an effect of patchy extinction and scattering. Similarities between the putative ULIG nuclei and QSO nuclei, the underlying host galaxies, and the apparent young age of the ULIGs (as evidenced by their compact star-forming knots) support the evolution of "warm" ULIGs into optical QSOs.

References

Surace, J.A., Sanders, D.B., Vacca, W.D., Veilleux, S., Mazzarella, J.M. 1998, *ApJ*, **492**, 116
McLeod, K.K. & Rieke, G.H. 1994, *ApJ*, **431**, 137

EVOLUTION OF VERY LUMINOUS INFRARED GALAXIES

H. WU AND Z.L. ZOU
Beijing Astronomical Observatory, P.R. China
X.Y. XIA
Dept. of Physics, Tianjin Normal University, P.R.China
AND
Z. G. DENG
Dept. of Physics, Graduate School, USTC, P.R.China

We have completed spectroscopic observations (Wu et al. 1997a) of a sample of 73 very luminous infrared galaxies ($\log(L_{\rm IR}/L_{\odot}) \geq 11.5; H_0 = 50$ km s^{-1} Mpc^{-1}) from the 2-Jy catalogue (Strauss et al. 1992) using the 2.16m telescope at the Beijing Astronomical Observatory. Spectral and interacting classifications are performed for the sample (Wu et al. 1997b). These statistical results provide strong evidence for the idea that interactions trigger nuclear activity and enhance the infrared luminosity. With the decrease of nuclear separation, relative velocity and specific angular momentum decrease rapidly, while on the contrary, both infrared luminosity and Hα equivalent width increase. Dynamical friction plays an important role even when two galaxies have large separation. This provides a favorable condition for strong star formation. We construct a simple merger sequence, from interaction class 1 to 4, to class 5 and 6 and then to the class 0 regarded as being in the stage of advanced merger. Along this sequence, spectral types change from HII-like to AGN-like. Considering the strong correlation of very luminous infrared galaxies in spectral classification schemes, it is reliable that infrared luminous galaxies evolve from HII-like galaxies to AGN-like galaxies. The different properties of infrared luminous Seyfert 1s and optically selected Seyfert 1s suggest that infrared luminous galaxies could evolve into optical Seyfert 1s in the last stage.

References

Strauss, M.A., et al. 1992, *ApJS*, **83**, 29
Wu, H., Zou, Z.L., Xia, X.Y., & Deng, Z.G. 1997a, *A&AS*, accepted
Wu, H., Zou, Z.L., Xia, X.Y., & Deng, Z.G. 1997b, in preparation

ASCA OBSERVATIONS OF THE TYPE-2 QUASAR RXJ13434+0001 AT $z = 2.35$

T. YAMADA
Astronomical Institute Tohoku University,
Aoba-ku, Sendai 980-77, Japan

Y. UEDA AND T. TAKAHASHI
Institute of Space and Astronautical Science,
3-1-1, Yoshinodani, Sagamihara, Kanagawa 229, Japan

K. OHTA
Department of Astronomy, Faculty of Science,
Kyoto University, Sakyo-ku, Kyoto, 606-01, Japan

M.CAPPI AND C.OHTANI
Institute of Physical and Chemical Research (RIKEN),
Wako, Saitoma, 351-01, Japan

AND

Y. ISHISAKI
Department of Physics, Tokyo Metropolitan University,
Hachioji, Tokyo 192-03, Japan

RXJ 13434+0001 is a rare example of radio-quiet type-2 quasars at high redshift. It was discovered through deep *ROSAT* observations and identified with a galaxy with a strong but narrow Lyα emission line at $z = 2.35$. In order to constrain the hard-X-ray properties we observed RXJ 13434+0001 with *ASCA*. The main purpose is to study the origin of the X-ray emission observed with *ROSAT*. If it is a scattered component from a strongly absorbed AGN, we could see it much brighter in the hard X-ray band.

ASCA results show that it is also very faint in the *ASCA* images, and the spectrum is consistent with the power-law extrapolation from the *ROSAT* PSPC results. RXJ 13434+0001 must be a partially absorbed ordinary type-1 quasar, which is consistent with the detection of broad Hα emission from the galaxy recently reported by Georgantopoulos et al.

COMPACT GROUPS OF GALAXIES

P. HICKSON
University of British Columbia, Dept. Physics & Astronomy
2219 Main Mall, Vancouver, B.C. V6T1Z4 Canada

Abstract. This paper reviews some of the outstanding questions concerning compact groups of galaxies. These relate to the physical nature and dynamical status of the groups, their formation and evolution, and their role in galaxy evolution. The picture that emerges is that compact groups are generally physically dense systems, although often contaminated by optical projections. Their evolution is likely a continuous process of infall, interaction and merging. As new galaxies are added, and previous ones merge, the membership of the group evolves. I suggest that while the size of the group changes little, other physical properties such as total mass, gas mass, velocity dispersion, fraction of early-type galaxies increase with time. This picture is at least qualitatively consistent with observations and provides a natural explanation for the strongest correlations found in compact group samples.

1. Introduction

One hundred and twenty years have passed since the discovery of the first compact group, Stephan's celebrated Quintet [41]. We now know of several hundred small groups of galaxies which have similar properties. They contain typically four or five galaxies in close proximity in the sky. With apparent space densities as high as 10^6 Mpc^{-3}, they are the densest known galactic systems. From the start, these systems have been enigmatic. When the first redshifts were obtained for galaxies in Stephan's Quintet, Seyfert's Sextet and VV 172, a remarkable chain of five galaxies, difficulties arose. All of these systems contain a galaxy whose redshift is discordant [5] [6] [40]. Even if the discordant redshifts are ignored, the velocity dispersion of the remaining galaxies is larger than that expected from the visible mass [5]. This discrepancy is now attributed to the presence of dark matter permeating the group and dominating its dynamics. Later, however, it became apparent that, even if the groups are dynamically bound, they would be unstable. Orbital decay due to dynamical friction should lead to stripping of galaxy halos, and eventual mergers, on relatively short time-scales [18] [7]. One must then explain how compact groups continue to form, and identify their progenitors and progeny. This leads to the larger question of the role of compact groups in galaxy evolution in general.

These difficulties would be alleviated if the groups themselves are not physically dense, and several alternative explanations for their high *apparent* density have been suggested. The question can in principle be settled by observations, and much progress in this area has been made over the past 15 years. It is timely to now ask what has been learned, and what is the status of the main questions regarding compact groups. The definition and selection of compact groups, and the main observational and theoretical results have recently reviewed by Hickson [14]. This paper briefly summarizes the current situation with regard to the most outstanding issues.

2. Physical Nature

It is generally accepted that the discordant redshifts seen in many compact groups are cosmological, the result of chance alignments of unrelated background or foreground galaxies. Statistical analysis (eg. [23]) suggests that the surprisingly large fraction of groups containing discordant redshifts (half of all quintets!) is consistent with chance alignments, although the subject continues to cause debate [1] [43].

Turning now to the accordant galaxies, we find several hypotheses that explain the apparent high space densities in compact groups:

Dense bound configurations. The natural explanation is that compact groups are indeed physically dense. Even after allowing for the modest increase in apparent density caused by selection biases [14], the typical physical separation between galaxies is only of order $50h^{-1}$ kpc. This hypothesis is supported by a variety of observational evidence [19]. Foremost is the high fraction of interactions seen both in the morphology and kinematics of compact group galaxies [37] [27]. More recently, strong evidence for this hypothesis has come from the detection of diffuse X-rays from a large fraction of compact groups (eg. [33]).

Transient configurations. Rose [38] first suggested that compact groups may be transient configurations caused by the accidental coincidence of galaxies passing unusually close to each other in their orbits. While otherwise attractive, the hypothesis is statistically improbable as the chance of four or more galaxies momentarily occupying the same small volume is very low [19].

Alignments within loose groups. Long championed by Mamon [25], the hypothesis remains a viable explanation for at least some compact groups.

The presence of at most two interacting galaxies may be explained by including a physical binary. However, the hypothesis fails to account for the many groups which contain more than two interacting galaxies or extended X-ray emission.

Alignments within filaments. Hernquist et al. [12], have recently suggested that compact groups may arise as projections along filaments. Their numerical simulations indicate that motions of widely-separated galaxies in filaments can conspire to produce apparent velocity dispersions comparable to those seen in compact groups. It offers an explanation of the relatively low X-ray luminosities of most compact groups, but appears now to be inconsistent with observations [33].

3. Relation to Environment

From the preceding discussion, we are faced with the probability that many, perhaps most, compact groups are dense bound systems. It is then natural to ask how they relate to their environment. Although many compact groups appear to be quite isolated, they generally trace the overall galaxy distribution, and are found in low-density associations, filaments, and loose groups. The fact that few compact groups are found in rich clusters is in part due to the selection criteria employed (eg. [13]), but dynamical disruption of small groups during cluster collapse may also play a role. There are, however, a number of important differences between galaxies in compact groups and those in their environment. For example, the fraction of elliptical galaxies is significantly higher in compact groups than in the field and correlates with velocity dispersion [16].

Several authors have investigated the formation of compact systems by gravitational collapse within loose groups (eg [9]). This seems to be a plausible mechanism, although it remains to be seen whether or not the timescales are consistent with the properties of the neighborhoods of observed compact groups [43].

If compact groups condense from loose groups, they appear to have lost all dynamical memory of their origins. Neither the shapes and orientations of the groups, nor the orientation of the galaxy spin axes, correlate with the distribution or orientation of surrounding galaxies [32].

4. Interactions

Because of their high density and relatively low velocity dispersions (next section), interactions in compact groups should be both frequent and strong. As expected, there are many cases of strong interactions leading to star formation and possible mergers in these systems [36] [24] [34] [28] [35] [45] [31] [46] [21] [22]. While the level of star formation in compact groups is enhanced with respect to that of isolated galaxies, it is lower than that found in strongly-interacting galaxy pairs [30] [44].

An interesting question is how interactions affect the environment of the galactic nuclei. Do they trigger nuclear star formation and/or nuclear (AGN) activity? A study of the nuclear regions of compact-group spiral galaxies indicates an order-of-magnitude enhancement in the star-formation rates compared to isolated spiral galaxies [29]. From spectra of 67 galaxies in compact groups, Coziol et al. [8] find 22% are starburst and 40% are (mostly low luminosity) AGN. The probability of finding an AGN is higher in luminous galaxies and early-type galaxies.

At higher luminosities, there is much circumstantial evidence that nuclear activity may be triggered by interactions in small groups of galaxies. A large fraction of ultraluminous infrared galaxies contain Seyfert nuclei, and essentially all are interacting [39]. Many QSOs appear to be interacting with close companions (eg. [42] [20] [3]). These may be extreme examples of interactions in compact groups, not found in local samples because of their rarity.

5. Dynamical Status

The observed velocity dispersions of compact groups range from a few tens to several hundreds of km sec^{-1}, with a median value of order 200 km sec^{-1}. When combined with their small sizes, this gives crossing times t_c which range from 10^{-3} to ~ 1 Hubble time, the median being $0.02\ H^{-1}$. Galaxies lose energy when moving through a background of lower-mass dark matter particles. The resulting dynamical friction timescale t_d depends on the amount and distribution of the dark matter, but for typical groups is of order $10t_c$. One expects orbital decay and merging to occur on this timescale, as is confirmed by dynamical simulations which indicate that groups typically evolve to form a single massive remnant, resembling an elliptical galaxy, within a few crossing times [4].

As it is unlikely that the present epoch is special this process has likely continued for of order a Hubble time, so one is led to the conclusion that the population of remnants could exceed the current population of compact groups by an order of magnitude. Since current estimates place the luminosity density of present-day compact groups at a few percent of the luminosity density of the universe, it follows that compact-group remnants should comprise a substantial fraction of all galaxies. While this may not at first seem implausible, it is not at all clear that the properties of such remnants are consistent with those of elliptical galaxies, or of any other identifiable population [44]. Just considering the luminosities, a difficulty arises because the total luminosity of individual compact groups exceeds that of typical elliptical galaxies by several magnitudes [26].

The problem may be avoided if individual compact groups can survive for a Hubble time or more. Recent numerical results indicate that it is possible for groups to survive this long given either special initial conditions [10] or massive dark matter halos [2]. Alternatively, groups may be replenished by the addition of new galaxies. Governato et al [11] point out that in a high-density ($\Omega \simeq 1$) universe, infall of surrounding galaxies onto a group can rejuvenate the system, prolonging its life. This reasonable picture may have observational support. Moles et al [31] present evidence for an example of recent infall in Stephan's Quintet.

6. Evolution

It is interesting to explore the logical consequences of the infall picture, and to see if it leads to testable predictions. Suppose, for example, that compact groups result from a continuous process of infall, interaction and merging. Assume that the infalling galaxies are gas-rich, and that some fraction of the interactions and mergers lead to elliptical and S0 galaxies. At each stage in the process we apply the compact group selection criteria to define the group members. One finds [15] that the membership of the group changes with time, as galaxies merge, and as new galaxies join the group. The radius of the group, however, remains roughly constant - it is determined more by the selection criteria than by any intrinsic size of the group.

In this simple model we find that, for any individual group, the following parameters increase with time:

- mass - grows as galaxies fall into the group.
- velocity dispersion - increases with mass.
- elliptical and S0 fraction - grows due to the effect of interactions and mergers
- halo mass - grows as individual galaxy halos are disrupted
- diffuse gas mass - grows due to stripping of galactic gas
- X-ray luminosity - increases with gas mass
- X-ray temperature - increases with velocity dispersion

The rate at which this evolution occurs depends on the initial density contrast of the perturbation which gives rise to the group. As this is likely to have a wide range of possible values, we expect to find groups in various stages of dynamical evolution. Those groups with the highest fraction of elliptical galaxies would be the most evolved.

From this picture we can infer the following observable consequences:

- Physical properties should not correlate strongly with group radius
- Spiral fraction should correlate inversely with velocity dispersion

- X-ray luminosity and temperature should correlate inversely with spiral fraction
- The mass and luminosity of the largest elliptical galaxy should correlate with velocity dispersion
- We should find a wide range of ages of stellar populations within compact group galaxies
- If globular clusters and/or dwarf galaxies are produced in interactions and mergers, the specific frequencies of these objects should correlate with group velocity dispersion

Comparing with observational results, we find that indeed, compact group parameters do not correlate strongly with radius. The morphology-density relation is absent or weak, velocity dispersion does not correlate with radius, etc. In contrast, the strong inverse correlation between spiral fraction and velocity dispersion is nicely explained. Looking at the X-ray results, we see that the relatively weak emission seen from spiral rich groups can result from the earlier evolutionary phase of these groups. An examination of the data reported by Hickson et al. [17] reveals that the luminosity of the brightest elliptical galaxy does indeed correlate with velocity dispersion, as predicted. The last two predictions have not yet been tested, but can in principle be determined by future observations.

In summary, we find that a self-consistent picture of the formation and evolution of compact groups emerges which is at least qualitatively in agreement with current observational data. Moreover, predictions are made which are amenable to observational tests.

In conclusion, compact groups have emerged as mostly dense physical systems, although contamination by optical projections is common. They are centres for gravitational interaction and the dynamical evolution of galaxies. They are by no means rare objects, but involve several percent of the total galactic population. Their physical properties, which cover a large range of values, are a function of the selection criteria by which they are identified, as would be expected if they represent the high-density tail of the clustering hierarchy.

References

1. Arp C, 1997, ApJ, 474, 74
2. Athanassoula E, Makino J & Bosma A, 1997, MNRAS, 286, 825
3. Bahcall JN, Kirharos S, Saxe DH & Schneider DP, 1997, ApJ, 479, 642
4. Barnes JE, 1989, Nature, 338, 123
5. Burbidge EM, & Burbidge GR, 1959, ApJ, 130, 23
6. Burbidge EM, & Burbidge GR, 1961, ApJ, 134, 244
7. Carnevali P, Cavaliere A & Santangelo P, 1981, ApJ, 249, 449
8. Coziol R, Ribeiro ALB, de Carvalho RR & Capelato, 1997, preprint
9. Diafario A, Geller MJ & Ramella M, 1994, AAJ, 107, 868

10. Governato F, Chincarini G, Bhatia R, 1991, ApJ, 371, L15
11. Governato F, Tozzi P, Cavaliere A, 1996, ApJ, 458, 18
12. Hernquist L, Katz N, Weinberg DH, 1995, ApJ, 442, 57
13. Hickson P, 1982, ApJ, 255, 382
14. Hickson P, 1997, ARAA, 35, 357
15. Hickson P, 1998, in preparation
16. Hickson P, Kindl E & Huchra JP, 1988, ApJ, 331, 64
17. Hickson P, Mendes de Oliveira C, Huchra JP & Palumbo GGC, 1992, ApJ, 399, 353
18. Hickson P, Richstone DO & Turner EL, 1977, ApJ, 213, 323
19. Hickson P & Rood HJ, 1988, ApJ, 331, L69
20. Hutchings JB, 1995, AJ, 109, 928
21. Iglesias-Paramo J, Vilchez JM, 1997, ApJ, 479, 190
22. Iglesias-Paramo J, Vilchez JM, 1997, ApJ, 489, L13
23. Iovino A & Hickson P, 1997, MNRAS, 287, 21
24. Longo G, et al, 1994, A&Ap, 281, 418
25. Mamon GA, 1986, ApJ, 307, 426
26. Mendes de Oliveira C & Hickson P, 1991, ApJ, 380, 30
27. Mendes de Oliveira C & Hickson P, 1994, ApJ, 427, 684
28. Menon TK, 1995, AJ, 110, 2605
29. Menon TK, 1995, MNRAS, 274. 845
30. Moles M, et al, 1994, A&Ap, 285, 404
31. Moles M, Sulentic JW & Márquez I, 1997, ApJ, 485, L69
32. Palumbo GGC, et al., 1993, ApJ, 405, 413
33. Ponman TJ, Bourner PDJ, Ebeling H, & Bohringer H, 1996, MNRAS, 283, 690
34. Rodrigue M, et al, 1995, AJ, 109, 2362
35. Ribeiro ALB, et al, 1996, ApJ, 463, L5
36. Rubin VC, Ford WKJ & Hunter DA, 1990, ApJ, 365, 86
37. Rubin VC, Hunter DA & Ford WKJ, 1991, ApJ Suppl, 76, 153
38. Rose JA, 1977, ApJ, 211, 311
39. Sanders DB & Mirabel IF, 1996, ARAA, 34, 749
40. Sargent WLW, 1968, ApJ Lett, 153, L135
41. Stephan ME, 1877, MNRAS, 37, 334
42. Stockton A, 1982, ApJ, 257, 33
43. Sulentic JW, 1997, ApJ, 482, 640
44. Sulentic JW & Raba ca 1994, ApJ, 429, 531
45. Verdes-Montenegro L et al., 1997, A&Ap, 321, 409
46. Yen MS, Verdes-Montenegro L, del Olmo A & Perea J, 1997, ApJ, 475, L21

VLA OBSERVATIONS OF NEUTRAL HYDROGEN IN COMPACT GROUPS

B.A. WILLIAMS
Physics and Astronomy Dept., University of Delaware
Newark, Delaware, 19716, USA

J.H. VAN GORKOM
Astronomy Dept., Columbia University
538 W. 120th Street, New York, NY 10027, USA

MIN YUN
NRAO
P. O. Box 0, Soccorro, NM 87801, USA

AND

LOURDES VERDES-MONTENEGRO
Instituto de Astrofisica de Andalucia
CSIC, Apda 3004, 18080 Granada, Spain

Abstract. VLA images of the neutral hydrogen (HI) in the direction of HCG 2, 16, 33, 88, and 92 (Stephan's Quintet) are examined. In HCG 2 and 16, the HI gas is bound to the individual galaxies but shows definite signs of tidal interaction; while in HCG 92, the more compact configuration, the HI gas is contained within a few prominent cloud features well displaced from the optical positions of any of the spiral members. In every case, the motions of the gas are consistent with the motions of the galaxies within the Hickson groups. A range of kinematical properties is observed for the HI gas, from well-ordered rotation, to small-scale systematic gradients within the cloud features.

1. Introduction

We have successfully imaged about a dozen of the Hickson (1982; HCG) groups in a long-term program to observe with the Very Large Array (VLA) all the Hickson groups detected by Williams and Rood (1987). The debate over the true nature of compact groups still continues. Their existence, age, and state of dynamical evolution have been and continue to be the subject of controversy. If these systems are real, then they are truly interesting objects because of their dynamical properties, i.e., short crossing and collision times, large space densities, low velocity dispersions, and short time scale for loss of orbital energy due to dynamical friction. Compact groups would be ideal places to study the evolution of galaxies as well as the entire group.

Two models have been proposed to explain their existence; they are truly compact systems/configurations, or they are chance alignments within larger systems (Rose 1977; Mamon 1986) or unbound galaxies along filaments (Hernquist, Katz, and Weinberg 1995). It is reasonable to assume that the objects in the Hickson catalog are not of a single nature but represent some heterogeneous mixture including physically dense systems, the cores of large groups, (HCG 58; Williams 1985), chance alignments in either loose groups or along edge-on filaments of galaxies, and even single galaxies (HCG 18; Williams and van Gorkom 1988). Given this potpourri of objects, what observational tests can be made to distinguish between the two basic models presented above? Diffuse gas trapped within the gravitational potential of a group provides a means of determining which systems are real and which are chance superpositions (Ostriker, Lubin and Hernquist 1995). Since all of the compact groups with an extended intragroup medium have high percentages of early-type galaxies (Mulchaey et al. 1996), X-ray emission is a poor observational test for the spiral-dominated Hickson groups. Clearly a better observational test for the spiral-rich groups would be the distribution and kinematics of the neutral hydrogen (HI) gas in the spiral galaxies. HI aperture synthesis could be used to distinguish between the real and illusory objects. Disk galaxies are normally more extended in HI than in starlight, so that the HI gas is suspectable to tides and direct collisions and ought to be more sensitive to recent interactions. In most cases, observations with the VLA can provide the spatial resolution needed to confirm the dynamical interactions suspected in the spiral-rich groups if they are indeed physically dense. The distribution and kinematics of the HI gas can give clues about the dynamical state and evolution of the spiral-rich groups.

2. Distribution and Kinematics of the Neutral Hydrogen Gas

Based on their HI morphology, the Hickson groups that we have imaged thus far fall into three basic categories. There is a class of Hickson groups that display evidence of physical compactness. The HI distribution shows signs of tidal tails, severely warped HI disks, and complex HI velocity fields, all suggestive of ongoing interactions between the members of the group.

The second category includes groups that display normal HI morphologies and kinematics. The HI emission is distributed about the optical centers of the spiral members and the overall patterns of the velocity field are consistent with motions generated by rotating disks of gas viewed at various inclinations. Most of the spiral galaxies in these groups have global HI kinematics, i.e., velocity widths, consistent with their absolute blue luminosities. The third category is the one that includes compact groups that have been misidentified and are in fact single galaxies. The new images that we present here of HCG 2, 16, 33, 88, and 92 (Stephan's Quintet) have HI morphologies that are consistent with the first two categories described above.

The integrated HI emission detected in the direction of HCG 2 *(fig. 1a)* can be clearly associated with the two spiral galaxies, *a* and *c*. In at least eight channel maps, the HI is elongated along a line connecting galaxies *a* and *b*. The HI contours of galaxy *a* in figure 1a are distorted in the direction toward galaxy *b*. This may be suggestive of some mild tidal interaction between the two members. In general, the velocity fields of both galaxies are well-ordered and are consistent with the motions associated with rotating disks. Figure 1b shows the global profiles that have been reconstructed from the channel maps. The total integrated flux measured

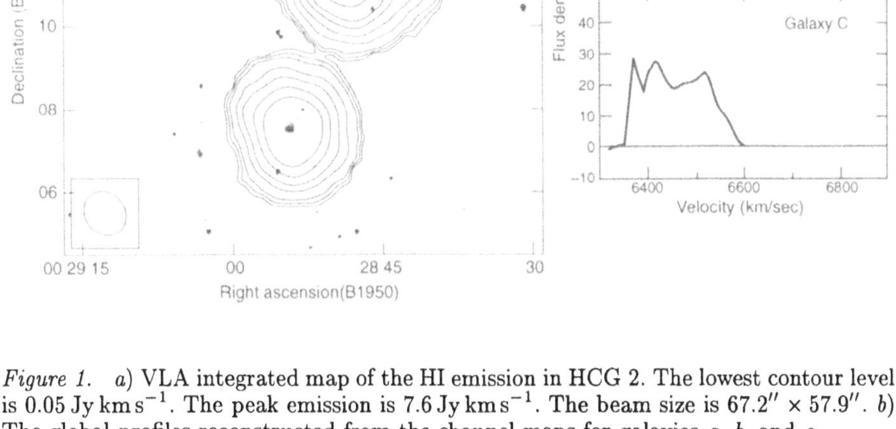

Figure 1. *a)* VLA integrated map of the HI emission in HCG 2. The lowest contour level is 0.05 Jy km s^{-1}. The peak emission is 7.6 Jy km s^{-1}. The beam size is 67.2″ × 57.9″. *b)* The global profiles reconstructed from the channel maps for galaxies *a*, *b*, and *c*.

by the D-array is 22.6 Jy km s^{-1} which is larger than that measured, by the NRAO 300-ft telescope (Williams and Rood 1987). Galaxies *a* and *c* have a hydrogen mass of $1.5 \times 10^{10} M_\odot$ and $3 \times 10^{9} M_\odot$, respectively. We have assumed that most of the neutral hydrogen in the direction of galaxies *a* and *b* is contributed by galaxy *a*. The velocity field appears too uniform to be produced by more than one galaxy. There is no strong evidence of tidal interaction between the two hydrogen-rich galaxies, *a* and *c*.

HI emission *(fig. 2a)* detected in the direction of HCG 16 is contained within a single cloud feature that is centered between galaxies *c* and *d* and elongated toward the southeast companion, NGC 848. At the resolution of the D-array, the emission from the individual galaxies cannot be determined but a comparison of the VLA integrated flux with that measured at Green Bank (Williams and Rood 1987) shows good agreement. Systematic motions are observed along two different gradients, none of which aligns with the major axis of any galaxy. It is difficult to interpret the D-array results with the present angular resolution. HCG 16 is clearly one group that needs to be reobserved with the higher angular resolution provided by the C-array. The global profiles of the three features resolved by the D-array are shown in figure 2b. NGC 848, the companion detected to the southeast,

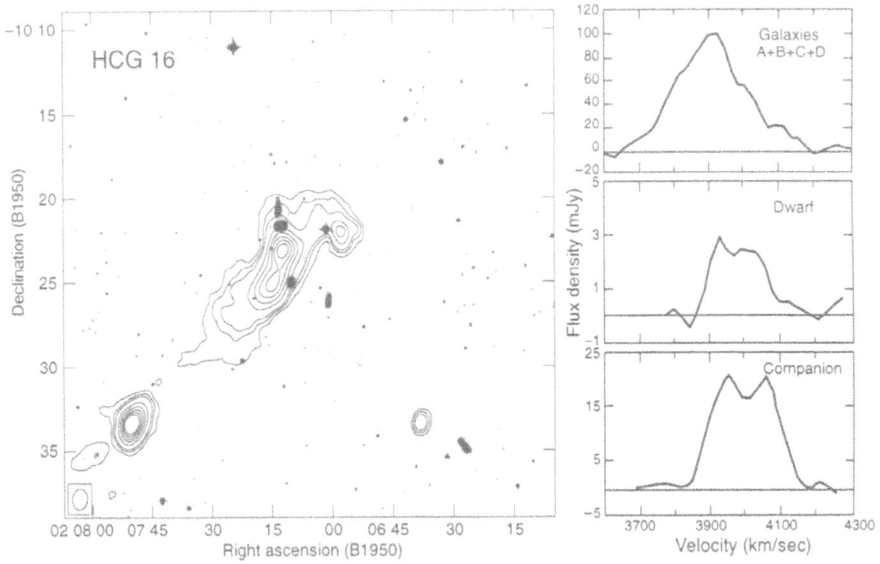

Figure 2. *a)* VLA integrated map of the HI emission in HCG 16. The lowest contour level is 0.25 Jy km s^{-1}. The peak emission is 3.9 Jy km s^{-1} and the beam size is $73.3'' \times 54.6''$. *b)* The global profiles reconstructed from the channel maps for the group, NGC 848, and the dwarf member.

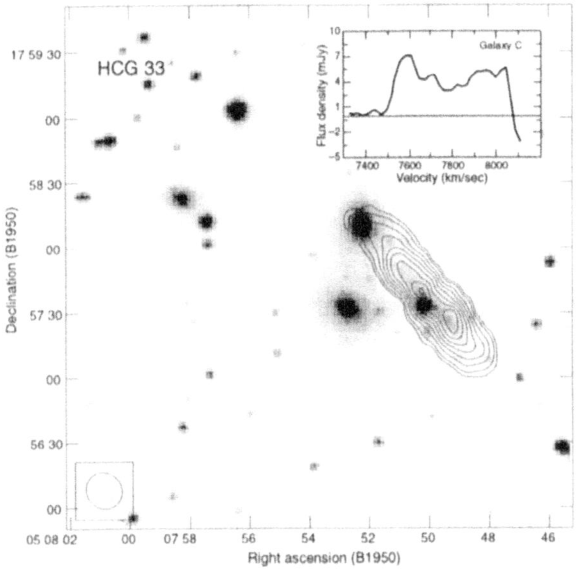

Figure 3. *a*) VLA integrated map of the HI emission in HCG 33. The lowest contour level is 0.05 Jy km s^{-1}. The peak emission is 0.83 Jy km s^{-1} and the beam size is $17.8'' \times 16.4''$. *b*) The global profiles reconstructed from the channel maps for galaxy *c*.

has a total hydrogen mass of $5 \times 10^9 M_\odot$, and the newly discovered dwarf member, southwest of the group, has a total hydrogen mass of $3 \times 10^8 M_\odot$.

The HI emission detected in the direction of HCG 33 is clearly identified with the only spiral member in the group, galaxy *c*. Measurements of the integrated flux with the C-array observations *(fig. 3a)* yield a value that is 40% of the integrated flux measured by the NRAO 300-ft telescope. Large amounts of extended emission are missing from the map shown in figure 3a. A lower limit of $6 \times 10^9 M_\odot$ can be set on the amount of HI in galaxy *c*. Given its luminosity, galaxy *c* is hydrogen-rich. The systematic motions observed in the velocity maps occur along the optical disk and are consistent with optical velocities measured by Rubin et al. (1991). As noted by Rubin et al. (1991), this low luminosity spiral has a peculiarly large internal velocity dispersion. The global profile shown in figure 3b has a velocity width (at 20% of peak emission) of 570 km s^{-1}. Given the luminosity of galaxy *c*, the predicted velocity dispersion is about half the measured value.

The HI emission *(fig. 4a)* is clearly associated with the individual galaxies in HCG 88. The total integrated flux (C-array) is consistent with the NRAO 300-ft measurement (Williams and Rood 1987). There are no obvious signs of tidal interaction in the HI distribution or the velocity field.

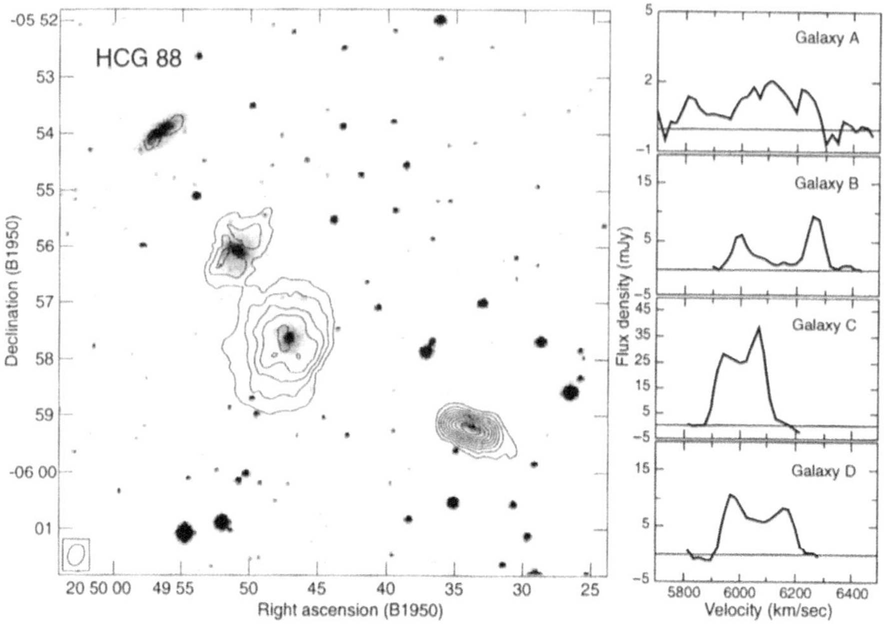

Figure 4. *a*) VLA integrated map of the HI emission in HCG 88. The lowest contour level is 0.074 Jy km s^{-1}. The peak emission is 0.74 Jy km s^{-1} and the beam size is 23.5″ × 17.3″. *b*) The global profiles reconstructed from the channel maps for galaxies *a*, *b*, *c*, and *d*.

Note that the HI disk associated with galaxy *a* is comparable in size to that of the optical disk. In the other galaxies, the HI disks are clearly much larger than the stellar ones. The integrated properties of galaxies *a* and *b* are worth noting. The hydrogen mass-to-light ratio of galaxy *a* and *b* is ~ 0.007 and ~ 0.05, respectively. For Sb-type galaxies, both are relatively deficient in neutral hydrogen. The measured velocity widths (20% peak) of their global profiles *(fig. 4b)* exceeds 560 km s^{-1} which are much broader than would be predicted from their luminosities. The amounts of hydrogen measured in galaxies *a*, *b*, *c*, and *d* is $3 \times 10^8 M_\odot$, $2 \times 10^9 M_\odot$, $9 \times 10^9 M_\odot$, and $4 \times 10^9 M_\odot$, respectively.

HI emission detected in the direction of HCG 92 (Stephan's Quintet), shown in *fig. 5a*, is displaced from any galaxies identified as members of the group. A similar distribution was observed by Shostak, Sullivan and Allen (1984). Where their observations detect emission between NGC 7320 and NGC 7319 and between NGC 7319 and NGC 7318A/B, ours do not. Most of the extended emission detected by Shostak et al. (1984) is missing from the C-array observations. The integrated flux detected with the VLA represents only $\sim$ 40% of the integrated flux measured by Shostak et al.

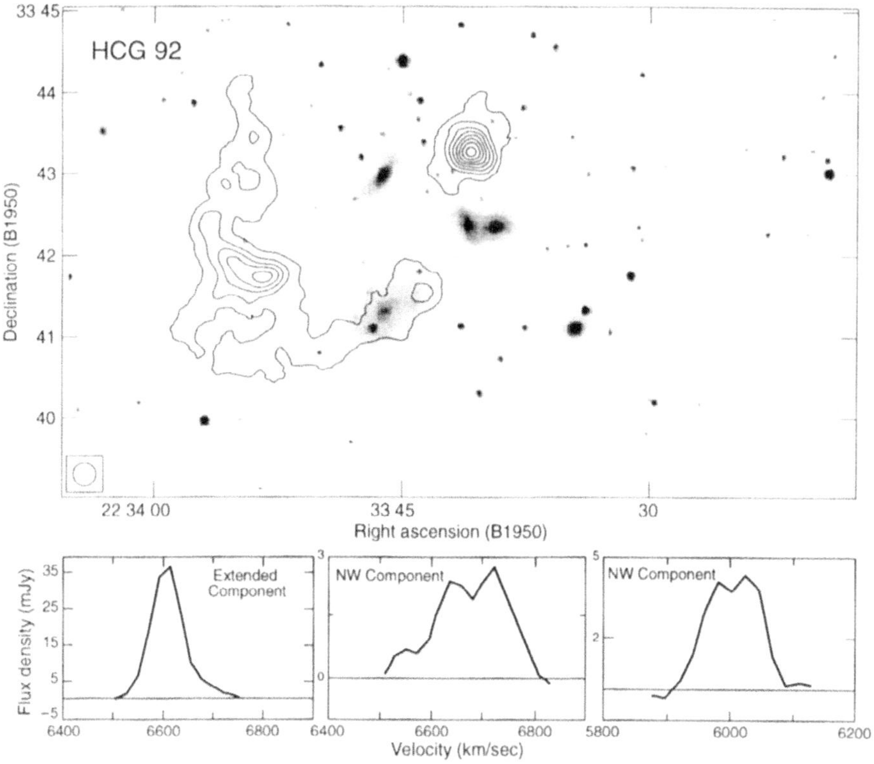

Figure 5. *a*) VLA integrated map of the HI emission in HCG 92. The lowest contour level is 0.030 Jy km s^{-1}. The peak emission is 0.48 Jy km s^{-1} and the beam size is 17.5$''$ × 16.8$''$. *b*) The global profiles reconstructed from the channel maps for the eastern cloud, and the northwest components.

(1984). HI is detected mainly in three cloud features, one near the end of the southeast tidal tail of NGC 7319 and two others north of NGC 7318A/B where Arp (1973) identified Hα regions at the same redshift as the group. The bulk of the HI is located in the curved eastern feature where the radial velocity of the gas is nearly constant *(fig. 5b)*. The compact emission north of NGC 7318A/B separates kinematically into two components, one at a systemic velocity of 6000 km s^{-1} and another at 6700 km s^{-1} *(fig. 5b)*. The peak emission in the northwest feature is 6 × 10^{20} H cm^{-2}, just at the threshold for OB star formation (Gallagher and Hunter 1984). There may be a physical connection between the compact HI emission and the underlying Hα objects identified by Arp (1973).

3. Summary

The HI morphology of HCG 2, 16, and 92 shows evidence that at least two or more galaxies in the groups are physically close. The galaxies are close enough for their gravitational fields to perturb the distribution and motions of the neutral hydrogen gas. In the case of HCG 92, it is the combination of the optical evidence of interactions both collisional and tidal, and the extraordinary HI distribution that present the most compelling case for physical compactness. If it is assumed that the HI detected in HCG 92 originally resided in the spiral members, then violent events must have caused the removal of large amounts of neutral hydrogen from these galaxies. HCG 2, 16, and 92 appear to be really compact systems. In contrast, the distribution and kinematics of HI in HCG 33 and 88 appear normal and show no signs of perturbations generated by the galaxies' tidal fields. As described by Hernquist, Katz, and Weinberg (1995), HCG 33 and 88 could be chance projections of unbound filaments of galaxies well separated along the line of sight.

If compact groups are in the process of merging, then we might expect to find them at different stages of the merging process. In the early stages, their properties should more closely resemble those of their progenitors, looser groups of galaxies. HCG 2 and 16 would be representative of an early stage in the merging process. The more evolved groups might appear more compact and more disordered in their optical and HI properties as a result of the dynamical interactions and major restructuring that is occurring within these systems, as observed in HCG 92. The imaged groups that show physical evidence of compactness have HI morphologies that are consistent with the model that describes compact groups as evolving objects.

References

Arp, H. 1973, Ap. J., **183**, 411.
Gallagher, J. S., and Hunter, D. A. 1984, A.R.A.A, **22**, 37.
Hernquist, L., Katz, N., and Weinberg, D. H. 1995, Ap. J., **442**, 57.
Hickson, P. 1982, Ap. J., **255**, 382.
Mamon, G. A. 1986, Ap. J., **307**, 426.
Mulchaey, J. S., David, D. S., Mushotzky, R. F., and Burstein, D. 1996, Ap. J., **456**, 80.
Ostriker, J. P., Lubin, L. M., and Hernsquist, L. 1995, Ap. J., **44**, L61.
Rose, J. A. 1977, Ap. J., **211**, 311.
Rubin, V. C., Hunter, D. A., Ford, W. K., Jr. 1991, Ap. J. Suppl., **76**, 153.
Shostak, G. S., Sullivan, W. T., and Allen, R. J. 1984, A.& A., **139**, 15.
Williams, B. A. 1985, Ap. J., **290**, 462.
Williams, B. A., and Rood, H. J. 1987, Ap. J. Suppl., **63**, 265.
Williams, B. A., and van Gorkom, J. H. 1988, A. J., **95**, 351.

RADIO DIAGNOSTICS OF GALAXY INTERACTIONS

T.K. MENON
Department of Physics and Astronomy
University of British Columbia
Vancouver, B.C. Canada. V6T 1Z4

The Hickson Compact Groups (Hickson 1982) are relatively isolated systems of galaxies with projected separations comparable to the diameters of the galaxies themselves and are an ideal laboratory for the study of the effects of interactions on the various properties of galaxies. Roughly one third of the galaxies in HCG show clear signs of interaction such as tidal distortion, truncation, and peculiar rotation curves. Galaxies that are not now interacting may have suffered past interactions or mergers. My study is based on a VLA survey of galaxies of different morphologies in 65 Compact Groups from Hickson's (1982, 1993) catalogue and only groups with minimum of 4 accordant velocities were included in the study. This final sample consisted of 298 galaxies (80E, 92S0, 126S) . The observations were carried out using the VLA in various configurations at wavelengths of 20cm and 6cm over a period of several years. The angular resolutions ranged from $20''$ to $0.3''$ and non-detection sensitivities ranged from 1 mJy to 0.3 mJy for different observations. A total of 86 galaxies were detected as radio sources (32ES0, 54S).

It is found (Menon 1992) that radio-loud ES0 galaxies in the HCG are predominantly the optically brightest galaxies in a group while the detected spirals are more or less uniformly distributed among the top three brightest members of a group. For the vast majority of radio-loud HCG galaxies the radio radiation is mainly from the central regions of those galaxies. I have shown (Menon 1995a) that the median radio luminosity of an isolated spiral (ISPL) sample is about 5 times greater than that of the HCG sample while the median value of the radio to optical luminosity ratio R is about 3 times greater for the ISPL sample. Hence it would appear that statistically the HCG galaxies have lower radio luminosity for a given optical luminosity. For both the ISPL and HCG samples the late type spirals have higher values of R than the earlier types. Quantitatively the median values of R are higher by factors of 2.4 and 5.5 respectively. For the early morphological types the difference between the samples is only marginal while for the later types the difference is particularly significant. It should be emphasized that the above comparison is for the total radiation from the two samples. The radio radiation from the HCG sample is in most cases from the nuclear regions only while for the ISPL sample the contribution is mostly from the disks of the galaxies. Hence if we consider the disc emission alone the HCG sample is significantly deficient compared to the ISPL sample. The median value of the ratio R_c of the nuclear radio luminosity to optical luminosity of the HCG spiral sample is about 12 times higher than for the isolated sample. Quantitatively the median values of R_c are about a factor of 3 higher for the earlier spiral types and a factor of 14 higher for the later spiral types. The structure and the spectral indices of the radio emission from these central regions suggest that the central regions of most HCG spirals may be undergoing bursts of star formation.

The results of numerical simulations suggest than during interactions there is both an inflow of gas into the center from the inner regions as well as an outflow of gas from the outer regions. The increase in the central gas density due to the inflows can lead to formation of molecular clouds, increase the frequency of cloud collisions leading to subsequent starburst activity. Such an enhancement of CO emission has been observed in interacting galaxies by Braine & Combes (1993) and Combes et al. (1994). Recently Leon et al. (1997) have found significant CO emission in HCG spirals similar to that in interacting spirals. The starbursts can then lead to the increase in supernovae resulting in the enhancement of the radio radiation. The greater degree of enhancement of the central component in the case of the later type spirals can be understood on the basis of the fact that the later type spirals have larger amount of gas available in their central regions. For a mean luminosity of 4×10^{20}W/Hz at 1.4GHz the star formation rate turns out to be about $0.1 M_{\odot}$ per year and a supernova rate of 0.02 per year. Even though these rates are not very large for a whole galaxy it may be recalled that in the case of HCG spirals the rate refers only to the nuclear regions. On the other hand the outflow of gas from the outer parts can be expected to lower the magnetic field intensities there due to the strong coupling between the gas and the field. The reported deficiency of neutral hydrogen in a number of HCG spirals by Williams and Rood (1987) may be attributed to a gas outflow of the type implied by the above results. This in turn may be the cause of the decrease in the total nonthermal radio radiation from these interacting systems.

A useful index of the strength of interaction is found to be a combination of the tidal force between the galaxies and the duration of the encounter. The tidal force itself is proportional to $M_p D_p^{-3}$ where M_p is the mass of the perturbing galaxy and D_p is its perigalactic distance. The duration of the encounter can be estimated from the perigalactic distance and the velocity difference between the two galaxies. Parameters of this type have been used by Dahari (1984), Byrd et al. (1986,1987) and Elmegreen et al. (1991) in their study of the effect of interactions among galaxies. The binned values of the sum of the tidal parameters due to all the other members of the group on each galaxy are plotted against $\log R$ for ES0 and S samples in Figure 1. It appears that in the case of ES0 galaxies there is a strong dependence of the radio emission on the total tidal force while in the case of spirals there is no such dependence.

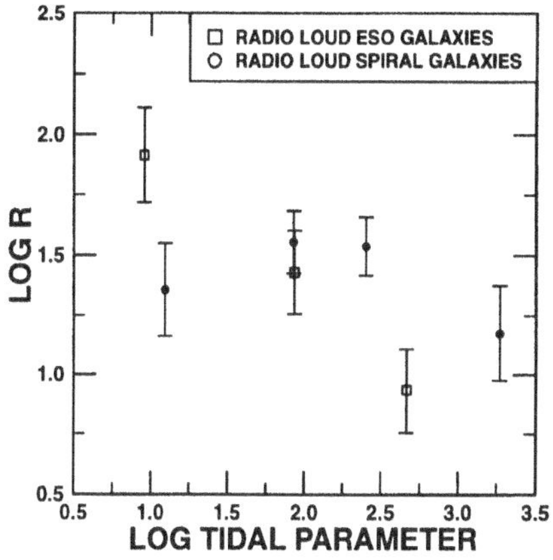

Figure 1. Dependence of $\log R$ on Tidal Parameter

The velocity dispersion of ES0–dominant groups is significantly higher than that of S–dominant groups and this can influence the duration or time scale of the interactions among the galaxies. It is possible that in the case of ES0 galaxies the duration is short enough that recognizable activity manifests itself only when the perturbing galaxy is very close to them producing the close correlation between the separation and activity. On the other hand in the case of S galaxies the lower velocity dispersion as well as the geometry of disc-halo systems extend both the range and duration over which the tidal forces are effective in initiating the sequence of processes starting with disc instabilities and ending up with inflows into the nucleus and star formation.

Even though the main consequence of interactions among galaxies appears to be the triggering of nuclear activity, numerical simulations suggest that in the case of spiral galaxies specific orbital orientations during the encounters can lead to major starburst activity in the spiral arms themselves. The radio maps of the HCG spiral 47a (Menon 1995b) suggest that in the case of this spiral the starburst activity in the spiral arms has been initiated by the tidal action of the companion elliptical galaxy 47b. Since the non-thermal luminosity is most likely produced by relativistic electrons originating in supernova remnants we can calculate the star formation rate to be SFR($M \geq 5M_{\odot}$) $\approx 3.7M_{\odot}\mathrm{yr}^{-1}$. This may be compared to the SFR($M \geq 5M_{\odot}$) $\approx 2.2M_{\odot}\mathrm{yr}^{-1}$ estimated for the whole galaxy M82 by Condon (1992). Since in the present case of HCG47a the volume of the spiral arms producing the radio luminosity is small compared to the galaxy the above rate is indeed very large.

In summary the radio properties of galaxies of different morphological types in high density regions suggest that tidal interaction is most likely the dominant physical process in the initiation of nuclear activity in these systems. In the case of ES0 systems the radio sources generated are generally of the low luminosity FR I type and in the case of spirals nuclear starbursts appear to be the main outcome of the interactions. The absence of any FRII type radio sources in any compact galaxy groups suggests that galaxy interactions alone may not be sufficient to explain the origin of such sources at earlier epochs.

This investigation was supported by a grant from the Natural Sciences and Engineering Research Council of Canada. National Radio Astronomy Observatory is operated by Associated Universities, Inc. under cooperative agreement with the National Science Foundation.

References

Braine J., & Combes F., 1993, A & A, 269, 7
Byrd, G.G., Sundelius, B.& Valtonen, M.J., 1987. A & A., 171,16
Byrd, G.G., Valtonen, M.J., Sundelius, B. & Valtaoja, L.,1986. A & A., 166,75
Combes F., Prugniel P., Rampazze R., & Sulentic J., 1994, A & A, 281, 725
Condon, J.J. 1992, Annual Review of Astronomy and Astrophysics, 30, 575
Dahari, O., 1984. AJ., 89,966
Dahari, O., 1985. ApJS., 57, 643
Elmegreen, D.M., Sundin, M., Elmegreen, B., Sundelius, B. 1991, A & A, 244, 52
Hickson P., 1982, ApJ., 255, 382
Hickson P., 1993, Ap Lett & Comm, 29, 1
Menon, T.K., 1991. ApJ., 372,419
Menon T.K., 1992, MNRAS, 255, 41
Menon, T.K., 1995a. MNRAS, 219, 305
Menon, T.K., 1995b. AJ, 110,2605
Williams B.A., & Rood H.J., 1987, ApJS, 63, 265

ENVIRONMENTAL EXTREMISTS IN THE VIRGO CLUSTER

JEFFREY KENNEY AND REBECCA KOOPMANN
Yale University, Astronomy Department,
P.O. Box 208101, New Haven, CT 06520-8101 USA

1. Introduction

Many types of galaxy interactions have been posited to occur in clusters, although it remains unclear which processes actually occur, and which ones might help explain the tendency for early type galaxies to inhabit high density environments, or cause the rapid evolution of cluster galaxies (e.g., Dressler et al. 1997). With these questions in mind, we have been conducting an environmental inventory of galaxies in the Virgo Cluster. Our approach is to combine surveys of spirals and S0s with detailed studies of the most interesting and peculiar galaxies. In this paper, we describe two main points. 1.) There is a population of spiral galaxies in the Virgo cluster with the small central light concentrations (bulge-to-disk ratios, or B/D's) characteristic of isolated Sb and Sc galaxies, but global star formation rates lower than those of isolated spirals of any Hubble class (Sa-Sc). These Virgo galaxies are generally classified as "early type" (e.g. Sa), and thus contribute to the morphology-density relationship. 2.) There are *several types* of environmental interactions occurring in Virgo, including low velocity tidal interactions and mergers, high velocity tidal interactions and collisions, HI accretion, and ICM-ISM stripping. We discuss examples of some of these interactions.

2. The Comparative Study of Cluster and Isolated Galaxies

It is important to compare Virgo and isolated galaxies in a manner independent of subjective galaxy classification. From a data set of R and Hα images of nearly 100 Virgo cluster and isolated spiral galaxies, we have measured objective parameters to test how well the Hubble system succeeds in distinguishing between galaxies with different physical characteristics in

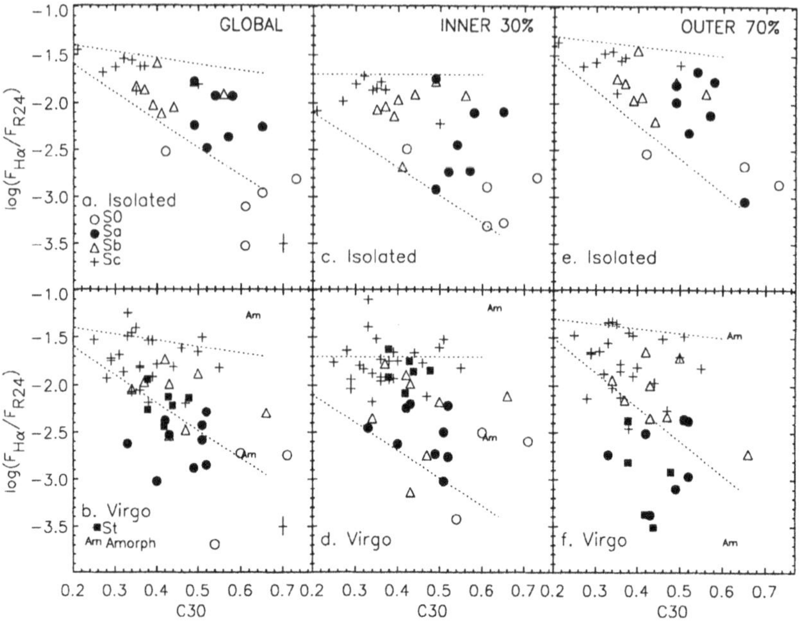

Figure 1. Hα to R flux ratio vs. central R light concentration for isolated (top) and Virgo (bottom) galaxies. Vertical axis shows flux ratio for (left) entire galaxy, (middle) central 30% of optical disk, and (right) outer 70% of optical disk. There are good correlations for isolated Sa-Sc's, but poor correlations for Virgo Sa-Sc's. Lines bound the regions in which isolated Sa-Sc's lie. Peculiar Virgo spirals are labeled St. Most isolated Sa's tend to be high concentration systems, while most Virgo Sa's are low concentration systems with low star formation rates. Star formation is normal or enhanced in the centers of Virgo galaxies, and strongly depleted in the outer disks (Koopmann & Kenney 1998a,b).

the two environments (Koopmann & Kenney 1998a,b). These results are important for understanding the morphology-density relationship, and also demonstrate that Hubble classification is meaningful for isolated spirals, but not for most cluster spirals.

We measure parameters which trace two of the criteria used for Hubble classification: the central light concentration in R (C30), a tracer of the B/D, and the Hα to R flux ratio ($\frac{F_{\rm H\alpha}}{F_{\rm R24}}$), a tracer of the star formation rate and the "knottiness" of spiral arms. These parameters are well correlated for isolated galaxies, as shown in Fig. 1a. Assigned Hubble types, indicated by different symbols, correlate well with both parameters. Note especially that Sa and Sc galaxies are distinguished by both their central light concentrations and their star formation rates in a manner consistent with their Hubble classifications. The situation is very different for the Virgo galaxies (Fig. 1b). C30 and $\frac{F_{\rm H\alpha}}{F_{\rm R24}}$ are virtually uncorrelated, and there is a greater

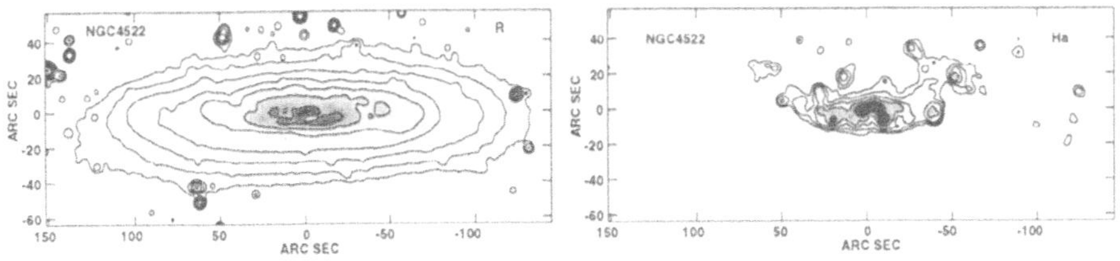

Figure 2. (left) R image of NGC 4522 shows fairly regular isophotes at large radii, indicating that the stellar disk is undisturbed. (right) Hα image shows truncated star-forming disk, and extraplanar HII regions. This combination suggests that NGC 4522 is presently experiencing a strong ICM-ISM interaction. From Kenney & Koopman (1998).

range of star formation rates at every central light concentration, compared to the isolated sample. *While the Sa and Sc galaxies are still well separated, they are distinguished principally by their star formation rates rather than their central light concentrations.* Many of the Virgo Sa galaxies are actually systems with the low central light concentrations characteristic of isolated Sb-Sc galaxies. Fig. 1c-f show that it is primarily the outer disks of Virgo galaxies which have low star formation rates. Most Virgo galaxies are truncated rather than globally anemic. This strongly suggests that environmental processes remove gas from the outer disks of cluster spirals of all B/Ds, and those that lose gas become classified as early type spirals due to the resulting decrease in their star formation rate. This has implications for the morphology-density relationship, as well as other comparisons of cluster and field spirals. *Part of the excess of early type galaxies in nearby clusters is due to small B/D systems with reduced SFRs. It is not all due to a systematic change in B/D with environmental density.*

3. Examples of Different Environmental Processes

3.1. NGC 4522 AND ICM-ISM STRIPPING

The stripping of gas from the interstellar medium (ISM) of galaxies due to interactions with the gas in the intracluster medium (ICM) may be one of the most important types of interactions which occur in clusters of galaxies. Although many galaxies have properties consistent with having been stripped, there have been no clear examples of gas actively being stripped from the disks of spirals. The highly inclined Virgo cluster spiral galaxy NGC 4522 may be such a case. Our R-band and Hα images from the WIYN telescope, shown in Figure 2, reveal a relatively undisturbed stellar disk and a bow-shock-shaped Hα morphology, which strongly suggest that the ISM of NGC 4522 is being stripped by the gas pressure of the ICM. The presence of HII regions apparently located above the disk plane suggests

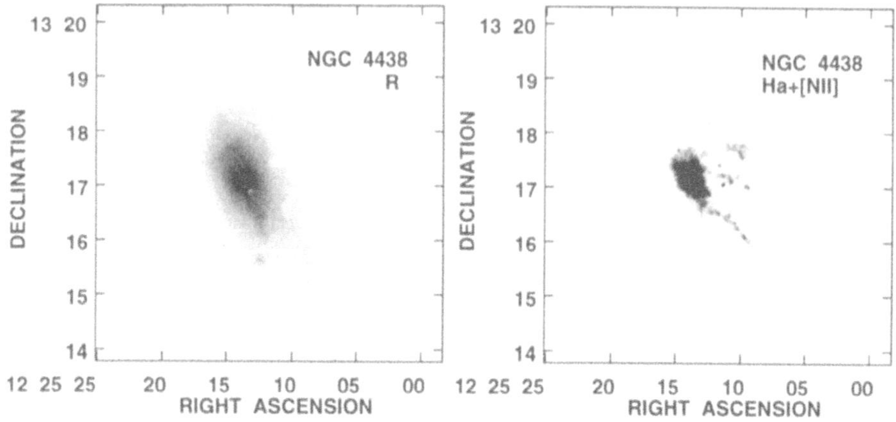

Figure 3. R (left) and Hα+[NII] (right) maps of NGC 4438, showing disturbed stellar disk and remarkable ionized gas filaments. The morphology and kinematics of this gas suggest that the gas is cooling and falling into the main body of NGC 4438. We propose that this type of disturbed ISM results from the aftermath of a high-velocity ISM-ISM collision. From Kenney et al. (1995).

that star formation is occurring in the stripped gas, and that newly formed stars will enter the galaxy halo and/or intracluster space.

3.2. NGC 4438 AND HIGH-VELOCITY COLLISIONS

Close interactions which occur at velocities too high to result in merging also disturb a galaxy's structure. Broadband images of NGC 4438 (Fig 3) show a highly disturbed edge-on stellar disk, with stellar debris displaced to the west of the galaxy's main disk. Simulations are able to produce this type of disturbed stellar morphology with a high velocity ($\sim$1000 km s^{-1}) interaction between galaxies of similar mass. Combes et al. (1988) suggest a collision with a closest approach of $\sim$5 kpc with the nearby galaxy NGC 4435, although Moore et al. (1996) suggest a more distant encounter at $\sim$30 kpc with an unidentified galaxy. Our multiwavelength study of NGC 4438 shows that the ISM in this galaxy is also severely disturbed (Kenney et al. 1995). The Hα+[NII] image (Fig. 3) reveals several remarkable ionized gas filaments which extend $\sim$5-10 kpc from a gas-rich nuclear disk to a second gas-rich region with strong [NII], x-ray, radio continuum, CO and HI emission. Spectroscopy shows that these filaments have low velocities and line ratios similar to those in "cooling flow" galaxies, which suggests that the filament gas is falling back into the disk and cooling. The ISM in NGC 4438 is much more heavily disturbed than in NGC 4522, and we propose that this results from a *high velocity ISM-ISM collision* between NGC 4438 and

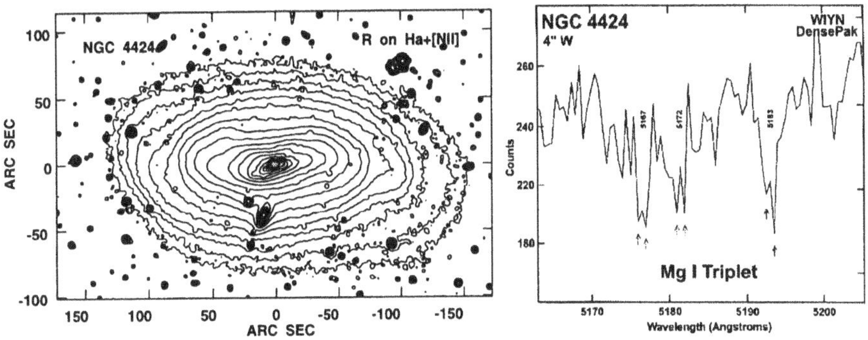

Figure 4. (left) R contour and Hα+[N II] grayscale plot of NGC 4424, a likely intermediate-mass-ratio merger remnant. Note heart-shaped contours in R and confinement of Hα emission to the central 10″. From Kenney et al. (1996). (right) Portion of optical spectrum 4″ W of nucleus in NGC 4424, obtained with the WIYN Densepak fiber array spectrograph. Note that each component of the Mg triplet is doubled, indicating cospatial, counterrotating stellar components. From Winnick & Kenney (in prep).

another galaxy. Therefore closest approach was probably less than 30 kpc, although the lack of any obvious ring-like structure in NGC 4438 suggests that closet approach may have been greater than 5 kpc.

3.3. NGC 4424 AND MERGERS

Mergers do not occur at the high velocities which characterize many of the interactions that occur in clusters. Yet there may be enough low velocity interactions in clusters to make merging an important process for transforming spirals into S0s. Our studies of the peculiar (Sa pec) Virgo cluster galaxy NGC 4424 indicate that this system is the product of a merger (Kenney et al. 1996). The R image reveals banana-shaped isophotes, shell-like features and other complex structure generally associated with mergers (Fig. 4a). The only Hα emission arises from a few bright HII complexes distributed along a bar-like feature located within 500 pc of the nucleus, inside the bulge-dominated region. Spectra reveal double-peaked stellar absorption lines (Fig. 4b), indicating the presence of cospatial, counterrotating stars in the circumnuclear region. These peculiar properties strongly suggest a recent, intermediate mass ratio (0.1-0.5) merger, and we propose that the galaxy will become a small-bulge S0 within $\sim$1 Gyr.

References

Combes, F., Dupraz, C., Casoli, F., & Pagani, L. (1988), AA, 203, L9.

Dressler, A., Oemler, A., Couch, W. J., Smail, I., Ellis, R. S., Barger, A., Butcher, H., Poggianti, B. M., & Sharples, R. M. 1997, ApJ, in press.

Kenney, J. D. P, Rubin, V. C., Planesas, P., & Young, J. S. (1995), ApJ, 438, 135.
Kenney, J.D.P. & Koopmann, R.A. (1998), ApJL, in prep.
Kenney, J.D.P., Koopmann, R. A., Rubin, V., & Young, J. S. (1996), AJ, 111, 152.
Koopmann, R.A. & Kenney, J.D.P. (1998a), ApJL, 000, submitted.
Koopmann, R.A. & Kenney, J.D.P. (1998b), ApJ, 000, in prep.
Moore, B., Katz, N., Lake, G., Dressler, A., & Oemler, A., Jr. (1996), Nature, 379, 613.

GALAXY HARASSMENT—INTERACTIONS FOR THE 90s

GEORGE LAKE
Department of Astronomy, University of Washington
Box 351580, Seattle WA 98195, USA

AND

BEN MOORE
Department of Physics, University of Durham
Durham City, DH1 3LE, UK

1. Introduction

The origin of the Hubble sequence remains a long-standing puzzle in astronomy. Giant galaxies range from slowly-rotating dense ellipticals to thin late-type spiral disks. At the faint end, there are two distinct classes of "ellipticals/spheroids" that are easily separated in plots of nearly any two of their properties, such as central surface brightness versus luminosity (Ferguson and Binggeli 1994; Kormendy 1985). The elliptical class includes the bright giants and extends to the rare high surface brightness "dwarf ellipticals", M32 being the prototype. The "spheroidal" galaxies have low surface brightnesses and are all $\gtrsim$3 magnitudes fainter than L_*, the characteristic break in the luminosity function. The dwarf spheroidal galaxies (dSph) in our Local Group of galaxies with magnitudes in the range $-8 \gtrsim M_B \gtrsim -12$ are often considered to be the low luminosity extreme of this sequence, but nearly all other known galaxies in this class reside in clusters.

There is no shortage of theories for galaxy formation and the origin of the Hubble sequence. Several speakers at this conference have described formation via merging (Toomre 1977) and "chaotic collapse" (Lake and Carlberg 1988). Spheroidal formation theories have a shorter history and have focused on incremental changes to the giant galaxy theory. There are many problems with the scheme that combines the notion of lower amplitude peaks in the hierarchical model with the use of stellar winds or supernovae to expel gas from small galaxies (Dekel and Silk 1986, Vader 1986). The clustering properties of dwarfs is opposite to the expectations of the Dekel and Silk model (Ferguson and Binggeli 1994). The model also has the seemingly impossible chore of explaining the general properties of both rapidly-rotating gas-rich dwarfs and gas free dwarf spheroidals.

Hubble Space Telescope (*HST*) observations reveal that the morphologies of galaxies in clusters changed dramatically since $z \sim 0.4$. Over 20 years ago, Butcher and Oemler (1978, 1984) discovered a large population of "blue galaxies" in clusters at $z \sim 0.4$. Giant ellipticals are already in place at $z \sim 0.4$, but the ubiquitous "blue galaxies" are distorted spirals that have vanished from present-day clusters (Dressler et al. 1994a). The population difference is greatest for galaxies fainter than $L_*/5$: 90% are bulgeless "Sd" disk systems in distant clusters, whereas 90% are spheroidals in nearby clusters (Sandage et al. 1985). Couch et al. (1994) present spectroscopic evidence that the distorted blue galaxies at $z \sim 0.3$ have undergone multiple burst events separated by 1–2 Gyr. In hierarchical clustering models, the influx of field galaxies into clusters peaks at $z \sim 0.4$ (Kauffmann 1995). So, we need to transform these galaxies when they enter clusters.

At speeds of several thousand kilometers per second, close encounters with bright galaxies cause impulsive gravitational shocks that can severely damage the fragile disks of Sc–Sd galaxies. Our earlier analytical work revealed that these collisions are frequent enough that disk galaxies would be harassed throughout a cluster (Moore et al. 1996a). Moore et al. (1996b) used numerical simulations to compare harassed galaxies to *HST* frames of galaxies in clusters at $z \gtrsim 0.3$. They stated that the cumulative effect of such encounters changes a disk galaxy into a spheroidal galaxy, thus identifying the present-day remnants of the disturbed blue galaxies and explaining the change in galaxy morphologies in clusters since $z \sim 0.4$. Moore et al. (1998) provide detailed comparisons of the harassed remnants with the photometric and kinematical properties of dwarf spheroidal galaxies. Lake et al. (1998) consider the feeding of quasars by galaxy harassment. We will review this work adding a few recent results.

2. Modeling Galaxy Harassment

We take a "minimalist" approach in our simulations. It is difficult to imagine how any galaxy could avoid the effects that we simulate. Our cluster models are based on properties of the Coma cluster with galaxies drawn from a Schechter luminosity function, assigned dispersions based on the Faber-Jackson relation and then tidally limited based on the pericenter of their cluster orbits. Galaxy harassment is slightly more effective at removing mass than tides alone. Our "victims" lose as much as half of their mass over a period of 3–5 Gyr. We reduced the initial galaxy masses by the time average of 25%. We expected that this was overly conservative as the largest galaxies do the harassing and are rather immune to it themselves.

Recently, we simulated the evolution of the dark matter in clusters of galaxies. The final state of the simulation has over a thousand identifiable "galactic halos" with masses that demonstrate that our assumptions about the masses of galactic halos within clusters were indeed conservative (Moore et al. 1998). We were also conservative in our choice of orbits for the harassment victims. At a fixed mean orbital radius, galaxies on elongated orbits experience greater harassment. We follow galaxies that have apo/peri ratios of 2 (*e.g.* apocenter at 600 kpc, pericenter at 300 kpc), whereas the typical value in a cluster with isotropic dispersions is ~ 6. As a result, our model galaxies avoid extremes of the cluster distribution and start with large dark halo masses determined by the tidal limit at their atypically large pericenters. The full details of the simulations can be found in Moore, Lake and Katz (1998).

If we assume that spiral disks follow the Tully-Fisher relationship ($L \propto v_{\rm circ}^4$) and are experiencing impulsive fly-by collisions from other galaxies that are tidally limited within a larger virialized system, we find a remarkable result. *The timescale to shake a disk into a spheroidal system is independent of the mass of the larger virialized system and independent of the orbital radius of the spiral disk within that virialized system.* Realistic conditions limit the validity of such universal statements. Galaxies with larger circular velocities are earlier type systems. High density bulges are effective at protecting disks from damage owing to encounters. Low surface brightness galaxies are more easily harassed. Even if they follow the Tully-Fisher relationship, the slower inner rise of their rotation curves increases the response to impulsive shocks. The impact parameters become too large at the edges of rich clusters for the collisions to be impulsive. Similarly, the velocities can be too slow in smaller groups, leading to merging rather than harassment. However, galaxy harassment is not just for rich clusters. It will occur anytime that galaxies are moving past one another at speeds that are much larger than their circular velocities. The three-dimensional dispersion velocity of a group with a total luminosity of just 10 L_* is ~ 700 km s^{-1}. Harassment will certainly occur in such an environment.

3. The Harassment Drama

The evolution proceeds in a violent, chaotic fashion that is best appreciated by watching the published video (Moore, Lake and Katz 1998). Typically, the first encounters create "disturbed barred spirals" with sharp and dramatic features drawn out from the dynamically cold disk. Tails of material can be pulled out and distorted by the tidal field of the cluster (Figure 1). The gas distribution often forms ring structures that tumble within the

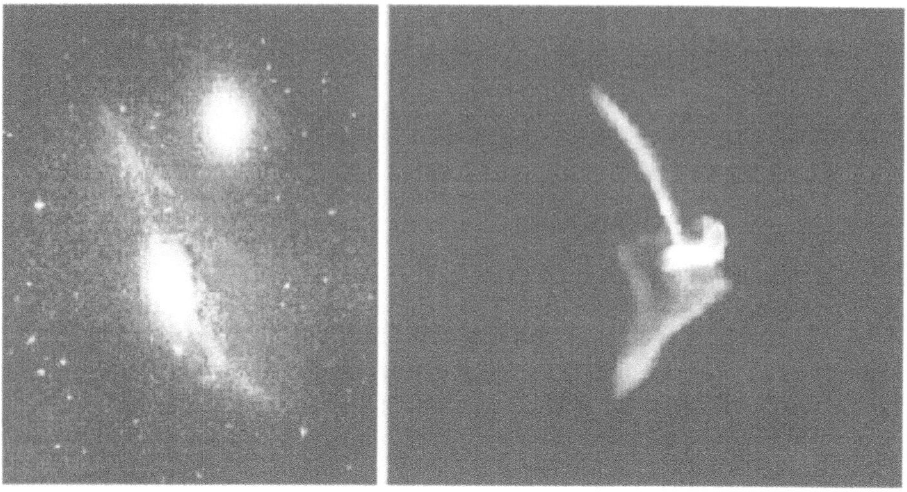

Figure 1. NGC 4438 is a Virgo cluster galaxy (left) with strong tidal tails. In our simulations, one of the first strong collisions often makes features such as those seen on the right. Combes (et al. 1988) constructed a model where the distortions owe to the optical companion. In their model, the true separation of the two galaxies is 100 kpc. There are many other galaxies that are closer and more massive, making them better candidates for the disturbance. Harassment is not too gentle to explain NGC 4438, as asserted by J. Kenney at this conference.

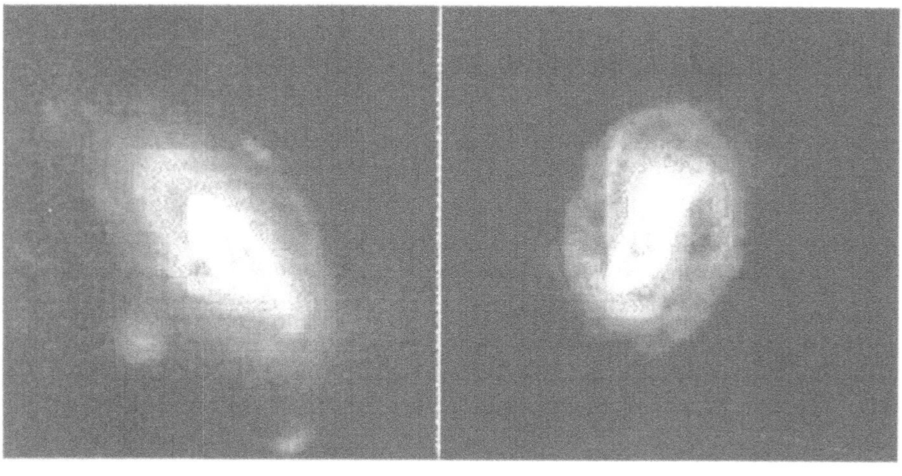

Figure 2. The left image is a spiral galaxy with a prominent ring in the distant rich cluster CL0939. The ring structure on the right is common in our simulations.

stellar bar (Figure 2).

The evolution is driven by just a few close encounters. These drive the multiple starbursts inferred from *HST* data (Barger et al. 1996). Another observational puzzle has been the ubiquity of disturbed galaxies with no sign of current interaction (Dressler et al. 1994b). Over the course of 3 Gyr, the closest approach of another galaxy is normally greater than 30 kpc. Since the relative velocity of strong encounters is ~ 1500 km s^{-1}, and the velocity impulse internal to the galaxy is $\lesssim 50$ km s^{-1}, the perturbing galaxy moves $\sim$ 100 kpc by the time the disk's response is noticeable. The galaxy delivering the shock is an L_* or brighter elliptical and barely noticed that it happened, eliminating the concern that one must simulate the internal response of the harasser (Joseph 1996).

4. The Spheroidal Remnants

After several strong encounters, angular momentum loss combined with impulsive heating, leads to a prolate figure supported equally by random motions and rotation. The gas sinks to the very center of the galaxy and the stellar distribution is heated to the extent that it closely resembles a dwarf elliptical, although some remnants retain very thick stellar disks and would be classed as dwarf lenticulars. At this stage in the evolution, encounters cease to create sharp distortions and fail to remove any more material from the compact remnant.

Moore, Lake and Katz (1998) make extensive comparisons of the harassed remnants to spheroidal galaxies in nearby clusters (Ferguson and Binggeli 1994; Kormendy 1985). We found good agreement with the luminosity function, surface brightness profiles, flattening, internal kinematics, mass-to-light ratios, stellar populations and clustering properties. This can't be much of a surprise. Disks exist in clusters at $z \sim 0.3$ and are mostly gone in present-day clusters, replaced at the faint end of the luminosity function by spheroids. We've shown that gravitational interactions with large galaxies drives such a transformation. The fact that the properties match suggests that other physical processes like ram pressure are unlikely to be important.

One might hope that radial gradients of the spheroidal populations would provide interesting tests of the model. However, the most important effect is independent of how the spheroidals formed: global tides coerce the lowest density (or surface brightness) objects into the diffuse stellar background (Ciardullo et al. 1997). Most radial correlations are projections of the fundamental correlation between density and survivability:

- only the densest spheroidals survive in the inner parts of the clusters, creating a paucity of faint spheroids there (Bernstein et al. 1995)
- correlations between density and color/metallicity create color gradients in the surviving ensemble (Secker 1996)
- the fraction of the more robust nucleated spheroidals increases towards the center (Binggeli et al. 1987)

- selective destruction of non-nucleated spheroidals with small pericentric radii can lead to a central deficiency in radial orbits causing a dip in the cluster's line-of-site velocity dispersion
- spiral disks seen in the central regions of clusters owe to projection, their velocity fields won't show virialization (Tonry, Ajhar and Luppino 1990; Bernstein et al. 1994)
- in the outer parts of the clusters, spirals on radial orbits are transformed faster than those on nearly circular orbits (Dressler 1986)

5. Feeding Quasars

When we first simulated a harassed galaxy with gas, we were aghast to see up to 90% of the gas was driven into the inner 500 pc in a few Gyr. Up to half of that mass can be transferred in a burst lasting just 100-200 Myr. This transport of gas to the center of a galaxy is far more efficient than any mechanism proposed before.

There are two observations that suggest that harassment could be important for feeding quasars at intermediate redshifts $0.2 \lesssim z \lesssim 0.8$ (Lake, Moore and Katz 1998). Quasars at intermediate redshifts are in Abell richness class 0-1 clusters of galaxies—an environment that is considerably richer than that of lower redshift quasars (Yates, Miller and Peacock [1989] find the break occurs at $z \sim 0.3$, while Yee and Ellingson [1993] state that it occurs at $z \sim 0.6$). There is evidence that many quasar hosts are less luminous than L_* at $z \sim 0.3$ (Bahcall, Kirhakos and Schneider 1995).

After observations of additional quasars, Bahcall et al. (1997) conclude that, "the luminous quasars studied in this paper occur preferentially in luminous galaxies". They reject the "null hypothesis" that all galaxies are equally likely to have quasars (e.g. a hypothesis that states that Draco and M87 are equally likely to host quasars). Their conclusion results because at least half of all galaxies are $\gtrsim 2$ magnitudes fainter than L_* whereas the dividing line for their sample of quasar hosts is $\sim L_*$ within their errors. Popular luminosity functions diverge at the faint end ($N \propto L^{-x}$, $1.5 > x > 1$), requiring a cutoff to define an "average luminosity" that is always 2-3 magnitudes brighter than the cutoff or $\gtrsim 2$ magnitudes fainter than L_*.

However, galaxies brighter than $\sim 0.75 L_*$ contain half of all the luminosity. This dividing line of luminosity is consistent with the Bahcall et al. midpoint of quasar hosts within their errors. The simplest summary of the observations to date is that quasars and galaxies may be related in the same way as stars and galaxies: the probability of finding either in a galaxy is proportional to the galaxy's luminosity but their individual luminosities are not determined by the luminosity of their host. *We need a mechanism at $z \sim 0.3$ that triggers quasars with a frequency that is roughly* *proportional to galaxy luminosity and prefers clusters. A mechanism that only operates in bright galaxies in the field such as mergers can not be the dominant trigger at* $z \sim 0.3$.

If galaxy harassment triggers quasars, we make four clear predictions:

1. QSO hosts will be found in systems where harassment occurs;
2. AGN frequencies are enhanced in clusters undergoing harassment;
3. resolved hosts should appear disturbed;
4. black holes should exist in some nucleated spheroidal galaxies.

Detailed discussions of these points can be found in Lake et al. (1998). Most if not all of the quasars with sub-L_* hosts are in high density environments. The *HST* images of the host candidates show tantalizing evidence of distortions (Bahcall, Kirhakos and Schneider 1995). There is an ongoing controversy with respect to the frequency of AGNs in Butcher-Oemler clusters, but we note that quasars at intermediate redshifts could not lie in clusters rich enough to be classified by Abell if nuclear activity were not enhanced in clusters. The final prediction suggests black hole hunting should be undertaken in some new places. In our original paper, we pointed to NGC 4486B as an interesting place to look, though one might argue whether it is an appropriate galaxy to consider in the context of harassment. Since then, Kormendy et al. (1997) have detected a substantial black hole in this galaxy.

To summarize, disk galaxies are seen in clusters at $z \sim 0.3$. We simulate the gravitational shocks that these galaxies feel when other galaxies in the clusters pass by them. The only thing that we need to know about the other galaxies are their masses. We adopted conservative values for these masses and the orbital distributions of the "victims". *We see absolutely no way that galaxies in clusters can avoid the gravitational interactions that we call harassment.* These interactions produce the distorted galaxies seen with *HST*. The collision frequency matches the interval between starburst events (Barger et al. 1995). The galaxies are transformed into spheroidal systems like those observed in clusters today. Quasar feeding depends on the flow of gas into the center; this could easily be stopped by star formation. As for the rest of our harassment results, the greatest uncertainty that remains in the model is the strength of tides in the very center of clusters of galaxies (*cf.* Moore et al. 1998). For this reason, we avoid simulating galaxies with orbits that have pericenters less than 150 kpc.

Galaxies are metamorphosed by their mutual interactions. "Merging" of spirals in groups creates bright ellipticals. In a cluster, one of these "cannablizes" its neighbors to become the giant central elliptical. The spheroidal galaxies are created by the harassment of low luminosity spirals. Our work to date has only touched on some of the most dramatic changes, the aetiology of harassment promises to be even richer than that of merging and cannibalism.

References

Bahcall, J. N., Kirhakos, S., Saxe, D. H. and Schneider, D. P. 1995, *Ap. J.*, **479**, 642.
Bahcall, J. N., Kirhakos, S. and Schneider, D. P. 1995, *Ap. J.*, **450**, 486.
Barger, A. J., Aragon-Salamanca, A., Ellis, R. S., Couch, W. J., Smail, I. and Sharples, R. M. 1996, *M.N.R.A.S.*, **279**, 1.
Bernstein, G. M., Nichol R. C., Tyson J. A., Ulmer M. P. & Wittman D. 1995, *A.J.*, **110**, 1507.
Binggeli B., Tammann, G. A. and Sandage, A. 1987, *A.J.*, **94**, 251.
Butcher H. and Oemler A. 1978, *Ap.J.*, **219**, 18.
Butcher H. and Oemler A. 1984, *Ap.J.*, **285**, 426.
Ciardullo, R., Jacoby, G., Feldmeier, J. and Bartlett, R. 1997, *Ap.J.*, in press.
Combes F., Dupraz C., Casoli F. and Pagani L. 1988, *Astr.Ap.*, **203**, L9.
Couch, W. J., Ellis R. S., Sharples R. and Smail I. 1994, *Ap.J.*, **430**, 121.
Dekel, A. and Silk, J. 1986, *Ap.J.*, **303**, 39.
Dressler, A. 1986, *Ap. J.*, **301**, 35.
Dressler, A., Oemler A., Butcher H. and Gunn J.E. 1994a, *Ap.J.*, **430**, 107.
Dressler, A., Oemler A., Sparks W.B. and Lucas R.A. 1994b, *Ap.J.Lett.*, **435**, L23.
Ferguson, H.C. and Binggeli B. 1994, *Astr. Ap. Rev.*, **6**, 67.
Joseph, B. 1996, *Nature*, **379**, 586.
Kauffmann, G. 1995, *M.N.R.A.S.*, **274**, 153.
Kormendy, J. 1985, *Ap.J.*, **295**, 73.
Kormendy, J., Bender, R., Magorrian, J., Tremaine, S., Gebhardt, K., Richstone, D., Dressler A., Faber, S. M., Grillmair, C. and Lauer-T-R. 1997, *Ap.J.Lett.*, **482** L139.
Lake, G. and Carlberg, R. G. 1986a, *A.J.*, **96**, 1581.
Lake, G., Moore, B. and Katz, N. 1998, *Ap. J.*, in press.
Moore, B., Governato, F., Quinn, T., Stadel, J. and Lake, G. 1998, *Ap. J.*, submitted.
Moore, B., Katz, N. and Lake, G., 1996, *Ap. J.*, **457**, 455.
Moore, B., Lake, G. and Katz, N. 1998, *Ap. J.*, in press.
Moore, B., Katz, N., Lake, G., Dressler, A. and Oemler, A. 1996, *Nature*, **379**, 613.
Sandage, A., Binggeli, B. & Tammann, G.A. 1985, *A.J.*, **90**, 1759.
Secker, J. 1996, *Ap.J.Lett.*, **469**, L81.
Tonry, J. L., Ajhar, E. A. and Luppino, G. A. 1990, *A.J.*, **100**, 1416.
Toomre, A. 1977, In *The Evolution of Galaxies and Stellar Populations*, ed. B.M. Tinsley and R.B. Larson, p. 401, (New Haven: Yale University Observatory).
Vader, P. 1991, *Ap.J.*, **305**, 669.
Valluri, M. and Jog, C. J. 1991, *Ap.J.*, **374**, 103.
Yates, M. G., Miller, L. and Peacock, J. A., *M.N.R.A.S.*, 240, 129.
Yee, H. K. C. and Ellingson, E. 1993 *Ap. J.*, 411, 43.

THE X-RAY PROPERTIES OF NEARBY ABELL CLUSTERS FROM THE ROSAT ALL-SKY-SURVEY

MICHAEL J. LEDLOW
University of New Mexico, Institute for Astrophysics
Dept of Physics & Astronomy, Albuquerque, NM USA

WOLFGANG VOGES
Max-Planck-Institut für Extraterrestrische Physik
Garching bei München, Germany

FRAZER N. OWEN
National Radio Astronomy Observatory
Socorro, NM USA

AND

JACK O. BURNS
University of Missouri, Office of Research
Columbia, MO USA

Using the *ROSAT* All-Sky-Survey (RASS), we examine the X-ray properties of a statistically complete sample of 294 nearby ($z < 0.09$) Abell clusters from our VLA 20cm survey (Ledlow & Owen 1995) and 49 Poor Groups ($z < 0.03$) (Burns et al. 1996). Our analysis includes a catalog of all significant ($> 3\sigma$) X-ray peaks, an analysis of the X-ray extents, identification of ICM emission, comparison to optical cluster properties, and a cross-correlation with our radio galaxy catalogue. We will make optical/X-ray overlays of the cluster fields available over the WWW in the near future (see http://astro.nmsu.edu/~mledlow for updates).

References

Burns, J.O., et al. 1996, *ApJ*, **467**, L49
Ebeling, H., et al. 1997, *ApJ*, **479**, 101
Ledlow, M.J., & Owen, F.N. 1995, *AJ*, **110**,1959

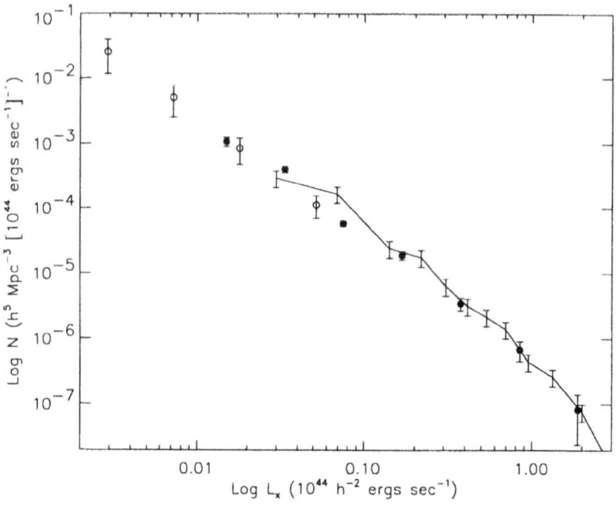

Figure 1. The X-ray luminosity function for our nearby rich clusters sample (solid circles), the poor clusters from Burns et al. (1996) (open circles), and the data points from the Brightest Cluster Sample of Ebeling et al. (1997) - solid line. We use $H_0 = 100$ km sec^{-1} Mpc^{-1}.

INVESTIGATIONS OF ENVIRONMENTAL EFFECTS IN CLUSTERS OF GALAXIES USING N-BODY SIMULATIONS

N.A. POPESCU AND M.D. SURAN
Astronomical Institute of the Romanian Academy,
75212 Bucharest 28, ROMANIA
e-mail: suran@roastro.astro.ro

In order to investigate cluster evolutionary effects we present a new chemo-dynamical model. The model takes into account N-body simulations, modified for galaxy-galaxy binary encounters. The model incorporates also the codes for spectrophotometric and chemical evolution of galaxies. In the same time the calculations include the interaction between two merging subclusters.

As initial parameters, the model uses $[(M_\mathrm{i}, \mathbf{x}_\mathrm{i}, \mathbf{v}_\mathrm{i}), (T_\mathrm{i}, Z_\mathrm{i})]$, taking into account different scales of interactions: galaxy–galaxy, galaxy–cluster and subcluster–subcluster (or cluster–cluster). The model is of N-body type, with normal collisions for $d > d_0$ and close binary interactions for $d \leq d_0$. The parameters of interactions are represented by $[\Delta M_\mathrm{i}(\Delta Z_\mathrm{i}), \Delta\mathbf{v}_\mathrm{i}, \Delta E_\mathrm{i}]$. In our simulations for the momentum transfer we use a simple Chandrasekhar formula with a constant $\mathcal{K}$ input parameter. Different types of interactions were taken into account: mergers and cannibalism ($|v| \leq V_0$), tidal stripping and harassment ($|v| > V_0$), tidal shaking ($T = T(M, Z)$), blocking of SFR ($M_\mathrm{g} = 0$), cool fluxes ($+\Delta M$), and evaporation (galactic winds).

The calculations have been done for *a Coma-like cluster, composed of two interacting substructures.* The initial configuration used in our calculations consists of 173 galaxies in two nucleated subclusters (A,B) with 119 galaxies and 54 galaxies, respectively. The initial cluster diameter was chosen to be ~3 Mpc. The two interacting substructures consist of 119 E+S galaxies in subcluster A: 40 galaxies in the nucleus (50% E galaxies) and 79 galaxies in the envelope (31% E galaxies). The subcluster B is composed of 54 E+S galaxies: 30 galaxies in the nucleus (50% E galaxies) and 24 galaxies in the envelope (45% E galaxies). The initial masses for galaxies were chosen to be $100 \times 10^{10}\ M_\odot$ for E-galaxies and $30 \times 10^{10}\ M_\odot$ for S-galaxies. The parameters for interactions where chosen as: $d_\mathrm{o} = 0.15$ Mpc; $\mathcal{K} = 0.001$, $\Delta M = 0.3 M_\odot$.

The chemo-spectro-photometric model is a Stellar Population Synthesis model, which takes into account supernova I and II rates, galactic winds and chemical mass transfer $\Delta M_{\rm i}(Z_{\rm i}(t))$.

The results of our simulations are compared with the observational data in the field of:

1. intrinsic galactic evolution (galactic tracks in C-C and C-M diagrams);
2. cluster evolution:

(a) photometric evolution (Suran, Popescu this volume): C-R effect; photometric evolution of first galaxies in cluster;

(b) dynamical evolution: existence of binary clusters with dumbbell galaxies; dependence of the properties of first galaxies on compactness and richness classes; profile and dynamics of cD and D galaxies;

3. complex chemical evolution: $SFR = SFR[Z(z)]$, $IMF = IMF[Z(z)]$;
4. morphological evolution: $T = T(z)$.

In our simulations, after a frontal interaction between the two subclusters, the core of the merged cluster contains two D or cD galaxies (which are the remnants of each central galaxy). This mode results in a cluster with dumbbell galaxies (separation $< 0.2h^{-1}$ Mpc).

We identify three different time scales: the interaction time scale between subclusters ($\sim 1-2$ Gyr); the relaxation time scale for the whole system (~ 10 Gyr) and the post-relaxation time scale (from which, up to now were inferred fortuitous interactions). The number of interactions for the entire process ($0 < z < 4$) increases as $\sim A \times (1+z)^{\gamma}$, where $A = 0.03, \gamma = 3.6 \div 4$. For short time scales, both A and γ depend on z.

The model can also reproduce colour-colour, colour-magnitude diagrams and chemical abundances in concordance with observations for all z ranges.

The time scale inferred in these effects (photometric time scale + abundance time scale + dynamical time scale) are of the same order (~ 10 Gyr), much higher than the needed time scale for the cluster evolution in the Standard Cold Dark Matter model (SCDM).

The estimated stellar mass for a D or cD galaxy has grown by a factor $\sim$ 3.2 for $z < 1$, in good agreement with other results (see Aragon-Salamanca et al., this volume, model with $\Omega_0 \sim 0.5$).

COLOUR GRADIENTS IN CLUSTERS OF GALAXIES

M.D. SURAN AND N.A. POPESCU
Astronomical Institute of the Romanian Academy,
75212 Bucharest 28, ROMANIA
e-mail: suran@roastro.astro.ro

In this paper we present a new indicator of the evolution of galaxies in clusters (both galactic and intrinsic cluster evolutionary phenomena) – the colour gradient in clusters of galaxies.

Our calculations were made for 16 clusters of galaxies and the results are presented in Table 1.

The individual values of the colour gradient (the slope $a = |\partial C/\partial R|$ of the C-R diagram for each cluster) present a cosmological dependence:

$$| a_{\rm B-V} | = 0.522(\pm 0.071)z + 0.264$$

Our results are in good agreement with other evolutionary effects for clusters of galaxies. So, our indicator could be:

- *directly related to Butcher-Oemler effect*

$$f_{\rm B} = 0.533(\pm 0.15)z + 0.043(n = 14)$$

- *related to the environmental effect* for different types of galaxies - the red galaxies prevail in the cores of clusters and blue galaxies in the outside regions;
- *related to the new environmental effect for galaxies of the same type* (Oemler et al. 1997, this Conference) – a high frequency of burst galaxies (blue and E+A galaxies) outside the cluster density peak;
- *possibly related to the aperture, and with classes of compactness.*

At the same time, the mean $< B-V >$ colour per cluster indicates a direct connection with the galactic evolutionary process (on the "burst" track in the Rocca-Volmerange (1988) model for galactic evolution and an enhanced stellar activity between $0.4 < z < 0.6$).

These results seem to be in good agreement with the numerical simulations (Popescu & Suran 1997, this volume), indicating an evolutionary time scale of $\sim$ 10 Gyr. The obtained frequency of "E+A like" galaxies (post-starburst phenomena) in 6 clusters with $0.35 < z < 0.55$ is $\sim$ 3%.

References

Rocca-Volmerange, B., & Guiderdoni, B. 1988, *A&A.Suppl*,**74**, 185
Popescu, N.A., & Suran, M.D. 1996, *Romanian Astron. J.*, **6 no.2**, 105

TABLE 1. f_B, a_{B-V} and the mean colour in 16 clusters of galaxies

Cluster	z	f_B	a^E_{B-V}	a^S_{B-V}	$\langle B-V \rangle_{roi}$	$\langle B-V \rangle^E$	$\langle B-V \rangle^S$
Virgo	0.0033	0.02	-0.019	-0.023	0.756	0.873	0.639
Coma	0.0235	0.011	-0.076		0.974	1.039	0.874
A401	0.074	0.071	-0.101	-0.091	1.042	1.104	0.901
3C 206	0.198	0.381	-0.417		1.443	1.607	0.87
A1942	0.224	0.196	-0.397	-0.433	1.178	1.40	1.001
1455+2232	0.259	0.	-0.409		1.695	1.695	0.896
1358+6245	0.32	0.152	-0.49	-0.514	1.469	1.709	1.159
0024+1654	0.3909	0.182	-0.416	-0.447	1.868	2.102	1.337
0939+4713	0.4069	0.321	-0.38	-0.448	1.797	2.036	1.33
0303+1706	0.4181	0.4	-0.495	-0.486	1.743	2.059	1.184
3C 295	0.46	0.263	-0.524	-0.624	1.677	2.071	1.22
1601+4253	0.5391	0.301	-0.548	-0.463	1.715	2.109	1.194
0016+1609	0.544	0.175	-0.456	-0.439	1.868	2.205	1.26
1305+2952	0.62	0.476	-0.579		1.506	1.68	0.916
3C 352	0.8	0.333	-0.723		1.284	1.405	0.75
3C 265	0.81	0.625	-0.726		1.504	1.762	0.937

THE K-BAND HUBBLE DIAGRAM FOR THE BRIGHTEST CLUSTER GALAXIES: A TEST OF GALAXY FORMATION MODELS.

A. ARAGÓN-SALAMANCA
Institute of Astronomy
Madingley Road, Cambridge CB3 0HA, England

C.M. BAUGH
Department of Physics
Science Laboratories, South Road, Durham DH1 3LE, England

AND

G. KAUFFMANN
Max-Plank-Institut für Astrophysik
D-85740 Garching bei München, Germany

We analyze the K-band Hubble diagram for a sample of brightest cluster galaxies (BCGs) in the redshift range $0 < z < 1$. We confirm that the scatter in the absolute magnitudes of the galaxies is small (0.3 magnitudes). The BCGs exhibit very little luminosity evolution in this redshift range: if $q_0 = 0.0$ we detect *no* luminosity evolution; for $q_0 = 0.5$ we measure a small *negative* evolution (i.e., BCGs were about 0.5 magnitudes fainter at $z = 1$ than today). If the mass in stars of these galaxies had remained constant over this period of time, substantial positive luminosity evolution would be expected: BCGs should have been *brighter* in the past since their stars were younger. A likely explanation for the observed zero or negative evolution is that the stellar mass of the BCGs has been assembled over time through merging and accretion, as expected in hierarchical models of galaxy formation. The colour evolution of the BCGs is consistent with that of an old stellar population ($z_{\rm form} > 2$) that is evolving passively. We can thus use evolutionary population synthesis models to estimate the rate of growth in stellar mass for these systems. We find that the stellar mass in a typical BCG has grown by a factor $\simeq 2$ since $z \simeq 1$ if $q_0 = 0.0$ or by factor $\simeq 4$ if $q_0 = 0.5$. These results are in remarkably good agreement with the predictions of semi-analytic models of galaxy formation and evolution set in the context of a hierarchical scenario for structure formation.

DISTRIBUTION OF STOCHASTIC FORCES IN GRAVITATIONALLY CLUSTERED SYSTEM OF GALAXIES

ELIANI ARDI[1,2] AND SHOGO INAGAKI[1]
[1]*Department of Astronomy, Kyoto University*
Kyoto 606-01, Japan
[2]*Department of Astronomy, Institute of Technology*
Bandung, Indonesia

Motivation:

We are interested in examining the influence of nearest neighbor galactic encounters in gravitationally clustered systems by using cosmological N-body simulation.

Research:

We investigate the evolution of the stochastic force distribution which comes from all the galaxies in the system and the force from the nearest galaxy before and during gravitational clustering of galaxies in the expanding universe, through numerical experiments. N-body simulations were done by using a COMOVEV code.

Conclusions:

The force acting on each galaxy in gravitationally clustered systems in the expanding universe is almost entirely due to the gravitational attraction of the nearest neighbor. Nearest neighbor galactic encounters play the main role in the dynamics of galaxy clustering in the expanding universe.

Implications:

In a gravitationally clustered system, each encounter can be treated as a two-body encounter representing the perturber galaxy and its nearest neighbor galaxy. The use of a softening parameter in the collisionless N-body simulation method may not be well suited to study the dynamics of galaxy clustering because the forces from the nearest neighbors may be neglected.

ENVIRONMENTAL EFFECTS

The Molecular Gas in Spiral Galaxies

D.F. DE MELLO
STScI, Baltimore, MD USA
T. WIKLIND
Onsala Space Observatory, Sweden
AND
M. MAIA
Observatório Nacional, Rio de Janeiro, Brazil

We have started a survey of the molecular gas content of spiral galaxies in high and low density regions (HDS and CS), selected according to well–defined criteria (Maia et al. 1994, *ApJS*, **93**, 425). The HDS sample is formed by galaxies that are in groups of three or more members. The groups are defined such that they have a density contrast $\delta\rho/\rho \geq 500$. This is equivalent to densities larger than 18 galaxies/Mpc^3. The CS sample is made up of galaxies which are not members of any group and which are situated in a region with a density contrast $\delta\rho/\rho \leq 0.01$, i.e. less than 0.0004 galaxies/Mpc^3.

We have detected CO emission in all 35 galaxies observed with the SEST radiotelescope on La Silla; 19 from the HDS and 16 from the CS. Although galaxies in the CS on average are more luminous than those in the HDS (a possible distance bias in our small subsample), the blue luminosity surface density of the two samples are indistinguishable from each other. Hence, the blue luminosity is a 'good' observable to use for normalization of the CO and far–infrared luminosities. We find no statistically significant difference in the molecular gas surface density of the two samples, nor of the $L_{\rm FIR}/M_{\rm H_2}$ ratio. Why are the two samples so similar? It is possible that although the HDS galaxies are in a higher density environment the tidal interaction has not yet taken place. Our results suggest that tidal forces which will affect galaxy evolution become important only when galaxies are in a very strong interaction, such as in close pairs of galaxies. Due to the relatively small sample, these results should be viewed with caution. For instance, the presence of an AGN in HDS galaxies could increase the $L_{\rm B}$, artificially lowering the average $L_{\rm FIR}/L_{\rm B}$ ratio.

GALAXY ORIENTATION IN SOME ABELL CLUSTERS

W. GODŁOWSKI AND F. BAIER
Obserwatorium Astronomiczne, Uniwersytet Jagielloński,
ul. Orla 171 30-244 Kraków, Poland
WIP Project Galaxienhaufen Universitat at Podsdam
An der Sternwarte 16 Podsdam Babelsberg Germany

We analyze a sample of galaxies in a region of three rich Abell Clusters. The data are taken from the Edinburgh Catalogue of the cluster. First, we divide the whole cluster area into different parts according to the assumed subclusters. Now we can find the position angles of the cluster and subclusters. We find strong evidence that the position angles of galaxies within our clusters are aligned to a large extent. For the cluster A754, position angles of galaxies tend to be perpendicular to the direction of the position angle of the cluster. Consequently, the angular momentum of galaxies are preferentially perpendicular to the cluster plane. For the cluster A14, position angles of galaxies tend to be parallel to the direction of the position angle of the cluster. Consequently, the angular momentum of galaxies are preferentially parallel to the cluster plane. For the cluster A3667 we obtain a more complicated picture suggesting that the alignment of galaxies in this cluster may have a different shape. From the distribution of the positions angles of galaxies we also found evidence for possible subclustering inside the whole cluster. This result is confirmed by the investigation of the distribution of the vectors normal to the galactic planes. Moreover we confirm the existence of a "line of sight" effect, originally found by Godłowski & Ostrowski (1996) for galaxies belonging to the clusters in the Tully Catalogue (1988), for the clusters in our basic catalog.

References

Godłowski, W. 1994, *MNRAS*, **271**, 19
Godłowski, W., Ostrowski, M. 1997, in press
Tully, R.B. 1988, *Nearby Galaxies Catalog*, Cambridge U. Press

ENVIRONMENTAL INFLUENCE ON STAR FORMATION OF GALAXIES IN THE LAS CAMPANAS REDSHIFT SURVEY

Y. HASHIMOTO AND A. OEMLER
Yale/OCIW
Dept. of Astronomy, Yale University, New Haven, CT 06511;
Carnegie Obs., 813 Santa Barbara St. Pasadena, CA 91101

1. Method

We have used a sample of 15749 galaxies taken from the Las Campanas Redshift Survey to investigate the effects of environment on the rate of star formation (SFR) in galaxies. For each galaxy we measure SFR by [OII] emission, while a concentration index (C) is used to decouple the effect of the "morphology-environment" relation from the SFR. Galactic environment is characterized *both* by the 3-space local density (ρ) of galaxies and by membership in groups and clusters.

2. Results

Cluster galaxies exhibit *reduced SFR for the same C.* A further division of clusters by "richness" reveals a new possible excitation of "starbursts" in poor clusters. Meanwhile, the SFR of a given C shows a continuous correlation with the ρ, in such a way that galaxies show higher levels of SFR in lower density. Interestingly, this trend is also observed both inside and outside of clusters, implying that physical processes responsible for this correlation might not operate intrinsically in the cluster environment. Galaxies with differing levels of SFR appear to respond differently to the local density. Low levels of star formation are more sensitive to environment inside than outside of clusters. In contrast, high levels of star formation, identified as "starbursts", are at least as sensitive to local density in the field as in clusters.

We conclude that at least two separate processes are responsible for the environmental sensitivity of the SFR.

A NEW AUTOMATED SAMPLE OF COMPACT GROUPS OF GALAXIES

A. IOVINO AND E. TASSI
Oss. Astron. di Brera, Milano, Italy
C. MENDES DE OLIVEIRA
USP, Sao Paolo, Brazil
P. HICKSON
UBC, Vancouver, Canada
AND
H. MACGILLIVRAY
ROE, Edinburgh, Great Britain

Compact Groups (CGs) of galaxies represent an extreme class of objects. They typically contain 4-8 galaxies with high space density (as in the centers of rich clusters) but with low velocity dispersions, being an excellent laboratory for the study of galaxy interactions and their effects. Up to now no unbiassed sample of CGs was available in the literature, and therefore several of the classical problems and paradoxes involving CGs could not be properly addressed.

We exploited the availability of large digitized galaxy catalogs to develop an algorithm for the detection of CGs, implementing clearly defined and rigidly applied selection criteria. The new algorithm is optimized for work also at fainter magnitudes, where both incompleteness and contamination become serious problems. We applied the new algorithm to an area of $\sim$ 4500 sq. deg. in the Southern Sky, obtaining $\sim$ 300 candidate CGs up to magnitude $b_j \sim 16.0$ for the brightest galaxy of the group.

Using the ESO 1.5m telescope at LaSilla, we measured redshifts for all the members of 60 candidate CGs, defining a brighter subsample (up to $b_j \sim 14.5$ for the brightest galaxy of the group). Assuming $1000\,\mathrm{km\,s^{-1}}$ as the velocity cut-off for group membership, 48 candidates have three or more concordant members, corresponding to an increase of $\sim$ 4 times in surface density with respect to Hickson's sample of CGs!

LARGE SCALE GRADIENT IN THE VELOCITY FIELD OF THE COMA CLUSTER AND A STUDY OF THE SPIN ORIENTATION OF GALAXIES IN THE VIRGO CLUSTER

M. IYE
National Astronomical Observatory, Tokyo 181, Japan

AND

T. OZAWA
Graduate University for Advanced Studies, Tokyo 181, Japan

This paper reports two studies on the distribution of the spin angular momenta of galaxies in physical clusters of galaxies.

i) Gradient in the Velocity Field of Coma Cluster

We studied the distribution of the line-of-sight velocity of galaxies in the Coma cluster by evaluating the dipole moment vector of the distribution. We compared the observed dipole moment vector with those obtained for 1000 Monte Carlo simulation runs where the observed line-of-sight velocities were reshuffled among the observed galaxies. We conclude that the Coma main cluster has a significant large scale line-of-sight velocity gradient across the cluster at the confidence level of 97%. The identified dipole moment vector, showing the large scale velocity gradient, is pointing toward the east and does not coincide with that of the X-ray temperature gradient vector observed by *ASCA*, which is actually pointing toward the southeast.

ii) Spin Vector Distribution in the Virgo Cluster

We looked for any systematic tendency in the distribution of spin vectors of galaxies in the Virgo cluster. We examined the correlation in the 5D phase space for 60 galaxies in the Virgo cluster for which the distances derived from the Tully-Fisher relation and the spin vector orientations are known. We tentatively conclude that there is no striking correlation in the 5D phase space for this sample. It is important to check whether the apparent lack of correlation is consistent with each scenario of galaxy formation. Numerical studies to derive estimates on expected residual correlation from simulations of various scenarios are under way.

MOLECULAR GAS IN HICKSON COMPACT GROUPS

S. LEON AND F. COMBES
DEMIRM, Observatoire de Paris, France

AND

T.K. MENON
Dep. of Physics & Astronomy, Univ. of British Columbia, Vancouver, Canada

Compact groups are ideal sites to study the influence of strong dynamical evolution due to environment on molecular cloud formation and star formation efficiency. We have observed 70 galaxies belonging to 45 Hickson compact groups (HCGs) in the $^{12}CO(1\rightarrow0)$ and $^{12}CO(2\rightarrow1)$ lines, in order to determine their molecular content. We compare the gas content relative to blue and L_{FIR} luminosities of galaxies in compact groups with respect to other samples in the literature, including various environments and morphological types. We find that there is some hint, of enhanced M_{H_2}/L_B and M_{dust}/L_B ratios in the galaxies from compact group with respect to our control sample, especially for the most compact groups, suggesting that tidal interactions can drive the gas component inwards, by removing its angular momentum, and concentrating it in the dense central regions, where it is easily detected. The threshold at 20-30 kpc in mean galaxy separation for the enhancement of H_2 suggests that it must correspond to an acceleration of the merging process and a significant inward gas flow. The molecular gas content in compact group galaxies is similar to that in pairs and starburst samples. However, the total L_{FIR} luminosity of HCGs is quite similar to that of the control sample, and therefore the star formation efficiency appears lower than in the control galaxies. However this assumes that the FIR spatial distributions are similar in both samples which is not the case at radio frequencies. Higher spatial resolution FIR data are needed to make a valid comparison. Given their short dynamical friction time-scale, it is possible that some of these systems are in the final stage before merging, leading to ultraluminous starburst phases. We also find for all galaxy samples that the H_2 content (normalized to blue luminosity) is strongly correlated with L_{FIR}, while the total gas content (H_2+HI) is not.

A K-BAND LUMINOSITY FUNCTION OF HICKSON COMPACT GROUPS OF GALAXIES

S. NISHIURA, T. MURAYAMA AND Y. TANIGUCHI
Astronomical Institute, Tohoku University

Y. SATO
ISAS and VILSPA

AND

D.B. SANDERS
Institute for Astronomy, University of Hawaii

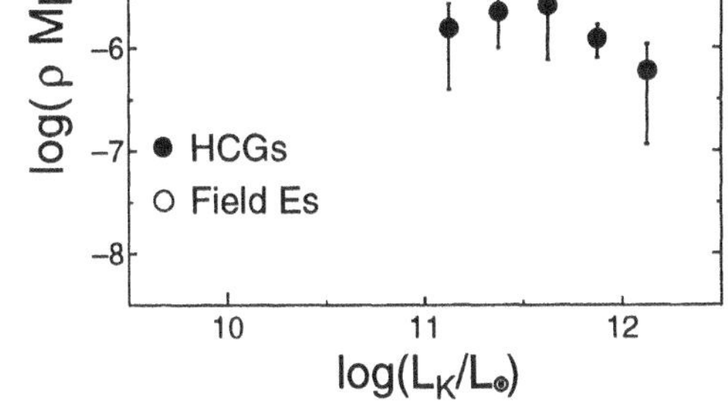

Figure 1. A K-band luminosity function of Hickson Compact Groups of Galaxies. The data were obtained at the University of Hawaii Planetary Patrol 24 inch telescope with a NICMOS3 camera. The K-band luminosity function of field ellipticals are derived from the data in Faber et al. (1989, *ApJS*, **69**, 763). We adopted 4.19 mag. as $(B-K)^0_T$ of ellipticals (as estimated from Govazzi & Boselli 1996, *Astro. Lett. & Comm.*, **35**, 1)

EFFICIENT STAR-FORMATION IN THE TIDAL ARMS OF THE STEPHAN'S QUINTET GROUP OF GALAXIES

Y. OHYAMA, S. NISHIURA, T. MURAYAMA AND Y. TANIGUCHI
Astronomical Institute, Tohoku University
Aramaki, Aoba, Sendai 980-77, JAPAN

NGC 7318B in Stephan's Quintet has two optical arms (toward N and S) emanating from the eastern part of the main body. Since these arms are similar morphologically to the tidal tails of merging galaxies such as NGC 4038/9, it is considered that NGC 7318B itself is a major merger with a retrograde orbit. In order to study the emission-line activity in the tidal arms of NGC 7318B, we took CCD narrow-band (Hα ON and OFF) images and then found a large-scale arc in Hα emission which traces closely the arms. This Hα arc resembles both the radio and the soft X-ray arcs morphologically (van der Hulst & Rots 1981; Pietsch et al. 1997), suggesting that a single physical mechanism is responsible for all of these kinds of emission. Our optical spectroscopic observations of the shell-like feature at the southern tip of the arc reveal both broad Hα emission and stronger-than-normal [NII] and [SII] emission lines, which are typical of supernova remnants (SNRs). The required number of SNRs is estimated to be as much as $\sim 10^6$.

The proposed scenario for the arc formation is the following (Ohyama et al. 1997): The two tidal tails were formed during a past merging event between two gas-rich disk galaxies. Giant HII regions containing numerous massive stars ($\sim 10^6$) were formed almost simultaneously along the tails (e.g., Barnes & Hernquist 1992). After $\sim 10^{6-7}$ years, supernovae exploded almost simultaneously and formed the emission arc observed in Hα, radio, and soft X-ray.

References

Barnes, J. E., & Hernquist, L. 1992, *Nature*, **360**, 715
Ohyama, Y., Nishiura, S., Murayama, T., & Taniguchi, Y. 1997, *ApJL*, in press
Pietsch, W., Trinchieri, G., Arp, H., & Sulentic, J. W. 1997, *A&A*, **322**, 89
van der Hulst, J. M., & Rots, A. H. 1981, *AJ*, **86**, 1775

MEASURING SUBCLUSTERS IN GALAXY CLUSTERS

Z.Y. SHAO
Shanghai Astronomical Observatory
80 Nandan Road, Shanghai 200030, China
E-mail: zyshao@center.shao.ac.cn

We assume that there are $K_{\rm c}$ subclusters and $K_{\rm f}$ fields (foreground or background) in a cluster region. Then, the distribution of all galaxies in this region can be described as follow:

$$\Phi = \sum_{c=1}^{K_{\rm c}} \Phi_c + \sum_{f=1}^{K_{\rm f}} \Phi_f = \sum_{c=1}^{K_{\rm c}} n_c \phi_c \mu_c + \sum_{f=1}^{K_{\rm f}} n_f \phi_f \mu_f. \tag{1}$$

where, n_c and n_f are normalized numbers of subcluster members and field galaxies. ϕ_c, ϕ_f, are their normalized distributions in radial velocity space. Both of them can be assumed as Gaussian. μ_c and μ_f are normalized distributions in the projected surface of the celestial sphere. For field galaxies, it's uniform, and for subcluster members, usually we use the King's approximate formulae. Distribution parameters and their uncertainties can be found by using the standard maximum likelihood method. And membership probabilities of the ith galaxy belonging to the cth subcluster can be calculated as $P_c(i) = \Phi_c(i)/\Phi(i)$.

Furthermore, we introduce a new index E_c to measuring the effectiveness of membership determination of the cth subcluster:

$$E_c = 1 - N \sum_{i=1}^{N} \{P_c(i)\,[1 - P_c(i)]\} \left\{ \sum_{i=1}^{N} P_c(i) \cdot \sum_{i=1}^{N} [1 - P_c(i)] \right\}^{-1} \tag{2}$$

By using the approach mentioned above. The Virgo cluster area is divided into 10 subcluster and 1 background field successfully (Shao 1996).

References

Shao, Z.Y. 1996, Ph.D. thesis, Shanghai Astronomical Obs.

NUCLEAR ACTIVITY IN THE HICKSON COMPACT GROUPS OF GALAXIES

M. SHIMADA, S. NISHIURA, Y. OHYAMA, T. MURAYAMA AND Y. TANIGUCHI
Astronomical Institute, Tohoku University
Aramaki, Aoba, Sendai 980-77, JAPAN

In order to study environmental effects on the nuclear activity in galaxies, we have been conducting a spectroscopic study of Hickson Compact Groups of galaxies (HCGs, Hickson 1982) which are the densest agglomeration of galaxies. We obtained nuclear spectra of 62 galaxies in 29 HCGs in the spectral range 6200-7000Å with the 188cm telescope at Okayama Astrophysical Observatory. These spectra were classified into the three types by using the emission line ratio [NII]λ6583/Hα; (1) AGN: [NII]λ6583/Hα >0.6, (2) HII nuclei: [NII]λ6583/Hα <0.6, and (3) Absorption: no emission line. We compared the nuclear activity of galaxies in HCGs with that of nearby galaxies (Ho 1996; Ho, Filippenko & Sargent 1997) which provides a representative sample of field galaxies. In early-type spirals (Sa-Sbc), the fraction of HII nuclei in HCGs is smaller than that in the field galaxies, while the fraction of absorption in HCGs is larger than that in field galaxies. On the other hand, in early-type galaxies (E-S0a) and late-type spirals (Sc-P), we found little difference in the nuclear activity between HCGs and field galaxies.

References

Hickson, P. 1982, *ApJ*, **255**, 382
Ho, L.C. 1996, in *The Physics of LINERS in View of Recent Observations, ASP Conf. Ser.*, **103**,103
Ho, L.C., Filippenko, A.V., & Sargent, W.L.W. 1997, *ApJ*, **487**, 568

EVOLUTION OF SUBSYSTEMS DURING COLLAPSE OF A CLUSTER

TOSHIO TSUCHIYA
Department of Astronomy, Kyoto University
Kyoto 606-01, Japan

Motivation: In the hierarchical clustering scenario, objects with smaller masses collapse and virialize earlier. I am interested in the effects of the existence of virialized subsystems on the structures of a collapsing larger mass object, and the fate of subsystems.
Comparison to previous works: Most work, both analytic and numerical, is concerned with the evolution of subsystems in a virialized cluster. Some numerical works have investigated dynamical effects in collapsing clusters, but these are usually not quite systematic.
Aim of this work: We try to study numerically the evolution of collapsing clusters with clumpy initial conditions. We make a systematic survey of two dimensional parameters: the number and the relative size of the subsystems.
Simulations: Numerical N-body simulations are made with the GRAPE-3A supercomputer. No cosmological expansion is included. Each subsystem is an equilibrium Plummer sphere. The initial radius of the cluster of the Plummer sphere is given by

$$R_0 \equiv 2r_\mathrm{v}\left(\frac{M_\mathrm{total}}{m}\right)^{(n_\mathrm{k}+5)/6} = N_\mathrm{c}^{(n_\mathrm{k}+5)/6}, \qquad (1)$$

where r_v is the initial radius of the cluster and the subsystems; M_total and m are the masses of the cluster and the subsystems and N_c is the number of the subsystems.
Results: We obtained a condition for the subsystems to survive the first violent phase of a collapsing cluster;

$$n_\mathrm{k} \geq -1, \qquad \text{and} \qquad N_\mathrm{c} \gg 1. \qquad (2)$$

ROSAT OBSERVATIONS OF CLUSTERS CL0500-24 & CL0939+4713

JOACHIM WAMBSGANSS
Astrophysikalisches Institut Potsdam (Germany)
e-mail: jwambsganss@aip.de

AND

SABINE SCHINDLER
MPI für extraterrestrische Physik, Garching (Germany)
e-mail: sas@mpa-garching.mpg.de

Cluster CL0500-24: HRI (Schindler & Wambsganss 1997, *A&A*, **322**, 66)

- *ROSAT*/HRI observation (Feb.1995): 37.8 ksec
- Relatively low X-ray luminosity, high gas mass fraction: $L_{X,\mathrm{ROSAT}} = 3.1^{+0.6}_{-0.4} \times 10^{44}$erg/s, $L_{X,\mathrm{bol}} \approx 5.6^{+4.2}_{-2.0} \times 10^{44}$erg/s, $f_{\mathrm{gas}} \approx 30^{+30}_{-10}$
- X-ray emission correlated with Northern subclump
- Derived total mass (≤ 1 Mpc) very low: $M_X \approx (1.5 \pm 0.8) \times 10^{14} M_{\odot}$

Cluster CL0939+4713: PSPC (Schindler & Wambsganss 1996, *A&A*, **313**, 113)

- *ROSAT*/PSPC observation (Nov. 1991): 14.4 ksec
- Low X-ray luminosity and Temperature: $L_{X,\mathrm{ROSAT}} = (7.9 \pm 0.3) \times 10^{44}$ erg/s, $T_X = 2.9^{+1.3}_{-0.8}$ keV
- Unusual properties of CL0939-24 can be explained if it is a merger

Cluster CL0939+4713: HRI (Schindler et al. 1997, in preparation)

- *ROSAT*/HRI observation (Oct. 1996): 45.6 ksec; reveals dramatically different features:
  Luminosity of former Maximum M3 decreased by more than a factor of 10 within 5 years! A background quasar is clearly visible: very luminous $L_{X,Q}=1.4 \cdot 10^{45}$erg/sec!
- Cluster emission can be traced out to 2.5 arcmin (≈ 1 Mpc) with 1060 source counts, corresponding to $L_X = (6.9 \pm 0.4) \times 10^{44}$erg/s.

A MULTI-MERGING GALAXY MRK 273 WITH HOT EXTENDED GASEOUS HALO AND AN EXTENDED SOFT X-RAY COMPANION

X.Y. XIA
Dept. of Physics, Tianjin Normal University, P.R.China
Z.G. DENG AND H. WU
Beijing Astronomical Observatory, P.R. China
AND
T. BOLLER
Max-Planck-Institut für Extraterrestrische Physik, Germany

The ultraluminous *IRAS* galaxy Mrk 273 is a Seyfert 2, multi-merger system with a triple-nucleus and a 20 kpc ($H_0 = 100$ km s^{-1} Mpc^{-1}) jet to the South. Also, a plume and faint tail extends ~50 kpc Northwest and there are more than 10 dwarf galaxies within 100 kpc of Mrk 273 as determined from the POSS II R and J film copy. Therefore, the precursor of Mrk 273 is probably connected with a group of galaxies.

ROSAT PSPC observations show that the soft X-ray emission of Mrk 273 is extended up to 50 kpc. Moreover there exists an extended soft X-ray companion at a distance of about 30 kpc that is certainly connected with Mrk 273 as revealed by our follow-up optical spectroscopic observation. The soft X-ray luminosity of the companion is 6.3×10^{41} erg s^{-1} and its optical counterpart is an irregular dwarf galaxy as determined from its optical image and estimated apparent magnitude.

The soft X-ray luminosity for known irregular dwarfs is not as high as 10^{41} erg s^{-1} and the extended hot gaseous halo is not the established character of a Seyfert 2 galaxy. As Ponman et al. (1994, *Nature*, **369**, 462) pointed out, an elliptical galaxy formed by the merger of a group will retain its halo. Therefore the hot gaseous halo and the soft X-ray companion of Mrk 273 very likely are a result of a multi-merger process and this may be evidence to support the merger origin for the intergroup or intercluster hot gas.

THE HIGH REDSHIFT POPULATION OF FIELD GALAXIES

DAVID C. KOO
UCO/Lick Observatory
University of California, Santa Cruz, CA 95064, USA

Abstract. In principle, observations of the high redshift population of field galaxies should provide powerful probes of interactions and mergers. As initial examples of such explorations, we highlight recent results from the DEEP program that take advantage of Keck and *HST*, including median redshifts for samples fainter than $I \sim 22$, morphologies and colors to $z \sim 1$, and especially the nature of $z \sim 3$ galaxies in the Hubble Deep Field. Although tantalizing, the results will need much larger samples to be confirmed and better quantified.

1. Introduction

High redshift observations of field galaxies have the potential to offer diverse and independent evidence for the rate and importance of interactions and mergers in the evolutionary history of galaxies. A sampling of indirect methods include:

∗ *morphological* evidence: such as in the frequency of close galaxy pairs or satellites

∗ *photometric* evidence: such as a decrease in the past of the volume density of bright elliptical galaxies with $r^{1/4}$ radial light profiles that are commonly assumed to arise from mergers of disks

∗ *global number count* evidence: such as an overall increase of the volume density of galaxies with lookback time, presumably due to the presence of more pre-merger components in the past

∗ *color* diagnostics: such as the colors of distant bulges, which some astronomers predict would be very red if they were formed prior to disks, while others claim bulge formation was coeval with disk formation and thus should have roughly similar colors at all redshifts

∗ *fundamental plane* deviations: e.g., this might be observed in the case of an unusually low rotation amplitude in the optical Tully-Fisher diagram for two galaxies colliding roughly head-on in the plane of the sky, resulting in an elongated system that mimics an highly inclined disk system

∗ *theoretical* expectations versus observations: e.g. substantial merging would result in lower median redshifts versus brightness or a different change of the bulge to disk ratio with lookback time

On the other hand, direct evidence for mergers and interactions can also be gathered. Compelling cases for such events would include images of galaxies that show tidal tails or multiple nuclei or spectral evidence for distorted or complex velocity structures. As discussed by Abraham (see proceedings), techniques are being developed which provide quantitative measures of such in-situ evidence for interactions and mergers.

2. Highlights from DEEP

As examples of the potential impact of such high redshift observations, we present some early results that are part of a much more comprehensive Keck survey of very faint galaxies, known as DEEP [1].

2.1. MEDIAN REDSHIFTS

One relatively robust test for whether mergers play a major role in galaxy formation is to compare the predicted median redshift for continually fainter samples of galaxies. Carlberg (1996), for example, estimated that a standard Cold Dark Model (SCDM) model with a substantial rate of mergers would yield a median redshift that remained constant between $20 < I < 28$ with a value near the $z \sim 0.6$ median actually measured by the $I \sim 22$ Canada France Redshift Survey (CFRS: Lilly et al. 1995a). This prediction is, however, found to be inconsistent with our Keck observation of a median $z \sim 0.81$ to 1.00 (95% confidence level), based on a small, but very faint ($22 < I < 24$), sample of 33 field galaxies (Koo et al. 1996). A larger pool of such faint galaxy redshifts observed in the HDF and other fields continue to support this measurement of a high median redshift. Although the "maximal merger rate" model of Carlberg (1996) is no longer compatible with the latest Keck redshifts, mergers may still play a significant, but less prominent, role. Promising matches to the $22 < I < 24$ data can be found, e.g., among more recent and improved merger models that predict median redshift values closer to $z \sim 0.67$ for SCDM and $z \sim 0.91$ for a low $\Omega \sim 0.3$ universe (Baugh et al. 1997).

[1]Information about the Deep Extragalactic Evolutionary Probe project is available on the WWW at URL **http://www.ucolick.org/~deep/home.html**

2.2. MORPHOLOGIES AND COLORS OF $z \sim 1$ GALAXIES

Although the CFRS provides a large pool of very bright galaxies at redshifts $z \sim 1$, the DEEP survey reaches galaxies fainter than $I \sim 22.5$, and thus with luminosities more comparable to typical (L^*) galaxies today. Several tantalizing hints support the view that a large fraction of such galaxies are participating in further agglomerations and in continued star formation due to interactions, major mergers, and infalling satellites by redshifts $z \sim 1$ (Koo et al. 1996). The samples are, however, still small and need to be expanded before firmer conclusions can be drawn.

As examples, we find:

* four of the six very red, high-redshift, early-type *field* galaxies show apparent tidal/merger "tails" or close projected neighbors (Koo et al. 1996). We speculate that we are watching the infall of dwarf galaxies, some quite gas rich, assuming the observed blue colors are due to active star formation. If so, perhaps ellipticals form early and yet can be built up with many more minor mergers over a much longer time period. This process would counter the expected dimming with time of a single-burst stellar population and thus be one possible explanation for the apparent lack of luminosity or density evolution among red galaxies seen in the CFRS (Lilly et al. 1995b) or the Mg II absorber sample (Steidel, Dickinson, & Persson 1994).

* several cases for massive-disk systems (Vogt et al. 1996, 1997) at redshifts $z \sim 1$. Some appear well-formed with thin disks and old bulges while others appear to be less well-organized and perhaps in the earlier phases of disk formation. One well-formed system also shows hints of very blue, very faint satellites that might well settle into it (Koo et al. 1996). This form of evolution has already been proposed for the Milky Way (Majewski 1993).

* Koo et al. (1995), Guzmán et al. (1996, 1997), and Phillips et al. (1997) discuss various kinematic surveys which demonstrate that some compact galaxies at intermediate redshifts are likely to have very low masses. The existence of such systems provide some evidence for the possible presence of pre-merger components or star-bursting dwarfs at high z, but our finding of more massive systems with high star formation rates at higher redshifts appears consistent with the "down-sizing" suggestion of Cowie et al. (1996), which at face-value seems counter-intuitive in a universe in which more massive systems are built up by mergers later in time.

2.3. NATURE OF $z \sim 3$ GALAXIES

An exciting recent advance in observational cosmology has been the pioneering work of Steidel et al. (1996b) in spectroscopically confirming with Keck a relatively large population of high redshift $z \sim 3$ field galaxies. Such

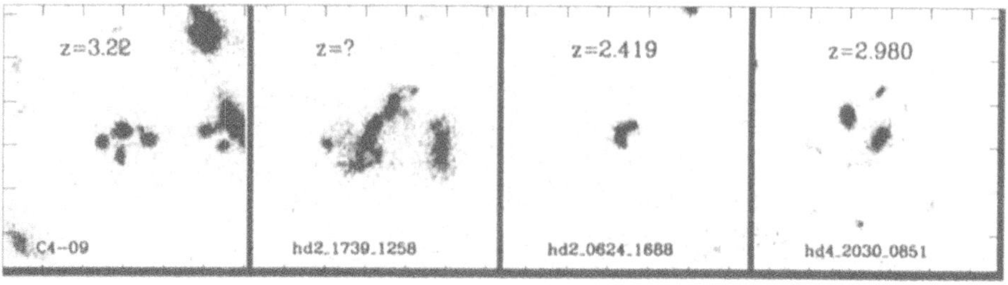

Figure 1. A selection of high-redshift candidate galaxies – three of them confirmed – from the HDF (Steidel et al. 1996a; Lowenthal et al. 1997). Each image is 6 arcsec on a side and the objects are typically $I = 25$. Note the complex morphologies, including linear structures and multiple, spatially separated knots.

galaxies provide direct views of the very early histories of galaxies and certainly place powerful constraints on the role of interactions and mergers in galaxy formation and evolution.

Their candidate selection uses multicolor broadband photometry to look for a "drop-out" of flux due to the redshifted Lyman break in the stellar continuum. This technique was suggested at least three decades ago as a way to look for primeval galaxies by Partridge and Peebles (1967) and followed by numerous surveys since the 1970's (see e.g., review by Koo 1985; Majewski 1989; Cowie 1989; Guhathakurta, Tyson, and Majewski 1990; Lilly, Cowie, and Gardner 1991; Steidel and Hamilton 1992). But none of these earlier efforts had the advantage of a 10-m Keck telescope to confirm any candidates.

To complement the early, important discovery of five such high redshift galaxies by Steidel et al. (1996a) in the HDF, Lowenthal et al. (1997) pushed about one magnitude fainter, used B as well as U band drop-outs to select candidates (see Fig. 2), and secured higher spectral resolution and S/N data to attempt improved kinematic and thus mass measures.

Lowenthal et al. (1997) confirmed 11 more galaxies with high redshifts. Combined with the five from Steidel et al. (1996a), we find an *observed* comoving volume density that is comparable to that of L^* galaxies today, and thus already roughly 3-4 times higher than those from the brighter limits of Steidel et al (1996b). Adjusting for the unobserved or spectroscopically unconfirmed high-redshift candidates in the HDF would yield densities almost 3 times larger. Instead of mainly simple, compact systems for these high redshift galaxies (Giavalisco, Steidel, & Macchetto 1996), we find within our HDF sample of candidates a broad diversity of morphologies, ranging from small, single knots; two more more knots; elongated systems; to more complex, asymmetric morphologies (see Fig. 1).

The nature of these high redshift galaxies is still a mystery. Remain-

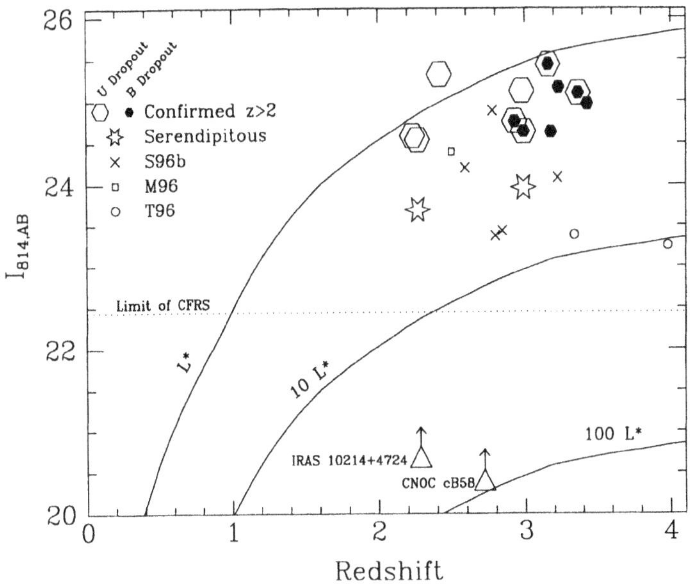

Figure 2. $I_{814\,AB}$ vs. redshift for the confirmed sample of Lowenthal et al. (1997). Also shown for comparison are the three lines corresponding to unevolved L*, 10 L*, and 100 L* blue star-forming galaxies; the five confirmed high-redshift objects in the HDF observed by Steidel et al. (1996b; × symbol); and high-z objects observed in other works as described by Lowenthal et al. (1997). Note that the B-dropouts do add galaxies not found via U-dropouts and have high redshifts, as expected.

ing evidence in support of their being the cores of massive ellipticals and spheroids as suggested by Steidel et al (1996b) are their recent findings of large-scale structure among these galaxies (Steidel et al. 1997), one candidate (C4-09) yielding a mass of roughly 2×10^{11} $M_{\odot}$ (Lowenthal et al. 1997), and strong theoretical support (e.g., Baugh et al. 1997). On the other hand, the original argument for large velocity widths based on the high equivalent widths of saturated interstellar absorption lines (Steidel et al. 1996a) has been countered by Conti et al. (1996); the smooth, compact morphologies being comparable in size to today's spheroids is no longer as compelling given our finding of large diversity; and arguments based on prior estimates of lower volume densities being consistent with massive galaxies today is now superseded by our fainter, *UB* drop-out survey of higher completeness.

Thus we suggest an alternative view (Lowenthal et al. 1997) that at least some, if not most, of these high redshift galaxies are either the progenitors of local dwarfs, such as spheroidals, with masses closer to 10^{10} $M_{\odot}$ or the components that will later merge to form more massive galaxies. In support of this view are: the low velocity width of $\sigma \sim 70$ km-s^{-1} measured for one galaxy in the near-IR (Pettini et al. 1997); the similarly low velocity widths of all six Lyα emission lines in our sample; the very high volume densities that would be inconsistent with each high redshift galaxy being an independent massive system today; the discovery of possible lower-redshift counterparts to the high-redshift galaxies among the very luminous, compact systems at intermediate redshift, some of which have very small masses and other characteristics consistent with being the progenitors of some spheroidals today (Koo et al. 1995; Guzman et al. 1996; Phillips et al. 1997); and the low star-formation rates of our fainter, high redshift galaxies that yield masses of only 10^{10} $M_\odot$ after several Gyr and assuming negligible dust extinction to convert UV luminosities to star formation rates.

Clearly a larger sample of reliable mass and star-formation measurements continues to be an important key to resolve the question of whether these systems are the cores of massive ellipticals and spheroids, the progenitors of dwarfs today, or the building blocks of larger galaxies.

3. Acknowledgements

The observations were obtained at the W. M. Keck Observatory, which is operated jointly by the University of California and the California Institute of Technology, and with the NASA/ESA *Hubble Space Telescope*, which is operated by AURA, Inc., under contract with NASA. The author thanks the IAU and UCSC Committee on Research for travel support to the IAU. The work highlighted here is a collaborative effort through DEEP and has been supported by NSF grants AST 91-20005 and AST 95-29098 and NASA grants AR05801.01-94A, AR06402.01-95A, AR06337.08-94A, AR06337.21-94A, and AR07532.01-96. I would like to especially note that J. Lowenthal should be credited for the vast bulk of the work on the $z \sim 3$ galaxies and for help with the figures.

References

Baugh, C. M., Cole, S., Frenk, C. S., & Lacey, C. G. 1997, *ApJ*, submitted
Carlberg, R. G. 1996, *Galaxies in the Young Universe*, eds. H. Hippelein, K. Meisenheimer, & H. J. Roser, p. 206
Conti, P. S., Leitherer, C., & Vacca, W. D. 1996, *ApJ*, **461**, L87
Cowie, L. L. 1989, *The Epoch of Galaxy Formation*, eds. C. S. Frenk et al., p. 31
Cowie, L. L., Songaila, A., Hu, E. M., & Cohen, J. G. 1996, *AJ*, **112**, 839
Giavalisco, M., Steidel, C. C.. & Macchetto, D. 1996, *ApJ*, **470**, 189
Guhathakurta, P., Tyson, J. A., & Majewski, S. R. 1990, *ApJ*, **357**, L9
Guzmán, R. et al. 1996, *ApJ*, **460**, L5
Guzmán, R. et al. 1997, *ApJ*, **489**, 559
Koo, D. C. 1985, *Spectral Evolution of Galaxies*, eds. C. Chiosi and A. Renzini, p. 419

Koo, D. C. et al. 1995, *ApJ*, **440**, L49
Koo, D. C. et al. 1996, *ApJ*, **469**, 535
Lilly, S. J., Cowie, L. L., & Gardner, J. P. 1991, *ApJ*, **369**, L79
Lilly, S. J., et al. 1995a, *ApJ*, **455**, 50
Lilly, S. J., et al. 1995b, *ApJ*, **455**, 108
Lowenthal, J. D. et al. 1997, *ApJ*, **481**, 673
Majewski, S. R. 1989, *The Epoch of Galaxy Formation*, eds. C. S. Frenk et al., p. 85
Majewski, S. R. 1993, *ARAA*, **31**, 575
Partridge, B. & Peebles, P. J. E. 1967, *ApJ*, **147**, 868
Pettini, M., et al. 1997, *Origins*, eds. J. M. Shull, C. E. Woodward, and H. Thronson, ASP Conf. Series, in press
Phillips, A. C. et al. 1997, *ApJ*, **489**, 543
Steidel, C. C., Dickinson, M., & Persson, S. E. 1994, *ApJ*, **437**, L75
Steidel, C. C., & Hamilton, D. 1992, *AJ*, **104**, 941
Steidel, C. C., et al. 1996a, *AJ*, **112**, 352
Steidel, C. C., et al. 1996b, *ApJ*, **462**, L17
Steidel, C. C., et al. 1997, *ApJ*, in press
Vogt, N. P. et al. 1996, *ApJ*, **465**, L15
Vogt, N. P. et al. 1997, *ApJ*, **479**, L121

DYNAMICS AND INTERACTIONS OF HIGH-REDSHIFT GALAXIES

M. NOGUCHI
Astronomical Institute, Tohoku University
Aoba, Sendai 980-77, Japan

Abstract. A large number of high redshift galaxies observed with the *Hubble Space Telescope* (*HST*) show anomalous morphology and photometric properties, which may be an indication of evolutionary process in young galaxies. We show here by means of numerical simulations that the copious interstellar gas existing in the disks of rapidly collapsing protogalaxies can bring about these peculiarities. Gravitational instability in a gas-rich disk leads to the formation of massive gas clumps with a typical mass of $10^9 M_\odot$. These subgalactic clumps make disk galaxy evolution a dynamically energetic and chaotic process, and give a natural explanation for peculiar morphology of high redshift galaxies. Moreover, the present model provides a new picture on the causal relationship between the emergence of quasar activities and the dynamical evolution of host galaxies. The clump-driven evolution model is also capable of explaining the correlations observed among present-day galaxies. Namely, the relative bulge dominance, existence of a thick disk, and a mass of the super-massive black hole situated at the galactic center should all be correlated positively. In contrast to their vigorous evolution in isolated state, primeval disk galaxies do not show any dramatic enhancement of activity or remarkable dynamical response in interaction with another galaxies.

1. Gravitational Instability in Young Galactic Disks

One of the most important characteristics of young galaxies is ample interstellar matter in their forming disks. The interstellar gas component is dynamically different from the stellar component in that the former is collisional and dissipative while the latter is collisionless. This difference manifests itself most remarkably in the presence of self-gravity. Therefore, the dominance of the gas in the disk galaxies at early cosmological epochs may bring about peculiarities in their structure and evolution.

In order to investigate the early phase of disk galaxy evolution, we simulated numerically the collapse of a rotating protogalaxy (Fig. 1, also Noguchi 1997). The protogalaxy was modeled as an ensemble of numerous gas clouds and dark matter particles distributed uniformly in a spherical volume. The masses of the gas component and the dark halo are identical. Inelastic collisions were introduced between gas clouds to simulate the dissipative nature of the gas, whereas the dark matter particles are assumed to move in a collisionless manner in the galaxy gravitational field. In the model depicted in Fig. 1, a uniform rotation was given so as to bring the system in a nearly centrifugal equilibrium initially. As seen in Fig. 1, the galaxy collapses nearly perpendicularly to the galactic plane within $\sim$ 1Gyr.

The most remarkable feature in this model is that several distinctive clumps are formed as the disk is built up by the gas infall to the galactic plane. The typical mass of individual clumps measured from the simulation is $10^9 M_\odot$. This is larger by three orders of magnitude than the masses of globular clusters and giant molecular clouds, which are most massive entities known in the present-day galaxies. These subgalactic clumps are created by the local gravitational instability of the mostly gaseous galactic disk in its early phase of formation. Star formation process is included in this simulation by converting gas clouds into stellar particles in accordance with a specified star formation law. The star formation rate reaches a maximum of $\sim 40 M_\odot yr^{-1}$ at the major epoch of disk formation ($t \sim 10$). This is also the epoch when the clumpy structure is most prominent. The clumpy and sometimes asymmetric disk seen in this model is strongly reminiscent of the images of many galaxies at large redshift obtained by the *HST* (e.g., van den Bergh et al. 1996).

2. Clump-driven Galaxy Evolution

The subgalactic clumps play a key role in the evolution of disk galaxies. They experience strong dynamical friction owing to their large masses. Resulting accumulation of clumps to the galactic center makes a spheroidal bulge (Fig. 1, $t = 30$). The bulge is not formed quickly in a single event as the protogalaxy collapses but is assembled gradually as individual clumps formed in the disk plane spiral into the galactic center. Fig. 2 shows the distribution of the age for the bulge stars at several epochs. At the final epoch indicated, $t = 28$, the age spread is as large as $\sim$ 1 Gyr. The nature and evolution of bulges remain one of the important unresolved problems in the galactic astronomy (e.g., Rich 1996). Recent observations suggest that the bulge of the Milky Way galaxy has a large age range and shows

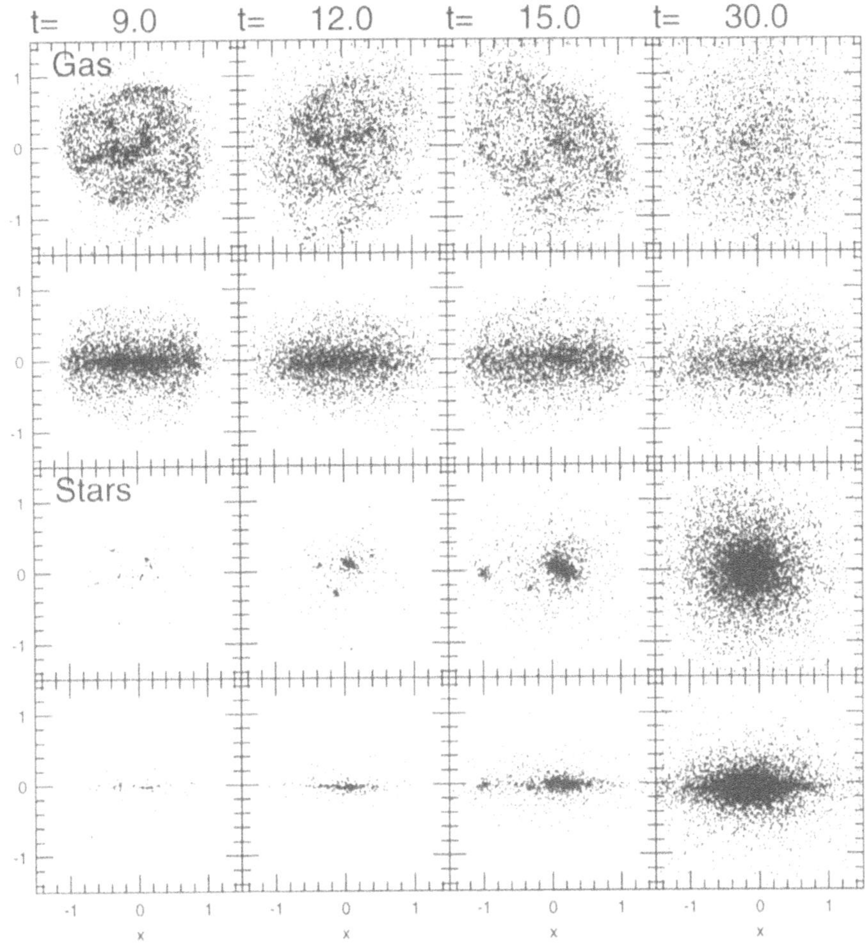

Figure 1. Morphological evolution of the disk galaxy model. The gas cloud particles and the stellar particles are projected onto the x-y plane (i.e., the disk plane) and the x-z plane. Time, t, is given in units of $\sim 10^8$ yr.

a correlation between abundance and kinematics of the constituent stars, which is difficult to explain by the conventional picture in which the bulge formation is a dynamically rapid process associated with the collapse of the protogalaxy. The clump-driven evolution model presented here may explain the enigmatic nature of the Milky Way bulge. The age distribution plotted in Fig. 2 indicates that the bulge sometimes contains a significant amount of young stars with ages less than 10^8 yr. This is owing to the starburst caused by the falling of one of massive clumps into the bulge region. This behavior may explain the blue nucleated galaxies, which Schade et al. (1995) noted

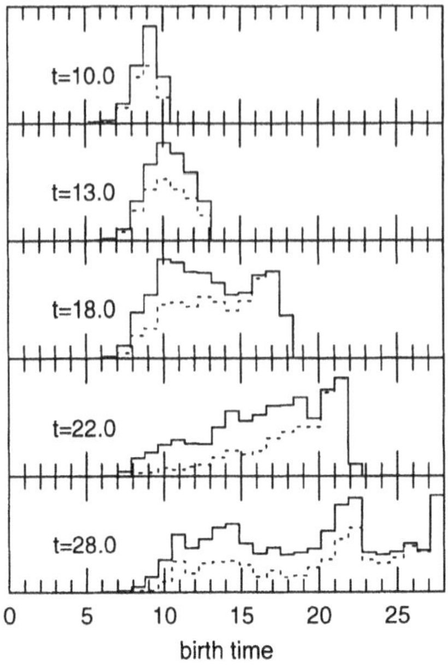

Figure 2. Distribution of the age for the stars contained in the bulge region. Each histogram shows, for the given epoch, the relative number of the stars which were born at the time indicated in the abscissa (unit time is $\sim 10^8$yr). Solid lines are distributions for stars located within 0.1 length units of the bulge center, whereas the dotted lines for stars located within 0.025.

among high redshift galaxy samples.

The clumpy disk has a tendency to acquire an exponential surface density distribution (Fig. 3), which is a universal feature of disk galaxies. This is understood naturally because a clumpy galactic disk in the present model can be viewed as a star-forming viscous disk (e.g., Saio & Yoshii 1990). As the bulge grows, the rotation curve of the disk component gradually steepens and the radial zone having a constant rotational velocity widens, leading to a flat rotation curve, which is commonly observed in disk galaxies, especially of early morphological types (e.g., Rubin et al. 1985). While the clumps are orbiting in the disk plane, stars and gas clouds are scattered off them and acquire random motions because of large masses of individual clumps. The old stellar disk component thus shows a progressively large scaleheight as the time elapses and evolves into what is called a thick disk.

3. Quasar Genesis

The clump-driven evolution model provides a unique prediction on the emergence of active galactic nuclei (AGN) in evolving galaxies. The stan-

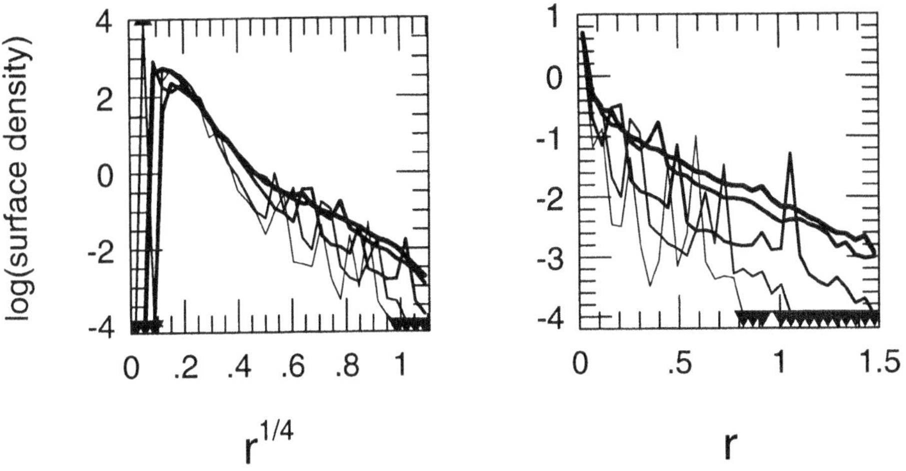

Figure 3. Growth of the bulge and disk components. Right panel plots the surface density distribution in the projection onto the x-y plane for the stars as a function of the galactocentric distance. The progressively thicker lines correspond to the epochs, $t =$ 11, 13, 15, 22, and 29. Left panel is the same as the right panel except that the abscissa denotes $r^{1/4}$ to facilitate the comparison with the de Vaucouleurs' density law.

dard picture of AGN (including quasars) presumes, as their engines, a super-massive black hole located at the galactic center and fed by a gas infall from the surrounding accretion disk at a rate of $\sim 1 M_{\odot} yr^{-1}$. The mergers and accumulation of the heavy subgalactic clumps may provide an efficient way to cause the required inflow. As Fig. 4 indicates, each merger event between two clumps induces mass accretion onto those clumps with infall rate up to $\sim 20 M_{\odot} yr^{-1}$. Each merging pair may be regarded as a potential site of quasar activity. According to this scenario, there is no single well-defined epoch of quasar activity, and one single galaxy can be activated many times repeatedly and intermittently. Also, the active site need not coincide with the galaxy nucleus, because a clump merger can happen in the outskirts of the galactic disk. There is even a possibility of multiple quasars in principle. For example, a double quasar is formed when *two pairs* of colliding clumps happen to merge at nearly the same instant. The growth of the black hole population within a galaxy is hierarchical, from numerous small seed black holes which are embedded in clumps and orbiting in the disk, to successively larger ones via mergers, and finally to one (or a few) super-massive black hole(s) located in the nucleus.

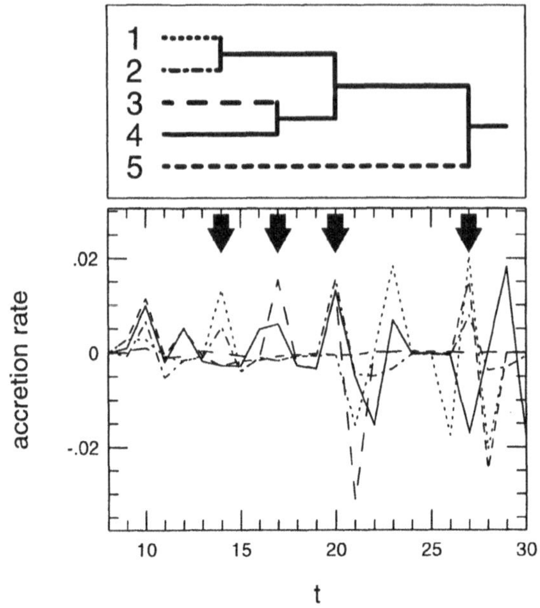

Figure 4. Merger history of the five selected clumps (upper panel) and the mass accretion onto individual clumps (lower panel). The same clump is indicated by the same line type in the upper and lower panels. Major accretion events associated with mergers between two clumps are indicated by arrows. A mass accretion of 0.01 corresponds to $\sim 20 M_{\odot} yr^{-1}$. Time, t, is in units of $\sim 10^8$yr.

4. Hubble Sequence

What is the meaning of the Hubble morphological sequence in the framework of the clump-driven evolution model? If the disk formation is slowed down by some reason, the clumping of the disk is weakened because the disk contains less gas at its major formation epoch (see Noguchi 1996 for details). Thus all the clump-driven dynamical processes are subdued. The slow collapse model develops a smaller bulge component, has a thinner stellar disk, and a rotation curve rising outwards more gently. These features are all characteristic of late-type disk galaxies. The clumpy galaxy model presented here predicts that the prominence of the bulge should be correlated with the existence of a thick disk and the mass of the supermassive black hole in the galaxy nucleus (e.g., Kormendy & Richstone 1995). Therefore, one possibility is that the Hubble sequence is largely that of disk formation time scale (Noguchi 1997).

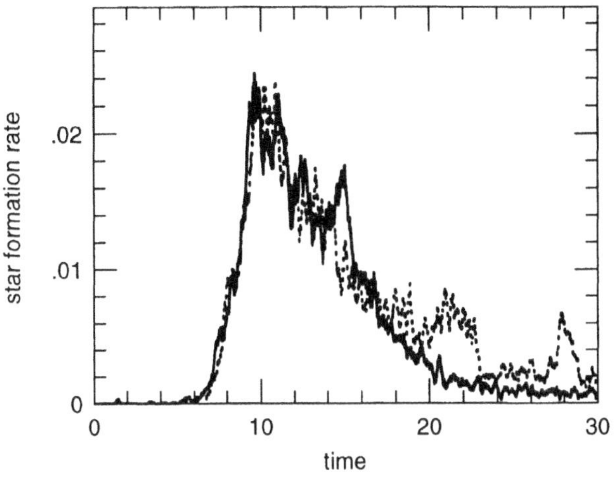

Figure 5. Star formation history of the perturbed galaxy model (solid) compared with that of the isolated evolution model (dotted). Star formation rate of 0.01 corresponds to $\sim 20 M_{\odot} yr^{-1}$.

5. Interactions of Primeval Disk Galaxies

In the nearby universe, many interacting and merging galaxies are observed to be sites of active star formation. Most of this starburst activity can be understood as a result of efficient gas flow to galactic centers induced either by tidally-created bars (Noguchi 1988) or merging process itself (e.g., Barnes & Hernquist 1992). Numerical simulations have been carried out in which the clumpy galaxy model in its early evolution stage was made to collide with a perturbing galaxy (Noguchi 1998). No significant enhancement in star formation rate was observed (Fig. 5). The clumpy nature of the gas-rich disk prevents any coherent structure (such as a bar) from developing (Fig. 6). A completed galactic disk supported mainly by well-ordered rotational motion (such as seen in nearby disk galaxies) is necessary for a galaxy to respond dramatically in tidal interactions.

References

Barnes, J.E., and Hernquist, L. (1992), *Ann. Rev. Astron. Astrophys.* **30**, 705-742

Kormendy, J., and Richstone, D. (1995), *Ann. Rev. Astron. Astrophys.* **33**, 581-624

Noguchi, M. (1988), *Astron. Astrophys.* **203**, 259-272

Noguchi, M. (1996), *Astrophys. J.* **469**, 605-622

Noguchi, M. (1997), submitted

Noguchi, M. (1998), in preparation

Rich, R.M. (1996), in *Unsolved Problems of the Milky Way* (eds. Blitz, L. & Teuben, P.) 403-410 (IAU Symp. No. 169, Kluwer Academic Publishers, Dordrecht)

Rubin, V.C., Burstein, D., Ford, W.K., and Thonnard, N. (1985), *Astrophys. J.* **289**,

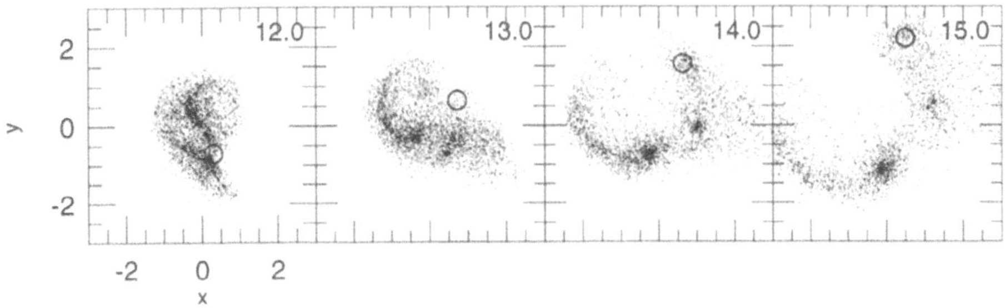

Figure 6. Morphological evolution of the perturbed galaxy model (to be compared with Fig. 1). Only gas clouds are shown. The circle indicates the position and effective radius of the Jaffe-type rigid perturber with a mass equal to that of the perturbed galaxy. Time in units of ~ 10^8yr is given in each panel.

81-104

Saio, H., and Yoshii, Y. (1990), *Astrophys. J.* **363**, 40-49

Schade, D., Lilly, S.J., Crampton, D., Hammer, F., Le Fèvre, O., and Tresse, L. (1995), *Astrophys. J.* **451**, L1-L4

van den Bergh, S., Abraham, R.G., Ellis, R.S., Tanvir, N.R., Santiago, B.X., and Glazebrook, K.G. (1996), *Astron. J.* **112**, 359-368

STRONG GRAVITATIONAL LENSING ON THE HUBBLE DEEP FIELD

R. D. BLANDFORD
130-33 Caltech,
Pasadena,
CA 91125, USA

1. Introduction

The Hubble Deep Field, henceforth HDF, (Williams et al 1996) is a unique data set for studies of faint galaxies. A small area was chosen well away from known local density enhancements and sources of obscuration. It was imaged for 30 complete orbits in each of four filters centered on wavelengths $\lambda 300, 450, 606, 814$ nm. There are three, contiguous WFPC2 frames and a solid angle 15,500 arcsec^2 is usable. This tiny region is teeming with galaxies. At least 2500 significant features can be identified down to $R = 30$, though it is not clear that these are all entire galaxies.

This data archive has, as intended, spawned a variety of follow up studies, including infrared and radio surveys and a redshift survey using the Keck telescope. This last program has garnered over 500 redshifts on the HDF and neighboring "flanking fields" (Cohen et al. 1996, Phillips et al. 1997), of which over 100 are on the HDF itself and therefore imaged with unprecedented depth. The redshift sample is essentially complete to $R \sim 23$.

2. Some Key Results

For completeness, let me list some developments that have come from analysis of the HDF. Some of these are original discoveries; others are verifications and elaborations of prior results. All of them are discussed in a forthcoming proceedings of a conference held at STScI ("The Hubble Deep Field", ed. Livio & Ealley).

- There appear to be $\sim$ 100 billion faint sources observable over the whole sky. These may be galaxies, but are surprisingly blue in color and they **appear** to be very compact with characteristic sizes $\sim 3h^{-1}$ kpc.

- The slope of the galaxy counts flattens fainter than $B \sim 25$ and most of the galaxian contribution to the brightness of the night sky comes from around this magnitude. (Again some caution is necessary because *HST* may not be sensitive to extended low surface brightness regions.)
- Perhaps the most significant discovery is of a substantial population of normal galaxies at redshifts $z \sim 3$. These are about 3 percent of the $\sim 25^m$ galaxies and represent a major breakthrough in the study of galaxy formation (Steidel et al. 1996, Lowenthal et al. 1997).
- It has also been found that just under half of galaxies with $z > 0.5$ are irregular morphologically, a far higher proportion than is the case at lower redshift (Abraham, these proceedings).
- The redshift surveys have shown that galaxies with $z \sim 0.5$ are less clustered than locally. In addition, about 40 percent of these are concentrated within narrow features in redshift space (Cohen et al. 1996). (As we only have a pencil beam survey, with linear width much smaller than the two point correlation function, this finding must be interpreted judiciously.) The redshift survey has also allowed a convincing demonstration of the reliability of the color-redshift method, at least for $z < 1$, $R < 24$ (Hogg et al. 1997 preprint).

3. The Number Problem

That there are too many faint galaxies may be demonstrated succinctly, if somewhat polemically. From the measured local luminosity function there are roughly $\sim 4\Phi^*$ galaxies with luminosity in excess of $\sim 0.01L^*$. This evaluates to ~ 0.4 billion galaxies (BG) per Hubble volume. Now, in an Einstein-De Sitter universe, there are roughly four Hubble volumes out to $z \sim 3$, which is as far as we are likely to see faint galaxies in the optical and therefore we only expect to see ~ 1.6 BG that are counterparts of the bright galaxies that we see around us. There are therefore roughly 60 times more faint objects than there ought to be.

There are a variety of resolution of this problem. In ascending importance these include:

- Denial. A surprisingly large number of cosmologists still see no problem.
- Incompetence. We may have underestimated the local luminosity function and may be missing many low surface brightness galaxies (McGaugh 1994). There is some evidence that this may be the case at the level of a factor 2, but this can hardly be the total explanation.
- Cosmography. Open universes give us more comoving volume before Lyman limit absorption extinguishes the galaxies. However, they do this at the expense of increasing the luminosity distance and mak-

ing the same galaxies much fainter. We have to appeal to a highly non-standard world model to find enough help here. This would be a wonderful discovery, if true.

– Local dwarfs. The problem here is that the local faint luminosity function appears to be quite flat, although it does appear to steepen at fairly modest redshifts. This is unlikely to be the whole explanation because cluster lensing (and related arguments) tell us that a sizable fraction of the faint galaxies have to be at high redshift.
– Fading population. There may be a large population of intermediate luminosity galaxies with $z \sim 1$ which form and have too shallow potential wells to retain their gas in the face of the supernova blast waves associated with the first generation of stars. These galaxies should still be visible locally in the infrared (Babul & Feguson 1996). Unfortunately, we see no evidence for them (Hogg et al. 1997).
– Mergers - the topic of this meeting. The faintest, 30^m objects may be the building blocks out of which galaxies are being formed. If so, then it is surprising that we do not see them more clustered (Colley et al. 1997). However, this may not be a fatal objection because the merging time can be quite short.

4. Strong Gravitational Lensing

These prefatory remarks provide a context for an ongoing investigation of strong gravitational lensing on the HDF. This is also motivated by radio and other surveys which are showing that a fraction ~ 0.002 of high redshift sources are multiply-imaged by intervening galaxies. On this basis, Hogg et al. (1996) argued that there should be $\sim 5 - 10$ strong lenses on the HDF and identified the most promising candidates on morphological grounds. Unfortunately, it has proved hard to follow this up as the candidates are very faint for spectroscopic study. There are still no convincing examples of multiply-imaged galaxies on the HDF. I have therefore tried another approach to attempt to locate these gravitational lenses or argue that they are not present, (which would immediately imply that the faint galaxy sources are predominantly relatively close by).

What I have tried to do is to combine the exquisite imaging of the potential lens galaxies on the HDF with their measured redshifts to identify the galaxies most likely to be lenses and then to use the faint galaxy counts to quantify the lensing probability. In addition I have performed simulations to see if it is possible to detect multiple imaging of a background against the light from the lens galaxy. As this research is ongoing, I can only report preliminary conclusions.

4.1. LENSES

As these are field galaxies with no discernible cluster along the line of sight, multiple imaging will be confined to the central regions of the most concentrated galaxies and least affected by dark matter in the halo. Now for every galaxy in the redshift sample, we can use the redshift to estimate the surface brightness in the frame of the source in the emitted B band. To accomplish this, we use the $F814W$ filter, (as this image will be least susceptible to the effects of dust), and make a spectral correction based upon the integrated color. (This is preferable to performing the spectral correction on a pixel by pixel basis because we are interested in the global mass distribution.) We next convert this surface brightness into a surface mass density by assuming a constant mass to light ratio in the B-band. Local measurements of $hM/L_{\rm B}$ suggest a value of 5 for spirals and 10 for ellipticals. Measurements of known gravitational lenses (which may be a biased sample) suggest twice this value (Keeton, Kochanek & Falco, 1997 preprint). (I included irregulars and unclassified galaxies in with the spirals.) Both observations and the expectations from passive stellar evolution indicate that galaxies were about twice as luminous at $z \sim 1$ and a simple, empirical evolutionary correction was made. We must next make some cosmographic assumptions. For illustration, I shall adopt an open universe with $\Omega_0 = 0.3$. (The outcome is independent of h.) I shall also place all the faint sources at $z = 3$. The next step is to adopt the surface brightness maps of each of the 114 galaxies with known redshifts $z < 1.5$ and solve Poisson's equation using a Fourier method to obtain the equivalent surface potential. This can, in turn, be converted into a deflection vector field and a magnification field, obtained from the gradient of the deflection field. Images with negative magnification are multiple and so we can compute the cross section for multiple imaging by projection back to the source plane. There were many surprises. The first was that most galaxies do not multiple image, at least at the resolution of HST. (Of course if every galaxy contains a massive black hole in its nucleus, they will each multiple image every other galaxy an infinite number of times (Rauch & Blandford 1994). However, this is is not a concern for observational astronomy!) Those that do are almost all elliptical galaxies that have higher central surface brightness and are assumed to have higher mass to light ratio. In fact, three-quarters of the cross section is contributed by four galaxies. Secondly, the propensity for multiple imaging is strongly dependent upon the mass to light ratio. For example, increasing the central mass to light ratio for local galaxies from 10 solar units to 20 solar units increases the cross section of the strongest lens from ~ 0.3 arcsec^2 to ~ 0.9 arcsec^2. As always, finding gravitationally lensed sources of known redshift is a very sensitive measurement of the central mass to light ratio of the lensing galaxy. Thirdly, the four strongest lenses all have $z \sim 1$ and $L \sim L^*$ and are therefore quite ineffectual if the sources have redshifts much less than $z \sim 2$. Finding multiple images systematically from a complete sample of lens galaxies is a potentially powerful probe of the faint source redshift distribution. The integrated cross section over the whole HDF is surprisingly low, only ~ 1 arcsec^2 for the "local" choice of M/L increasing to ~ 3 arcsec^2 for the lens-normalized choice.

4.2. SOURCES

The sources were classified by size rather than integrated flux. Size was defined operationally as the number of contiguous 0.04 arcsec pixels with counts above 2.5σ. There are $\sim$ 1600 galaxies with sizes between 5 and 50 pixels (about two-thirds of the total number of galaxies). The median galaxy in this sample has a size of 13 pixels. As we extrapolate to smaller images, we do not expect to find a large increase in the potential source density unless there is a sharp upturn in the magnitude source counts below $\sim 30^m$. The total density of source galaxies is therefore ~ 0.1 arcsec^{-2} and the total probability of finding one of them multiple-imaged if they are at redshift $z \sim 3$ is ~ 0.1 assuming the local M/L and ~ 0.3 for the lens-normalized M/L. For the individually strongest galaxies, the probabilities per lens are therefore ~ 0.03 and ~ 0.1 respectively.

4.3. DETECTABILITY

What is clear from the foregoing discussion is that it may pay to seek multiple images very carefully in judiciously chosen locations, rather than by inspecting a large sample of galaxies. Now, it goes without saying that this is a daunting task. What we are looking for are faint, inclined arc-like features within ~ 5 kpc. They are likely to be patchy and blue. In other words, they look just like spiral arms! Now, I believe that the prospects are not so bleak if one has images with a resolution and depth comparable with those on the HDF in several filters. In particular, ellipticals are clearly preferred and can be identified on spectral and not just morphological grounds. They have relatively smooth light profiles against which arcs can be brought out by image processing techniques. Furthermore the image pattern that is expected, and can be easily simulated, is quite distinctive. The magnified images should have similar colors although variable reddening by the lens galaxy can confound this expectation.

A quick inspection of the images of all of the galaxies in the redshift survey reveals several features that could be formed by galaxy lensing. However, none of these is convincing, based upon the criteria listed above. Looking at the four best candidates, only the strongest of these exhibits a feature that is worth considering further and this is a large, tangential arc at a radius of ~ 1 arcsec. It has a similar color to the rest of the galaxy and would require $hM/L \sim 40$ solar units. It is far more likely to be a shell produced by tidal stripping of a merging galaxy.

5. Conclusion

Using the observed surface brightnesses of galaxies on the HDF, we estimated a strong lensing frequency at least an order of magnitude lower than that based on radio surveys and advertized in Hogg et al. 1996. If this is the correct approach to estimating the frequency of multiple imaging, then it should not be a surprise that we have no good examples to offer as yet. Put another way, the best multiple image candidates selected from the HDF on morphological grounds, required M/L ratios far in excess of those measured in local, normal galaxies. Perhaps, in retrospect, this is quite reasonable. Many of the secure examples of gravitational lensing are unusual because they belong to a rich compact group or cluster which magnifies the focusing power of an individual galaxy and the HDF was selected, as far as possible, to avoid such concentrations. Furthermore, the source population in the radio surveys is known to be at high redshift, whereas not all of the faint optical sources may be distributed in this manner.

Nevertheless, in spite of this somewhat negative, interim report, I hope that I have demonstrated that scrutiny of deep images in a larger sample of bright, high redshift, field ellipticals should tells us much about both the evolution of their mass to light ratios and the nature and distances of the very faintest optical sources on the sky.

Acknowledgements

I thank my colleagues Tereasa Brainerd, David Hogg and Tomislav Kundić for collaboration on various parts of this investigation and Roland van der Marel for consultation on mass to light ratios. Special thanks are due to Judith Cohen who has led the Caltech-Keck redshift survey and colleagues at UCSC and UH who have also contributed many of the redshifts. Acknowledgement must also be made of Bob Williams who had the foresight to see the long term value of the HDF project. Support under NSF grant AST 95-29170 is gratefully noted.

References

Babul, A. & Ferguson, H. C. ApJ 458 100
Cohen, J. G. et al. 1996 ApJ 471 L5
Colley, W. N. et al. 1997 ApJ 488 579
Hogg, D. W. et al. 1996 ApJ 467 L73

Hogg, D. W. et al. 1997 AJ 113 474
Lowenthal, J. D. et al. 1997 ApJ
McGaugh, S. S. 1994 Nature 367 538
Phillips, A. C. et al. 1997 ApJ in press
Rauch, K. P. & Blandford, R. D. ApJ 421 46
Steidel, C. C. et al. 1996 AJ 112 352
Williams, R. E. et al. 1996 AJ 112 1335

MEASURING THE EVOLUTION OF THE MASS-TO-LIGHT RATIO FROM $z = 0$ TO $z = 0.6$ FROM THE FUNDAMENTAL PLANE

MARIJN FRANX AND PIETER VAN DOKKUM
Kapteyn Astronomical Institute
Groningen, The Netherlands

DAN KELSON AND GARTH ILLINGWORTH
University of California, Santa Cruz, USA

AND

DAN FABRICANT
Center for Astrophysics, Cambridge, USA

Abstract. Galaxy evolution is probably a complex process. Mergers, infall, and starbursts may change galaxy properties systematically with time. As a result, the interpretation of the luminosity function is ambiguous, and information on the mass evolution of galaxies is needed. Such information can be retrieved from the evolution of the Tully-Fisher relation, Faber-Jackson relation, or the Fundamental Plane with redshift.

Observations of this kind have recently become possible. We present the Fundamental Plane relation measured for galaxies in the rich clusters out to $z = 0.58$. The galaxies satisfy a tight Fundamental Plane, with relatively low scatter (17%). The M/L ratio evolves slowly with redshift, $\ln M/L_V \propto 0.8z$. This result is consistent with simple evolutionary models if the bulk of the stellar population formed at high redshift.

It is not clear yet how these results can be made consistent with the rapid evolution of galaxies in intermediate redshift clusters as indicated by the Butcher-Oemler effect. Observations of post-starburst galaxies ("E+A" galaxies) indicate that these systems are dominated by disks. They may evolve into galaxies which are underrepresented in most "normal" Fundamental Plane samples.

1. Measuring the evolution of mass: $F(M_*, z)$ or $F(v_c, z)$

Galaxy evolution may be a complex process, with possibly a large role for mergers, interactions, infall, and starbursts triggered by these events. Such processes complicate the interpretation of observations of high redshift galaxies, as galaxies can change rapidly in luminosity (due to starbursts), and can change morphology due to mergers, infall of gas, and enhanced star formation. The progenitors of certain types of galaxies at some redshift may be of different type at some other redshift, and their luminosities may be quite different.

In order to quantify these effects, more information is needed than the evolution of luminosity and color of galaxies, such as measured by the evolution of the luminosity function. Detailed information on the morphological evolution, and the evolution of the mass function is essential. The evolution of the mass function is possibly the most important, as it gives direct insight into the mass evolution of individual galaxies, and can directly determine when typical galaxies were assembled.

Unfortunately, the total masses of galaxies are notoriously difficult to measure. However, there exist good relations between circular velocity, and velocity dispersion, and photometric parameters: the Tully-Fisher relation for spirals (Tully & Fisher 1977), the Faber-Jackson relation (Faber & Jackson 1976), and the Fundamental Plane for early-types (Djorgovski & Davis 1987, Dressler et al. 1987). These relations are very suitable for evolutionary studies, because their intrinsic scatter is low at $z = 0$.

The general purpose of such observational studies will be a measurement of the evolution of the Tully-Fisher relation, Faber-Jackson relation, and Fundamental Plane with redshift. The combination of the observations with the evolution of the luminosity function of galaxies, can be used to constrain the evolution of the distribution of circular velocities $F(v_c, z)$, which is less sensitive to starbursts than the equivalent $F(L, z)$. Similarly, the stellar mass locked up in early type galaxies can be measured in a similar way. Such results will provide direct constraints on theories of galaxy formation and evolution.

2. Evolution of the Fundamental Plane

Here we present new results on a program to measure the evolution of the Fundamental Plane relation with redshift. Early results can be found in Franx (1993a,b, 1995). The Fundamental Plane is a relation between effective radius r_e, effective surface brightness I_e, and central velocity dispersion σ of the form $r_e \propto \sigma^{1.24} I_e^{-0.82}$ (e.g., Bender et al. 1992, Jørgensen et al 1996, JFK). Its scatter is low, at 17% in r_e (Lucey et al. 1991, JFK). The implication of the Fundamental Plane is that the M/L ratio of galaxies is well behaved (e.g., Faber et al. 1987). Under the assumption that galaxies are a homologous family, the implied M/L scaling is $M/L \propto r_e^{0.22}\sigma^{0.49} \propto M^{0.24}$. Such scaling is sufficient for the existence of the Fundamental Plane, and vice versa. The cause of the variation in M/L with mass is not well understood, but it is thought to be mainly due to variations in metallicity (see also, e.g., Renzini & Ciotti 1993).

Observations at higher redshifts will yield the evolution of the M/L ratio as a function of redshift. Below we explore the expected variation of M/L.

2.1. MODELS FOR THE EVOLUTION OF THE M/L RATIO

The luminosity of a co-eval stellar population is expected to evolve with time. Tinsley (1980) showed that the luminosity evolves like

$$L \propto 1/(t - t_{\rm form})^{\kappa}$$

where $\kappa = 1.3 - 0.3x$, and x is the slope of the IMF. The Miller–Scalo IMF implies $x = 0.25$, and $\kappa \approx 1.2$. Recent studies indicate that the value of κ depend on passband and metallicity (Buzzoni 1989, Worthey 1994). These authors find $0.6 < \kappa < 0.95$ for the V band.

To first order, this evolution implies that the M/L ratio evolves like

$$\ln M/L(z) = \ln M/L(0) - \kappa(1 + q_0 + 1/z_{\rm form})z,$$

where $z_{\rm form}$ is the formation redshift (Franx 1995). Hence the logarithm of the M/L ratio is expected to decrease linearly with redshift, and the coefficient depends on κ(IMF), q_0, and $z_{\rm form}$. This equation is valid for $q_0 \approx 0$, and high $z_{\rm form}$. The equation implies that the rate at which the M/L ratio decreases is a function of several unknown variables, and a direct interpretation of the observed decrease of the M/L ratio may not be very straightforward.

Fig. 1a shows the expected evolution of the L/M ratio if all galaxies form at the same redshift. As can be seen, the evolution depends strongly on the formation redshift. It is unlikely that galaxies formed in such a simple way. For Fig. 1b we explored models in which galaxies form at a range of redshifts. As a result, scatter is introduced in the L/M ratio, which increases with look back time. This is due to the fact that the relative age difference increases with look back time.

2.2. COMPLEX EVOLUTION

Even the last model is probably an over-simplification of the formation of early types. There is no good reason to assume that all stars in an early-type galaxy formed in a very short burst. A single galaxy may have had a

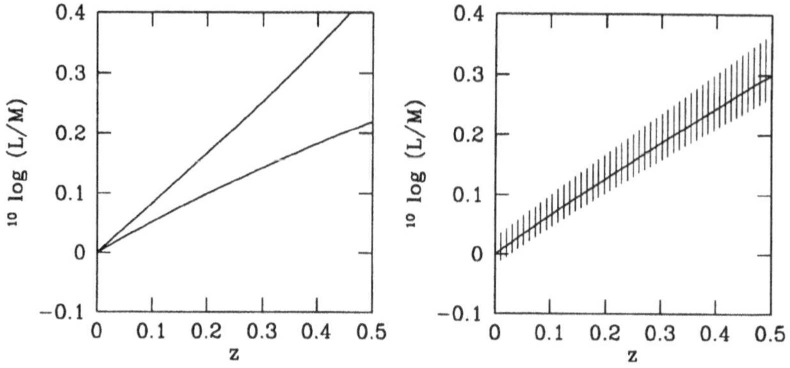

Figure 1. The evolution of galaxies with a simple star formation history. a) shows the luminosity evolution for galaxies with co-eval populations. Galaxies which formed recently evolve faster. b) the evolution of the mean L/M ratio for a sample of early-types which formed at a random time between $z = 1$ and $z = 2$. The scatter in the relation increases with redshift, as the relative age difference increases with lookback time.

complex formation history, with star formation extending over a long time. The evolution of the M/L ratio will be more complex if such age differences are taken into account.

We have created models in which early type galaxies form by transformation of galaxies with continuous star formation. It is assumed that the progenitors form stars in a continuous way, until a burst of star formation occurs, and the galaxies are transformed into non-star forming galaxies. These will appear as post starburst galaxies for another 1.5 Gyr, and then appear to be early types.

This type of evolution implies that the morphologies of galaxies evolve with time, from spiral, to post star burst galaxy, to early-type. This has important consequences, since the set of early-types at higher redshifts will be a special subset of the set of early-types at $z = 0$. If we select early-types at higher and higher redshift, we are selecting a subsample that is more and more biased towards the oldest early-types. In short, we may be selecting the oldest galaxies, and find that they are old.

The effect is illustrated in Fig. 2. Fig. 2a shows the typical evolution of 3 galaxies. The solid curve is the phase in which they appear as early-types. Clearly, the oldest early-types appear as early-type for the longest time. Fig. 2b demonstrates the effect on the observed L/M ratios of a large sample. At low redshifts, all galaxies appear as early types, and the evolution of the median L/M ratio remains normal. The scatter around the mean increases rapidly with redshift. Around $z = 0.2$, some of the galaxies appear as post star burst galaxies, and they would be excluded.

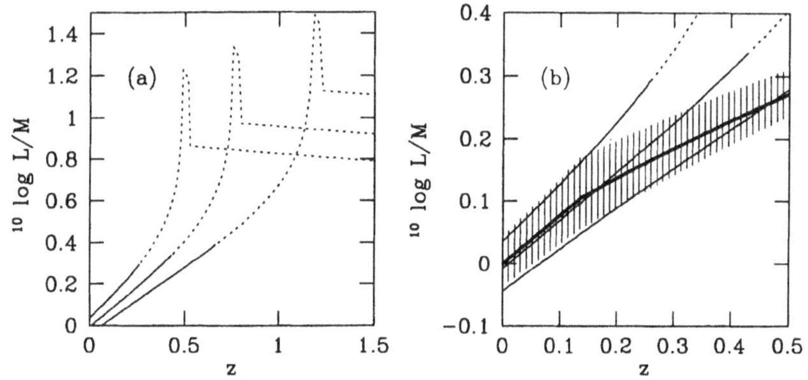

Figure 2. The evolution of galaxies which undergo three distinct phases: I — regular star formation in a disk, II — starburst, III — quiescent evolution. a) shows the luminosity evolution for three such galaxies. The galaxies are classified as regular early-types after 1.5 Gyr after the burst. This epoch is indicated by the solid curve. b) the evolution of the mean L/M ratio for a sample of early-types which formed in this complex way. The starburst is assumed to occur at a random time between $z = 0.5$ and $z = 2$. The thick line indicates the median L/M, the shaded area is bounded by the upper and lower quartile of the sample. The median L/M ratio bends at $z = 0.2$, as more and more galaxies drop out from the sample. The sample becomes more and more biased to the oldest early-types at higher redshifts.

The median L/M ratio is biased towards low values. This effects increases at higher redshifts. The bias is as strong as 30% at $z = 0.5$. As galaxies disappear from the sample, the scatter in the L/M ratio may decrease at higher redshifts.

2.3. THE FUNDAMENTAL PLANE IN CL0024+16 AT $z = 0.39$

CL0024 is a rich cluster at $z = 0.39$, and has been extensively observed (e.g., Dressler et al. 1985). We have obtained a deep, 19 hour integration at the MMT to measure the internal velocity dispersions of luminous galaxies in the cluster. *HST* images were used to measure the structural parameters of the galaxies. Full details of the observations and the analysis can be found in van Dokkum and Franx (1996).

Fig. 3a shows the resulting Fundamental Plane. There is a very clear relation, with relatively low scatter (15%). The slope is very similar to that for nearby cluster galaxies (e.g., JFK). In short, *early-type galaxies exist at* $z = 0.4$ *which are very similar to galaxies at* $z = 0$.

Fig. 3b shows the observed M/L ratios for Coma and CL0024 against the parameter $r_e^{0.22}\sigma^{0.49}$. The Fundamental Plane implies that galaxies lie along a line in the plot. We see a clear offset between the two data sets. The

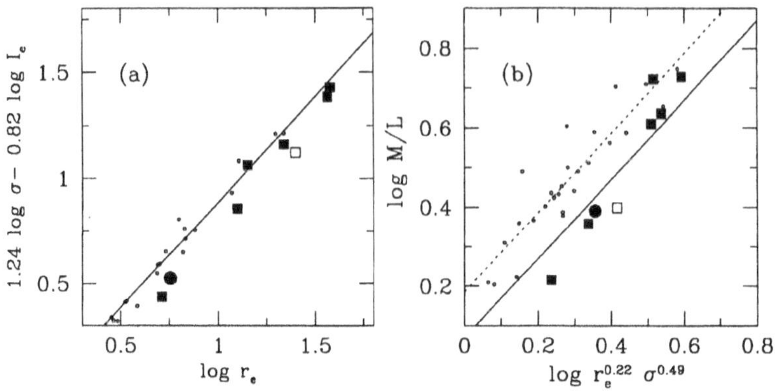

Figure 3. a) The Fundamental Plane for galaxies in CL0024+16 at $z = 0.391$ in the redshifted V band. The small symbols are galaxies in Coma. The Fundamental Plane in CL0024 is very similar to that in Coma, with similar low scatter (15%). b) The M/L ratio against $r_e^{0.22}\sigma^{0.49} \propto M^{0.24}$, for CL0024 and Coma. The lines are fits to the data points. The M/L ratio in CL0024 is lower by 31± 12%.

lines indicate fits to both data sets. The mean difference in the M/L ratio is 31%. The error is dominated by systematic effects, and is estimated at 12%. It is clear that the sample for CL0024 is biased towards the most massive galaxies, and this selection bias is partly the cause for the systematic uncertainty.

3. Using Keck to extend to $z = 0.58$

With modern telescopes and efficient instrumentation it is possible to extend the Fundamental Plane work out to higher redshifts. We have recently used Keck to measure the Fundamental Plane in two clusters at $z = 0.33$ and $z = 0.58$, CL 1358+65 and MS 2053-05 respectively. A full description can be found in Kelson et al. (1997). Typical integration times were 2 hours on the Keck telescope. Fig. 4a shows the resulting Fundamental Planes from $z = 0$ to $z = 0.58$. The figure demonstrates how well the relation is defined at each redshift interval.

Surprisingly, the scatter in the relation remains low. We have now 22 galaxies with Fundamental Plane parameters, and we find a scatter of 17%. This is quite comparable to the scatter in nearby rich clusters.

The evolution of the M/L_V ratio is shown in Fig. 4b. The data are consistent with a slow evolution of $\ln M/L_V \propto 0.8z$. Both the low evolution, and the small scatter are consistent with high formation redshifts for cluster early types ($z_{\rm form} > 2$).

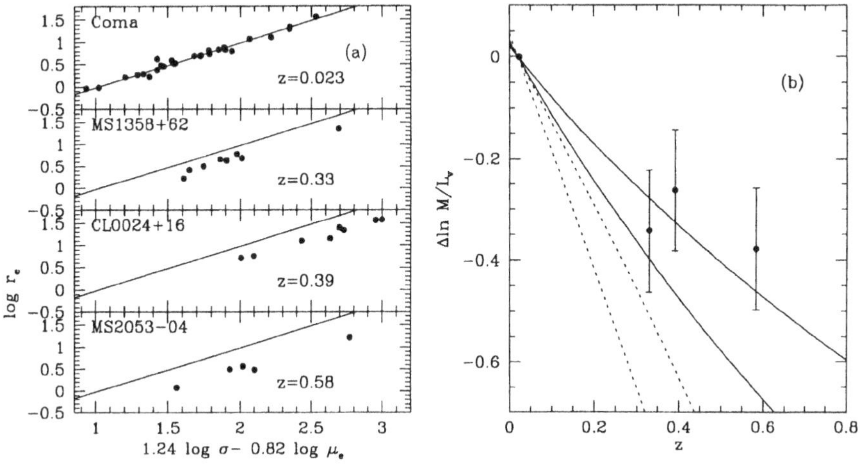

Figure 4. The evolution of the Fundamental Plane from $z = 0$ to $z = 0.58$, based on data from Kelson et al 1997. Panel a) shows the individual Fundamental Plane for the clusters. The solid line is the relation for Coma. The offset is mostly due to surface brightness dimming. Panel b) shows the evolution of the mean M/L_V ratio, after correction for surface brightness dimming ($q_0 = 0.5$). The M/L_V ratio decreases slowly with redshift. The solid lines indicate stellar population models with formation redshifts of ∞. The dashed lines indicates models with a formation redshift of 1. The data are consistent with high formation redshift. The datapoints move downward if lower values for q_0 are used.

4. How about the Butcher–Oemler effect and E+As?

It has been well established that distant clusters have a high proportion of blue members (Butcher & Oemler, 1984). Furthermore, some galaxies have post starburst spectra (Dressler and Gunn, 1983, 1992). These galaxies have spectra which can be modeled as the superposition of a young component and an old component, and were named "E+A" by Dressler and Gunn. The "E" stands for early type, and "A" for A-star. These galaxies were defined to have no emission lines, i.e., little or no star formation. The population models invoked a peak in the star formation rate 1 Gyr before their light was emitted, and a subsequent drop in the star formation rates.

The relatively high fraction of such post starburst galaxies in clusters, and the short lifetime of the phenomenon implies that many galaxies in clusters underwent such a phase at intermediate redshifts. Is this consistent with the slow evolution and high formation redshift indicated by the Fundamental Plane?

To answer this question, it is necessary to determine the morphologies of the "E+A" galaxies. This can be done on the basis of imaging and kinematics. We discuss two samples below.

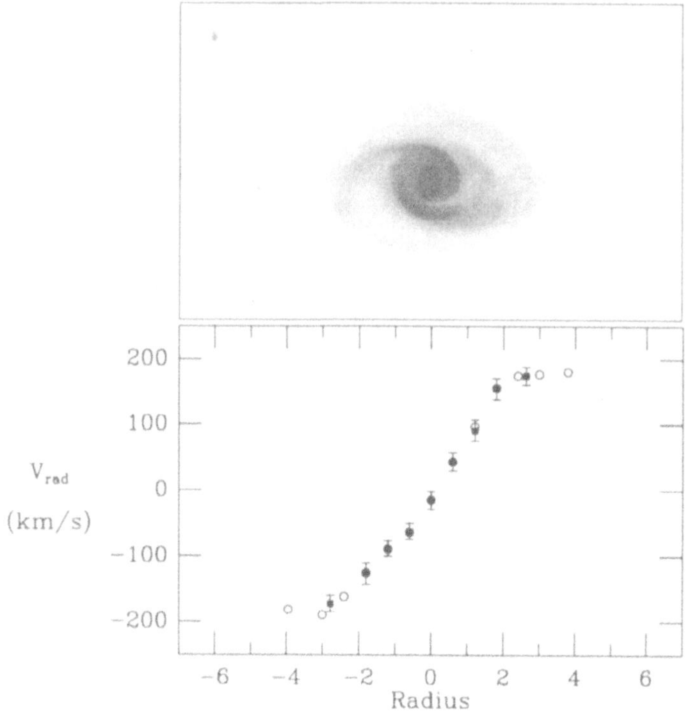

Figure 5. A bright E+A galaxy in Abell 665 at $z = 0.18$. The upper panel shows the *HST* image. The galaxy has smooth spiral arms, without regions of star formation. The lower panel shows the rotation curve of the galaxy, at the same scale. The galaxy is dominated by rotation at large radii. This E+A galaxy has virtually no bulge.

4.1. A GROUP OF E+A GALAXIES IN ABELL 665 AT $z = 0.18$?

Abell 665 contains a very bright, blue E+A galaxy. It was included by chance in the study of the kinematics of cluster members (Franx 1993a,b). The rotation curve of the galaxy proved that the galaxy was supported by rotation, and that it was likely a disk. Recent *HST* imaging has confirmed this. Fig. 5 shows the *HST* image, in combination with the rotation curve. The galaxy is strongly dominated by the disk, and has a very small bulge. It shows weak, smooth spiral arms, without any signs of star formation. The colors along the arms are also very smooth. The rotation curve is typical of a disk, and symmetric.

These data demonstrate that the galaxy is a disk. It will probably evolve into a disk dominated S0, given the lack of star formation. We notice that the optical morphology is very rare in the nearby universe: smooth spiral arms without star formation. This may be related to the recent cutoff in star formation.

There are two galaxies with the same morphology very close to the E+A in Abell 665. Furthermore, there is a fourth galaxy with strong star formation in the same area. This suggests that the other two galaxies might have similar E+A spectra, and that the fourth galaxy is still in the star forming stage – possibly shortly before the cutoff? Clearly, spectroscopic confirmation is required. The data are very suggestive that we are dealing with a small group of galaxies that are undergoing the transformation from the star formation phase into a early type phase.

The mechanism of the transformation is not clear: it could be triggered by the small group itself, or it could be triggered by infall into the rich cluster. It is clearly not triggered by a major merger, as such a merger would not produce disk dominated galaxies. Minor mergers, or tidal interactions, can certainly not be ruled out.

4.2. E+A GALAXIES IN CL 1358+65

We have obtained a large *HST* mosaic of 7x7 arcmin on the cluster CL 1358+65 at $z = 0.33$. We have spectra of 190 cluster members in the mosaic, and have selected E+A galaxies on the basis of this spectroscopy. Again, as in Abell 665, the E+A galaxies are disk dominated. Some of the galaxies have very strong Balmer lines, with a mean equivalent width of $\langle H_{\beta,\gamma,\delta} \rangle \approx 8$ Å. The colors are generally smooth across the galaxies, indicating that the young component is smoothly distributed. More detailed spectroscopy is needed to characterize the galaxies better. This will be presented in Franx et al, in preparation.

4.3. WHERE DO E+AS GO?

The above evidence implies that E+As may evolve into disk dominated S0s. We have to note that we cannot be certain about this: the E+As may also be in groups which merge to form ellipticals, or bulge dominated S0s. We can think of several ways to explain the low evolution and low scatter in the Fundamental Plane and the evidence for recent bursts and star formation from E+As:

1. The E+As may evolve into disk dominated S0s which are under-represented in the current sample for the Fundamental Plane.
2. The E+As may have generally low central dispersions, and such galaxies are usually excluded from Fundamental Plane analysis.
3. The stellar population models may need to be adapted.
4. The E+As may merge with older systems so that the influence of the burst is "diluted".

More studies based on larger samples are needed to distinguish between these possibilities. Such studies are now in progress.

5. Discussion

We have shown that it is possible to determine the Fundamental Plane at intermediate redshifts, all the way up to $z = 0.58$. The relation is well defined at these redshifts, with a relatively low scatter of 17%. The mean evolution of the M/L_V ratio is low, at about 45% at $z = 0.58$. This evolution is consistent with an early star formation epoch for our galaxies. We note, however, that the current measurement may be biased, mostly due to the fact that we only use non-star forming red galaxies. We may therefore exclude the star forming progenitors of current-day early type galaxies.

We have analyzed the structure of E+A galaxies in our program clusters. We find that most of our E+As are strongly disk dominated. They would likely evolve into strongly disk dominated S0s, unless they merge with other galaxies. It is still not quite clear how to explain the low scatter in the Fundamental Plane on the one hand, and the high fraction of E+As on the other hand. It is possible, however, that most of the E+As have a low central velocity dispersion, and are mostly omitted from Fundamental Plane samples in nearby clusters.

These observations demonstrate that information on galaxy masses can be obtained from deep, ground based spectroscopy. The next step is to extend this work to higher redshift, and to the field. Furthermore, similar studies have demonstrated that the Tully-Fisher relation can be used (Vogt et al 1996, Rix, Colless, & Guhathakurta 1996). The new generation optical telescopes will allow a rapid extension of this type of work to larger samples, higher redshifts, and lower galaxy masses.

Eventually, these observations can be used to determine the evolution of the distribution of circular velocities for galaxies $F(v_c, z)$, and the evolution of stellar masses locked up in early type galaxies $F(M_*, z)$. This requires accurate observations of the evolution of the luminosity function and the "mass relations" in the field. The final outcome of such a program can be expected to provide strong constraints on the models of galaxy formation and evolution.

It is a pleasure to thank the organizers for a stimulating conference.

References

Bender R., Burstein D., Faber S. M., 1992, ApJ, 399, 462
Buzzoni A., 1989, APJS, 71, 817
Djorgovski S., Davis M., 1987, ApJ, 313, 59
Dressler A., Gunn J. E., Schneider D. P., 1985, ApJ, 294, 70

Dressler A., Lynden-Bell D., Burstein D., Davies R. L., Faber S. M., Terlevich R. J., Wegner G., 1987, ApJ, 313, 42
Faber S. M., Dressler A., Davies R. L., Burstein D., Lynden-Bell D., Terlevich R., Wegner G., 1987, Faber S. M., ed., Nearly Normal Galaxies. Springer, New York, p. 175
Faber S. M., Jackson R. E., 1976, ApJ, 204, 668
Franx M., 1993a, ApJ, 407, L5
Franx M., 1993b, PASP, 105, 1058
Franx M., 1995, van der Kruit P. C., Gilmore G., eds, Proc. IAU Symp. 164, Stellar Populations. Kluwer, Dordrecht, p. 269
Jørgensen I., Franx M., Kjærgaard P., 1996, MNRAS, 280, 167 [JFK]
Kelson, D. D., van Dokkum, P. G., Franx, M., Illingworth, G. D., Fabricant, D., 1997, ApJL, in press, astroph-9701115
Lucey J. R., Guzmán R., Carter D., Terlevich R. J., 1991, MNRAS, 253, 584
Renzini A., Ciotti L., 1993, ApJ, 416, L49
Rix, H. W., Colless, M. M., Guhathakurta, P., 1996, Bender, R., Davies, R. R., eds, Proc. IAU Symp. 171, New Light on Galaxy Evolution. Kluwer, Dordrecht, p. 241
Tinsley B. M., 1980, Fundamentals of Cosmic Physics, 5, 287
Tully R. B., Fisher J. R., 1977, AA, 54, 661
van Dokkum P. G., Franx M. 1996, MNRAS, 281, 985
Vogt, N. P., et al, 1996, ApJ, 465, L15
Worthey G., 1994, APJS, 95, 107

POPULATION SYNTHESIS IN A UNIVERSE OF INTERACTING GALAXIES

GUSTAVO A. BRUZUAL
Centro de Investigaciones de Astronomía (C.I.D.A.)
Apartado Postal 264, Mérida, Venezuela

1. Introduction

The traditional view in population synthesis and evolutionary synthesis models assumes that galaxies can be considered closed systems. Thus, the evolution of the stellar population in galaxies which do not interact with the environment (including other galaxies) is described by the so called Pure Luminosity Evolution (PLE) model. As we have heard several times during this conference (and before), galaxies do interact and very few galaxies are expected to evolve passively. Early-type galaxies seem to behave closer to the PLE model than later types, but E galaxies in the bright phase predicted by this model at the galaxy formation epoch have not been observed. Most likely E galaxies do not form all their stars in short lived initial bursts (several talks in this conference). Number, size, and luminosity evolution is required to understand late-type and Irr galaxies. The PLE model is unable to reproduce the number counts, and color and redshift distributions of the galaxies in the HDF (Pozzetti et al. 1997).

Under these circumstances, we may ask ourselves: *are population synthesis models still useful?* I think that the answer is *yes*. In this conference we have seen already several applications of population synthesis models to interacting galaxies. Population synthesis models provide tools to understand the properties (colors, luminosity, evolutionary age, initial mass function, star formation rate, etc.) of the dominant population in star clusters and galaxies. The models also allow us to predict the effects of non-passive events (mergers, starbursts) in the evolution of these systems. In this paper I will show some results from the Bruzual & Charlot (1997, BC97 hereafter) evolutionary models suitable to study interacting galaxies. I make no attempt to review here work by other authors, notably Arimoto and collaborators, Bressan, Chiosi and collaborators, Buzzoni and collaborators, Fritze-v. Alvensleben and collaborators, Leitherer and collaborators, Rocca-Volmerange and collaborators, Vazdekis and collaborators, Worthey and collaborators. The reader is referred to various papers in this volume which illustrate the use of different sets of evolutionary models.

2. Population Synthesis Models

BC97 have extended the Bruzual & Charlot (1993, BC93 hereafter) evolutionary population synthesis models to provide the evolution in time of the spectrophotometric properties of SSPs for a wide range of stellar metallicity. The BC97 models are based on the stellar evolutionary tracks computed by Alongi et al. (1993), Bressan et al. (1993), Fagotto et al. (1994a, b, c), and Girardi et al. (1996), which use the radiative opacities of Iglesias et al. (1992). This library includes tracks for stars with initial chemical composition $Z = 0.0001, 0.0004, 0.004, 0.008, 0.02, 0.05$, and 0.10, with $Y = 2.5Z + 0.23$, and initial mass $0.6 \leq m/M_{\odot} \leq 120$ for all metallicities, except $Z = 0.0001$ ($0.6 \leq m/M_{\odot} \leq 100$) and $Z = 0.1$ ($0.6 \leq m/M_{\odot} \leq 9$). This set of tracks will be referred to as the Padova or P-tracks hereafter. To allow for uncertainties in the stellar evolution theory, BC97 consider the evolutionary tracks computed by Schaller et al. (1992) ($m \geq 2\,M_{\odot}$) and Charbonnel et al. (1996) ($0.8 \leq m/M_{\odot} < 2$) for solar metallicity as an alternative to the P-tracks. In this case the abundances are $X = 0.68$, $Y = 0.30$, and $Z = 0.02$, and the opacities are from Iglesias et al. (1992) (for $m \geq 2\,M_{\odot}$) and Iglesias & Rogers (1993) (for $0.8 \leq m/M_{\odot} < 2$). This set of tracks will be referred to as the Geneva or G-tracks hereafter. For $Z_{\odot} = 0.02$ both sets of tracks are normalized to the temperature, luminosity, and radius of the Sun at an age of 4.6 Gyr. The published tracks go through all phases of stellar evolution from the zero-age main sequence to the beginning of the thermally pulsing regime of the asymptotic giant branch (AGB, for low- and intermediate-mass stars) and core-carbon ignition (for massive stars), and include mild overshooting in the convective core of stars more massive than 1 $M_{\odot}$ (Padova set) and 1.5 $M_{\odot}$ (Geneva set). The Post-AGB evolutionary phases for low- and intermediate-mass stars were added to the tracks by BC97 from different sources (see BC97 for details).

The BC97 models use the library of synthetic stellar spectra compiled by Lejeune et al. (1997a,b, LCB97 hereafter) for all the metallicities listed above. This library consists of Kurucz (1995) spectra for the hotter stars (O-K), Bessell et al. (1989, 1991) and Fluks et al. (1994) spectra for M giants, and Allard & Hauschildt (1995) spectra for M dwarfs. There are two versions of this atlas. One version (LCB97-O hereafter) contains the model spectra as published, but rebinned in a homogeneous fundamental parameter (effective temperature, gravity, metallicity) and wavelength scale (Lejeune et al. 1996). In the second version (LCB97-C hereafter), LCB97 corrected the original model spectra for systematic deviations that become apparent when color-temperature relations computed from the models are compared to empirical ones at $Z_{\odot}$. The corrections are especially important for the M star models. These semi-empirical blanketing corrections are defined for every wavelength, and should be a function of the fundamental model parameters: temperature, gravity, and Z. Due to the lack of calibration standards at various metallicities, LCB97 applied the corrections derived for $Z_{\odot}$ at all metallicities. Since the blanketing correction functions are multiplicative, the differential properties of the libraries are nearly conserved (see LCB97a,b for details). For instance, well known photometric differential metallicity indicators for F-G-K stars, such as the ultraviolet excesses, $\delta(U-B)$ in UBV photometry, and $\delta(C-M)$, $\delta(C-T_1)$ in Washington photometry, are very well reproduced for dwarf and giants by both the original and the corrected versions of the Kurucz (1995) library (Lejeune & Buser 1996). For $Z = Z_{\odot}$, BC97 also use an extended version of the Gunn & Stryker (1983) atlas, assembled from mostly empirical stellar data (EGS atlas hereafter, see BC97 for details).

Regardless of the specific computational algorithm used, all evolutionary synthesis models depend on three adjustable parametric functions: *(1)* the stellar initial mass function, $f(m)$, or IMF; *(2)* the star formation rate, $\Psi(t)$, or SFR; and *(3)* the chemical enrichment law, $Z(t)$. For a given choice of $f(m)$, $\Psi(t)$, and $Z(t)$, a particular set of evolutionary synthesis models provides: *(1)* Galaxy spectral energy distribution (SED) *vs.* time, $F_{\lambda}(\lambda, Z(t), t)$; *(2)* Galaxy colors and magnitude *vs.* time; *(3)* Line strength and other spectral indices *vs.* time.

Fig 1. shows the evolution in time of the $B-V$ and $V-K$ colors, and the M/L_V ratio predicted by BC97 for chemically homogeneous simple stellar populations (SSPs) of the indicated metallicity. In an SSP all the stars form at $t = 0$ and evolve passively afterward. In all the examples shown in this paper I assume that stars form according to the Salpeter (1955) IMF in the range from $m_L = 0.1$ to $m_U = 125\ M_{\odot}$. The total mass of the model galaxy is $1\ M_{\odot}$.

From Fig 1 it is apparent that there is a uniform tendency for galaxies to become redder in $B-V$ as the metallicity increases from $Z = 0.0004$ ($\frac{1}{50} Z_{\odot}$) to $Z = 0.05$ ($2.5 \times Z_{\odot}$). The tendency reverses at the highest metallicity shown, $Z = 0.10$ ($5 \times Z_{\odot}$). After 14 Gyr this model becomes as blue as the lower Z models. One reason for this behavior is the appearance of AGB-manqué stars at $Z = 0.10$ (Greggio & Renzini 1990). These stars skip the AGB phase and instead go through a long lived hot HB phase. However, this particular result should be taken with caution. The opacities,

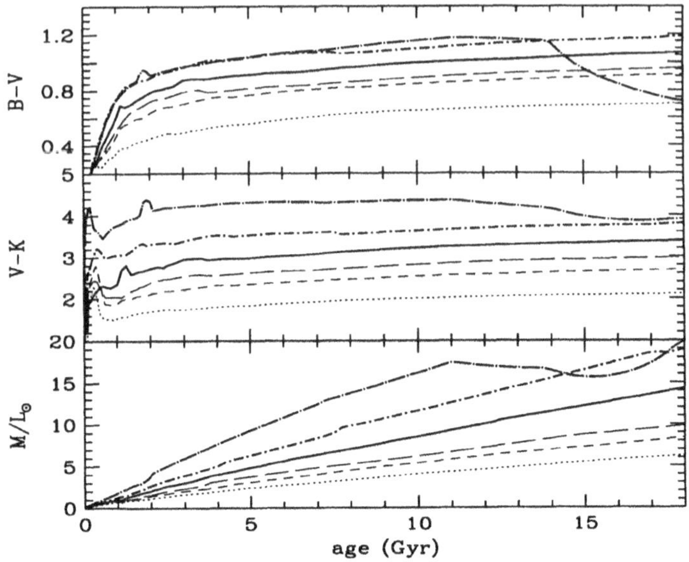

Figure 1. Evolution in time of the $B - V$, and $V - K$ colors, and the M/L_V ratio for BC97 models of different metallicities. Each line represents a different metallicity, as indicated in the middle panel in Fig 2. BC97 models constructed with the Padova tracks and the LCB97-C stellar library. See text for details.

and hence both the evolutionary tracks and the stellar model atmospheres, are quite uncertain at this high Z. There are very few, if any, examples of galactic stars which such a high Z. On the contrary, the $V - K$ color and the M/L_V ratio show the expected tendency with metallicity, i.e. $V - K$ becomes redder and M/L_V becomes higher with increasing Z. However, even in these two quantities there is a trend at $t > 12$ Gyr for the $Z = 0.10$ model to approach the $Z = 0.05$ model.

Fig. 2 shows the evolution in time of the Mg_b, H_β, and Ca spectral indices as defined by Worthey (1994) for the same BC97 SSP models shown in Fig. 1 Again, except for the $Z = 0.10$ model, the models show the expected tendency with Z and match the values computed by Worthey (1994). It should be remarked that the time behavior of the line strength indices at constant Z is due to the change in the number of stars at different positions in the HR diagram produced by stellar evolution and is not related to chemical evolution. The indices change also in chemically homogeneous populations. The H_β index is less sensitive to the stellar metallicity than the Mg_b and Ca index. Instead, the H_β index is high when there is a large fraction of MS A-type stars ($t < 1$) Gyr. The behavior of the 3 indices for $Z = 0.10$ and $t > 12$ Gyr is dominated by the presence of the hot

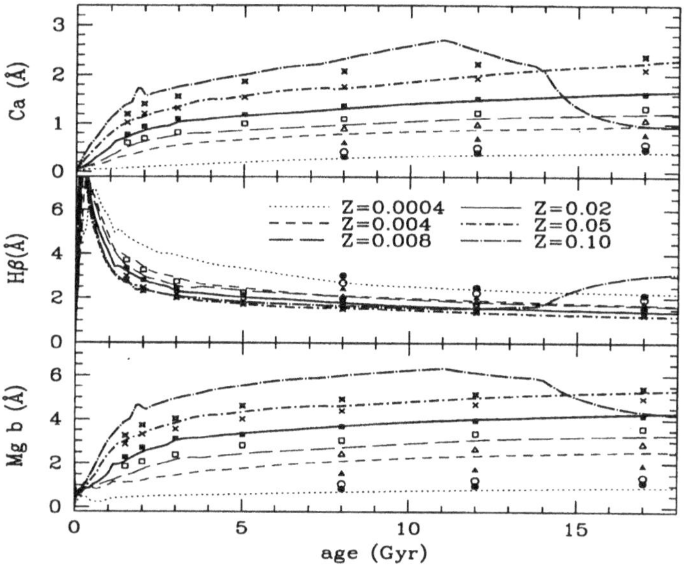

Figure 2. Evolution in time of the Mg_b, H_β, and Ca spectral indices as defined by Worthey (1994) for the BC97 models shown in Fig. 1. The different symbols represent the values of the indices computed by Worthey for the same range in Z. Each line represents a different metallicity, as indicated in the middle panel. BC97 models constructed with the Padova tracks and the LCB97-C stellar library.

HB (AGB-manqué) stars. Again, this prediction is uncertain and should be taken with caution.

3. Application to Interacting Galaxies

As an example of the use of population synthesis to model galaxies with a more complex star formation history than traditionally assumed, I discuss here the case of galaxies described by the following functions $\Psi(t)$ and $Z(t)$.

$$\Psi(t) = \begin{cases} 3 \times \Psi_0, & Z(t) = 0.0004[-1.65], \quad 0 \le t < \ 5 \text{ Gyr;} \\ 1 \times \Psi_0, & Z(t) = 0.0040[-0.64], \quad 5 \le t < 11 \text{ Gyr;} \\ 8 \times \Psi_0, & Z(t) = 0.0080[-0.33], \quad 11 \le t < 12 \text{ Gyr,} \end{cases} \tag{1}$$

and

$$\Psi(t) = \begin{cases} 1 \times \Psi_0, & Z(t) = 0.02[+0.09], \quad 0 \le t < 0.8 \text{ Gyr;} \\ 1 \times \Psi_0, & Z(t) = 0.05[+0.56], \ 0.8 \le t < 0.9 \text{ Gyr;} \\ 1 \times \Psi_0, & Z(t) = 0.10[+1.01], \ 0.9 \le t < 1.0 \text{ Gyr.} \end{cases} \tag{2}$$

The numbers inside the square brackets in (1) and (2) represent the value of [Fe/H] for the corresponding Z and $\Psi_0 = 1\ M_\odot\ \text{yr}^{-1}$. The complex nature of $\Psi(t)$ and $Z(t)$ may result from interactions with the environment

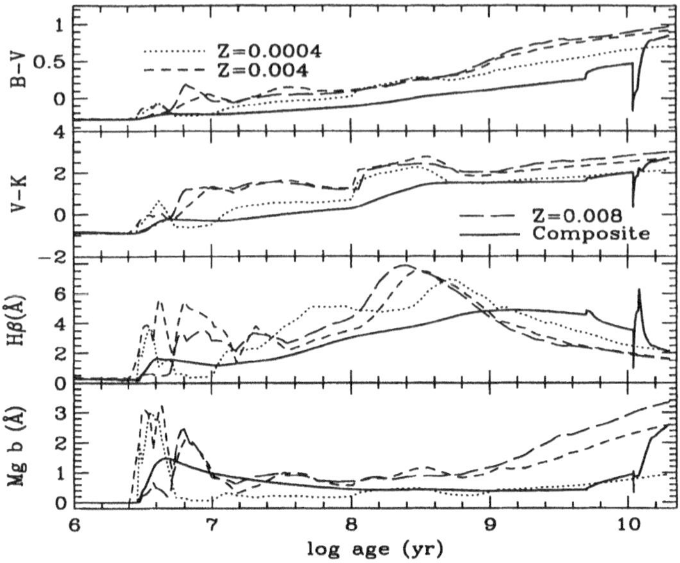

Figure 3. Evolution in time of the $B-V$ and $V-K$ colors and the H_β and Mg_b spectral indices for the different SSPs used to build the composite population described by eq. (1). The heavy solid line represents the composite population. BC97 models constructed with the Padova tracks and the LCB97-C stellar library. Notice the logarithmic scale in the horizontal axis.

(other galaxies, IGM, dark matter) or from the galaxy formation process itself. The multiple bursts of star formation in (1) are typical of dIr galaxies characterized by episodic star formation (Smecker-Hane et al. 1995). The SFR (2) represents the history of star formation in E galaxies when metal enrichment above solar values during the last fifth of the initial burst is taken into account. Following BC93 we can compute the properties of the composite population described by (1) and (2) by convolving $\Psi(t)$ with the evolving spectrum of the SSPs for the corresponding $Z(t)$. For simplicity $f(m)$ is assumed to follow the Salpeter power law at all times.

The evolution in time of the B-V and V-K colors and the H_β and Mg_b spectral indices for the populations described by (1) and (2) are shown in Figs 3 and 4, respectively, together with the evolution of the same properties for the different SSPs used to build each composite population. A few things should be remarked from these figures. The properties of the composite population evolve more smoothly than for the SSPs due to the convolution with the SFR. The discontinuities in $\Psi(t)$ are also present in the integrated properties. The population becomes bluer following an increases in $\Psi(t)$. There is a delay in the corresponding increase of the H_β index lasting until the newly formed O-B stars leave the main sequence. Once passive

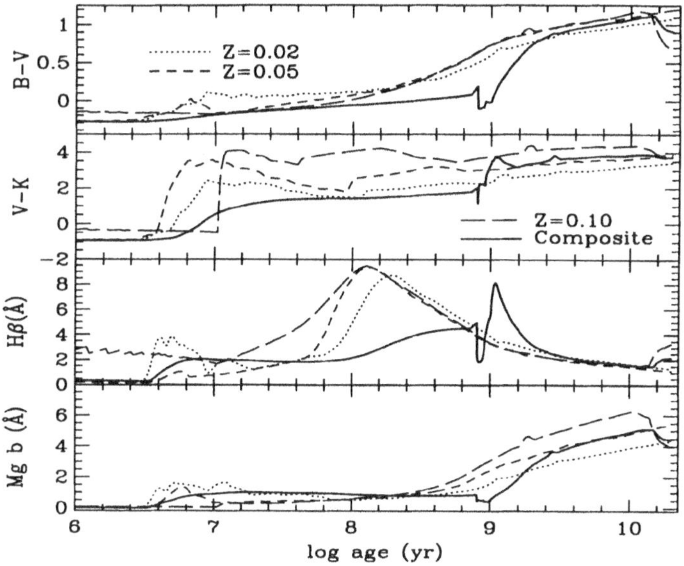

Figure 4. Evolution in time of the $B-V$ and $V-K$ colors and the H_β and Mg_b spectral indices for the different SSPs used to build the composite population described by eq. (2). The heavy solid line represents the composite population. BC97 models constructed with the Padova tracks and the LCB97-C stellar library. Notice the logarithmic scale in the horizontal axis.

evolution sets in, the population reddens, H_β decreases, and Mg_b increases. This tendency reverses in Fig 4 once the highest metallicity stars, $Z = 0.10$ in (2) are formed, and the AGB-manqué stars appear: the integrated colors become slightly bluer, H_β increases, and Mg_b decreases. As remarked above, this particular result should be taken with caution.

4. Conclusions

In this paper I have illustrated by means of two examples how to use simple stellar population synthesis models to study the observational properties of composite stellar populations subject to star formation laws plausible for interacting or forming galaxies of various kinds. Even though these models do not follow the dynamical processes involved in galaxy formation and interactions, they provide insight into the problem of spectral amd photometric evolution of interacting or forming galaxies. The results from population synthesis can be input into dynamical models to build self-consistent models which include chemical and spectral evolution (see, for instance, several papers in this volume).

References

Allard, F., & Hauschildt, P.H. 1995, ApJ, 445, 433

Alongi, M., Bertelli, G., Bressan, A., Chiosi, C., Fagotto, F., Greggio, L., & Nasi, E. 1993, A&AS, 97, 851

Bessell, M.S., Brett, J., Scholtz, M., & Wood, P. 1989, A&AS, 77, 1

———. 1991, A&AS, 89, 335

Bressan, A., Fagotto, F., Bertelli, G., & Chiosi, C. 1993, A&AS, 100, 647

Bruzual A., G. & Charlot, S. 1993, ApJ, 405, 538 (BC93)

———. 1997, ApJ, in preparation (BC97)

Charbonnel C., Meynet G., Maeder A., & Schaerer D., 1996, A&AS, 115, 339

Fagotto, F., Bressan, A., Bertelli, G., & Chiosi, C. 1994a, A&AS, 100, 647

———. 1994b, A&AS, 104, 365

———. 1994c, A&AS, 105, 29

Fluks, M. et al. 1994, A&AS, 105, 311

Girardi, L., Bressan, A., Chiosi, C., Bertelli, G., & Nasi, E. 1996, A&AS, 117, 113

Greggio, L., & Renzini, A. 1990, ApJ, 364, 35

Gunn, J.E. & Stryker, L.L. 1983, ApJS, 52, 121

Iglesias, C.A., Rogers, F.J., & Wilson, B.G. 1992, ApJ, 397, 717

Iglesias, C.A., & Rogers, F.J. 1993, ApJ, 412, 752

Kurucz, R. 1995, private communication

Lejeune, T., Buser, R. 1996, Baltic Astronomy, 5, 399

Lejeune, T., Cuisinier, F., & Buser, R. 1996, in "A Data Base for Galaxy Evolution Modeling", eds. C. Leitherer et al., PASP, 108, 996

———. 1997a, A&AS, 125, 229

———. 1997b, A&A, in preparation (LCB97b)

Pozzetti, L., Madau, P., Zamorani, G., Ferguson, H.C., & Bruzual A., G. 1997, MNRAS, in press

Salpeter, E.E. 1955, ApJ, 121, 161

Schaller G., Schaerer D., Meynet G., & Maeder A. 1992, A&AS 96, 269

Smecker-Hane, T.A., Stetson, P.B., Hesser, J.E., and VandenBerg, D.A. 1995, in "From Stars to Galaxies: The Impact of Stellar Physics on Galaxy Evolution", eds. C. Leitherer, U. Fritze-von Alvensleben, and J. Huchra, ASP Conf. Ser., 98, 328

Worthey, G. 1994, ApJS, 95, 107

EMISSION LINE GALAXIES AT $1 < z < 1.5$

K. GLAZEBROOK
Anglo-Australian Observatory, P.O. Box 296, Epping, NSW 2121, Australia

R.G. ABRAHAM
Institute of Astronomy, Madingley Road, Cambridge CB3 0HA, United Kingdom

AND

C.A. BLAKE
Department of Physics, Astrophysics, 1 Keble Road, Oxford OX1 3NP, United Kingdom

1. Why study emission line galaxies at high redshift?

In this paper we wish to introduce the first results on two new projects aimed at detecting emission lines in galaxies at $z > 1$. There are two primary motivations for doing this: Firstly to try and measure the cosmic star-formation rate at these redshifts. The combination of $z < 1$ redshift surveys and the discovery of the $z \sim 3$ Hubble Deep Field ultraviolet dropout objects has led to a 'first draft' history of the cosmic SFR (Fig. 1). These results are based on UV continuum fluxes, it is highly desirably to confirm these studies with line diagnostics and extend the work to the redshift of the inferred peak ($z \sim 1.5$).

Secondly like the drunk looking for his keys under the lamp-post one reason for looking for emission lines galaxies is that the line/continuum contrast renders them easier to find at high-z! Since we don't yet have *any* significant samples of normal field galaxies in the $z \sim 1.5$–2 regime to study (primarily because the optical lines are redshifted in to the high-background near-infrared and the strong UV lines have yet to make an appearance in the optical) this is not just a throwaway point. Also for some studies, such as the cosmic SFR, samples selected down to a constant line luminosity can be regarded as complete.

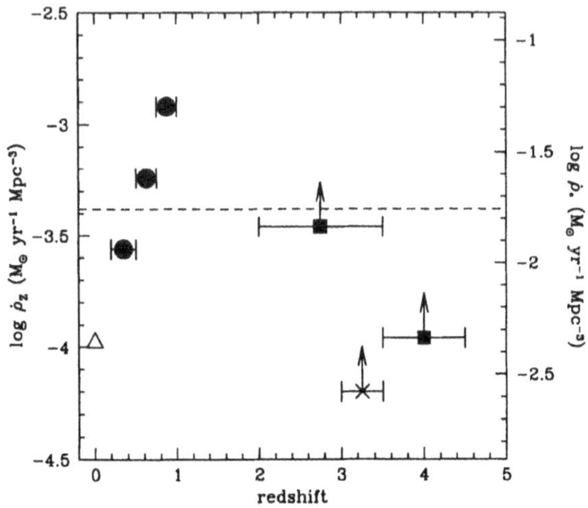

Figure 1. Star-formation density limits from the compilation of [6]. The points are: triangles — local Hα survey [2], filled dots — the CFRS survey[5], diagonal cross — lower limits from Lyman limit galaxies observed by [7], filled squares — from HDF [6]. The dashed line shows the average rate required to produce the current abundance of metals for $\Omega = 1$ and $H_0 = 50$ km s^{-1} Mpc^{-1}.

2. The WHT/TTF survey

The first survey to be presented here is a narrow-band imaging survey for emission line objects in the Hubble Deep Field. This is unique in our use of the 'TAURUS Tunable Filter' [1] to image very narrow (10Å) slices. Unlike standard narrow-band glass filters the narrower, wavelength tunable, bandwidth is perfectly matched to the typical linewidths ($\sim 300\,\mathrm{km\,s^{-1}}$) of normal galaxies. TTF uses Fabry-Perot technology: the TTF etalon has an unusually small band-gap — between 2–15μm which gives $\Delta\lambda$ = 10–60Å (c.f. 100–200 μm and $\Delta\lambda$ = 1–2Å for a conventional FP). In our HDF survey at the William Herschel Telescope in La Palma we scan for [OII] emission over $0.9 < z < 1.5$ in 3 OH-line free regions of the I-band: I1: z = 0.894–0.908 (0.2 hr/slice), I5: z =1.173–1.191 (1 hr/slice) and I8: z =1.426–1.448, (1.5 hr/slice). We cover a volume of $\sim 3000\,\mathrm{Mpc}^3$ down to fluxes of $10^{-21}\,\mathrm{W\,m^{-2}}$ which corresponds to SFRs as low as $1 - 2M_\odot\,\mathrm{yr}^{-1}$ at these redshifts. While we are still analyzing this data our initial work shows emission line objects appearing at about the rate expected from an extrapolation of Figure 1. Some examples are shown in Figure 2.

Figure 2. Candidate emission line objects (indicated by circles) at different redshifts in the Hubble Deep Field). These images show two slices at different wavelengths which have been continuum subtracted (removing over 98% of the objects).

3. CGS4 Observations of Hα in known $z \sim 1$ galaxies

A related project we are pursuing is to try and detect the Hα line in a sample of *known* $z \sim 1$ normal field galaxies from the CFRS/LDSS2 samples ([4], [3]). At this redshift the line is in the J-band. Our strategy is to observe at high spectral resolution ($R > 2000$) and select from the 74 galaxies with $0.9 < z < 1.4$ those with redshifted Hα lying in regions between the night sky OH lines — this greatly reduces our effective background and lets us reach fluxes of 10^{-19} W m^{-2} in just a few hours.

By measuring the Hα line we get a independent, well-calibrated measure of the SFR in these systems which we can compare with the UV-derived measurements of [5]. Moreover at these redder wavelengths extinction is less and comparison can address the big question of the importance of dust in these objects. This sample has already been subject to a great deal of morphological analysis from *HST* data so we can also look at the SFR in different systems, and correlations with parameters such as asymmetry and bimodality (candidate mergers).

In our survey so far (using CGS4 on the UK Infrared Telescope in Hawaii) we have observed 13 galaxies and reliably detected Hα in 8 of these. Most of the galaxies also have the continuum detected in J which indicates reliable object acquisition. From the Hα fluxes we can make a first pass at estimating the SFR — this is shown in Figure 3. Our preliminary, tentative, conclusion is that we find none of these $z \sim 1$ galaxies have greatly

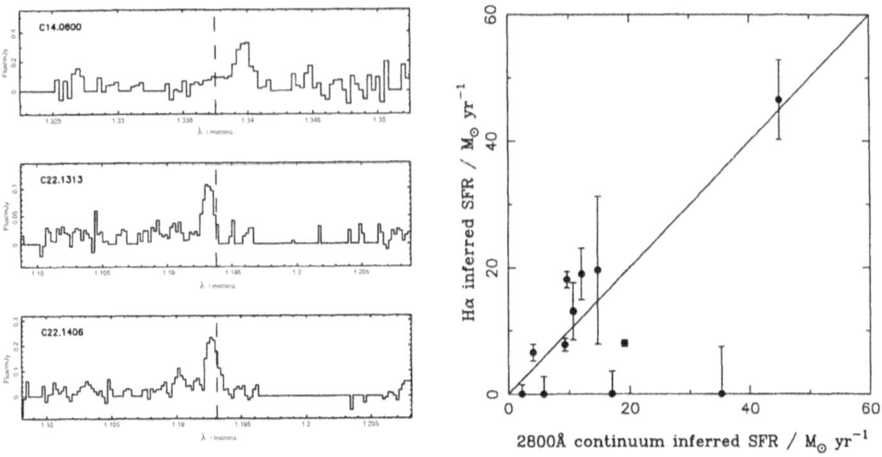

Figure 3. Left: some sample Hα spectra. Right: SFR per object as inferred from Hα and UV indices.

enhanced SFR from Hα c.f. UV estimates. Rather we find good agreement. This suggests to us that dust extinction is not of major importance in this sample of objects (though of course we can not rule out hidden, 'dark galaxies'), confirming the Madau et al point at $z \sim 1$. We do note that some galaxies show UV but not Hα flux. Probably the most likely explanation is the differing lifetimes of young stars contributing at 2800Å ($\sim 10^9$ yr) and Hα ($\sim 10^7$ yr), or perhaps some of the UV flux does not come from hot young main sequence stars (e.g. AGN). We are currently investigating this further by means of comparing SED fits to populations vs Hα strength.

In conclusion we observe that the detection of emission lines in normal galaxies with modest SFRs is now possible at $z > 1$, even with 4m telescopes. The technique of OH avoidance augurs well for future near-IR studies of normal galaxies at high redshift.

References

1. Bland-Hawthorn J., Jones D. H.,1997, MNRAS, in preparation (**astro-ph/9707315**).
2. Gallego J., Zamorano J., Arágon-Salamanca A., Regg M., 1995, ApJ, 455, L1
3. Glazebrook K., Ellis R. S., Colless M. M, Broadhurst T. J., Allington-Smith J. R., Tanvir N. R., 1995A, MNRAS, 273, 157
4. Lilly S. J., Le Fevre O., Crampton. D., Hammer F.,Tresse L., 1995, ApJ, 455, 50
5. Lilly S. J., Le Fevre O., Hammer F., Crampton. D., 1996, ApJ, 460, L1
6. Madau P., Ferguson H. C., Dickinson M. E., Giavalisco M., Steidel C. C., Fruchter A., 1996, MNRAS, 283, 1388
7. Steidel C. C., Giavalisco M., Dickinson M. E., Adelberger K. L., 1996, AJ, 112, 352

DISTANT RADIO GALAXIES: PROBES OF THE FORMATION OF MASSIVE GALAXIES

HUUB RÖTTGERING, PHILIP BEST, LAURA PENTERICCI AND GEORGE MILEY
Leiden Observatory,
P.O. Box 9513, 2300 RA Leiden, The Netherlands

Abstract. In this paper we review some of the evidence that the hosts of powerful high redshift ($1 \lesssim z \lesssim 5$) radio galaxies (HzRGs) are the progenitors of present day brightest cluster galaxies (BCGs). On the basis of *HST* imaging we argue that the scenario describing the formation of BCGs consists of at least two important stages. At $z > 2$ a significant fraction of the stellar mass of a BCG is formed during a massive burst of star-formation. By $z \sim 1$, well developed massive ellipticals are already observed and from then until the present epoch, the total mass in stars of the BCG will further grow by a factor of order 3, mainly through the accretion of cluster ellipticals.

1. Introduction

During the last decade, evidence has been accumulated that powerful radio galaxies at large distances ($1 \lesssim z \lesssim 5$) are hosted in galaxies that are the most massive for their epoch and are situated in a dense, cluster type environment (Best et al. 1997a). There are now more than 140 radio galaxies with $2 < z < 4.5$ known (e.g. Röttgering et al. 1997). This sample of $z > 2$ radio galaxies combined with the large number of known $z \sim 1$ radio galaxies therefore forms an ideal tool to study the evolution of brightest cluster galaxies from $z = 5$ down to the present epoch.

In this paper we will first review briefly some of the basic properties of distant radio galaxies, with particular emphasis on the question of whether or not these high redshift radio sources are located in cluster-like environments. Subsequently, two *HST* imaging projects are discussed, one focusing on $z \sim 1$ radio galaxies and the other on $z > 2$ radio galaxies. Finally, we will show that the observations do indeed sketch a picture of how the assembly of massive galaxies from $z = 5$ to the present epoch has taken place.

2. Properties of $z > 2$ radio galaxies

Since in this paper we would like to discuss the use of distant radio galaxies as probes of the formation of massive galaxies, two properties of $z > 2$ HzRGs are of particular relevance: their total mass in stars, and the rate at which they form stars.

There are two fundamental difficulties in obtaining a robust estimate for the number of stars in these systems. Firstly, the best wavelength region to probe the bulk of the stellar population is around $1 - 2$ μm rest-frame. For the HzRGs this region is red-shifted towards the thermal infrared, a region in which it is difficult to carry out observations with sufficient sensitivity. Secondly, the radio galaxies have an emission component, with a blue spectrum, which is aligned with the radio source. This blue light is likely to be due to a combination of young stars, scattered light from a hidden quasar and nebular continuum emission from the emission line gas. Despite these difficulties, the spectrum of HzRGs indicates that they contain a stellar mass which is generally within a factor of a few of 5×10^{11} $M_{\odot}$ (McCarty et al. 1992; Eales and Rawlings 1996).

In several HzRGs the UV continuum is probably dominated by young stars, and the estimated star-formation (SF) rates can then be directly estimated (Madau et al. 1996). SF rates of up to a 1000 $M_{\odot}$ yr^{-1} are found (Dey et al. 1997). In three distant radio galaxies submillimeter continuum emission has been detected (Hughes et al. 1997). Since this dust is likely to be heated by young stars, this submillimeter detection directly implies SF rates of well over 1000 $M_{\odot}$ yr^{-1}. These two estimates give similar high rates for the SF indicating that distant radio galaxies can indeed form stars at such a high rate. It further indicates that dust and young stars are distributed in such a way that despite the copious amount of dust a large fraction of the UV light emitted by the young stars can nevertheless escape the galaxy.

It is now well established that at least some HzRGs at $z \sim 1$ are located in fairly massive clusters (Dickinson 1997). At higher redshifts the evidence is more tentative, but clearly building up; in a number of cases companion galaxies to the HzRGs are detected (for a review see Dickinson 1997). Of order 20% of the radio galaxies have large rotation measures (RM), strongly suggesting that the radio sources are surrounded by hot magnetized cluster gas (Carilli et al. 1997). And indeed, the HzRGs with the highest RM, 1138-262 at $z = 2.156$ (Pentericci et al. 1997), was recently detected with

ROSAT, indicating an amount of hot gas present similar to that of the cluster around Cygnus A (Carilli et al. 1998).

3. *HST* imaging of 28 $z \sim 1$ 3CR radio Galaxies

During the last 4 years we have carried out a detailed study of a virtually complete sample of 28 $z \sim 1$ 3CR radio galaxies using the *HST*, UKIRT and the VLA (Best et al. 1997a;1997b). From a detailed analysis, we concluded that the magnitudes, colours and profiles are all consistent with the host of the radio galaxy being a massive ($M_\star \sim 5 \times 10^{11} M_\odot$) elliptical galaxy that formed at high redshift ($z > 3 - 5$). For a number of reasons, we suggest that a significant fraction of the excess blue light is due to young stars, indicating that that these systems must be forming stars with a rate of order $f \times 50 - 100 M_\odot$ yr^{-1}, where f is the fraction of the excess blue light that is due to stars. If this star burst last for 10^7 years, then the total mass in the stars being formed is $\sim f \times 10^9 M_\odot$, a small fraction of the mass in stars already present.

4. *HST* imaging of $z > 2$ radio galaxies

To study the formation of massive galaxies using radio galaxies as probes we are observing the continuum emission of 10 of the most distant radio galaxies with WFPC2 on the *HST*. Our first impression is that there is a large diversity in the morphologies of the objects. The degree of clumpiness and the overall alignments with the axis of the radio source vary greatly from object to object. An extreme case is that of 1138-262 that contains of order 10 clumps. These clumps have typical sizes in the range of $2 - 10$ kpc, have profiles that can not be well fit by either an $r^{1/4}$ or an exponential law and have star formations rates of order $1 - 5 M_\odot$ yr^{-1}. Interestingly, all these characteristics are similar to those of the UV-dropout galaxies at $z \sim 3$ (Giavalisco et al. 1996). A first comparison with the *HST* images of the $z \sim 1$ 3CR radio galaxies, indicates that in general the most distant radio galaxies contain a factor of $2 - 3$ more of these star forming clumps.

5. The assembly of massive galaxies

The observations discussed suggest a scenario for the formation of BCGs that contains at least two important stages. At $z > 2$, the high star-formation rates as inferred from the UV-continuum and dust emission measurements, suggest the presence of a starburst that is so vigorous that we are indeed witnessing the formation of major parts of the galaxies. The stars in the galaxy then settle in the potential well and at $z \sim 1$ we can observe the fully developed ellipticals hosting the powerful 3CR radio sources.

Interestingly, the brightest cluster galaxies at $z \sim 0$ are about one magnitude more luminous at K-band than the $z \sim 1$ galaxies. If the hosts of distant 3CR galaxies are indeed the precursors of present day BCGs, then the 3CR galaxies must grow by about a factor 3 in stellar mass between $z \sim 0$ and $z \sim 1$. This can not be done by actively forming a significant amount of new stars during this period, basically because the $z \sim 0$ would be much bluer than observed. The most natural way of doing this is by accreting faint cluster ellipticals onto the 3CR galaxy. From the observations of a sample of BCG associated with known rich clusters up to $z \sim 1$, Aragón-Salamanca et al. (1997) also suggest that vigorous accretion is a dominant process during the formation of BCGs. They further show that such a scenario can be understood quantitatively using semi-analytical modeling techniques.

The naive interpretation from the observations therefore indicate that a vigorous starburst and the subsequent accretion of dwarf ellipticals are both important processes in the formation of BCGs. However, in a number of $z \sim 1$ radio galaxies and in the case of 1138-262 X-ray emission has been observed, presumably from hot cluster gas. If in this gas a cooling flow has developed, then the amount of gas cooling must be enormous and will have an impact on the evolution of the BCGs.

Concluding, if all these three mechanisms contribute in the formation of BCGs, then there is no "epoch of formation" of BCGs, but the formation of BCGs is an ongoing process during the history of the universe.

References

Aragón-Salamanca A., Baugh C., Kauffmann G., 1997, in S. D'Odorico, A. Fontana and E. Giallongo (eds.), The Young Universe, A.S.P. Conf. Ser, in press, astro-ph/9711146

Best P., Longair M. S., Röttgering H. J. A., 1997a, MN, in press, astro-ph/9709195

Best P., Longair M. S., Röttgering H. J. A., 1997b, MN, in press, astro-ph/9707337

Carilli C. L., Harris D. E., Pentericci L., Röttgering H., Miley G. K., Bremer M. N., 1998, ApJL: submitted

Carilli C. L., Röttgering H., van Ojik R., Miley G. K., van Breugel W., 1997, ApJS, 109, 1

Dey A., Van Breugel W., Vacca W. D., Antonucci R., 1997, ApJ, 490, 698

Dickinson M., 1997, in L. Da Costa (eds.), the ESO/VLT meeting: Galaxy Scaling Relations: Origins, Evolution and Applications, in press, astro-ph/9612178

Eales S. A., Rawlings S., 1996, ApJ, 460, 68

Giavalisco M., Steidel C. C., Macchetto F. D., 1997, ApJ, 470, 189

Hughes D. H., Dunlop J. S., Rawlings S., 1997, MNRAS, 289, 766

Madau P., Ferguson H. C., Dickinson M. E., Giavalisco M., Steidel C. C., Fruchter A., 1996, MNRAS, 283, 1388

McCarthy P. J., Persson S. E., West S. C., 1992, ApJ, 386, 52

Pentericci L., Röttgering H., Miley G. K., Carilli C. L., McCarthy P., 1997, A&A, 326, 580

Röttgering H., van Ojik R., Miley G., Chambers K., van Breugel W., de Koff S., 1997, A&A, 326, 505

LOW-IONIZATION BALQSOS: WARM ULTRALUMINOUS GALAXIES AT HIGH REDSHIFTS

E. EGAMI
Max-Planck-Institut für extraterrestrische Physik
Postfach 1603
D-85740 Garching
Germany

A decade ago Sanders et al. (1988a, b) proposed a possible QSO formation scenario as a sequence of Galaxy Mergers → Ultraluminous IR Galaxies (ULIRGs) → Warm ULIRGs → QSOs. Since then, this proposal has become a paradigm against which all the subsequent observations will be compared. Fortunately for us, this paradigm makes two specific predictions: that is, 1) ULIRGs are powered by hidden AGNs, and 2) ULIRGs should be more abundant at $z \sim 2$, where the comoving density of optically-selected QSOs is sharply peaking up.

The observations testing the first prediction had been producing confusing results for a long time, but now the results seem to be finally converging. Most recently, the *ISO*-SWS spectroscopic study by Genzel et al. (1998) showed that 20–30% of ULIRGs they looked at seem to be predominantly powered by AGNs. In the meantime, based on their optical and near-IR spectra, Veilleux et al. (1997) concluded that 25–30% of ULIRGs show clear signs of AGN activities. Therefore, as long as we define ULIRGs as galaxies with $L_{\rm ir} > 10^{12}\ L_{\odot}$, the majority of these galaxies do not seem to be predominantly powered by AGNs.

The next question is whether the QSO formation scenario above is still valid when these new results are taken into account. Although not all of ULIRGs as originally defined seem to fit in this picture, it is possible that only a subset of them, namely higher-luminosity ones, are the progenitors of QSOs. Veilleux et al. (this volume) indicate that the AGN-dominated fraction increases rapidly at $L_{\rm ir} > 10^{12.3} L_{\odot}$. If this is the case, it might be only these higher-luminosity objects that will later become QSOs. Furthermore, these objects might be much more abundant at high redshifts.

Here, we try to test the validity of the paradigm by checking the second prediction, i.e. whether there is a large number of ULIRGs at high redshifts, especially at $z \sim 2$. To address this question properly, we need a large-area sensitive IR survey, which is not available for the moment. Therefore, what we are doing is to see if we can find a ULIRG-like population at high redshifts in the existing optically-selected samples.

To state the conclusion first, it now seems that a subset of Broad Absorption Line QSOs (BALQSOs) called Low-ionization BALQSOs are actually high-redshift analogues of Warm ULIRGs such as Mrk 231. Low-ionization BALQSOs are a class of QSOs which show broad UV absorption lines of low-ionization ions (Mg II, Al III) as well as those of the high-ionization species (C IV, Si IV) regularly seen in BALQSOs (Weymann et al. 1991). Since this classification is based on the restframe UV spectra, the majority of this population known so far are at high redshifts. These QSOs are known to be rare, comprising only $\sim$ 15% of BALQSOs and $\sim$ 1.5% of all QSOs. Over the past years, pieces of evidence have been accumulating that Low-ionization BALQSOs are dusty: they are common in *IRAS*-selected samples (Low et al. 1989), their UV continua show signs of moderate reddening (Sprayberry & Foltz 1992), and the absence or the extreme weakness of their [O III] 5007 Å line might be suggesting that the ionizing radiation is blocked before reaching the outer low density regions, where the formation of the forbidden line is possible (Boroson & Meyers 1992). Voit et al. (1993) suggested that Low-ionization BALQSOs are probably "young quasars in the act of casting off their cocoons of gas and dust," which is identical to the interpretation of ULIRGs by Sanders et al. (1988).

A direct link was made between Low-ionization BALQSOs and Warm ULIRGs when space UV spectroscopy showed that two of the nearby Warm ULIRGs are actually Low-ionization BALQSOs (IRAS 07598+6508 by Lipari et al. (1994b); Mrk 231 by Smith et al. (1995)). What we would like to do here is to make this connection at high redshifts by comparing the restframe optical spectra of Low-ionization BALQSOs with those of nearby Warm ULIRGs. We performed near-IR spectroscopy of Low-ionization BALQSOs, and found that they have typical characteristics of Warm ULIRGs in the sense that (1) they are dusty, and (2) they emit strong restframe-optical Fe II emission.

The dustiness of these QSOs is inferred from the large values of their Balmer decrement. Table 1 compares the Balmer decrement values of Low-ionization BALQSOs with those of normal QSOs. The measurements are from Egami et al. (1996) and Egami (1998). It is clear from the comparison that the Balmer decrement of a Low-ionization BALQSO is roughly a factor of two larger than that of a normal QSO. This is also in line with the absence of the [O III] 5007 Å line, probably indicating strong internal absorption of ionizing radiation. The most interesting case is Hawaii 167, whose Balmer decrement is as large as 13. Egami et al. (1996) interpret this object as a heavily dust-enshrouded young QSO whose internal reddening is so large as to completely extinguish the QSO light in the restframe UV, leaving only the UV light from its starbursting host galaxy.

TABLE 1. Balmer decrements of low-ionization BALQSOs

Object	Object type	z	$H\alpha/H\beta$
Q0059-2735	Low-ionization BALQSO	1.59	7.6
Q0335-3339	Low-ionization BALQSO	2.26	8.1
Q1011+0910	Low-ionization BALQSO	2.30	6.9
Hawaii 167	Low-ionization BALQSO	2.36	13
Q1246-0524	Normal BALQSO	2.25	4.0
Q1428+0202	Normal QSO	2.12	4.0

Another characteristic which links Low-ionization BALQSOs and Warm ULIRGs is their strong restframe-optical Fe II emission. Figure 1 shows that one of the low-ionization BALQSOs emits a very strong Fe II emission at 4924 Å; on the other hand, this line is much weaker or even completely absent in the spectra of normal QSOs at similar redshifts we have looked at. In fact, such a strong optical Fe II 4924 Å emission is a common property among Warm ULIRGs (Lipari 1994b). If the strong optical Fe II emission is coming from Type II supernovae as Lipari (1994a, b) suggests, then this might mean that the host galaxies of these objects are undergoing massive starbursts, which is consistent with the picture of Hawaii 167 suggested by Egami et al. (1996).

Based on the large Balmer decrements, the absence of the [O III] 5007 Å line, and the strong Fe II 4924 Å emission, we suggest that high-z Low-ionization BALQSOs are Warm ULIRGs. This indicates that at least a subset of the ULIRG population (i.e. the "Warm" variety) does exist at high redshifts. At present only a small number of Low-ionization BALQSOs are known, but it might simply be a selection effect of the optical surveys. A large-area sensitive IR survey is essential to properly access the abundance of such objects. The more important goal of such a survey is to pick up bona fide ULIRGs at high redshifts; these objects would not have been detected in the previous optical surveys because of their much redder (i.e. cooler) color. These surveys, however, were able to detect high-z Warm ULIRGs simply because they have flatter SEDs due to their less obscured central AGNs. It will be especially interesting if we can find more of Hawaii 167-like objects in which we can directly see the starlight from the starbursting host galaxy.

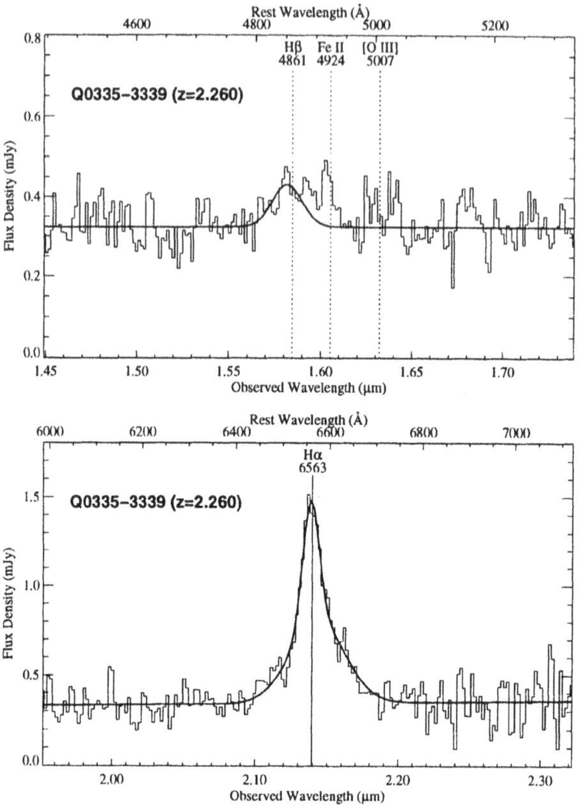

Figure 1. Near-IR spectra of Q0335-3339 taken with IRS on CTIO 4m telescope.

References

Boroson, T.A., & Meyers, K.A. 1992, ApJ, 397, 442
Egami, E. 1998, in preparation
Egami, E., Iwamuro, F., Maihara, T., Oya, S., & Cowie, L.L. 1996, AJ, 112, 73
Genzel, R., Lutz, D., Sturm, E., Egami, E., Kunze, D., Moorwood, A.F.M., Rigopoulou, D., Spoon, H.W.W., Sternberg, A., Tacconi-Garman, L.E., Tacconi, L., & Thatte, N. 1998, Ap J, in press.
Lipari, S. 1994, ApJ, 436, 102
Lipari, S., Colina, L., & Macchetto, F. 1994, ApJ, 427, 174
Low, F.J., Cutri, R.M., Kleinmann, S.G., & Huchra, J.P. 1989, ApJ, 340, L1
Sanders, D.B., Soifer, B.T., Elias, J.H., Madore, B.F., Matthews, K., Neugebauer, G., & Scoville, N.Z. 1988a, ApJ, 325, 74
Sanders, D.B., Soifer, B.T., Elias, J.H., Neugebauer, G., & Matthews, K. 1988b, ApJ, 328, L35
Smith, P.S., Schmidt, G.D., Allen, R. G., & Angel, J.R.P. 1995, ApJ, 444, 146
Sprayberry, D., & Foltz, C.B. 1992, ApJ, 390, 39
Veilleux, S., Sanders, D.B., & Kim, D.C. 1997, ApJ, 484, 92
Voit, G.M., Weymann, R.J., & Korista, K.T. 1993, ApJ, 413, 95
Weymann, R.J., Morris, S.L., Foltz, C.B., & Hewett, P.C. 1991, ApJ, 373, 23

A DEEP, LARGE-AREA K-BAND SURVEY FOR HIGHLY REDSHIFTED Hα EMISSION

P.P. VAN DER WERF
Leiden Observatory
P.O. Box 9513
NL - 2300 RA Leiden
The Netherlands
(pvdwerf@strw.leidenuniv.nl)

1. Introduction

The most significant unknown in searches for high-z starburst galaxies is the role and importance of absorption by dust. Emission line searches have mostly targeted the Lyα line, which, due to resonant scattering, will be strongly suppressed even if only small amounts of dust are present. Indeed, most of the high redshift starburst galaxies identified by Steidel et al. (1996) have *no* Lyα emission. The U-band and V-band dropout techniques developed by these authors and by Madau et al. (1996) are likewise sensitive to dust absorption, both because of absorption of the observed ultraviolet (UV) radiation and because UV-obscured galaxies will not be accounted for in these dropout studies.

Dust absorption is largely avoided by searching for Hα emission. We have carried out a deep, large-area near-IR survey for Hα emission at $z = 2.1 - 2.4$, using the near-IR camera IRAC2B at the ESO/MPI 2.2 m telescope at La Silla, to image in total about 42 $\Box'$ in 2% narrow-band filters to an r.m.s. noise of typically $10^{-16}\,\mathrm{erg\,s^{-1}\,cm^{-2}}$, and in a broad-band K$'$ filter. We selected objects with significant excess narrow-band flux as emission-line candidates.

2. Results and conclusions

The limits of our survey are presented in Fig. 1, which shows that this survey for the first time probes all reasonable (evolved) Hα luminosity functions to some extent. Two serendipitous emission line objects were found. After correcting for the part of the luminosity function not sampled by our survey, a comoving star formation density of $0.12\,M_\odot\,\mathrm{yr}^{-1}$ at $z =$

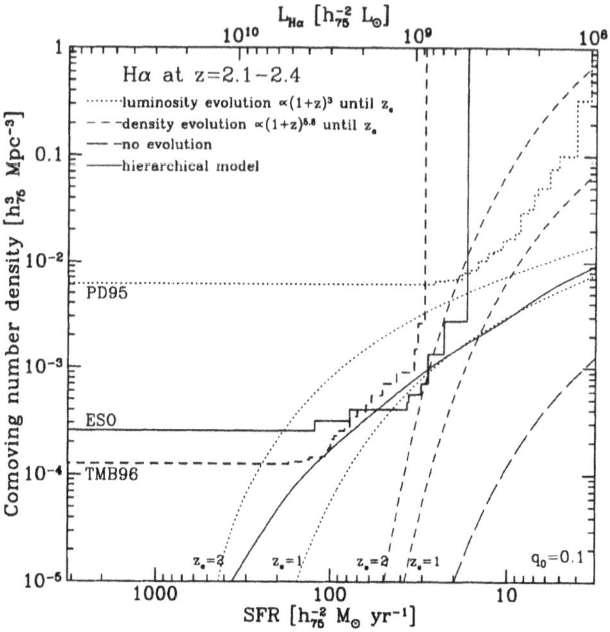

Figure 1. Limiting comoving volume densities at 90% confidence level implied by our survey (ESO) and the surveys by Thompson et al. (1996, TMB96) and Pahre & Djorgovski (1995, PD95), for $H_0 = 75\,\mathrm{km\,s^{-1}\,Mpc^{-1}}$ and $q_0 = 0.1$, as assumed throughout this paper. The curves indicate integrated Schechter-type luminosity functions based on the local Hα luminosity function (Gallego et al. 1995), labeled "no evolution", and evolved in luminosity or in density as indicated. The continuous curve results from a semi-analytical calculation of galaxy formation and evolution in a hierarchically clustering CDM universe (Baugh, priv. comm.).

2.25 results, where we have adopted the continuous curve in Fig. 1 as the luminosity function, a Salpeter Initial Mass Function, and the Kennicutt (1983) conversion factor from Hα luminosity to star formation rate. This survey shows the potential of future work with 8 m class telescopes, which will allow a determination of the Hα luminosity function at the peak of the AGN era, giving a reliable census of the star formation properties of the universe in this important epoch.

References

Gallego, J., Zamorano, J., Aragón-Salamanca, A., & Rego, M. 1995, *ApJ*, **455**, L1
Kennicutt, R.C. 1983, *ApJ*, **272**, 54
Madau, P. 1996, *MNRAS*, **283**, 1388
Pahre, M.A., & Djorgovski, S.G. 1995, *ApJ*, **449**, L1
Steidel, C.C., Giavalisco, M., Pettini, M. et al. 1996, *ApJ*, **462**, L17
Thompson, D., Mannucci, F., & Beckwith, S.V.W. 1996, *AJ*, **112**, 1794

ARE PRESSURE-CONFINED CLOUDS IN GALACTIC HALOS POSSIBLE MODELS OF LYMAN ALPHA CLOUDS?

KEIKO MIYAHATA AND SATORU IKEUCHI
Department of Earth and Space Science, Faculty of Science, Osaka University, Machikaneyama-cho, Toyonaka, Osaka 560

Our understanding of the Lyα forest has changed considerably following observations by *HST* and Keck. Lyα clouds at low redshifts ($z < 1.7$) observed by *HST* showed two unexpected features: Lanzetta et al. (1995) found that most luminous galaxies at such redshifts produce Lyα absorptions at mean impact parameter $\sim 160h^{-1}$kpc, and established the association between Lyα clouds and galaxies. Ulmer (1996) pointed out the strong clustering of Lyα clouds in this redshift range. Motivated by the above, we propose a two-component protogalaxy model for the Lyα clouds based upon our previous work (Miyahata & Ikeuchi 1995). In our model, the Lyα clouds are stable cold clouds confined by the pressure of ambient hot gas in a galactic halo. We determine the properties of these cold clouds and hot gas on the basis of theoretical and observational constraints. We take into account the stability of a cold cloud in the galactic halo in addition to the general stability conditions in a two-component medium (e.g. Ikeuchi & Ostriker 1986), and compare the derived quantities of Lyα clouds in the galactic halo and in the intergalactic medium at both high and low redshifts. We conclude that the ciondition that a cloud is stable against both evaporation and tidal disruption by a hot galactic halo is very restrictive. In the most noteworthy example at $z \sim 0.5$, a pressure-confined, stable spherical Lyα cloud with $N_{\rm HI} = 10^{14}$cm^{-2} cannot survive in the galactic halo, although much higher column density clouds of $N_{\rm HI} = 10^{17}$cm^{-2} can. Miyahata & Ikeuchi (1997) discuss how these results constrain an alternative model for Lyα clouds associated with the galaxies observed by Lanzetta et al. (1995).

References

Ikeuchi,S. & Ostriker,J.P. 1986, *ApJ*, **301**,522
Lanzetta, K.M., Bowen, D.V., Tytler, D., & Webb, J.K. 1995, *ApJ*, **442**, 538
Miyahata,K. & Ikeuchi,S. 1995, *PASJ*, **47**, L37
Miyahata,K. & Ikeuchi,S. 1997, *PASJ*, submitted, (astro-ph/9705040)
Ulmer, A. 1996, *ApJ*, **473**, 110

METAL ENRICHMENT OF LY ALPHA CLOUDS AND THE INTERGALACTIC MEDIUM

I. MURAKAMI
National Institute for Fusion Science, Toki, 509-5292, Japan
AND
K. YAMASHITA
Information Processing Center, Chiba Univ., Chiba 263, Japan

We have considered the metal enrichment of the intergalactic medium (IGM) and Lyα clouds at high redshifts. The CIV absorption lines associated with the Lyα forest have been observed (e.g. Songaila & Cowie 1996) and the metallicity is estimated as $\sim 0.01 Z_{\odot}$. Here, based on a galactic wind model, we have examined how the IGM is metal-enriched by the outflows driven by supernova explosions in primeval galaxies and minihalos.

In a spherical cloud model for a minihalo which experiences star formation, supernova explosions cause the expanding gas shell which propagates towards the collapsing envelope. The expanding velocity depends on the star formation timescale. The diffuse hot gas behind the shell is polluted by metal and expands with the shell.

The propagation of metallic gas into the IGM, where diffuse void regions and dense gas walls exist, has been examined in a 3D toy model of grid geometry. The outflow towards the dense wall is prevented as expected. Metallic gas is accumulated mostly at the dense shell, but the metallicity is nearly uniform at the shell and in the hot cavity behind the shell.

The 3D hydrodynamical simulation for the IGM in the CDM model is performed and we examine the metallicity distribution of the IGM. At $z = 3$ only 10 % of the IGM has metallicity $Z > 10^{-2} Z_{\odot}$, and this volume fraction increases up to 30 % at $z = 0$. The void region is almost metal free and the metallic gas traces the filamentary structures of the matter distribution.

References

Songaila, A. & Cowie, L.L. 1996, *AJ*, **112**, 335

EVOLUTION OF DWARF GALAXIES IN HIGH PRESSURE ENVIRONMENTS

I. MURAKAMI
National Institute for Fusion Science, Toki, 509-5292, Japan

AND

A. BABUL
Dept. of Physics & Astronomy, Univ. of Victoria, P.O.Box 3055, Victoria, BC V8W 3P6, Canada

We use 2D hydrodynamical calculations to examine the effect of the external medium on evolution of supernova-driven outflows from dwarf galaxies. Babul & Rees (1992) have suggested a high external pressure may be able to prevent the outflows from escaping beyond the galaxy and that this material, as it cools and falls back into the galaxy, would serve as fuel for a second epoch of star formation. When thermal pressure is dominant, such evolution of the outflows is seen in our simulations and the gas falls back into the galaxy. Babul & Rees, however, did not take into account the possibility that in high pressure environments such as clusters, galaxies are moving and therefore, subject to ram pressure. In our simulations, we find that ram pressure causes the mass shell associated with the outflow to fragment into clumps. These clumps remain in the vicinity of the galaxy for a few tens of million years before being swept away. The distribution of the clouds gives the galaxy in our simulations a characteristic "head-tail" appearance. If the clouds experience star formation during this epoch, we would expect that the light distribution would also show this "head-tail" feature. The tail-like structure is a transient feature that will eventually disappear. We speculate that galaxies observed by Dickinson (1996) in the $z = 1.15$ cluster around 3C324 are such galaxies.

References

Babul, A.,& Rees, M.J. 1992, *MNRAS*, **255**, 346

Dickinson, M. 1996, in *HST and the High Redshift Universe*, eds. N. Tanvir N., A. Aragon-Salamanca A., J.V. Wall, (London: World Scientific) (astro-ph/9612178)

THE EFFECTS OF SPATIAL CORRELATIONS ON MERGER TREES OF DARK MATTER HALOS

MASAHIRO NAGASHIMA AND NAOTERU GOUDA
Department of Earth and Space Science, Graduate School of Science, Osaka University, Toyonaka, Osaka 560, Japan

The effects of spatial correlations of density fluctuations on merger histories of dark matter halos (so-called '*merger trees*') are analyzed (Nagashima & Gouda 1997). We compare the mass functions of dark haloes derived by a new method for calculating merger trees, that proposed by Rodrigues & Thomas (1996), with those given by other methods such as the Block model, the Press-Schechter formula and our own formula in which the mass functions are analytically expressed in a way that takes into consideration the spatial correlations (Yano et al. 1996). It is found that the mass functions given by the new method are well fit by those given by our formula. We believe that the new method *naturally* and correctly takes into account the spatial correlations of the density fluctuations due to a calculated, grid-based realization of the density fluctuations, and so is very useful for estimating the merger tree accurately in a way that takes into consideration spatial correlations.

Moreover, by applying our formula, we present an analytic expression which reproduces the mass function derived by the Block model. We therefore show clearly why and how the mass functions given by the new method and the Block model are different from each other. Furthermore, we note that the construction of merger trees is sensitive to the criterion of collapse and merging of overlapped halos in cases in which two or more halos happen to overlap. In fact, it is shown that the mass function is very much affected when the criterion of overlapping is changed.

References

Nagashima, M. & Gouda, N. 1997, *MNRAS*, **233**, 637
Rodrigues, D.D.C., & Thomas, P.A. 1996, *MNRAS*, **282**, 631
Yano, T., Nagashima, M. & Gouda, N. 1996, *ApJ*, **466**, 1

ON THE ANGULAR CORRELATION FUNCTIONS OF THE HUBBLE DEEP FIELD

BOUDEWIJN F. ROUKEMA
National Observatory of Japan, Mitaka, 181 Tokyo, Japan
email: roukema@iap.fr

Roukema & Valls-Gabaud (1997) reinforce the conclusion of Colley et al. (1996, 1997) that the Hubble Deep Field (HDF) "galaxies" are probably star-forming regions, not "building-blocks".

Consider a "building-block" hypothesis:

(1) all (colour-selected high z) HDF galaxy-like objects are galaxies;

(2) these objects have a spatial correlation function $\xi(r, z) = b^2(r_0/r)^\gamma (1+z)^{-(3+\epsilon-\gamma)}$ where $b \gg 1$ is a strong bias factor at high z (e.g., Ogawa et al. 1997; see Groth & Peebles 1977 and Roukema & Valls-Gabaud 1997 for other parameters) and $b \geq 1, \partial b/\partial r < 0 \; \forall r, z$;

such that the projection of ξ (3-D) into w (angular correlation; 2-D), via Limber's equation (Limber 1953), matches Figs 1a,1d of Colley et al. (1996).

Since $w(1'') \gtrsim 1$ in Figs 1a,1d of Colley et al. (1996), at least 50% of the $1''$ object pairs can be considered "excess pairs". Table 1 of Roukema & Valls-Gabaud (1997) therefores shows, *conservatively*, that of all the $1''$ object pairs, and under the above hypotheses, 25% are spatially separated by a median of only $3 - 7h^{-1}$ kpc (proper units), and 45% are spatially separated by a median of $12 - 30h^{-1}$ kpc[1], *taking into account projection effects.* Many excess pairs have $\theta \sim 0.25''$. Hence, for a pure "building-block" model, galaxy formation models (e.g., Roukema et al. 1998) would have to post-dict the existence of *many* $R_{\rm halo} \ll 2$ kpc *(proper units), very highly biased galaxies* at $2.5 \lesssim z \lesssim 5$. This result is not very sensitive to ϵ, Ω_0, λ_0 or $z_{\rm median}$ (Roukema & Valls-Gabaud 1997).

References

Colley, W.N., Rhoads, J.E., Ostriker, J.P. & Spergel, D.N. 1996, *ApJ*, **473**, L63
Colley, W.N., Gnedin, O.Y., Ostriker, J.P. & Rhoads, J.E. 1997, *ApJ*, **488**, 579
Groth, E.J. & Peebles, P.J.E. 1977, *ApJ*, **217**, 385
Limber, D.N. 1953, *ApJ*, **117**, 134
Ogawa, T., Roukema, B.F., Yamashita, K. 1997, *ApJ*, **484**, 53
Roukema, B.F., Peterson, B.A., Quinn, P.J. & Rocca-Volmerange, B. 1998, *MNRAS*, in press (astro-ph/9707294)
Roukema, B.F. & Valls-Gabaud, D. 1997, *ApJ*, **488**, 524

FAR-IR GALAXY COUNTS EXPECTED IN THE IRIS SURVEY

T.T. TAKEUCHI, H. HIRASHITA, T. G. HATTORI AND K. OHTA
Dept. of Astronomy, Kyoto Univ.
Kitashirakawa Oiwake-cho, Sakyo-ku, Kyoto, 606-01, Japan

AND

H. SHIBAI
Division of Particle and Astrophysical Sciences, Nagoya Univ.
Furo-cho, Chikusa-ku, Nagoya 464-01, Japan

1. Introduction

Infrared Imaging Surveyor (*IRIS*, officially Astro-F) is a satellite which will be launched in the winter of 2003. The main purpose of the *IRIS* mission is an all sky survey in the mid- and far-IR with a flux limit much deeper than that of *IRAS*. In order to examine the performance of the survey and to find a suitable set of bandpasses for tracing galaxy evolution and picking up protogalaxy candidates as effective as possible using *IRIS*, we estimated the FIR galaxy counts based on a simple model with various sets of cosmological parameters and evolution types.

2. Method and Conclusion

We adopted a multicomponent model consisting of cirrus and starburst components for galaxy spectra, and the nearby FIR luminosity function derived from that of *IRAS* galaxies. We derived the $\log N$-$\log S$ and N-z relations for: 1) no evolution, 2) pure luminosity evolution, and 3) pure density evolution with various sets of cosmological parameters (H , q_0). The results are consistent with those obtained with *IRAS*, *ISO*, and *COBE*. Picking up protogalaxy candidates only by FIR colors is possible, although the contamination of nearby faint objects would be significant.

CLUSTERING OF RED GALAXIES NEAR A RADIO-LOUD QUASAR AT $z = 1.086$

I. TANAKA AND T. YAMADA
Astronomical Institute, Tohoku University,
Aoba-ku, Sendai 980-77, Japan

A. ARAGÓN-SALAMANCA AND T. KODAMA
Institute of Astronomy,
Madingley Road, Cambridge, CB3 0HA, UK

K. OHTA
Department of Astronomy, Kyoto University,
Kyoto, 606-01, Japan

AND

N. ARIMOTO
Institute of Astronomy, University of Tokyo,
Mitaka, Tokyo 181, Japan

We investigated the environment of the radio-loud quasar 1335.8+2834 at $z = 1.086$ where an excess surface number density of galaxies was reported by Huthings et al. (1993). By obtaining near-infrared and new deep optical images of the field, we found a clustering of objects with very red optical-NIR color, $4 < R - K < 6$ and $3 < I - K < 5$ near the quasar. The colors and magnitude of the reddest objects are consistent with those predicted for old (2–4 Gyr) passively evolving elliptical galaxies at $z = 1.1$.

There is a large fraction (> 60% after field correction) of objects with moderately red colors, $4 < R - K < 4.5$, which have a similar sky distribution as the reddest objects. They are interpreted as cluster galaxies with on-going star formation. For more details, see Yamada et al. (1997) and Tanaka et al. (1997, in preparation).

References

Hutchings, J. B., Crampton, D., & Persram, D. 1993, *AJ*, **106**, 1324
Yamada, T., Tanaka I., N., Aragón-Salamanca, A., Kodama, T., Ohta, K., & Arimoto, N. 1992, *ApJ*, **487**, L125

A SEARCH FOR EXTENDED OBJECTS WITH VARIABLE NUCLEI

D. TRÈVESE
Istituto Astronomico, Università di Roma "La Sapienza", via G.M. Lancisi 29, I-00161 Roma

M.A. BERSHADY
Department of Astronomy & Astrophysics, Pennsylvania State University, 525 Davey Lab, University Park, PA 16802

AND

R.G. KRON
Fermi National Accelerator Laboratory, MS 127, Box 500, Batavia, IL 60510

A large sample of QSOs/AGNs with detectable host galaxies is important to study the relations between the properties of the host and nucleus. Relatively faint active nuclei are of particular interest since they represent the low luminosity part of the QSO luminosity function, whose cosmological evolution is still poorly known. The non-stellar colors criterion cannot be applied to extended objects since most galaxies appear as non-stellar in color space. We have selected a sample of candidate AGNs with extended image structure on the basis of their variability, extending a previous survey for variability in objects with stellar structure (Trèvese et al. 1989, *AJ*, **98**, 108; 1994, *ApJ*, **433**, 494). The new sample allows a comparison with different selection techniques and increases the completeness of our previous survey. We add to the previously published spectroscopic observations new data providing the confirmation of some additional candidates. Presently we have 5 confirmed candidates from our primary sample of 16 objects brighter than $B_J = 22$ (Bershady, Trèvese, & Kron 1997, *ApJ*, in press). Since it is likely that our sample contains some additional bona fide AGNs, further spectroscopic work is desirable. We can put a lower limit of 106±20 deg^{-2} on the surface density of AGNs brighter than B_J=22.0 mag. The newly detected extended AGNs are at least ten times fainter than M_B=-23 mag and represent 13 % of the total number of AGNs.

GALAXY MORPHOLOGY, INITIAL CONDITIONS AND THE HUBBLE SEQUENCE

P.R. WILLIAMS AND A.H. NELSON
Department of Physics and Astronomy, University of Wales College Cardiff, PO Box 913, Cardiff, CF2 3YB

We have carried out over 120 galaxy formation N–body simulations modeling gravity, gas dynamics and star formation using TREESPH on a parallel computer. Our aims were to investigate whether or not numerical galaxies formed from idealized cosmological perturbations can account for the Hubble sequence and the diversity of disk galaxies observed in the field.

The initial conditions used were isolated uniform density spheres in solid body rotation and in Hubble expansion. One tenth of the total mass was initially in gaseous form and nine tenths in dark matter. Perturbations were further constrained to be bound, and have a dimensionless spin parameter λ, (a measure of rotational support,) $0.03 \leq \lambda \leq 0.11$.

Our parameter survey has shown that:

- The range of disk sizes, bulge sizes, bulge–to–disk ratios and arm definition observed in field galaxies is comparable to that found in the numerical galaxies.
- There is a well defined link between the physical parameters of the initial perturbation and the parameters defining the final galaxy morphology. But there is no direct one–to–one mapping from the parameters of the initial perturbation to Hubble type.
- Perturbations which decoupled from the Hubble flow at $z > 10$ formed galaxies which were found to be too compact, rotated too quickly and converted most of their mass into stars at redshifts > 5. Madau et al. (1996) have found that the global star formation rate of the Universe rises from $z \simeq 5$ to a peak at $z \simeq 1$ and then decays. Therefore a cosmology in which most galaxies form at late times is preferred.

References

Madau, P., Ferguson, H.C., Dickinson, M.E., Giavalisco, M., Steidel, C.C, Fruchter, A. 1996, *MNRAS*, **283**, 1388

THE FORMATION OF GALAXY DISKS AND BULGES

P.R. WILLIAMS AND A.H. NELSON
Department of Physics and Astronomy, University of Wales College Cardiff, PO Box 913, Cardiff, CF2 3YB

Using a TREESPH code modified to run on a parallel computer and a new algorithm for modeling star formation, we have investigated the formation of disk galaxies from idealized cosmological perturbations.

The initial conditions used were an isolated uniform density sphere in solid body rotation and in Hubble expansion. The total mass was $5 \times 10^{11}\ M_{\odot}$, 10% gas and 90% dark matter. The star formation law used was a Schmidt law of index 1.5. Our simulations have shown that isolated initial conditions with no small scale noise can parent numerical galaxies which are very similar to those observed in the field. The resultant numerical galaxies were found to have four mass components: a dark matter halo, a thin gas disk, a stellar disk, and a spherical bulge. Also, the numerical galaxies were found to have the properties found in typical field spirals:

- Rotation curves were found to be flat over the galaxy disk and have a maximum velocity of ~ 170 km s^{-1}.
- The surface density profile was found to be well modeled with two exponential distributions, representing the disk and bulge components. The bulge and disk scale lengths were found to be typically 0.25 kpc and 1.4 kpc respectively.
- The star formation rate increased rapidly to 1000 $M_{\odot}$yr^{-1} during collapse and decreased to 10 $M_{\odot}$yr^{-1} within 1 Gyr.
- A spherical bulge component, supported by an isotropic velocity dispersion, formed during the collapse.

Steinmetz & Müller (1995) have suggested that the size of the bulge component in a disk galaxy is determined by the initial level of small scale noise and the initial amount of random kinetic energy. Our simulations suggest that the properties of the bulge are more likely determined by the physical parameters of the initial perturbation.

References

Steinmetz M., & Müller E. 1995, *MNRAS*, **276**, 549

HUBBLE SEQUENCE AS A TEMPORAL EVOLUTION SEQUENCE

XIAOLEI ZHANG
Harvard-Smithsonian Center for Astrophysics
60 Garden St., MS 78
Cambridge, MA 02138, USA

The results from *Hubble Space Telescope's* Medium Deep Survey and Deep Fields indicate that there exists far more blue spiral galaxies at the intermediate and high redshifts than at the present epoch. A natural question therefore is: what have become of these excess late type galaxies as the Universe aged?

Recent theoretical work on the secular dynamical evolution of spiral galaxies (Zhang 1996, *ApJ*, **457**, 125; Zhang 1998, *ApJ*, **494**, in press) indicates that an average spiral galaxy undergoes significant morphological evolution during its lifetime, in the process of which its nuclear bulge increases in size through the radial accretion of *both* stellar and gaseous mass, and its Hubble type changes from late to early and eventually to elliptical galaxies of varying ellipticities. The underlying mechanism responsible for this evolution is a nonlinear and dissipative energy and angular momentum exchange process between the spiral density wave and the basic state of the galactic disk. During this process the angular momentum of the inner disk material gets loaded onto the spiral density wave, which transports this angular momentum outward, and eventually the angular momentum gets unloaded from the wave back to the material in the outer disk. As a result, the disk material in the inner disk spirals inward, and in the outer disk spirals outward, so with time a more and more centrally-concentrated configuration is obtained, together with an extended outer envelope.

Since the rate of spiral-induced secular evolution is proportional to the square of the wave amplitude and to the square of the pitch angle of the spiral, the interactions among galaxies generally accelerate the speed of the morphological evolution of the participating members through the excitation of large-amplitude, open spiral wave patterns. This appears to be the underlying process responsible for the rapid steepening of the galaxy morphology-environment relation with decreasing redshift.

NGST: SEEING THE FIRST STARS AND GALAXIES FORM

H.S. STOCKMAN

Space Telescope Science Institute
3700 San Martin Drive, Baltimore, MD 21218

AND

JOHN MATHER

NASA Goddard Space Flight Center
Code 685, Greenbelt, MD 20771

Abstract. The *Next Generation Space Telescope* (*NGST*) is a key element in NASA's Origins program. The primary goals for the *NGST* are observing the origins of stars, galaxies, and the elements that are necessary for life. To reach those goals, the telescope must work in the near and mid-infrared – at wavelengths where the Earth's atmosphere outshines the distant galaxies by up to 8 orders of magnitude. NASA, industry, US astronomers and international collaborators have completed the initial feasibility study and have begun the development of the technologies required to make the mission affordable and ready to launch by 2007.

We describe the three missions developed in the feasibility studies: a 6 m monolithic telescope in a 1×3 AU elliptical orbit, and two 8 m deployable telescopes put into L2 halo orbits. All three telescopes are radiation cooled to temperatures of 30–60 K. All three achieve point-source imaging sensitivities of 1 nanoJansky in a 3 hr exposure. Such sensitivity and diffraction-limited imaging can see the birth of stellar clusters at redshifts of $z = 10 - 20$ – even before the universe was reionized. We also illustrate *NGST*'s potential for studying the detailed interactions and mergers of galaxies during the peak of metal production at redshifts of $z = 1 - 2$, an epoch when dust may mask much of the galactic light and the regions of ongoing star formation.

1. Introduction

Today we enjoy unprecedented facilities with which to observe the interactions between galaxies and the galactic denizens of the universe. The *Hubble Space Telescope* (*HST*) brings us breathtakingly clear views of galactic train-crashes and the strands of resulting star formation regions. The *Infrared Space Observatory* (*ISO*) peers through the blanket of dust that surrounds the most recent stellar nurseries. Sub-millimeter and millimeter arrays penetrate the nuclear cores and find dense rings of heated dust and star formation within the inner hundreds of parsecs. This symposium is a celebration of our progress and a forerunner of many years of exciting research. Within the next five years, we look forward to the operation of a dozen 8 m-class groundbased telescopes and the Stratospheric Observatory For Infrared Astronomy (SOFIA), the construction of an international millimeter array (MMA), and the flight of the *Space Infrared Telescope Facility* (*SIRTF*). Each of these will expand on our current investigations, making new detections and more quantitative measures in our nearby universe and near the apparent peak of star formation at redshifts, $z \sim 1-2$. What will follow these ambitious programs? How might astronomers and the scientific community pursue the study of the formation of and interactions between galaxies at cosmological epochs? This paper describes the potential of a new space facility, a large passively cooled telescope optimized for operation in the near infrared (NIR), the *Next Generation Space Telescope* (*NGST*).

2. The NGST Project

The original idea for using radiative cooling to permit the construction of large aperture infrared telescopes traces at least as far back as the early European Space Agency (ESA) designs for the *Infrared Astronomy Satellite* (*IRAS*). It was developed further in the Edison proposal for the ESA M3 mission (Hawarden et al 1992, Thronson 1993); and it will be used for the cool-down phase of *SIRTF*. But the broad scientific case for such a facility was best made in the report of the *HST* and Beyond Committee (Dressler 1996). The charge to the committee was to consider the needs of the ultraviolet, optical, and infrared community after the end of the *HST* mission and the flights of other planned NASA and ESA facilities, such as *SIRTF*. In their recommendations, the committee identified key initiatives that would service the foreseeable scientific needs, engage the public interest, and provide a broad long range vision of ultraviolet-infrared astronomy in space. These recommendations and those of the Exploration of Neighboring Planetary Systems (ExNPS) board (Beichmann 1996) are the foundation of the NASA Office of Space Science (OSS) long range plan to search of the origins of the universe, stars, and planetary systems. The OSS initiated the

NGST Project at Goddard Space Flight Center (GSFC), in collaboration with the Space Telescope Science Institute (STScI), to study the feasibility of developing a large (>4 m dia.) space telescope optimized for the study of the origins of galaxies such as the Milky Way galaxy. The NIR wavelength region ($\lambda = 1 - 5\mu$m) is ideal for observing the redshifted optical light from existing stellar populations in spiral and elliptical galaxies during the peak star formation period at redshifts of $z = 1 - 2$; with sufficient sensitivity, an *NGST* can detect the ultraviolet light from the very earliest formation of galaxies and stellar systems at redshifts greater than $z > 10$. But the potential of a large, diffraction limited telescope in the optical and mid-IR ($\lambda = 5 - 30\mu$m) was not ignored and the *NGST* Project was to consider options for extending the capabilities in these directions.

Both NASA and the *HST* and Beyond Committee knew that developing and operating such a telescope for a fraction of the cost of *HST* would be a requirement and a challenge. The goal was to break not only the cryogenic paradigm but also the constraints of existing and planned rocket shroud diameters: build, launch and operate an 8 m class IR telescope for less than \$1B (in 1996 \$). The *NGST* Project and the NASA OSS are addressing this goal through the engagement of academia and industry and an extensive investment program in advanced technologies. In particular, the development of ultralightweight optical segments (< 20 kg m^{-2}), the ability to adjust them to submicron accuracy on orbit, and the availability of ultra-low noise IR detector arrays. Marshall Space Flight Center(MSFC) and the Jet Propulsion Laboratories (JPL), as well as Ames Research Center and Langley Research Center are leading many of the these technology development efforts. The goal is to bring technologies to sufficient maturity by 2003 to support an *NGST* construction and flight in 2007.

In the 1996 *NGST* Feasibility study, three government and industry teams proposed concepts that would satisfy the scientific and fiscal objectives. These concepts are shown in Figure 1. Both the GSFC and TRW teams developed designs (shown in the upper panels) for a deployable 8 m telescope that would be launched to halo orbits about the L2 Lagrange point. Large multilayer sunshields would shadow the optics and instrument sections, allowing them to cool to < 60 K and < 30 K respectively. The Lockheed Martin design incorporates a 6 m monolithic mirror based upon the thin mirror technology developed at the University of Arizona Mirror Laboratories. A Proton or Ariane 5 launcher with a special 7 m dia. shroud could inject the telescope into a 1×3 AU elliptical orbit, thereby providing very low zodiacal light background for most of the 2.7 yr orbital period. TRW and Ball Aerospace Corp. were selected in summer 1997 to consider further tradeoffs among these and other concepts. The GSFC-led team are developing detailed modeling tools to ensure that the original "Yard-

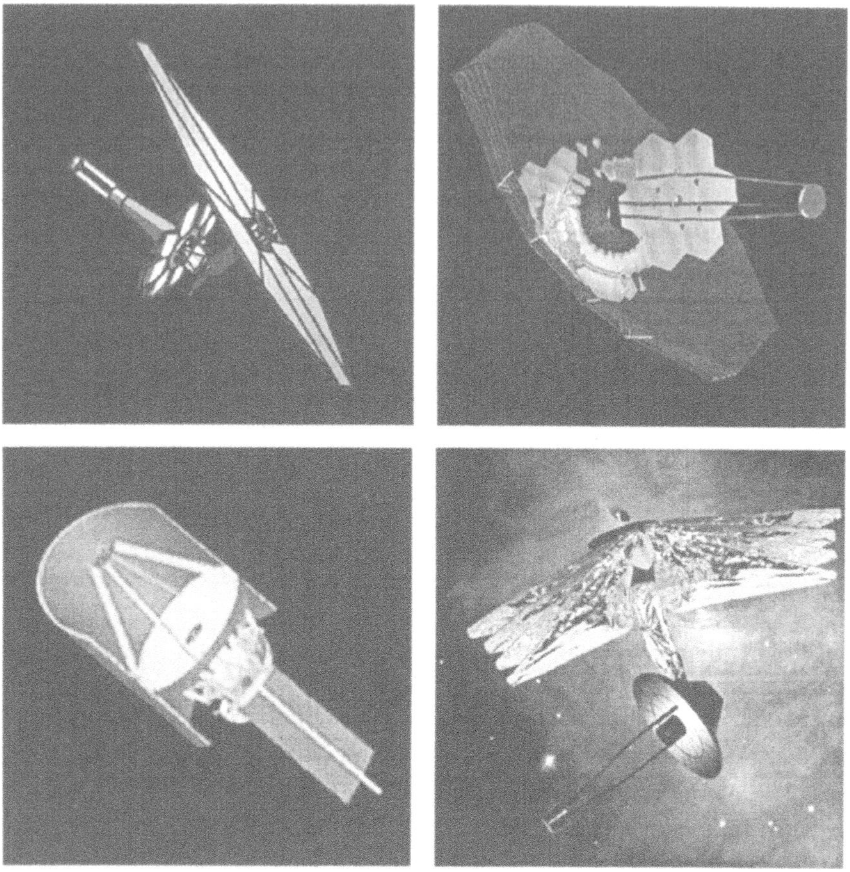

Figure 1. Four *NGST* Concepts. The upper left panel shows the "Yardstick" concept developed by the GSFC-led team. The upper right panel shows the TRW concept that uses a deployment technique developed by TRW for sub-mm antennae. The lower left panel displays the Lockheed Martin monolithic mirror design while the lower right panel shows a deployable concept developed by Ball Aerospace.(STScI/NASA)

stick" GSFC concept can meet the difficult thermal and pointing control requirements and to exercise different techniques for maintaining the optical alignment in space. NASA will also select five teams to develop designs for the scientific instruments. In November 1997, ESA and NASA agreed to a course that should lead to a major partnership for ESA in the *NGST* mission. ESA will study the feasibility of the 1×3 AU orbit, scientific instruments, and other technical areas in which ESA would lead the *NGST* development and operations efforts.

3. Studying Galaxy Formation

The 186 IAU Symposium has provided many fascinating examples of the merging of galaxies through collisions and subsequent gravitational interactions. While small prograde portions of the galaxies may become unbound by the tidal effects during the encounters, the great majority of stars and gas eventually will coalesce into a larger unit: one with a stellar population made of the merging stellar populations and peppered with new stellar systems spawned in the merger. With star formation rates ranging between 10–100 $M_\odot$ yr^{-1} during the 10^8 yr after the initial encounter, the number of stars formed by mergers can be comparable to the number of stars formed in quiescent disks at ~ 0.4 $M_\odot$ yr^{-1} for 10^{9-10} yr. In Color plate 5 (p. xxiii), we show a recent *HST* image of NGC 4038/4039, the Antennae, in which the star formation regions are clearly resolved into thousands of bright, revolved young globular cluster and associations. By studying the age of these young systems using their ultraviolet and optical colors, we can estimate how the star formation was triggered during the initial encounter (c.f. Whitmore and Schweizer 1995). Using *NGST*, we can detect these brilliant stellar nurseries at great distances and early epochs. While *NGST* observations of most star forming galaxies at high redshift will not achieve the resolution of the Antennae image, we expect that observations of serendipitous lensed galaxies, similar to those reported by Franx et al. (1997) will resolve the star formation regions in the transverse dimension over scales < 100 pc. The Franx et al. (1997) images of a lensed galaxy at redshift $z = 4.92$ clearly show bright star formation on scale corresponding to the clumps of globular cluster-like sources seen in Color plate 5 (p. xxiii). To demonstrate the sensitivity of *NGST*, we show in Figure 2 the spectrum of a newly formed, large globular cluster at different redshifts compared to the sensitivity of an 8 m *NGST* at L2 (1 AU) and a 6 m *NGST* at 3 AU. Both telescopes would be capable of detecting such a starburst at a redshift of $z \sim 10 - 20$ ($\Omega = 1$, $H_0 = 50$ km s^{-1} Mpc^{-1}). Even if some of the bursts are dust enshrouded like in the Antennae, the *NGST* would be capable of detecting interacting galaxy systems through their redshifted ultraviolet light to very great distances. Indeed, if the earliest form of star formation were through such collective star forming mechanisms, as has been suggested by Haiman and Loeb (1997), Loeb (1997), and Gnedin and Ostriker (1997), *NGST* would detect them.

Both far infrared and sub-millimeter observations demonstrate that many interacting systems display intense star formation tori in their remnant nuclei. Dust extinction in these systems is such that the systems are optically thick at mid-IR wavelengths (e.g. the contributions by Dave Sanders and Nick Scoville in this symposium). Except perhaps for low incli-

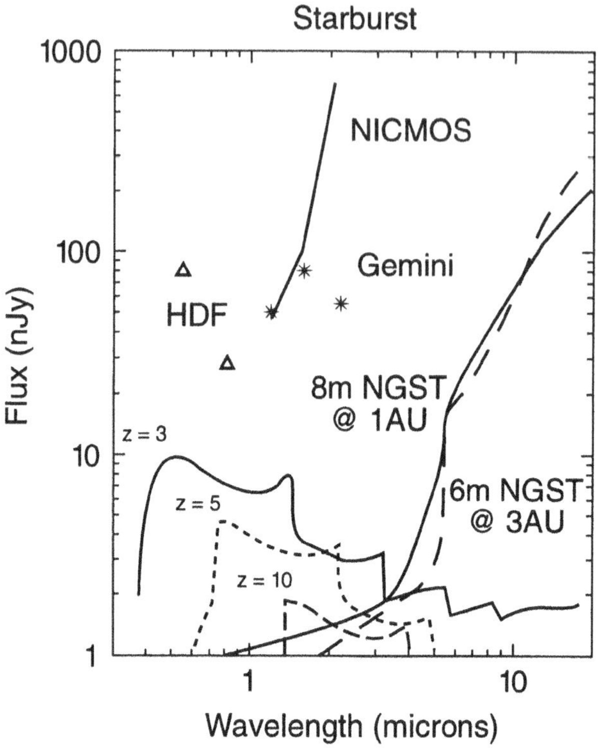

Figure 2. The spectrum from a massive starburst, $10^6 M_{\odot}$, after 5 Myr and observed at a number of different redshifts. The spectral features are due to wind nebulae around hot stars (Leitherer and Heckman 1995). The two imaging sensitivity curves are for a 8 m *NGST* at L2 and a 6 m *NGST* at 3 AU: 33% bandwidth, 10^4 s exposure, 10 σ point source detections. The other curves correspond to the sensitivity of a ground-based 8 m telescope and adaptive optics (Gillette 1997) and *HST*/NICMOS. The two triangles correspond to the faintest high redshift sources (U and B dropouts) detected in the Hubble Deep Field. (STScI)

nation views of these disks, neither *NGST* nor the IRAC camera on *SIRTF* will be capable of completely penetrating these regions but will be able to detect the outer skins of star formation in these tori. The recent NICMOS images of the ultraluminous IR galaxy Arp 220 easily resolves the surface of the inner torus or two separate star forming regions (Scoville et al. 1998). *NGST*, with 0.3 arcsec resolution at 10 μm, would be capable of resolving and penetrating 300 parsec dust tori at redshifts of $z \sim 1$. Because we expect that the metallicity and dust content of these systems will drop significantly at higher redshifts, *NGST* may be capable of detecting starforming nuclear tori to extremely large redshifts, $z > 2$. Through observations such as these, coupled to deep NIR and MIR spectroscopy of the brighter systems and dynamic studies of quiescent galaxies – if there are any – at $z = 2 - 3$, the *NGST* will be able to address the relative importance of merging on the growth and structures of galaxies such as our own. The recently selected Ad Hoc Science Working Group (ASWG) will review and develop programs such as these to direct the design trades now being studied. Other *NGST* scientific programs, such as the study of stellar populations in the Local Group and Virgo, will provide the fossil evidence with which to compare our understanding of the early universe to the records of steady growth and cataclysmic changes that we find around us.

We are pleased to acknowledge the strong support of *NGST* by Harley Thronson and Ed Weiler at NASA OSS, as well as institutional support and guidance at the GSFC and STScI. We thank Brad Whitmore and Nick Scoville for generously providing their superb *HST* images prior to publication. We direct interested readers to the booklet, *Next Generation Space Telescope, Visiting a Time When Galaxies Were Young* (Stockman 1997) and the proceedings of the 1997 Goddard Space Flight Center workshop, *Science with NGST.*

References

Beichmann, C.A. ed. (1996) *A Road Map for the Exploration of Neighboring Planetary Systems*. JPL Publ. 96-22.

Dressler, A. et al., (1996) *Exploration and the Search for Origins, A Vision of Ultraviolet, Optical, and Infrared Space Astronomy* . AURA, Space Telescope Science Institute.

Franx, M., Illingworth, G.D., Keson, D.D., van Dokkum, P. G., and Tran, K-V, (1997) *Astrophysical Journal,* **486**, pp. L75–78

Gillette, F. (1997) *Science with NGST*, ASP Conference Proceedings, ed. E. Smith and A. Koratkar, in press

Gnedin, N.Y. and Ostriker, J.P. (1997) *Astrophysical Journal,* **486**, pp. 581–598

Haiman, Z. and Loeb, A. (1997) *Astrophysical Journal,* **483**, pp. 21–37

Hawarden, T. G., Cummings, R.O., Telesco, C.M., and Thronson, H.A. (1992) *Space Science Reviews,* **61**, pp. 113-144

Leitherer, C. and Heckman, T. (1995) *Astrophysical Journal Supplement,* **96**, pp. 9-50

Loeb, A. (1997) *Science with NGST*, ASP Conference Proceedings, ed. E. Smith and A. Koratkar, in press.

Mather, J.C., Smith, E.P., Seery, B.D., Bely, P.Y., Stiavelli, M., Stockman, H.S., and Burg, R. (1997) *Science with NGST*, ASP Conference Proceedings, ed. E. Smith and A. Koratkar, in press.

Scoville, N. Z., Evans, A.S., Dinshaw, N., Thompson, R., Rieke, M., Schneider, G., Low, F.J., Hines, D., Stobie, B., Becklin, E., and Epps, H. (1998), *Astrophysical Journal,* in press.

Stockman, H.S. ed. (1997) *Next Generation Space Telescope, Visiting a Time When Galaxies Were Young*. Space Telescope Science Institute.

Thronson, H. A. ed. (1993) *Edison: an M3 Proposal to ESA*. Didcot: Rutherford Appleton Labs.

Whitmore, B.C. and Schweizer, F. (1995) *Astronomical Journal,* **109**, pp 960–980

AUTHOR INDEX

SUBJECT INDEX

OBJECT INDEX

GPSR Compliance

The European Union's (EU) General Product Safety Regulation (GPSR) is a set of rules that requires consumer products to be safe and our obligations to ensure this.

If you have any concerns about our products, you can contact us on ProductSafety@springernature.com

In case Publisher is established outside the EU, the EU authorized representative is:

Springer Nature Customer Service Center GmbH
Europaplatz 3
69115 Heidelberg, Germany

Batch number: 10761697

Printed by Printforce, the Netherlands